Foundations of Engineering Mechanics

Series Editors: V.I. Babitsky, J. Wittenburg

Foundations of Engineering Mechanics

Series Editors: Vladimir I. Babitsky, Loughborough University, UK
                    Jens Wittenburg, Karlsruhe University, Germany

Further volumes of this series can be found on our homepage: springer.com

(Continued after index)

D. Skubov · K.S. Khodzhaev

Non-Linear Electromechanics

With 96 Figures

Springer

Series Editors:

V.I. Babitsky
Department of Mechanical Engineering
Loughborough University
Loughborough LE11 3TU, Leicestershire
United Kingdom

J. Wittenburg
Institut für Technische Mechanik
Universität Karlsruhe (TH)
Kaiserstrasse 12
76128 Karlsruhe
Germany

Authors:

Prof.Dr. Dmitry Skubov
St. Petersburg Technical University
Dept. Mechanical and
Control Processes
Polytechnicheskaya ul. 29
St. Petersburg
Russia 195251
root@fmf.hop.stu.neva.ru

Prof. Kamil Shamsutdinovich Khodzhaev
Institutsky Av. 29
St. Petersburg
Russia 194021

ISSN print edition: 1612-1384

ISBN: 978-3-540-25139-2   e-ISBN: 978-3-540-44779-5

Library of Congress Control Number: 2008928144

Cover design: Mönnich, Max

Printed on acid-free paper

9 8 7 6 5 4 3 2 1

springer.com

Foreword

Books like their authors have their destinies. This book for the most part could have been written earlier, but it happened to be released as late as the 21st century. It is based on the numerous fundamental studies of my teacher Kamil Sh. Khodzhaev. His disease prevented him from implementing his ideas in the form that he had been contemplating for years. It was me who tried to convey his concepts and ideas with the least possible distortions.

This book contains a number of solutions worked out by Kamil Sh. Khodzhaev himself as well as problems solved by the authors jointly or separately.

Kamil Sh. Khodzhaev founded St. Petersburg school of electromechanics, with the focus on mechanics as a part, however distinctive, of general analytical mechanics, having been inspired by the desire of our common Teacher Anatoly I. Lurie "to set order" in this significant branch of science.

Khodzhaev's school has many followers and their work is an integral part of the book. Some original ideas and studies of our colleagues are provided with footnotes in the corresponding sections while Chapter 6 dealing with the motion of the charged particle in electromagnetic field was written in cooperation with another co-author Alexander G. Chirkov.

I would like to express my deep gratitude to the colleagues and students, without whose contribution many problems analyzed in this book would have been left unsolved. My deep gratitude goes to our friends and colleagues Iliya I. Blekhman, Robert F. Nagaev, Dmitry A. Indeitsev as well as the staff of the Department of Mechanics and Control Processes of State Polytechnical University of St. Petersburg for their helpful suggestions, understanding and unfailing support rendered in the process of working at this book. Sincere gratitude is expressed to Alexander K. Belyaev who not only took the trouble of translation into English and preparing the camera-ready manuscript, but who also made a number of important suggestions for its improvement.

Dmitry Yu. Skubov

Introduction

The systems with essentially interacting mechanical and electromagnetic processes are referred to as the electromechanical systems. There is a great amount of applications of such systems in modern technology. It is worth mentioning the converters of mechanical energy into electric energy (various types of generators, electromagnetic vibration exciters, induction sensors etc.) and the inverters of electric energy into mechanical one (electric motors, catapults, electric-type instrument etc.). For description of processes in the electromechanical system one needs prescribing the generalized coordinates and velocities of the mechanical system and the electromagnetic field created by magnets or currents at any time instant. The sources of the magnetic field can be a part of the mechanical system and move together with it. An alternative is that they do not appear in the mechanical system and in this case their motion and position are described independently. For example, in the electric motor or generator a part of conductors (the windings) moves together with the rotor whereas another part is located on the motionless stator.

Electromechanical systems can be split into two classes: the systems with (i) distributed and (ii) concentrated currents. The systems of the first type describe motions of a conducting rigid body in the magnetic field (magnetic suspensions, bearings, systems for orientation) whereas the second type of electromechanical systems deals with the systems with linear currents (conductors). There are various approaches to investigation of dynamics of the electromechanical systems and this difference is especially evident under consideration of the systems with distributed currents. The first approach is connected with solving of the coupled electrodynamical (Maxwell equations) and mechanical problems, an extensive bibliography being resulted in [80]. Another approach is based on the discrete description of the electromagnetic field by representing the distributed whirling currents in the form of a finite-dimensional or infinite-dimensional systems of conducting contours [90]. For such a description the vector of magnetic induction and the vector of intensity of the electric field are expressed in terms of a finite or countable set of other quantities. They are the effective charges and contour currents which are analogous to the generalized coordinates and generalised velocities. In the framework of this approach, the conditions of quasi-stationarity ignoring the electromagnetic waves generated by motion of the charges should be satisfied. The second approach is more natural for the case in which an electric subsystem consists of a set of linear conductors in the electric circuit.

Extensive literature is devoted to dynamics of the electromechanical systems. Among the literature in Russian it is necessary to mention, first of all, a chapter "Dynamics of nonholonomic systems and general theory of electric machines" in the classical book "Dynamics of nonholonomic systems" by Yu.I.Neimark and N.A.Fufaev. This work laid the groundwork for the fundamental principle of the investigation of the electromechanics considered nowadays as a part of general mechanics and analytical mechanics. An important role in the development of electromechanics played the book by Yu.G.Martynenko "Motion of a rigid body in electric and magnetic fields". One of the principal results of this book was development of the method of splitting the coupled problem of calculation of the field and motion of a suspended rigid body in two extreme cases: (i) high-frequency and (ii) quasi-static magnetic field. The book "Electromechanical systems" by A.Yu.Lvovich published in 1989 is more likely a textbook.

Among the books of the foreign authors it is worth noting the book "Electromechanical systems" by A.Lenk and "Electromechanical energy converters" by D.C.White and H.H.Woodson.

The works of the Nizhni Novgorod school of mechanics, namely the works by G.G.Denisov, N.V.Derendyaev, O.D.Pozdeev, Yu.M.Urman and others played a significant role in development of the theory of electromechanical systems and especially the theory of electromagnetic suspension. It is also necessary to mention a series of works by V.V.Beletsky on the theory of motion of the magnetised satellite and the works by A.A.Burov, V.V.Kozlov and V.A.Samsonov on the general problem of motion of a rigid body in the magnetic field.

Extensive literature of the mechanical-mathematical character is devoted to the theory of electric machines. Here, first of all, it is necessary to notice the results by G.A.Leonov and the groups of scientists of Petersburg State University on the qualitative theory of equations of the electric machines and the works of scientists of the Illinois Center of Electromechanics (P.V.Kokotovic, P.W.Sauer, M.A.Pai et al) on investigation of dynamics of synchronous machines and power engineering systems.

At the same time a lot of general questions of the theory of electromechanical systems have remained without attention. It is first of all the general problem of stability of equilibrium and motion of the electromechanical systems with linear and distributed currents, as well as the theory of periodic motions in electromechanical systems with a low electric dissipation. The priority belongs to K.Sh.Khodzhaev who is one of the authors of this book. The authors together with colleagues and post-graduate students have solved the important problems of simplification of the equations of synchronous electric machines by means of separating fast and slow electromechanical processes and some other problems of the qualitative analysis of their motions. K.Sh.Khodzhaev was the first who studied the oscillations of mechanical systems in the electromagnetic field which resulted in a series of engineering facilities.

The book consists of seven chapters. In the first chapter the basic features of the electromechanical systems are defined from positions of analytical mechanics. The theorems on motions in the electromechanical systems of various structures are formulated and proved.

In the second chapter the influence of a fast oscillating magnetic field on motion and equilibrium of the conducting rigid bodies is studied. Some particular problems presenting a certain interest for mechanical engineering are considered.

Some new simplified models for synchronous electric machines with various types of loadings are suggested in the third chapter. The qualitative dependences are studied and the general properties of motion of the electromechanical systems including the synchronous machines are determined.

The fourth chapter is concerned with formalization and further generalization of the problem of excitation of mechanical oscillations, in particular, in the electromechanical systems. Technical peculiarities of the electromechanical systems of vibration excitation allow one to formulate an integral criterion of stability of the periodic motions in these systems which differs from that in the theory of quasi-conservative synchronised systems [98].

In the fifth chapter the vibrations in mechanical systems caused by the electromagnetic vibration exciters of various types are studied in detail. The effects due to the interaction of processes in the exciter with the mechanical vibrations are discussed. The vibrations in the excitation systems with impacts and other types of mechanical nonlinearities are investigated. The influence of the magnetic nonlinearity on the vibration is determined.

The sixth chapter of the book is devoted to investigation of dynamics of relativistic and non-relativistic charged particle in fast oscillating and essentially inhomogeneous magnetic fields, this question being of a great importance for cosmophysics. Similar to the problems considered in the previous chapters this problem admits a correct and effective description and solution by means of the asymptotic methods of nonlinear mechanics. The main results are construction of higher approximations reduced to the equations of motion of a material point in the force field of a certain structure or to the equations of motion in canonical form, the latter having less physical meaning however being more convenient for calculations and establishing the adiabatic invariants.

The seventh chapter deals with the statement and solving some "simplest" problems of the linear-elastic magnetoelasticity, i.e. the cases in which it is necessary to consider the dependence of the field and ponderomotive forces on the motion in spite of the fact that the latter can be rather small. The term "magnetoelasticity" has already been used in the works on elastic waves in conducting body and their interaction with the electromagnetic field. The described theory has a little in common with these problems since the present approach is concerned with the elastica, that is, the forms of elastic equilibrium of conducting or ferromagnetic rigid bodies in a constant field in which the waves are impossible. The obtained nonlinear boundary-value problems require some other methods of analysis.

Finally, new methods of solving the systems of differential equations with a small parameter contributing to development of the averaging method for some special cases are suggested and proved in the appendices.

Contents

1

Description of electromechanical systems

1.1   Electrostatics and magnetostatics

Any physical theory is based upon the axiomatic definitions and notions, as well as auxiliary definitions and experimental facts. Relating these definitions, notions and facts is understood as the physical laws. The basis notions of the theory of electromagnetism are the charge, the current and the electromagnetic field, the latter describing the interaction between the charges or the currents. The electromagnetic field is described by two auxiliary vectorial quantities $\mathbf{E}$ and $\mathbf{H}$, referred to as the electric intensity (that is the intensity of the electric field caused by charges) and the magnetic induction (caused by the magnetic field due to the currents or motion of the charges). The auxiliary nature of these quantities is due to the fact that they characterize the measure of the force interaction of the electromagnetic field described by two experimental laws of Coulomb and Ampere.

The Coulomb law describes the force of interaction between the point charges and is written as follows

$$\mathbf{F}_{21} = \frac{1}{4\pi\varepsilon_0} \frac{g_1 g_2}{r_{21}^3} \mathbf{r}_{21} \,, \tag{1.1.1}$$

where $\mathbf{F}_{21}$ denotes the force acting on the second charge in the field of the first charge, $\mathbf{r}_{21}$ is the position vector of the second point relative to the first one and ε_0 denotes the dielectric permittivity of vacuum. Coulomb's law introduces the vector of electric intensity $\mathbf{E}$ of the electric field. At the second point this vector is given by

$$\mathbf{E}_{21} = \frac{1}{4\pi\varepsilon_0} \frac{g_1 \mathbf{r}_{21}}{r_{21}^3}, \tag{1.1.2}$$

provided that the electric field is caused by a point charge g_1. For a distributed system of charges with density $g(\mathbf{r})$ a natural generalization of eq. (1.1.2) has the form

$$\mathbf{E}(\mathbf{r}) = \frac{1}{4\pi\varepsilon_0} \int_V \frac{g(\mathbf{r}_1)(\mathbf{r} - \mathbf{r}_1)}{|\mathbf{r} - \mathbf{r}_1|^3} dv_1 . \tag{1.1.3}$$

By virtue of eq. (1.1.3), the vector of the electric intensity can be represented as a gradient of the potential $\varphi(\mathbf{r})$

$$\mathbf{E} = -\nabla\varphi, \quad \varphi = \frac{1}{4\pi\varepsilon_0} \int_V \frac{g(\mathbf{r}_1)}{|\mathbf{r} - \mathbf{r}_1|} dv_1 . \tag{1.1.4}$$

Direct calculation convinces that

$$\nabla \cdot \mathbf{E} = -\nabla^2\varphi = \frac{g}{\varepsilon_0}, \quad \nabla \times \mathbf{E} = 0, \tag{1.1.5}$$

that is, the potential φ satisfies Poisson's equation.

The volume force $\mathbf{f} = g\mathbf{E}$ is potential, and its potential is the energy of electric field

$$V = \frac{\varepsilon}{2} \int_V E^2 dv. \tag{1.1.6}$$

An elegant proof of the fact that the electrostatic forces are potential is given in [29]. The work done by the potential volume forces $\mathbf{f}$ on a virtual displacement $\delta\mathbf{u}$ must be equal to the variation of energy given by eq. (1.1.6)

$$\int_V \mathbf{f} \cdot \delta\mathbf{u}\, dv = -\delta V = -\varepsilon_0 \int_V \mathbf{E} \cdot \delta\mathbf{E}\, dv. \tag{1.1.7}$$

However

$$\mathbf{E} \cdot \delta\mathbf{E} = \nabla\varphi \cdot \nabla\delta\varphi = \underline{\nabla \cdot (\varphi\nabla\delta\varphi)} - \varphi\Delta\delta\varphi \Rightarrow \delta V = \int_V \varphi\delta g\, dv. \tag{1.1.8}$$

The underlined term in eq. (1.1.8) vanishes under integration over the total field since the field tends to zero at infinity. Using the equation of the charge balance

$$\delta g + \nabla \cdot (g\delta\mathbf{u}) = 0 \tag{1.1.9}$$

we obtain

$$\varphi\delta g = -\nabla \cdot (\varphi g\delta\mathbf{u}) + \nabla\varphi \cdot g\delta\mathbf{u}, \Rightarrow \delta V = -\int_V g\mathbf{E} \cdot \delta\mathbf{u}\, dv . \tag{1.1.10}$$

There are no free charges in dielectrics. One introduces the dipole moment or the polarization vector, which is equal to

$$\mathbf{P} = \sum_i e_i \mathbf{r}_i, \quad \sum_i e_i = 0 \tag{1.1.11}$$

for the unit volume of the dielectric, see [115]. In the case of the inhomogeneous polarization of the dielectric one can introduce the density of the dipole moment $\mathbf{p}(\mathbf{r})$, and the potential of the electric field will be equal to [115, 29]

$$\varphi(\mathbf{r}) = \int\limits_V \frac{\mathbf{p} \cdot (\mathbf{r} - \mathbf{r}_1)}{|\mathbf{r} - \mathbf{r}_1|^3} dv_1 \tag{1.1.12}$$

Applying the divergence theorem allows us to obtain another form for this integral

$$\varphi(\mathbf{r}) = \int\limits_S \frac{\sigma_*(\mathbf{r}_1)}{|\mathbf{r} - \mathbf{r}_1|} ds_1 + \int\limits_V \frac{g_*(\mathbf{r}_1)}{|\mathbf{r} - \mathbf{r}_1|} dv_1 \,, \tag{1.1.13}$$

where

$$\sigma_* = \mathbf{n} \cdot \mathbf{p}, \quad g_* = -\nabla \cdot \mathbf{p} \,. \tag{1.1.14}$$

Hence, the electric field in dielectrics is not changed if the dielectric is replaced by a system of charges with the volume density g_* and the surface density σ_*. Obviously, the total polarization charge is equal to zero.

Instead of the polarization vector it is convenient to enter a vector of the electric induction which is equal to [115]

$$\mathbf{D} = \varepsilon_0 \mathbf{E} + \mathbf{r} \,. \tag{1.1.15}$$

Experience suggests that, in the isotropic media, the polarization is proportional to the intensity $\mathbf{E}$, i.e.

$$\mathbf{D} = \varepsilon_0 \varepsilon \mathbf{E} \,. \tag{1.1.16}$$

Generalizing to the case of anisotropic media we have

$$\mathbf{D} = \varepsilon_0 \varepsilon \cdot \mathbf{E}, \tag{1.1.17}$$

where ε denotes the tensor of dielectric permittivity. The energy of electric field of the dielectric is determined by the formula

$$V = \frac{1}{2} \int\limits_V \mathbf{D} \cdot \mathbf{E} dv \,, \tag{1.1.18}$$

which generalises eq. (1.1.6). This expression provides one with a potential for the electrostatic force $\mathbf{f}$. For the dielectric in the linear case (1.1.16) this force is as follows, [115],

$$\mathbf{f} = \frac{1}{2} \varepsilon_0 (\varepsilon - 1) \nabla E^2. \tag{1.1.18}$$

Let us consider the problem of magnetostatics in which the magnetic field is created by the direct currents. The interaction force of the elements of currents is determined by the Ampere law

$$\mathbf{F}_{21} = \frac{\mu_0}{4\pi} \frac{\mathbf{dj}_2 \times (\mathbf{dj}_1 \times \mathbf{r}_{21})}{r_{21}^3}, \qquad (1.1.20)$$

where μ_0 denotes the magnetic permeability of vacuum[1]. Here it is necessary to make a comment as to what is understood under the element of current. Generally speaking, in real conducting bodies the distribution of current can be rather complex and determination of this distribution presents a separate problem. In the electrical engineering, the distribution of current is often assumed to be given *a priori*. This allows one to introduce the concept of a linear current which is a current (i.e. a flux of the charged particles) whose cross-sectional dimensions are small enough in comparison with the distance to the field point considered. For the linear current $\mathbf{dj} = J\mathbf{ds}$, where J and $\mathbf{ds}$ denote the total current and the oriented element of the current contour respectively.

Equation (1.1.20) can be set in the form

$$\mathbf{F}_{21} = \mathbf{dj}_2 \times \mathbf{B}_{21}, \qquad (1.1.21)$$

where

$$\mathbf{B}_{21} = \frac{\mu_0}{4\pi} \frac{\mathbf{dj}_1 \times \mathbf{r}_{21}}{r_{21}^3} \qquad (1.1.22)$$

determines the vector of induction of the magnetic field at point 2 caused by the current element $\mathbf{dj}_1$. Expression (1.1.21) can be naturally generalized to the case of distributed currents. In this case the vector of the volume forces is determined in terms of the density of the volume current

$$\mathbf{f} = \mathbf{j} \times \mathbf{B}. \qquad (1.1.23)$$

For the distributed currents, the expression for the magnetic induction can be written in the integral form

$$\mathbf{B}(\mathbf{r}) = \frac{\mu_0}{4\pi} \int_V \frac{\mathbf{j}(\mathbf{r}_1) \times (\mathbf{r} - \mathbf{r}_1)}{|\mathbf{r} - \mathbf{r}_1|^3} dv_1. \qquad (1.1.24)$$

By virtue of eq. (1.1.24) the vector of magnetic induction can be represented in the form

$$\mathbf{B} = \nabla \times \mathbf{A}, \qquad \mathbf{A} = \frac{\mu_0}{4\pi} \int_V \frac{\mathbf{j}(\mathbf{r}_1)}{|\mathbf{r} - \mathbf{r}_1|} dv_1. \qquad (1.1.25)$$

---

[1]It is obvious that relationship (1.1.20) does not satisfy the third law of Newton, i.e. $\mathbf{F}_{12} \neq -\mathbf{F}_{21}$. Ampere himself derived the law of interaction of current's elements from experiments and suggested another form, satisfying the third law, see [115]. Relationship (1.1.20) and Ampere's form yield the same result in the integral form, which requires calculation of the total force of interaction of the current contours.

As vector $\mathbf{A}$ (the vectorial potential) is determined up to gradient of a scalar function, an additional condition imposed on $\mathbf{A}$ is as follows

$$\triangle \cdot \mathbf{A} = 0. \tag{1.1.26}$$

One can prove by a direct substitution that vectors $\mathbf{B}$ and $\mathbf{A}$ satisfy the following relationships

$$\nabla \times \mathbf{B} = -\triangle \mathbf{A} = \mu_0 \mathbf{j}, \quad \nabla \cdot \mathbf{B} = \mathbf{0}. \tag{1.1.27}$$

Similar to the case of electrostatics, the magnetic forces are potential with the potential equal to the energy of magnetic field with the opposite sign, [115],

$$\int_V \mathbf{f} \cdot \delta \mathbf{u} dv = -\delta W, \quad W = -\frac{1}{2\mu_0} \int_V B^2 dv. \tag{1.1.28}$$

Let us vary the expression for the magnetic energy and take into account that

$$\mathbf{B} \cdot \delta \mathbf{B} = \nabla \times \mathbf{A} \cdot \delta \mathbf{B} - \nabla \cdot (\mathbf{A} \times \delta \mathbf{B}) + \nabla \times \delta \mathbf{B} \cdot \mathbf{A} \Rightarrow$$
$$\int \mathbf{B} \cdot \delta \mathbf{B} dv = \mu_0 \int \mathbf{A} \cdot \delta \mathbf{j} dv. \tag{1.1.29}$$

The virtual change in the current density can be determined from the condition that the current strength through an arbitrary surface does not change under the virtual displacement $\delta \mathbf{u}$ (provided that this surface is deformed or moves together with the medium)

$$\int_S \mathbf{j} \cdot \mathbf{n} ds = J = \text{const}. \tag{1.1.30}$$

Under displacement of the medium element $\delta \mathbf{u}$ the current density can change which results in a vector $\delta \mathbf{j}$ at different points of surfaces S and a deformation of contour L of this surfaces. Let $\mathbf{ds}$ be an element of contour L, then under displacement $\delta \mathbf{u}$ this element spans the area $\delta \mathbf{S} = \delta \mathbf{u} \times \mathbf{ds}$. Hence the total change in the current J through an arbitrary surface S is equal to

$$\delta J = \int_S \delta \mathbf{j} \cdot \mathbf{n} dS + \oint_L \mathbf{j} \cdot (\delta \mathbf{u} \times \mathbf{ds}) = 0. \tag{1.1.31}$$

The contour integral is transformed by means of the Stokes theorem

$$\oint_L \mathbf{j} \cdot (\delta \mathbf{u} \times \mathbf{ds}) = \oint_L (\mathbf{j} \times \delta \mathbf{u}) \cdot \mathbf{ds} = \int_S (\nabla \times (\mathbf{j} \times \delta \mathbf{u})) \cdot \mathbf{n} dS. \tag{1.1.32}$$

Due to the arbitrariness of surface S we obtain from eq. (1.1.31)

$$\delta \mathbf{j} = -\nabla \times (\mathbf{j} \times \delta \mathbf{u}). \tag{1.1.33}$$

Inserting into eq. (1.1.29) we arrive at the following equality

$$\mathbf{A}\cdot\delta\mathbf{j} = -\mathbf{A}\cdot\nabla\times(\mathbf{j}\times\delta\mathbf{u}) = -\nabla\cdot[(\mathbf{j}\times\delta\mathbf{u})\times\mathbf{A}]+(\nabla\times\mathbf{A})\cdot(\mathbf{j}\times\delta\mathbf{u})\,. \quad (1.1.34)$$

The first term vanishes after integration over the total field, that is, substitution of eq. (1.1.34) into eq. (1.1.29) yields

$$\mu_0\int_V \mathbf{A}\cdot\delta\mathbf{j}dv = \int_V \mathbf{B}\cdot(\mathbf{j}\times\delta\mathbf{u})dv = -\int_V (\mathbf{j}\times\mathbf{B})\cdot\delta\mathbf{u}dv\,. \quad (1.1.35)$$

Therefore, the volume force is $\mathbf{f}=\mathbf{j}\times\mathbf{B}$, and W, eq. (1.1.28), is the energy of magnetic field, indeed.

Like to the fact that dielectrics distort the external magnetic field because of their polarization, the magnet-sensitive materials (magnetics) in the magnetic field cause changes in this field due to their magnetization. However in contrast to the polarization of dielectrics which disappears when the external field is removed, the diverse classes of magnetics behave differently when the field has been removed. Paramagnetics and diamagnetics are completely demagnetized while ferromagnetics remain not only magnetized when the external field disappears, but they also can produce an own magnetic field.

The field of the magnetized magnetic is caused by the electric distributed (molecular) currents $\mathbf{j}_*$ circulating in it, hence

$$\int_V \mathbf{j}_*dv = 0\,, \quad (1.1.36)$$

that is, the currents circulating in magnetics produce no electromotive force (emf). The system of closed currents is characterized by the magnetic moment

$$\mathbf{m} = \frac{1}{2}\int_V \mathbf{r}\times\mathbf{j}_*dv\,. \quad (1.1.37)$$

The vector potential $\mathbf{A}$ of all molecular currents circulating in all elementary volumes of the magnetics is determined by the following integral [115]

$$\mathbf{A}(\mathbf{r}) = \frac{\mu_0}{4\pi}\int_V \frac{\mathbf{m}(\mathbf{r}_1)\times(\mathbf{r}-\mathbf{r}_1)}{|\mathbf{r}-\mathbf{r}_1|^3}dv\,, \quad (1.1.38)$$

where $\mathbf{m}(\mathbf{r})$ denotes the magnetic moment of the elementary volume dv. Using the following rule of vector analysis

$$\nabla\times\left(\frac{\mathbf{a}}{r}\right) = \frac{\mathbf{r}\times\mathbf{a}}{r^3} + \frac{1}{r}\nabla\times\mathbf{a}$$

we can transform this expression to the form

$$\mathbf{A}(\mathbf{r}) = \frac{\mu_0}{4\pi}\left(\int_O \frac{\mathbf{J}_*(\mathbf{r}_1)}{|\mathbf{r}-\mathbf{r}_1|}do_1 + \int_V \frac{\mathbf{j}_*(\mathbf{r}_1)}{|\mathbf{r}-\mathbf{r}_1)}dv_1\right)\,, \quad (1.1.39)$$

where

$$\mathbf{J}_* = -\mathbf{n} \times \mathbf{m}, \quad \mathbf{j}_* = \nabla \times \mathbf{m}.$$

Thus, the field of a magnetic is equivalent to the field caused by the surface current $\mathbf{J}_*$ and the volume current $\mathbf{j}_*$. Separating the external current $\mathbf{j}$ from the molecular current $\mathbf{j}_*$ we can lead the first equation in (1.1.27) to another form

$$\nabla \times \mathbf{B} = \mu_0(\mathbf{j} + \nabla \times \mathbf{m}) \Rightarrow \nabla \times \mathbf{H} = \mathbf{j}, \qquad (1.1.40)$$

where

$$\mathbf{H} = \frac{1}{\mu_0}\mathbf{B} - \mathbf{m}.$$

By analogy to the dielectrics, in order to close the system of field equations it is necessary to add an equation relating the medium magnetization $\mathbf{m}$ and magnetic induction $\mathbf{H}$. Depending upon the character of the dependence $\mathbf{m}(\mathbf{H})$ the magnetics can be divided into three classes. In isotropic paramagnetics and diamagnetics

$$\mathbf{m} = \kappa\mathbf{H} \Rightarrow \mathbf{B} = \mu_0\mu\mathbf{H}, \qquad (1.1.41)$$

where $\mu = 1 + \kappa$ denotes the magnetic permeability and κ is the magnetic susceptibility. The magnetic susceptibility of paramagnetic bodies is positive $\kappa > 0$, i.e. the direction of the magnetization vector coincides with the field direction. The magnetic susceptibility of diamagnetics is negative $\kappa < 0$, thus their magnetization is opposite in direction to the magnetizing field. The dependence $\mathbf{B}(\mathbf{H})$ for ferromagnetics has a character of the hysteresis loop [115], that is, it is essentially nonlinear and determined by the prehistory of magnetization of the material. The nonlinearity of dependence $\mathbf{B}(\mathbf{H})$ gives rise to the remanent magnetism of ferromagnetics determined by the coercive force. Unless stated otherwise, in what follows the linear-magnetic materials are considered, and in most cases μ for ferromagnetics is supposed to be equal to infinity.

Superconductors possess some specific magnetic properties [71]. For present purposes it is enough to consider the superconductor as an ideal diamagnetic, i.e. to suppose that $\mu = 0$. This means that the field does not penetrate into the superconductor.

The expression for energy of the magnetic field can be considered as a potential for the volume forces

$$W = -\frac{1}{2}\int_V \mathbf{B} \cdot \mathbf{H} dv. \qquad (1.1.42)$$

This results in the following expression for the ponderomotive forces [115]

$$\mathbf{f} = \frac{1}{2}\mu_0(\mu - 1)\nabla H^2. \qquad (1.1.43)$$

Hence, in contrast to paramagnetics which moves toward the increase in the magnetic field, the diamagnetics are pushed away from the regions of maximum induction.

1.2   The Lagrange-Maxwell equations

The electromagnetic field is said to be known if the vector of magnetic induction **B** and the vector of electric intensity **E** are prescribed at any point of the space. These vectors (or other quantities which can be expressed in terms of **B** and **E**) characterize the electrodynamic field. However dealing with technical electromechanical facilities one can limit the consideration to the case when the fields of **B** and **E** are expressed in terms of a finite number of other quantities appearing in the equations of electromechanics. Such a description is formally analogous to introducing generalized coordinates and generalized velocities in mechanics. To this end, the conditions of quasi-stationarity must hold. These conditions require that the electromagnetic waves can be neglected. In addition to this, the transverse dimensions of conductors are often much smaller than their longitudinal dimensions, thus the conductors can be considered as linear bodies (with the exception of the capacitor plates).

Linear conductors connected to each other, capacitor plates and external power sources form an electric circuit. Let us consider l parallel circuits connected with each other by means of inductors. This means that the conductors of each circuit are not connected to conductors of other circuits however all circuits are in a magnetic field and thus their electromagnetic processes are interrelated. Each circuit has non-parallel subcircuits consisting of conductors, and possibly, an emf and capacitors in series. The current in any cross-section of the linear conductor in the branch remains constant. Each parallel circuit has nodes where at least three subcircuits are connected. It is assumed that the circuit does not change topologically under mechanical motion, i.e. the initially connected conductors do not disconnect and the initially disconnected conductors do not connect.

This condition is not satisfied for the electromechanical facilities with interrupters, commutators etc. Some of these facilities are described by the equations with nonholonomic constraints, see for example [77, 90].

Let z_s and y_s denote respectively the number of branches and nodes in the $s-th$ circuit. Let us arbitrarily choose a positive direction of the current in any branch and designate the currents in $s-th$ branch as i_j ($j = 1..., N$, where $N = z_1 + z_2 + \ldots + z_l$). According to Kirchhoff's first law these currents are related as follows

$$\sum_{j=1}^{N} \gamma_{jk} i_j = 0, \qquad (1.2.1)$$

where $\gamma_{jk} = 0$ if $j - th$ branch is not adjacent to $k - th$ node; $\gamma_{ik} = 1$ if this branch is adjacent to $k - th$ node and the current enters the branch; $\gamma_{jk} = -1$ if this branch is adjacent to $k - th$ node and the current leaves the branch.

Relations (1.2.1) can be integrated and written in terms of the charges, hence relations (1.2.1) play the role of holonomic constraints for electromechanical systems. One has $y_s - 1$ independent equations (1.2.1) for $s - th$ circuit. Hence it is possible to choose m independent circuits ($m = (z_1 - u_1) + (z_2 - u_2) + \ldots + (z_l - u_l) + l$) from currents $i_1, \ldots, i_N$. For $s - th$ circuit, the number of independent currents is equal to $m_s = z_s - y_s + 1$. The parallel circuits can be presented in the form of an aggregate (superposition) of m closed non-parallel contours such that the currents in the contours are $i_1, \ldots, i_m$. The currents in branches will be the algebraic sums of currents in the contours in which the branch enters

$$i_j = \sum_{r=1}^{m} \beta_{jr} i_r \qquad (j = m+1, \ldots, N), \qquad (1.2.2)$$

where $\beta_{jr} = 0$ if $j - th$ branch is not included in $r - th$ contour; $\beta_{jr} = 1$ if the branch is included in the contour and the positive directions of the branch and the contour current i_r are coincident; $\beta_{jr} = -1$ in the case of the opposite directions.

The choice of currents $i_1, \ldots, i_m$ is not unique, however it is not arbitrary. In order to choose the appropriate currents, it is necessary to find a broken line in circuit's chart which passes through all nodes and forms no closed contour. Currents in the branches which are not included in this line are the sought currents $i_1, \ldots, i_m$. Adding a branch with current i_r to this line results in a closed contour. Different ways of choosing the broken line lead to different sets of currents $i_1, \ldots, i_m$ and different decompositions of the circuit into contours.

For the magnetically linear medium the magnetic induction $\mathbf{B}$ is a linear function of the currents

$$\mathbf{B} = \sum_{r=1}^{m} \mathbf{B_r}(x, y, z) i_r \qquad (1.2.3)$$

at every point of the space with the coordinates x, y, z.

The electric field is considered only in the space between the capacitor plates, thus vector $\mathbf{E}$ is a linear function of the capacitor charge g_j

$$\mathbf{E} = \mathbf{E_j}(x, y, z) g_j. \qquad (1.2.4)$$

The energy of magnetic field is deemed as the kinetic energy of the currents

$$W = \frac{1}{2} \int_V \frac{1}{\mu \mu_0} B^2 \, dx dy dz.$$

Substituting eq. (1.2.3) results in a quadratic function of contour's currents

$$W = \frac{1}{2} \sum_{r,s=1}^{m} L_{rs} i_r i_s = \frac{1}{2} i^{\mathrm{T}} L i. \qquad (1.2.5)$$

The coefficients L_{rs} form a symmetric positive-definite matrix L whose entries are the coefficients of self-induction and mutual induction.

Energy of the electric field between the capacitor plates is as follows

$$V = \frac{1}{2} \sum_{j=1}^{N} \int_{\Omega_j} \varepsilon E^2 dx dy dz = \frac{1}{2} \sum_{j=1}^{N} \frac{g_j^2}{C_j}, \qquad (1.2.6)$$

where ε is the dielectric permitivity and C_j denotes the capacitance.

For the branches without capacitors one should take $1/C_j = 0$. Integrating eq. (1.2.2) yields the following relation between the charges

$$g_j = \sum_{r=1}^{m} \beta_{jr}(g_r - g_{r0}) + g_{j0} \qquad (j = m+1, \dots, N), \qquad (1.2.7)$$

where g_{r0} and g_{j0} denote the initial values of the capacitor charges. This allows one to express V in terms of charges in the first m branches only, i.e.

$$V = \frac{1}{2} \sum_{r,s=1}^{m} \frac{1}{C_{rs}} g_r g_s + \sum_{r=1}^{m} b_r g_r + V_0 = \frac{1}{2} g^{\mathrm{T}} C^{-1} g + Bg + V_0. \qquad (1.2.8)$$

The quantities $1/C_{rs}$ given by the relationships

$$\frac{1}{C_{rs}} = \frac{1}{C_r} \delta_{rs} + \sum_{j=m+1}^{N} \frac{1}{C_j} \beta_{jr} \beta_{js} \qquad (1.2.9)$$

are termed the inverse capacitances and they are the elements of the symmetric positive-definite matrix C^{-1}. The quantities b_r and V_0 depend on the initial charges g_{r0} and their expressions are obtained by substituting relationships (1.2.7) into eq. (1.2.6)

$$b_r = \sum_{j=m+1}^{N} \left[\beta_{jr} g_{j0} - \sum_{s=1}^{m} \beta_{jr} \beta_{js} g_{s0} \right] \frac{1}{C_j},$$

$$V_0 = \frac{1}{2} \sum_{j=m+1}^{N} \frac{1}{C_j} \left[g_{j0}^2 + \sum_{r,s=1}^{m} \beta_{jr} \beta_{js} g_{r0} g_{s0} - 2 g_{j0} \sum_{r=1}^{m} \beta_{jr} g_{r0} \right]. \qquad (1.2.10)$$

Thermal dissipation of energy on resistances R_j is determined by the electric dissipative function Φ. For $R_j = \mathrm{const}$ this function is as follows

$$\Phi = \frac{1}{2} \sum_{j=1}^{N} R_j i_j^2. \qquad (1.2.11)$$

Using eq. (1.2.2) it is possible to express function Φ in terms of the contour currents

$$\Phi = \frac{1}{2} \sum_{r,s=1}^{m} R_{rs} i_r i_s = \frac{1}{2} i^T R i, \qquad (1.2.12)$$

where the coefficients

$$R_{rs} = R_r \delta_{rs} + \sum_{j=m+1}^{N} \beta_{jr} \beta_{js} R_j$$

are the elements of a symmetric positive-definite matrix of resistances R.

The algebraic sum of emf in $r - th$ contour is expressed in terms of the emf in branches E'_j in the following fashion

$$E_r = E'_r + \sum_{j=m+1}^{N} \beta_{jr} E'_j. \qquad (1.2.13)$$

In order to describe the mechanical motion one needs the generalized coordinates $q_1, \ldots, q_n$ where n denotes the number of degrees of freedom. When the system moves, the magnetic field as well as the position of linear conductors and capacitor plates can change. This leads to the fact that the coefficients of induction L_{rs} and the inverse capacities $1/C_{rs}$ become functions of the generalized coordinates of the mechanical system. The equations describing coupled electromagnetic and mechanical processes in the electromechanical system are referred to as the Lagrange Maxwell equations and have the form

$$\frac{d}{dt} \frac{\partial W}{\partial i_r} + \frac{\partial \Phi}{\partial i_r} + \frac{\partial V}{\partial g_r} = E_r \qquad (r = 1, \ldots, m),$$

$$\frac{d}{dt} \frac{\partial T}{\partial \dot{q}_k} - \frac{\partial (T + W)}{\partial q_k} + \frac{\partial (\Pi + V)}{\partial q_k} = Q_k \qquad (k = 1, \ldots, n). \qquad (1.2.14)$$

Here $T(q_1, \ldots, q_n, \dot{q}_1, \ldots, \dot{q}_n)$ and $\Pi(q_1, \ldots, q_n)$ are the kinetic and potential energy of the mechanical system respectively and the non-potential generalized forces are denoted as $Q_k(q_1, \ldots, q_n, \dot{q}_1, \ldots, \dot{q}_n)$. The constraints in the mechanical system are assumed to be holonomic.

System (1.2.14) contains $m+n$ equations of the second order for the variables $g_1, \ldots, g_m, q_1, \ldots, q_n$. Equations (1.2.14) have a structure of Lagrange equations if the variables $g_1, \ldots, g_m, q_1, \ldots, q_n$ are taken as the generalized coordinates of the electromechanical system with the Lagrange function $L_a = T + W - (\Pi + V)$. The currents $i_r = \dot{g}_r$ play the role of the generalized velocities whereas W and V can be formally attributed to the kinetic and potential energy respectively. The quantities

$$\Psi_r = \frac{\partial W}{\partial i_r} = \sum_{s=1}^{m} L_{rs} i_s \qquad (r = 1, \ldots, m), \qquad (1.2.15)$$

are referred to as the induction flux or the magnetic flux linkage. They are analogous to the generalized momenta. Further, Ψ_r is the flux of induction vector $\mathbf{B}$ through any surface spanned on contour r.

The elements $\dfrac{\partial W}{\partial q_k}$ and $-\dfrac{\partial V}{\partial q_k}$ are caused by the mechanical interaction of magnetic and electric fields and called ponderomotive forces.

If there are no capacitors in the electric circuit, that is, when the currents are closed, the charges g_r do not appear in the expression for the Lagrange function L_a and, thus, they are similar to mechanical quasi-cyclic coordinates. Expressing the currents in terms of the magnetic flux linkage and inserting the result in the expression for the magnetic energy

$$W = \frac{1}{2}\Psi^T L^{-1} \Psi \tag{1.2.16}$$

we obtain the equations which are similar to Routh's equations in mechanics

$$\dot{\Psi}_r + \sum_{s=1}^{m} R_{rs} \frac{\partial W}{\partial \Psi_s} = E_r \qquad (r = 1, \ldots, m),$$

$$\frac{d}{dt}\frac{\partial T}{\partial \dot{q}_k} - \frac{\partial T}{\partial q_k} + \frac{\partial W}{\partial q_k} + \frac{\partial \Pi}{\partial q_k} = Q_k \qquad (k = 1, \ldots, n). \tag{1.2.17}$$

Here $W = W(\Psi_1, \ldots, \Psi_m, q_1, \ldots, q_n)$ and the currents are expressed in terms of the magnetic flux linkage as follows

$$i_s = \frac{\partial W}{\partial \Psi_s} = \sum_{r=1}^{m} L_{rs}^{-1} \Psi_r. \tag{1.2.18}$$

This relationship is similar to the well-known relation between the velocities and momenta in mechanics where L_{rs}^{-1} denote the components of the matrix which is inverse to the inductivity matrix L.

1.3 Stability of stationary motions of systems with quasi-cyclic coordinates and the mechanical equilibrium under the action of magnetic field

In the problem of stationary motions of the electromechanical systems with constant emf the mechanical equilibrium is determined independently of the currents which are considered as parameters for determining the ponderomotive forces. While studying the stability it is necessary to take into account that the currents and the coordinates should be determined together (because one studies the stability of stationary motion rather than the stability of equilibrium of the system subjected to forces depending on parameters). As it will be shown in the present section, for stability of such a motion it is necessary and sufficient that the mechanical equilibrium is stable for non-varied currents, i.e. the currents can be considered as parameters. This conclusion simplifies the analysis of stability and allows one to judge the stability by analysing the change in solution under a change in the currents.

Let us consider a system with the holonomic stationary constraints described by m quasi-cyclic coordinates $(q_1, \ldots, q_m)$ and $n - m$ positional coordinates $(q_{m+1}, \ldots, q_n)$ (according to [76] a coordinate is termed quasi-cyclic if it does not appear in the expression for the kinetic energy and the generalized forces, however the corresponding generalized force is not identically equal to zero). Let us assume that the generalized forces corresponding to quasi-cyclic coordinates are of two types: (i) the dissipative forces depending only on the quasi-cyclic generalized velocities and (ii) constant forces. The generalized forces corresponding to the positional coordinates are considered as being potential forces (the question of the influence of dissipation on the positional coordinates will be discussed in what follows). The "positional" subsystem can be a system with distributed parameters.

The kinetic potential of the described system L and dissipative function F have the form

$$L = T - \Pi = T_1 + U + T_2 - \Pi,$$

$$T_1 = \frac{1}{2} \sum_{r,s=1}^{m} a_{rs}(q_{m+1}, \ldots, q_n)\dot{q}_r\dot{q}_s, \quad \Pi = \Pi(q_{m+1}, \ldots, q_n),$$

$$U = \sum_{r=1}^{m} \sum_{s=1}^{n-m} a_{rm+s}(q_{m+1}, \ldots, q_n)\dot{q}_r\dot{q}_{m+s}, \quad F = F(\dot{q}_1, \ldots, \dot{q}_m),$$

$$T_2 = \frac{1}{2} \sum_{r,s=1}^{n-m} a_{m+rm+s}(q_{m+1}, \ldots, q_n)\dot{q}_{m+r}\dot{q}_{m+s}. \tag{1.3.1}$$

The following stationary motions

$$\dot{q}_r = h_r = \text{const} \quad (r = 1, \ldots, m)$$

$$q_{m+r} = u_r = \text{const} \quad (r = 1, \ldots, n - m) \tag{1.3.2}$$

are possible in the considered systems. Here the constants h_r, u_r are determined from the equations

$$\frac{\partial F(h)}{\partial h_r} = e_r \quad (r = 1, \ldots, m),$$

$$\frac{\partial \Pi(u)}{\partial u_r} = \frac{\partial T_1(h, u)}{\partial u_r} \quad (r = 1, \ldots, n - m). \tag{1.3.3}$$

Here e_r denotes the constant generalized forces corresponding to quasi-cyclic coordinates $h = (h_1, \ldots, h_m)$. For systems with distributed parameters, the second group of equations in (1.3.3) is understood as a conditional notation for the equations of equilibrium of the position subsystem subjected to the forces produced by the quasi-cyclic subsystem.

By virtue of eq. (1.3.3), in the stationary motion the quasi-cyclic velocities do not depend on the positional coordinates. Changing the dissipation and the constant generalized forces we can achieve arbitrary values for these velocities (at least within some bounds). Therefore while determining the possible equilibrium positions of the positional subsystems it is possible to consider that the quasi-cyclic velocities are prescribed in advance. As a result we obtain the problem of equilibrium of the positional subsystem subjected to the forces depending on parameters whereas the quasi-cyclic system is excluded from consideration.

Let us show that the problem of stability of the stationary solutions (1.3.2) is reduced to analysis of stability of equilibrium of the positional subsystem under the assumption that the quasi-cyclic velocities are prescribed (non-varied) parameters.

Let us introduce the perturbations η_r and ζ_r by the relationships $\dot{q}_r = h_r + \dot{\eta}_r, r = 1, \ldots, m, q_{m+r} = u_r + \zeta_r, r = 1, \ldots, n-m$ and write down the equations in variations

$$\sum_{s=1}^{m}[a_{rs}(u)\ddot{\eta}_r + b_{rs}(h)\dot{\eta}_r] + \sum_{s=1}^{n-m}[a_{rm+s}(u)\ddot{\zeta}_s + g_{rm+s}(h, u)\dot{\zeta}_s] = 0,$$

$$(r = 1, \ldots, m),$$

$$\sum_{s=1}^{n-m}[a_{m+rm+s}(u)\ddot{\zeta}_s + (g_{m+rm+s}(h, u) - g_{m+sm+r}(h, u))\dot{\zeta}_s + c_{rs}(h, u)\zeta_s]+$$

$$+ \sum_{s=1}^{m}[a_{sm+r}(u)\ddot{\eta}_s - g_{sm+r}(h, u)\dot{\eta}_s] = 0, \qquad (r = 1, \ldots, n-m). \quad (1.3.4)$$

The following denotations for coefficients are entered here

$$g_{rm+s}(h, u) = \sum_{s=1}^{m} \frac{\partial a_{ri}(u)}{\partial u_s} h_i, \ g_{m+rm+s}(h, u) = \sum_{s=1}^{m} \frac{\partial a_{im+r}(u)}{\partial u_s} h_i,$$

$$b_{rs}(h, u) = \frac{\partial^2 F(u)}{\partial h_r \partial h_s}, \ c_{rs}(h, u) = \frac{\partial^2 \Pi(u)}{\partial h_r \partial h_s} - \frac{1}{2} \sum_{i,j=1}^{m} \frac{\partial a_{ij}(u)}{\partial u_r \partial u_s} h_i h_j. \quad (1.3.5)$$

Let us study stability with respect to variables $\dot{q}_1, \ldots, \dot{q}_m, q_{m+1}, \ldots, q_n$. The original system is obtained from the conservative system by introducing the dissipative forces with the partial dissipation. Therefore the motion (1.3.2) is stable if the variational equations have no unbounded solutions (i.e. undamped vibrations corresponding to pure imaginary roots are allowed) and unstable if the unbounded solutions exist. The same criterion is also applicable for systems with distributed parameters. We do not consider the case in which the solution $\dot{\eta}_1, \ldots, \dot{\eta}_m, \zeta_1, \ldots, \zeta_{n-m}$ is independent of time.

The full dissipation with respect to the quasi-cyclic coordinates and a positive definite matrix $\|b_{rs}\|$ are now assumed.

Let us assume that for the constant quasi-cyclic velocities the equilibrium of the positional subsystem is *unstable* or possesses *a temporal stability*. Let us also admit that none of the solutions ζ_ν of equations describing small vibrations of the positional subsystem for constant quasi-cyclic velocities

$$\sum_{s=1}^{n-m} \left[a_{m+rm+s}(u)\ddot{\zeta}_{\nu s} + (g_{m+rm+s}(h,u) - g_{m+sm+r}(h,u))\dot{\zeta}_{\nu s} + \right. $$
$$\left. + c_{rs}(h,u)\zeta_{\nu s} \right] = 0, \qquad (r = 1, \ldots, n-m) \quad (1.3.6)$$

satisfies simultaneously the following m conditions

$$\sum_{s=1}^{n-m} \left[a_{rm+s}(u)\ddot{\zeta}_{\nu s} + g_{rm+s}(h,u)\dot{\zeta}_{\nu s} \right] = 0 \quad (r = 1, \ldots, m). \quad (1.3.7)$$

Then the corresponding solution (1.3.2) is *unstable*.

Indeed, let us assume that it is stable. It follows from the relationship

$$dH_*/dt = -2F_* \quad (1.3.8)$$

where

$$H_* = T_1(u, \dot{\eta}) + U(u, \dot{\eta}, \dot{\zeta}) + T_2(u, \dot{\zeta}) + \Pi_*(u, \zeta),$$
$$F_* = \frac{1}{2} \sum_{r,s=1}^{m} b_{rs}\dot{\eta}_r\dot{\eta}_s, \qquad \Pi_* = \frac{1}{2} \sum_{r,s=1}^{n-m} c_{rs}\zeta_r\zeta_s,$$

that the velocities $\dot{\eta}_1(t), \ldots, \dot{\eta}_m(t)$ of the stable motion are such that the integral

$$\int_{t_0}^{t} F_*(\tau)d\tau$$

is bounded at $t \to \infty$.

We will show that in this case the system (1.3.4) can not have only solutions satisfying the condition $\dot{\eta}_r, \zeta_r \to 0$ at $t \to \infty$. Indeed, let us choose the initial conditions such that $H_{*0} < 0$. According to eq. (1.3.8) $H_* \le H_{*0}$ for all t. However if $\dot{\eta}_r, \zeta_r \to 0$ then $H_* \to 0$ which contradicts the requirement $H_* \le H_{*0}$.

Thus, system (1.3.4) must have a particular solution $\zeta_s = z_s \cos(\lambda t + \psi_s)$ and at least a part of z_s is not zero and integral of $F_*(t)$ is limited at $t \to \infty$ for such $\dot{\eta}_s$. Let us consider the second sum in one of the first m equations

(1.3.4). It is a linear form in $\dot{\zeta}_s, \ddot{\zeta}_s$ which is either identically equal to zero or equal to $Z_r \cos(\lambda t + \vartheta_r)$ for the above values of ζ_s. The first sum in the considered equation will have the same form. But this sum is a linear form in $\dot{\eta}_s, \ddot{\eta}_s$. Consequently, if both sums are not identically equal to zero then at least one of the functions $\dot{\eta}_s$ must contain a term $x_s \cos(\lambda t + \gamma_s)$. But for such $\dot{\eta}_s$ and positive-definite $\|b_{rs}\|$ the integral of F_* is not bounded. Therefore for the stable motion all $2m$ sums must be identically equal to the zero.

In the same manner we will prove that each of $2(n - m)$ sums in the second group of equations (1.3.4) must be identically equal to zero. As a result we obtain that ζ_r satisfy both equations (1.3.6) and conditions (1.3.7) which is impossible. Therefore system (1.3.4) can not have only the solutions for which integral of $F_*(t)$ is bounded and ζ_r are bounded or tend to zero at $t \to \infty$. This proves the above mentioned instability.

Equalities (1.3.7) correspond to the cases when some of the unknowns in eq. (1.3.4) are determined independently of the others. As before, instability of the positional subsystem leads to instability of the whole system (1.3.4). However in the case of temporal instability of the positional subsystem the complete system can be stable or unstable.

Let the positional subsystem be unstable for the constant quasi-cyclic velocities. If unbounded solution $\dot{\zeta}_\nu$ satisfies conditions (1.3.7) then system (1.3.4) has an unbounded solution of the form $\dot{\eta} \equiv 0, \zeta = \zeta_\nu$ which means instability. Let us assume that several bounded solutions ζ_ν, ζ_μ etc. satisfy conditions (1.3.7). The solutions of the sort $\dot{\eta} \equiv 0, \zeta = \zeta_\nu, \zeta_\mu, \ldots$, will be the particular solutions of system (1.3.4). Let us consider a set of solutions which is linearly independent of these solutions. The above proof of instability is valid for them.

Let us consider the following special case of *temporal stability* of a positional subsystem and let the particular solutions ζ_ν, satisfy the requirement that the inequality $T_2(\dot{\zeta}_\nu) + \Pi_*(\zeta_\nu) < 0$ holds for some t. We assume that all these ζ_ν satisfy conditions (1.3.7). Then for any solution $\dot{\eta}, \zeta$ of system (1.3.4) which is linearly independent of all solutions $\dot{\eta} \equiv 0, \zeta = \zeta_\nu$, the inequality $H_*(\dot{\eta}, \zeta, \dot{\zeta}) > 0$ holds for all t, that is, the previous proof of instability is invalid. In this case (and only in this case) the temporal stability of the total system (1.3.4) can be maintained. A trivial example of temporal stability in that in which all $a_{sm+r} g_{sm+r}$ are equal to zero and system (1.3.4) is split into two uncoupled subsystems.

If a positional subsystem possesses the *secular stability* for non-varied quasi-cyclic velocities, then the solution (1.3.2) is stable. Indeed, let us consider a system whose vibrations are described by the equations of perturbed motion. The following energy relationship

$$dH_{*1}/dt = -2F_{*1},$$

$$H_{*1} = H_* + H_{*2}(\dot{\eta}, \dot{\zeta}, \zeta), \qquad F_{*1} = F_* + F_{*2}(\dot{\eta}) \qquad (1.3.9)$$

holds, in which the expansion of H_{2*} and F_{2*} in power series in terms of their arguments begins with the terms of degrees greater than two. Therefore H_{*1} is positive-definite for sufficiently small absolute values of arguments whereas its time derivative calculated by means of the equations of perturbed motion is not positive.

If $g_{m+rm+s} - g_{m+sm+r} = 0, r, s = 1, \ldots, n - m$ (for example, in the case of the gyroscopically uncoupled system at $U = 0$) the equations of small vibrations of the positional subsystem for the constant quasi-cyclic velocities do not contain gyroscopic terms and the temporal stability is impossible. In this case the stability of stationary motions is uniquely determined by properties of the equilibrium and does not depend upon whether dissipation with respect to the position coordinates is taken into account. In general case, instability and secular stability are determined by the properties of equilibrium (regardless of dissipation in the positional subsystem). Also in this case the judgement of stability is unambiguous provided that the dissipative forces corresponding to the positional coordinates are taken into account.

An important example of systems of the considered class is the electromechanical systems with the closed currents (i.e. those without capacities and sliding contacts). In many cases such systems are rather exactly described in the quasi-stationary approximation. If, in addition to this, the cross-sectional dimensions of the explorers are small in comparison to their length, and the active resistances do not depend on the displacements, the system will have the kinetic potential and the dissipative function of the above kind. The part of the quasi-cyclic coordinates and velocities is played by charges and currents respectively, the constant "quasi-cyclic" generalized forces are the external emf, the dissipation with respect to quasi-cyclic coordinates is stipulated by the active resistances of explorers and the positional coordinates are the mechanical generalized coordinates. The energy of magnetic field corresponds to term T_1 in the expression for kinetic energy, and the derivatives of the energy of magnetic field with respect to q_{m+r} determine the ponderomotive forces. The expression for T usually contains no terms with U.

In the case of electromechanical systems the stationary solution describes the constant currents and the mechanical equilibrium in a constant magnetic field. The results obtained allow one to ignore the "electric" degrees of freedom (i.e. electric chart, details of power supply etc.). Additionally these allow one to judge the stability by means of dependences of the equilibrium forms on the currents by Poincaré's theory of bifurcation.

The above-said is also valid for the "magnetically nonlinear" electromechanical systems provided that it is possible to neglect the hysteresis. The expression for T will differ from the above one only in that another function of currents appears instead of the quadratic form T_1. Therefore it is sufficient to show that the form

$$\sum_{r,s=1}^{m} \frac{\partial T_1}{\partial h_r \partial h_s} \dot{\eta}_r \dot{\eta}_s \tag{1.3.10}$$

is positive definite. Let us restrict our consideration to the case when the induction vector $\mathbf{B}$ and magnetic induction vector $\mathbf{H}$ are parallel in the magnetics and the dependence $B(H)$ is an increasing function. For parallel $\mathbf{B}$ and $\mathbf{H}$ we have

$$T_1(h_1,\dots,h_m) = \int\limits_{V} dv \int\limits_{0}^{H} B(H)dH, \tag{1.3.11}$$

where the first integral is evaluated over the whole space. Let us obtain an increment $\Delta T_1 = T_1(h + \dot{\eta}) - T_1(h)$ keeping the squares of $\dot{\eta}_1,\dots,\dot{\eta}_m$. We designate an increment in the magnetic induction by $\Delta \mathbf{H} = \mathbf{H}(h + \dot{\eta}) - \mathbf{H}(h)$. Retaining the terms of second order in $\Delta \mathbf{H}$ yields

$$\Delta T_1 = \int\limits_{V} dv \left[\mathbf{B} \cdot \Delta \mathbf{H} + \frac{H^2(\Delta H)^2 - (\mathbf{H} \cdot \Delta \mathbf{H})^2}{2H^3} B + \right.$$

$$\left. + \frac{1}{2} \frac{dB}{dH} \left(\frac{\mathbf{H} \cdot \Delta \mathbf{H}}{H} \right)^2 \right]. \tag{1.3.12}$$

In the magnetically nonlinear case $\Delta \mathbf{H}$ includes not only the first but also the higher degrees of increase in the currents. It follows from the relationship [71]

$$\int\limits_{V} \mathbf{B} \cdot \Delta \mathbf{H} dv = \sum_{r=1}^{m} \Phi_r \dot{\eta}_r, \tag{1.3.13}$$

where Φ_r denotes the magnetic flux through the contour of $r - th$ current, that the integral of $\mathbf{B} \cdot \Delta \mathbf{H}$ is a linear form in $\dot{\eta}_1,\dots,\dot{\eta}_m$. Therefore the expression for the form (1.3.10) is obtained from the last two terms in eq. (1.3.12) if $\Delta \mathbf{H}$ contains only the linear terms in $\dot{\eta}_1,\dots,\dot{\eta}_m$. These terms vanish only at $\dot{\eta}_1,\dots,\dot{\eta}_m = 0$. Taking into account the structure of these terms in eq. (1.3.12) we conclude that the form (1.3.10) is positive definite.

In the case of cyclic coordinates it is possible to prove the stability of a stationary motion in terms of stability of equilibrium of the positional subsystem by means of Routh's theorem generalized by Lyapunov and its complements, cf. [85]. In this case rather the cyclic momenta than the velocities are considered to be constant. The stability conditions obtained with the help of Routh's theorem are broader than the stability conditions for the same system with the quasi-cyclic coordinates.

Let us prove this for the case $U = 0$. We designate the cyclic momenta and Routh's function by $p_1, \ldots, p_m$ and $V_R = \Pi + T_1(p, u)$ respectively. The solution $q_{m+r} = u_r = \text{const}, r = n - m$ is stable if the quadratic form

$$\sum_{r,s=1}^{n-m} \frac{\partial^2 V_R}{\partial u_r \partial u_s} v_r v_s \tag{1.3.14}$$

is positive definite and unstable if this form is negative for some $v_1, \ldots, v_{n-m}$. When the coordinates $q_1, \ldots, q_m$ are quasi-cyclic it is necessary to insert function $V = \Pi - T_1(h, u)$ in eq. (1.3.14) instead of V_R. Let $\dot{q}_r = h_r, r = 1, \ldots, m$ be the same for the systems with cyclic and quasi-cyclic coordinates. Let us construct the form (1.3.14) and an analogous form with V for these solutions by taking into account that the matrix of coefficients $\|a_{rs}^{(-1)}\|$ in $T_1(r, u)$ is inverse to matrix $\|a_{rs}\|$. As a result we obtain that the difference between eq. (1.3.14) and the second form is equal to the non-negative value

$$\sum_{r,s=1}^{m} a_{rs}^{(-1)} f_r f_s,$$

where

$$f_r = \sum_{i-1}^{n-m} \sum_{s-1}^{m} \frac{\partial a_{rs}}{\partial u_i} h_s v_i.$$

Routh's theorem for the electromechanical systems is also valid for the systems with superconducting contours. Let us ignore those exceptional cases when both quadratic forms are equal to zero for the same $v_1, \ldots, v_{n-m}$. The equilibrium forms in the magnetic field which are stable in the field due to the contours of finite conductivity will be also stable in the case of the superconductivity. However there exist forms which are stable only for the superconducting contours. Hence the systems with superconducting contours possess new qualitative features in a "pure mechanical" sense.

As an example of the problem of equilibrium of electromechanical system with closed currents we consider a system of two rings with currents. For the sake of simplicity we assume that the rings are identical and have a common vertical axis. The upper ring is fixed whereas the lower one hangs freely in the gravity field, cf. Fig. 1.1.

In order to ensure a magnetic attraction force between the rings with non-zero resistance (the case of superconducting rings is considered below) an external emf is needed in each of the conducting rings. One expects that the equilibrium is unstable in the case of constant emf in each contour. Stability of the equilibrium is possible to ensure in two ways: either by introducing a variable component in the external emf or by using superconducting rings, cf. [64].

Let us first consider the case of constant emf in the contours of current-conducting rings, that is $U_1, U_2 = \text{const}$. The magnetic energy of the system

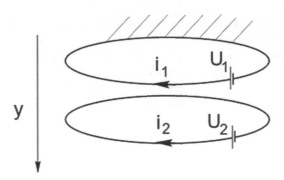

FIGURE 1.1.

of two rings can to be written in the form

$$W = \frac{1}{2}Li_1^2 + \frac{1}{2}Li_2^2 + M(y)i_1i_2 \, ,$$ (1.3.15)

where L and $M(y)$ denote the internal and mutual inductance of the rings respectively. It is assumed that the lower ring can move only along the vertical axis (axis y is directed downward and y denotes the vertical displacement). The coefficient of mutual induction is a rather complex function of the distance between the rings

$$M(y) = \mu_0 R \left[\left(\frac{2}{k} - k \right) K(k) - \frac{2}{k} E(k) \right] ,$$ (1.3.16)

where

$$k = \frac{1}{\sqrt{1 + \left(\dfrac{y}{2R} \right)^2}} < 1,$$

and functions

$$K(k) = \int\limits_0^{\pi/2} \frac{d\varphi}{\sqrt{1 - k^2 \sin^2 \varphi}}, \quad E(k) = \int\limits_0^{\pi/2} \sqrt{1 - k^2 \sin^2 \varphi}\, d\varphi$$

denote the complete elliptic integrals of the first and second kind respectively. Function $M(y)$ decreases monotonically from L to 0 which corresponds to reducing the magnetic-flux linkage of the rings with increase in the distances between them.

Let us construct the Lagrange-Maxwell equations for the described electromechanical system.

$$L\ddot{i}_1 + M(y)\ddot{i}_2 + \frac{\partial M}{\partial y}i_2\dot{u} + Ri_1 = U_1,$$

$$L\ddot{i}_2 + M(y)\ddot{i}_1 + \frac{\partial M}{\partial y}i_1\dot{u} + Ri_2 = U_2,$$

$$m\ddot{u} - \frac{\partial M}{\partial y}i_1 i_2 = mg \qquad (1.3.17)$$

and find the equilibrium position of system (1.3.17). The corresponding stationary values of the currents are given by

$$i_1 = \frac{U_1}{R}, \quad i_2 = \frac{U_2}{R}. \qquad (1.3.18)$$

The coordinate y_* of the equilibrium positions is determined from the third equation of system (1.3.17)

$$-\frac{\partial M}{\partial y}\frac{U_1 U_2}{R^2} - mg. \qquad (1.3.19)$$

The derivative $\dfrac{\partial M}{\partial y} < 0$ and varies from $-\infty$ to 0, thus for any values of U_1 and U_2 of the same signs there exists a unique equilibrium position. As shown above, the mechanical equilibrium in the problem of stationary motion is determined regardless of the problem of determining the currents which are simply the parameters in eq. (1.3.19). For stability of such a motion it is necessary and sufficient that the mechanical equilibrium is stable for the non-varied currents. For the stationary currents the equation in variations about the equilibrium position obtained from eq. (1.3.19) has the form

$$m\ddot{\xi} - \frac{\partial^2 M}{\partial u^2}\bigg|_{y=y_*}\frac{U_1 U_2}{R^2}\xi = 0. \qquad (1.3.20)$$

The derivative $\dfrac{\partial^2 M}{\partial u^2}$ is positive for any $y \in (0, \infty)$. Thus the characteristic equation for (1.3.20) has real-valued roots of different signs, i.e. the equilibrium position is a saddle.

Let us consider an equilibrium in the system of two superconducting rings. The magnetic-flux linkage which are introduced according to eq. (1.3.15)

$$\Psi_1 = Li_1 + M(y)i_2, \quad \Psi_2 = M(y)i_1 + Li_2 \qquad (1.3.21)$$

are constant momenta whilst the currents i_1 and i_2 are the cyclic velocities. Routh's equation for this system corresponding to the vertical displacement of the ring is written down as follows

$$m\ddot{y} + \frac{LM(\Psi_1^2 + \Psi_2^2) - (L^2 + M^2)\Psi_1\Psi_2}{(L^2 - M^2)^2}\frac{\partial M}{\partial y} = mg. \qquad (1.3.22)$$

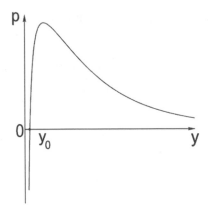

FIGURE 1.2.

The magnetic-flux linkage constants Ψ_1, Ψ_2 appear in the expression for ponderomotive forces as parameters. The quadratic form $LM(\Psi_1^2 + \Psi_2^2) - (L^2 + M^2)\Psi_1\Psi_2$ is not of a constant sign. It is easy to prove without Sylvester's criterion. Indeed, we have

$$LM(\Psi_1^2 + \Psi_2^2) - (L^2 + M^2)\Psi_1\Psi_2 = -(L^2 - M^2)^2 i_1 i_2 = \qquad (1.3.23)$$
$$= -(\Psi_1 L - \Psi_2 M)(\Psi_2 L - \Psi_1 M).$$

For any values of Ψ_1 and Ψ_2 of the same sign there is such a value of coordinate y_0 for which one of the parentheses, and consequently one of the currents is equal to zero (this is a result of monotonically decreasing function $M(y)$). This position corresponds to vanishing electromagnetic force. For $y < y_0$ the electromagnetic force is negative which corresponds to repulsion of the rings whereas for $y > y_0$ the force is positive which corresponds to the attraction of them. For $y \to \infty$ the attraction weakens and tends to zero. Thus, the electromagnetic force has a maximum in the range (y_0, ∞) corresponding to the point of vanishing rigidity of the suspension. It is easy to prove directly by taking derivative of the expression for the electromagnetic force with respect to coordinate y. A qualitative character of dependence of the electromagnetic forces on y is shown in Fig. 1.2.

When the weight of the lower ring is less than the value of maximum of the attraction force the system has two equilibrium positions. It is easy to prove that the equilibrium position with the smaller distance between the rings (y_1) is stable whereas that with the larger distance y_2 is unstable. The range $(0, y_2)$ is the domain of attraction of the stable equilibrium positions. For $y > y_2$ the magnetic attraction force is not sufficient for stability of the lower ring.

1.4   Motion of electromechanical systems due to constant external electromotive forces

Numerous technical facilities (some sensors, contactors, actuators etc.) are the systems with constant external emf. In these cases only frictional forces are the non-potential generalized forces. For such systems it is possible to classify the possible motions.

First we consider the systems with closed linear currents. Let us assume that the dissipative generalized forces $Q_1, \ldots, Q_n$ corresponding to the mechanical generalized coordinates $q_1, \ldots, q_n$ satisfy the requirement

$$P = \sum_{j=1}^{n} Q_j(q_1, \ldots, q_n, \dot{q}_1, \ldots, \dot{q}_n)\dot{q}_j \leq 0, \qquad (1.4.1)$$

the equality $P = 0$ being arrived only at $\dot{q}_1 = \ldots = \dot{q}_n = 0$. Coordinates $q_1, \ldots, q_n$ can also have potential generalised forces. We assume that the electromagnetic processes are quasi-stationary and the mechanical constraints are stationary and holonomic.

Let us consider the case of constant external emf. The active resistances are assumed to be independent of mechanical coordinates however, generally speaking, they depend on the currents, voltage drops across the resistances being increasing functions of the currents. Let us show that undamped bounded mechanical motions are impossible in the system.

We consider the general case of parallel electric circuits. Using denotations of Section 1.2 we write down the "electrical" subsystem of the Lagrange-Maxwell equations in the form

$$\dot{\Phi}_s + \sum_{k=1}^{N} \beta_{sk} R_k(I_k)I_k = e_s, \qquad s = 1, \ldots, m. \qquad (1.4.2)$$

Here $\Phi_s = \partial W/\partial i_s$ denotes the magnetic flux in $s-th$ contour, $W(i,q)$ is the energy of magnetic field, R_k is the resistance of $k-th$ branch and $e_s = $ const denotes the combination of external emf in the branches forming $s-th$ contour.

The second group of the Largange-Maxwell equations are the equations of motion

$$\frac{d}{dt}\frac{\partial T}{\partial \dot{q}_k} - \frac{\partial(T - \Pi)}{\partial q_k} - \frac{\partial W}{\partial q_k} = Q_k, \qquad k = 1, \ldots, n, \qquad (1.4.3)$$

where T and Π denote respectively the kinetic and potential energy of the system.

Next it is assumed, that functions T, Π, W, P are prescribed and bounded (as well as their first derivatives with respect to all arguments) in some region D of the spaces of variables $q, \dot{q}, i$, the points of $\dot{q}$ and $i = 0$ belonging to D. Let us also assume that there are a "sufficient number" of resistances R_k such that the circuits have no contours with zero resistances, however the resistances of some branches R_k can be equal to zero. Finally it is assumed that functions $U_k(I_k) = R_k(I_k)I_k$, which are not identically equal to zero, are prescribed for $i \in D$, have bounded derivatives with respect to I_k and are increasing functions of I_k.

Let us consider the equations for determining the stationary values of currents I_{k0}, i_{s0} in the case of no mechanical motions

$$\sum_{k=1}^{N} \beta_{sk} U_k(I_{k0}) = e_s, \qquad s = 1, \ldots, m,$$

$$I_k = \sum_{s=1}^{m} \beta_{sk} i_{s0}, \qquad k = 1, \ldots, N. \tag{1.4.4}$$

We will show that in the region of the values of emf, including the point $e_1 = \ldots = e_m = 0$, eqs. (1.4.4) are uniquely resolved. To this end, let us consider the dependences $I_{k0}(e_1, \ldots, e_m)$ and $i_{s0}(e_1, \ldots, e_m)$. We rewrite eqs. (1.4.4) in the form

$$F_k(I_{10}, \ldots, I_{N0}, i_{10}, \ldots, i_{m0}, e_1, \ldots, e_m) = 0. \tag{1.4.5}$$

For $e_1 = \ldots = e_m = 0$ eqs. (1.4.4) or (1.4.5) have a solution if all $I_{k0}, i_{s0} = 0$. We consider the functional determinant

$$\Delta = \frac{D(F_1, \ldots, F_{N+m})}{D(I_{10}, \ldots, I_{N0}, i_{10}, \ldots, i_{m0})}. \tag{1.4.6}$$

Let us designate $R_{k0} = (\partial U_k / \partial I_k)_0$ and consider the equations for $J_{10}, \ldots, J_{N0}$ $j_{10}, \ldots, j_{m0}$

$$\sum_{k=1}^{N} \beta_{sk} R_{k0} J_{k0} = e_{s0}, \qquad s = 1, \ldots, m,$$

$$J_{k0} = \sum_{s=1}^{m} \beta_{sk} j_{s0}, \qquad k = 1, \ldots, N. \tag{1.4.7}$$

Here e_{s0} are any numbers, $R_{k0} = 0$ if the corresponding resistance $R_k = 0$ and $R_{k0} > 0$ for $R_k \neq 0$. Equations (1.4.7) describe the distribution of direct currents created by emf e_{s0} in the circuit which differs from the considered in that the branches include the resistances R_{k0} (which do not depend on currents) instead of R_k. By assumption, this circuit does not contain independent contours with zero resistance. For such a circuit the equations in eq. (1.4.7) are uniquely resolvable for any e_{s0}.

Equations (1.4.7) form a linear system for J_{k0}, j_{s0} which is uniquely resolvable for any e_{s0}, hence its determinant is not zero. On the other side, this determinant coincides with the determinant Δ obtained at the "zero" point.

The obtained relationship $\Delta \neq 0$ and accepted assumptions about the smoothness of functions U_k allow one to apply the theorem on explicit functions (see for example [120]). Therefore, in a neighbourhood of the zero point there are unique dependences $I_{k0}(e_1, \ldots, e_m), i_{s0}(e_1, \ldots, e_m)$.

We will show that these dependences are continuously and uniquely extended on the whole domain where function $U_k(I_k)$ is determined and has the above mentioned properties. In other words, we will show that functions $I_{k0}(e_1, \ldots, e_m)$ and $i_{s0}(e_1, \ldots, e_m)$ have no points of bifurcation with respect to parameters $e_1, \ldots, e_m$. Let us suppose that such a point is revealed and consider the determinant Δ at this point. In the same manner as before we can establish that it differs from zero. However it is impossible at the point of bifurcation, hence the point of bifurcation does not exist at all.

Finally we show that the considered $m-$parametric branch of solutions of eq. (1.4.4) is unique. Let equations in (1.4.4) have two solutions $I_{k0}^{(1)}, i_{s0}^{(1)}$ and $I_{k0}^{(2)}, i_{s0}^{(2)}$ for the same values of $e_1, \ldots, e_m$. We rewrite the first equation in (1.4.4) in the form

$$e_s - \sum_{k=1}^{N} \beta_{sk} U_k(I_{k0}^{(1)}) = 0, \qquad s = 1, \ldots, m. \qquad (1.4.8)$$

We multiply both parts of this equation by the difference $i_{s0}^{(1)} - i_{s0}^{(2)}$ and sum the result over s. Taking into account that $I_{k0}^{(2)}, i_{s0}^{(2)}$ are also the solution of eq. (1.4.4) we obtain

$$\sum_{s=1}^{m} \left[e_s - \sum_{k=1}^{N} \beta_{sk} U_k(I_{k0}^{(1)}) \right] (i_{s0}^{(1)} - i_{s0}^{(2)}) =$$

$$= \sum_{s=1}^{m} \left[\sum_{k=1}^{N} \beta_{sk} U_k(I_{k0}^{(2)}) - \sum_{k=1}^{N} \beta_{sk} U_k(I_{k0}^{(1)}) \right] (i_{s0}^{(1)} - i_{s0}^{(2)}) =$$

$$= \sum_{k=1}^{N} [U_k(I_{k0}^{(2)}) - U_k(I_{k0}^{(1)})](I_{k0}^{(1)} - I_{k0}^{(2)}) = 0. \qquad (1.4.9)$$

Each member in the last sum in eq. (1.4.9) is not positive, therefore all members must vanish which is possible only if $I_{k0}^{(1)} = I_{k0}^{(2)}$.

Let us show that the electromechanical systems with the constant external emf and the dissipative generalized forces possess the property of dichotomy, i.e. all bounded motions tend to stationary ones. We use the following power relationship

$$\dot{H} = P + \sum_{s=1}^{m} \left[e_s - \sum_{k=1}^{N} \beta_{sk} U_k(I_k) \right] i_s, \qquad (1.4.10)$$

where $H = W + \Pi + T$ denotes the total energy of the electromechanical system. Let us assume that system (1.4.4) has a solution for the considered values of e_s. We construct the function

$$V = H - \sum_{s=1}^{m} i_{s0}\Phi_s. \tag{1.4.11}$$

Differentiating V and replacing Φ_s and e_s by means of eqs.(1.4.2) and (1.4.4) respectively we obtain

$$\dot{V} = P + \sum_{s=1}^{m}\left[e_s - \sum_{k=1}^{N}\beta_{sk}U_k(I_k)\right]i_s - \sum_{s=1}^{m}\left[e_s - \sum_{k=1}^{N}\beta_{sk}U_k(I_k)\right]i_{s0} =$$

$$= P + \sum_{s=1}^{m}\left[\sum_{k=1}^{N}\beta_{sk}U_k(I_{k0}) - \sum_{k=1}^{N}\beta_{sk}U_k(I_k)\right](i_s - i_{s0}) = P + K. \tag{1.4.12}$$

Here

$$K = \sum_{k=1}^{N}[U_k(I_{k0}) - U_k(I_k)](I_k - I_{k0})$$

and the terms P and K in the expression for $\dot{V}$ are non-positive.

We consider the bounded motions in the system, i.e. such motions that for all t the vector-columns $q, \dot{q}, i$ remain in a bounded domain of the space of these variables contained in D. Let us show that these motions are either stationary of the type $I_k = I_{k0}, i_s = i_{s0}, q_1, \ldots, q_n = $ const or tend to stationary ones.

Indeed, for a bounded motion function V is also bounded and consequently the integral

$$\int_{t_0}^{t}[P(\tau) + K(\tau)]d\tau$$

is bounded at all t. Forasmuch as P and K are non-positive, each of the integrals of P and K is bounded.

We will take into account that under the considered motion $\dot{P}$ and $\dot{K}$ are bounded because the accelerations $\ddot{q}_k$ and the derivative of currents i_m in the expressions for $\dot{P}$ and $\dot{K}$ are bounded by virtue of the Lagrange-Maxwell equations (1.4.2) and (1.4.3). Hence the integrals of the functions which have constant signs and bounded derivative with respect to t are bounded for all t. However it is possible only under the condition that these functions tend to zero at $t \to \infty$.

Let us notice that this statement is not correct for a function with an unbounded derivative. Such a function can have a bounded integral and does not tend to zero. For example, this is the case in which a function has an infinite number of "splashes" of decreasing area and not decreasing height.

Thus we proved that functions P and K tend to zero at $t \to \infty$ if the motion is bounded. This is possible only if $\dot{q}_1, \ldots, \dot{q}_n$ tend to zero whilst I_k and i_s tend respectively to I_{k0} and i_{s0}. Differently, any bounded motion is either stationary or tending to one of stationary motions.

Under the stationary motion the currents are direct and do not depend on the mechanical coordinates, thus the mechanical equilibrium is due to the electromagnetic forces. However such a solution of the Lagrange-Maxwell equations is a stationary motion rather than an equilibrium because the currents are analogous to the generalized velocities in mechanics.

These conclusions are valid for the systems with non-closed currents (e.g. the systems with capacitors). Let the energy of electric field E depend on $l \leq m$ independent charges $g_1, \ldots, g_l$ and probably mechanical coordinates. Then it is possible to take $i_1 = \dot{g}_1, \ldots, i_l = \dot{g}_l$ and the antiderivatives of the other $m - l$ currents do not appear in the Lagrange-Maxwell equations. The Lagrange-Maxwell equations for electromechanical systems with capacitors are as follows

$$\dot{\Phi}_s + \sum_{k=1}^{N} \beta_{sk} U_k(I_k) + \frac{\partial E}{\partial g_s} = e_s, \qquad s = 1, \ldots, l,$$

$$\dot{\Phi}_s + \sum_{k=1}^{N} \beta_{sk} U_k(I_k) = e_s, \qquad s = l+1, \ldots, m,$$

$$\frac{d}{dt}\frac{\partial T}{\partial \dot{q}_k} - \frac{\partial(T - \Pi - E)}{\partial q_k} - \frac{\partial W}{\partial q_k} = Q_k, \qquad k = 1, \ldots, n. \qquad (1.4.13)$$

As above, the total energy of the system $H = W + T + \Pi + E$ satisfies relationship (1.4.10). A motion under which the mechanical coordinates, velocities, currents and charges $g_1, \ldots, g_l$ change for $t \in [t_0, \infty)$ in a bounded domain is referred to as the bounded motion. Under such a motion the function

$$V = H - \sum_{s=l+1}^{m} i_{s0} \Phi_s \qquad (1.4.14)$$

is bounded. Here the stationary values of currents i_{s0} are determined from the system

$$\sum_{k=1}^{N} \beta_{sk} U_k(I_{k0}) = e_s, \qquad s = l+1, \ldots, m,$$

$$I_{k0} = \sum_{s=l+1}^{m} \beta_{sk} i_{s0}, \qquad k = 1, \ldots, N. \qquad (1.4.15)$$

The other currents $i_{s0}, s = 1, \ldots, l$ are equal to zero.

Differentiating eq. (1.4.14) and replacing $\dot{\Phi}_s$ $(s = l+1, \ldots, m)$ by means of eq. (1.4.13) and e_s $(s = l+1, \ldots, m)$ by means of eq. (1.4.15) we obtain

$$
\dot{V} = P + \sum_{s=1}^{l} \left[e_s - \sum_{k=1}^{N} \beta_{sk} U_k(I_k) \right] i_s + \sum_{s=l+1}^{m} i_{s0} \left[e_s - \sum_{k=1}^{N} \beta_{sk} U_k(I_k) \right] i_s -
$$

$$
- \sum_{s=l+1}^{m} i_{s0} \left[e_s - \sum_{k=1}^{N} \beta_{sk} U_k(I_k) \right] = P + \sum_{s=1}^{l} e_s \dot{g}_s - \sum_{s=1}^{l} i_s \sum_{k=1}^{N} \beta_{sk} U_k(I_k) +
$$

$$
+ \sum_{s=l+1}^{m} (i_s - i_{s0}) \left[\sum_{k=1}^{N} \beta_{sk} U_k(I_{k0}) - \sum_{k=1}^{N} \beta_{sk} U_k(I_k) \right]. \qquad (1.4.16)
$$

Let us add and subtract $\sum_{s=1}^{l} i_s \sum_{k=1}^{N} \beta_{sk} U_k(I_{k0})$ in the right hand side of expression (1.4.16). The result is

$$
\dot{V} = P + \sum_{s=1}^{l} e_s \dot{g}_s + \sum_{s=1}^{l} i_s \left[\sum_{k=1}^{N} \beta_{sk} U_k(I_{k0}) - \sum_{k=1}^{N} \beta_{sk} U_k(I_k) \right] +
$$

$$
\sum_{s=l+1}^{m} (i_s - i_{s0}) \left[\sum_{k=1}^{N} \beta_{sk} U_k(I_{k0}) - \sum_{k=1}^{N} \beta_{sk} U_k(I_k) \right] - \sum_{s=1}^{l} i_s \sum_{k=1}^{N} \beta_{sk} U_k(I_{k0}) =
$$

$$
= P + \sum_{s=1}^{l} e_s \dot{g}_s + \sum_{s=1}^{m} (i_s - i_{s0}) \left[\sum_{k=1}^{N} \beta_{sk} U_k(I_{k0}) - \sum_{k=1}^{N} \beta_{sk} U_k(I_k) \right] -
$$

$$
- \sum_{s=1}^{l} \dot{g}_s \sum_{k=1}^{N} \beta_{sk} U_k(I_{k0}) = P + K + \sum_{s=1}^{l} e_s \dot{g}_s - \sum_{s=1}^{l} \dot{g}_s \sum_{k=1}^{N} \beta_{sk} U_k(I_{k0}).
$$

$$
(1.4.17)
$$

Introducing the function

$$
V_1 = V - \sum_{s=1}^{l} e_s g_s + \sum_{s=1}^{l} g_s \sum_{s=1}^{l} \sum_{k=1}^{N} \beta_{sk} U_k(I_{k0}),
$$

we can rewrite eq. (1.4.17) in the form $\dot{V}_1 = P + K$.

In a similar fashion we come to the conclusion that $P \to 0, K \to 0$ at $t \to \infty$ and the bounded motions are either stationary or tend to them. If the solutions of equations (1.4.15) do not exist, the stationary and bounded motions do not exist as well.

If $W, T, \Pi, Q_1, \ldots, Q_n$ are bounded, for instance they are continuous-periodic functions of some mechanical coordinates, we can show by analogy that there are no motions for which these coordinates unboundedly increase or decrease whereas the corresponding generalized velocities are bounded and do not tend to zero at $t \to \infty$. For this reason, there are no undamped rotatory motions with bounded angular velocities. In particular, this provides us with the proof that the commutatorless direct current motor is not feasible.

The obtained results are obviously generalized to the pure mechanical systems with quasi-cyclic coordinates for which the dissipative and potential forces correspond to the positional coordinates whilst the dissipative and constant generalized forces correspond to the quasi-cyclic coordinates.

The above proofs are based on the energy relations (1.4.10) which can be constructed for the mechanical systems with distributed parameters. Therefore these results are also valid for such systems as well as for the systems with volume conductors. Indeed, by using the discrete description of the body currents [90] we arrive at the same relationships, however summation in equations (1.4.2) is carried out over a countable set of contours.

2

Motion of the conducting rigid body in alternating magnetic field

2.1 Asymptotic transformation of the equations of motion for conducting rigid body in alternating magnetic field

It is known that the alternating magnetic forces can destabilize a stable equilibrium position which results in self-oscillations. Mathematically, this effect is related to change in the structure of the Lagrange-Maxwell equations. In addition to the gyroscopic forces the circulational generalized forces due to the alternating external field appear in these equations.

Let us consider in detail the problem of the slow motion of a conducting rigid body in the alternating magnetic field. We assume that the distribution of eddy currents in the rigid body can be presented as an expansion in terms of a complete system of solenoidal vectorial functions $\mathbf{S_r}$, the latter being constant in the coordinate system fixed in the rigid body [90]

$$\mathbf{j}(t, h, u, z) = \sum_{r=1}^{\infty} i_r(t)\mathbf{S}_r(h, u, z). \tag{2.1.1}$$

The expression for the energy of magnetic field can be presented as the sum

$$W = \frac{1}{2}i^{\mathrm{T}}Li + J(t)^{\mathrm{T}}M(q)i. \tag{2.1.2}$$

Here the infinite-dimensional vector i of the coefficients of decomposition (2.1.1) is equivalent to the vector of currents in the "fictitious" contours of the rigid body. Thus, the rigid body is presented as a countable or finite system of conducting contours whose mutual position does not change under the motion. This corresponds to a matrix L of constant coefficients of self-induction and mutual induction. Next, $J(t)$ denotes the vector of currents prescribing the alternating external field and $M(q)$ is a rectangular matrix of the coefficients of mutual induction of the contours in the rigid body and the contours of external sources of the alternating field.

Let our consideration be limited by the case of periodic or quasi-periodic function $J(t)$, that is,

$$J(t) = \sum_k (I_k \cos \nu_k t + J_k \sin \nu_k t) \qquad (2.1.3)$$

and let us assume that the differences in frequencies $\nu_i - \nu_j$ have approximately the order of the smallest frequency ν_1. The electromagnetic forces $F_e = J^T \partial M / \partial q$ have the components with frequencies of order ν_1 and higher. These components cause mechanical oscillations with the amplitudes of the order of $[F_{er}]/\left(\nu_1^2[A_{rr}]\right), r = 1, \dots, n$, where the square brackets denote the characteristic value of the corresponding quantity. In the case of high-frequency current J these amplitudes are small in comparison with the characteristic value $[q_r]$ of the coordinate q_r.

Let us introduce a small parameter ε by means of the following relation: $\varepsilon^2 \sim [F_{er}]/(\nu_1^2[A_{rr}][q_r])$. In the problem of slow motions the values of momenta are assumed to be of the order of ε in comparison to value $\nu_1[A_{rr}][q_r]$. Hence, the generalized forces $Q\left([Q_r] = \varepsilon^2 \nu_1^2[A_{rr}][q_r]\right)$ should be considered as being small, i.e. of the order of forces $[F_{er}]$.

Let us now introduce the dimensionless time which is equal to the product of the dimensional time and ν_1, the dimensionless coordinates and currents as the ratios of the dimensional values to their characteristic values and the dimensionless momenta which are equal to the ratios of the momenta to $\varepsilon \nu_1[A_{rr}][q_r]$.

Keeping the previous denotation for the dimensionless momenta we arrive at the equations

$$\dot{q} = \varepsilon A^{-1} p, \quad \dot{p} = \varepsilon \left(-\frac{1}{2} p^T \frac{\partial A^{-1}}{\partial q} p + J \frac{\partial M}{\partial q} i + Q \right),$$
$$L\ddot{i} + R i + (J M^T)^{\cdot} = 0. \qquad (2.1.4)$$

The above assumptions and equations (2.1.4) are used in a number of technical applications: orientation of details by magnetic field, suspension of a rigid body for the non-crucible process of melting, magnetic suspensions etc. The problem of rapid rotation of a rigid body in the magnetic field is studied in [79]. In this case, the equations are similar to eq. (2.1.4) and the small parameter appears only in front of the electromagnetic forces.

According to Section A.1 we seek the solution of eq. (2.1.4) in the form of an asymptotic series

$$q = \xi + \varepsilon u_1(\xi, \eta, t) + \dots, \quad r = \eta + \varepsilon v_1(\xi, \eta, t) + \dots,$$
$$i = i_0(\xi, t) + \varepsilon i_1(\xi, \eta, t) + \dots. \tag{2.1.5}$$

Functions u_1, v_1 are assumed to have zero mean value

$$\langle u_1 \rangle = \lim_{T \to \infty} \frac{1}{T} \int_{t_0}^{t_0+T} u_1(\xi, \eta, t) dt = 0, \tag{2.1.6}$$

which ensures the uniqueness of second approximations in the averaged equations. We construct the equations for ξ, η in the form

$$\dot{\xi} = \varepsilon \Xi_1(\xi, \eta) + \varepsilon^2 \Xi_2(\xi, \eta) + \dots,$$
$$\dot{\eta} = \varepsilon H_1(\xi, \eta) + \varepsilon^2 H_2(\xi, \eta) + \dots, \tag{2.1.7}$$

and limit the analysis by the terms of order $O(\varepsilon^2)$. It is sufficient to find i_0, i_1 satisfying the equations

$$L\ddot{i}_0 + R i_0 + M J = 0,$$
$$L\ddot{i}_1 + R i_1 + L \frac{\partial i_0}{\partial \xi} A^{-1} \eta + J \frac{\partial M}{\partial \xi} A^{-1} \eta - 0. \tag{2.1.8}$$

Here $\xi, \eta = \text{const}$ and it was taken into account that $\Xi_1 = A^{-1} \eta$. To obtain eq. (2.1.7) it is sufficient to determine only the periodic or quasi-periodic solutions of equations (2.1.8)

$$i_0 = \sum_{k=1}^{n} (I_{0k} \cos \nu_k t + J_{0k} \sin \nu_k t),$$
$$i_1 = \sum_{k=1}^{n} (I_{1k} \cos \nu_k t + J_{1k} \sin \nu_k t). \tag{2.1.9}$$

Inserting eq. (2.1.9) into eq. (2.1.8) yields

$$-\nu_k L I_{0k} + R J_{0k} = \nu_k M I_k, \quad \nu_k L J_{0k} + R I_{0k} = -\nu_k M J_k,$$
$$-\nu_k L I_{1k} + R J_{1k} + L \frac{\partial J_{0k}}{\partial \xi} A^{-1} \eta + J_k \frac{\partial M}{\partial \xi} A^{-1} \eta = 0,$$
$$\nu_k L J_{1k} + R I_{1k} + L \frac{\partial I_{0k}}{\partial \xi} A^{-1} \eta + I_k \frac{\partial M}{\partial \xi} A^{-1} \eta = 0. \tag{2.1.10}$$

Resolving the system of linear equations (2.1.10) for the unknowns I_{0k}, J_{0k}, I_{1k}, J_{1k} we obtain

$$I_{0k} = -\nu_k^2 I_k U_k M - \nu_k J_k V_k M, \quad J_{0k} = -\nu_k^2 J_k U_k M + \nu_k I_k V_k M,$$

$$I_{1k} = \left(\nu_k U_k L \frac{\partial J_{0k}}{\partial \xi} + \nu_k J_k U_k \frac{\partial M}{\partial \xi} - V_k L \frac{\partial I_{0k}}{\partial \xi} - I_k V_k \frac{\partial M}{\partial \xi}\right) A_{-1}\eta,$$

$$J_{1k} = -\left(\nu_k U_k L \frac{\partial I_{0k}}{\partial \xi} + \nu_k I_k U_k \frac{\partial M}{\partial \xi} + V_k L \frac{\partial J_{0k}}{\partial \xi} + J_k V_k \frac{\partial M}{\partial \xi}\right) A^{-1}\eta.$$

$$(2.1.11)$$

Here the positive definite symmetric matrices U_k, V_k are given by the formulae

$$U_k = (\nu_k^2 L + R L^{-1} R)^{-1}, \quad V_k = (R + \nu_k^2 L R^{-1} L)^{-1},$$
$$L V_k = R U_k, \quad V_k L = U_k R. \tag{2.1.12}$$

The averaged equations of second approximation are as follows

$$\dot{\xi} = \varepsilon A^{-1}(\xi)\eta,$$

$$\dot{\eta} = \varepsilon \left(-\frac{1}{2}\eta^{\mathrm{T}}\frac{\partial A^{-1}}{\partial \xi}\eta + \left\langle J\left(\frac{\partial M}{\partial \xi}\right)^{\mathrm{T}} i_0\right\rangle + \langle Q_1\rangle\right) +$$

$$+ \varepsilon^2 \left(\left\langle J\left(\frac{\partial M}{\partial \xi}\right)^{\mathrm{T}} i_1\right\rangle + \langle Q_2\rangle\right). \tag{2.1.13}$$

Equations (2.1.13) are the equations of motion of the original mechanical system subjected to the forces

$$\varepsilon\langle Q_1\rangle + \varepsilon^2\langle Q_2\rangle, \quad \varepsilon\langle J(\partial M/\partial \xi)^{\mathrm{T}} i_0\rangle + \varepsilon^2\langle J(\partial M/\partial \xi)^{\mathrm{T}} i_1\rangle.$$

We will study the properties of these forces. Taking into account eqs. (2.1.10) and (2.1.12) we obtain after some transformations

$$P_1(\xi) = \left\langle J\left(\frac{\partial M}{\partial \xi}\right)^{\mathrm{T}} i_0\right\rangle = \frac{1}{2}\sum_k I_k \left(\frac{\partial M}{\partial \xi}\right)^{\mathrm{T}} I_{0k} + \frac{1}{2}\sum_k J_k \left(\frac{\partial M}{\partial \xi}\right)^{\mathrm{T}} J_{0k}$$

$$= -\frac{1}{2}\sum_k \left\{\left(\frac{\partial I_{0k}}{\partial \xi}\right)^{\mathrm{T}} L I_{0k} + \left(\frac{\partial J_{0k}}{\partial \xi}\right)^{\mathrm{T}} L J_{0k}\right\} +$$

$$+ \frac{1}{2}\sum_k \frac{1}{\nu_k}\left\{\left(\frac{\partial J_{0k}}{\partial \xi}\right)^{\mathrm{T}} R I_{0k} - \left(\frac{\partial I_{0k}}{\partial \xi}\right)^{\mathrm{T}} R J_{0k}\right\}. \tag{2.1.14}$$

It follows from eq. (2.1.11) that the second sum in eq. (2.1.14) is equal to zero. Then $P_1(\xi)$ can be set in the form

$$P_1(\xi) = -\frac{\partial}{\partial \xi}\langle W_0\rangle, \quad W_0 = \frac{1}{2}i_0^{\mathrm{T}} L i_0. \tag{2.1.15}$$

Thus, the averaged electromagnetic forces calculated in the first approximation are potential and the potential is the mean value of the energy of magnetic field of Foucault currents provided that the energy is obtained by the same (first) approximation.

Using eq. (2.1.11) we transform the second approximation to the expressions for the mean electromagnetic forces to the form

$$
P_2(\xi, \eta) = \left\langle J \left(\frac{\partial M}{\partial \xi} \right)^{\mathrm{T}} i_1 \right\rangle =
$$

$$
= \frac{1}{2} \sum_k I_k \left(\frac{\partial M}{\partial \xi} \right)^{\mathrm{T}} I_{1k} + \frac{1}{2} \sum_k J_k \left(\frac{\partial M}{\partial \xi} \right)^{\mathrm{T}} J_{1k} = \qquad (2.1.16)
$$

$$
= -\frac{1}{2} \sum_k (I_k^2 + J_k^2) \left(\frac{\partial M}{\partial \xi} \right)^{\mathrm{T}} (V_k - 2\nu_k^2 U_k R U_k) \frac{\partial M}{\partial \xi} A^{-1} \eta.
$$

Taking into account that $\varepsilon A^{-1} \eta = \dot{\xi}$ and the matrices

$$
\Phi_k = \frac{1}{2} (I_k^2 + J_k^2) \left(\frac{\partial M}{\partial \xi} \right)^{\mathrm{T}} (V_k - 2\nu_k^2 U_k R U_k) \frac{\partial M}{\partial \xi} \qquad (2.1.17)
$$

are symmetric, we can set the expression $\varepsilon^2 P_2(\xi, \dot{\xi})$ in the following form

$$
\varepsilon^2 P_2(\xi, \dot{\xi}) - \frac{\partial}{\partial \dot{\xi}} \left(\frac{\varepsilon}{2} \dot{\xi}^{\mathrm{T}} \sum_k \Phi_k \dot{\xi} \right) = -\frac{\partial \Psi}{\partial \dot{\xi}}. \qquad (2.1.18)
$$

The expression for the dissipative function Ψ can be written as a quadratic form of the rates of change of the induction coefficients

$$
\Psi = \frac{\varepsilon}{2} \dot{M}^{\mathrm{T}} D \dot{M}, \quad D = \frac{1}{2} \sum_k (I_k^2 + J_k^2)(V_k - 2\nu_k^2 U_k R U_k). \qquad (2.1.19)
$$

Thus the second approximations to the averaged electromagnetic forces describe the "formally dissipative" forces. The sign of the dissipative function depends on the relationship between the time constant of attenuation of the proper field of the conducting body and the period of change of the external field.

By virtue of relationships (2.1.15), (2.1.18) and (2.1.19) the equations for second approximation (2.1.13) can be written down in the form

$$
\dot{\xi} = \varepsilon A^{-1}(\xi) \eta,
$$

$$
\dot{\eta} = \varepsilon \left(-\frac{1}{2} \eta^{\mathrm{T}} \frac{\partial A^{-1}}{\partial \xi} \eta - \frac{\partial \Lambda}{\partial \xi} + \langle Q_1 \rangle \right) + \varepsilon^2 \left(P_2(\xi, \dot{\xi}) + \langle Q_2 \rangle \right), \qquad (2.1.20)
$$

where $\Lambda = \langle W(i_0) \rangle$.

In the most important case in which forces $\varepsilon\langle Q_1 \rangle$ are potential with the potential function $\Pi(\xi)$ (for example, the moments due to the gravity force) the first approximation in the averaged equations (2.1.13) describes the conservative system whose motions can be qualitatively changed by forces of the next order of smallness. Therefore the second approximations are necessary in order to have the averaged equations for qualitative description of some motions, e.g. attenuating or increasing oscillations. In this case the oscillations decrease to small amplitudes within the time $t \sim T/\varepsilon^2$.

However it follows from the general theorems of the averaging method that the expressions $q = \xi + \varepsilon u_1, r = \eta + \varepsilon v_1$ approximate the solution of the original systems (3.1.5) with the error $O(\varepsilon^2)$ in the interval $t \sim T/\varepsilon$. Therefore these theorems are not sufficient for the qualitative analysis of motion of the considered system, for example, these do not allow one to assert, that the system tends to quasi-stationary motions, when the forces $\varepsilon^2(P_2 + \langle Q_2 \rangle)$ are dissipative.

If the frictional forces in $\langle Q_1 \rangle$ are sufficiently large and can be deemed as quantities of the order of ε, the first approximation is sufficient for analysis of stability of quasi-equilibria and motions. However in a number of applications [86, 104] one observes increasing oscillations of a rigid body which can be explained only by instability due to swinging character of forces $\varepsilon^2 P_2$. In these cases the forces of external friction are naturally associated with the forces of order ε^2 and higher. Because the forces are potential in the first approximation, the motion of the system depends drastically on the forces of next order of smallness, in particular $\varepsilon^2 P_2$.

By virtue of eqs. (2.1.18) and (2.1.19) we have

$$\varepsilon^2 P_2(\xi, \dot{\xi}) = \varepsilon B(\xi)\dot{\xi}, \qquad (2.1.21)$$

where, generally speaking, the symmetric matrix B of the "formally dissipative" forces P_2 depends on the generalized coordinates. Thus the influence of magnetic field on slow motions results in the appearance of potential and "dissipative" forces in the first and second approximations, respectively.

Let us consider system (2.1.20) in the first approximation, i.e. without terms $O(\varepsilon^2)$. This system is conservative and has the Hamilton function

$$H = \frac{\varepsilon}{2}\eta^{\mathrm{T}} A^{-1}(\xi)\eta + \varepsilon\Lambda + \varepsilon\Pi. \qquad (2.1.22)$$

Following Appendix D we take function V in the form $V = H/\varepsilon$.

Let the system of first approximation have a stable equilibrium position. It is encircled by closed surfaces $V = C, C = \text{const}$. Let $\Lambda = \Pi = 0$ in the equilibrium position and let D_1 denote the region bounded by two surfaces $S(C_1), S(C_2), C_1 > C_2, C_1, C_2 = O(1)$, the first enclosing the second. By virtue of eq. (2.1.20) the derivative $\dot{V}$ is equal to

$$\dot{V} = -\varepsilon^2(A^{-1}\eta)^{\mathrm{T}}(B + G)(A^{-1}\eta), \qquad (2.1.23)$$

where G denotes the matrix of coefficients of forces of external viscous damping. We will consider the case when matrix B is positive definite. It is possible to show that the inequality

$$V(\xi(t^{(0)} + T/\varepsilon)) - V(\xi(t^{(0)})) \leq -\varepsilon^{m-1} W_0 \qquad (2.1.24)$$

holds true. Indeed, if $|\eta(t^{(0)})| = O(1)$, then by virtue of eq. (2.1.22) an increase in function V caused by the solution $\xi(t), \eta(t)$ in any time interval Δt is not greater than $-\varepsilon^2(\text{const}\,\Delta t)$. If $|\eta(t^{(0)}|$ is a small value, we have $|\eta(t^{(0)})| = O(1)$ after the time span of the order

$$\Delta t = \text{const}\left(\varepsilon \inf_{D_1} \left|\frac{\partial(\Lambda + \Pi)}{\partial \xi}\right|\right)^{-1}$$

according to the second equation in (2.1.20). Choosing $T = \varepsilon k \Delta t, k > 1$ which affects only the constant c_m in the estimate $|\xi_m - \xi^{(m)}| \leq c_m \varepsilon^m$ we obtain the required inequality.

Inequality (2.1.24) and relationship (2.1.22) are satisfied for any non-small value of C_2. This allows us to apply Theorem 3 of Appendix D. As a result we obtain that for sufficient small ε any phase trajectory $\xi(t), \eta(t)$ in the above region D_1 eventually get in a small neighbourhood of the equilibrium position and will remain there. This corresponds to oscillations of the original system which are qualitatively similar to damped oscillations tending to quasi-statical ones.

The divergent oscillations are also observed in the case of the negative definite matrix of the total friction $B + G$.

Matrix B depends on ξ. Therefore it is possible that the total non-potential forces $\varepsilon^2 Q_2 - \varepsilon B \dot{\xi}$ are destabilizing near the equilibrium position and dissipative far away from it. In these cases a limit cycle is possible. At least, for a system with one mechanical degree of freedom it is possible to show in the similar manner that $\xi(t)$ reaches a small neighbourhood of this cycle. For a periodic function $J(t)$ and large t the oscillations are qualitatively similar to quasi-periodic ones and the motions of the system look like decreasing or increasing oscillations tending to quasi-periodic ones. However it seems that the existence of exact quasi-periodic solutions has not been proved for the cases when a limit cycle is revealed in the higher approximations.

Let us consider the dependence of forces $\varepsilon^2 P_2$ on the field frequency and values of electric resistances. We will assume a harmonic current $J(t)$, that is, $J(t) = I \cos \nu t + J \sin \nu t$. There is an orthogonal transformation of vector i which reduces matrices L and R to the unit and diagonal form, i.e. $R = \text{const}(\lambda_1, \lambda_2 \ldots)$, where $\lambda_1 < \lambda_2 < \ldots$. Matrices U and V also become diagonal and matrix D expressed in terms of them is as follows

$$D = \text{const}\left[\frac{(\lambda_1^2 - \nu^2)\lambda_1}{(\lambda_1^2 + \nu^2)^2}, \ldots\right]. \qquad (2.1.25)$$

One can see that for $\nu < \lambda_1$ matrix D is positive definite, i.e. if the field frequency is smaller than the first eigenvalue of the problem of attenuating electromagnetic processes in the body, the average electromagnetic forces in the second approximation are dissipative. Thus the dissipation is complete when the coefficients of the matrix of mutual inductions M depend on all generalized coordinates.

If $\nu > \lambda_1$, force $\varepsilon^2 P_2$ can be of the divergent character. These forces can be dissipative or destabilising since they depend on the coordinates. This fact indicates that self-oscillations are possible. Because of a small external friction, the self-oscillations are possible also if forces $\varepsilon^2 P_2$ are destabilising for all values of coordinates.

It follows from eq. (2.1.20) that the destabilizing effect of forces $\varepsilon^2 P_2$ can be eliminated by decreasing the body conductivity. Indeed, the coefficients of matrix R and the absolute values of eigenvalues $|\lambda_1|, |\lambda_2|, \ldots$ increase as conductivity reduces in a similar way to increasing eigenfrequencies as rigidity increases. In a number of cases the body heating can eliminate instability of the considered type.

Having reduced the matrices R and L to the diagonal form one can set the expression for forces εP_1 in the form

$$
\begin{aligned}
P_1^T(\xi) &= -\frac{\partial}{\partial \xi}[\nu_1^2(I^2 + J^2)M^T(\nu^2 ULU + VLV)M] = \\
&= -(I^2 + J^2)\frac{\partial}{\partial \xi}\left[M^T \text{const}\left(\frac{\nu^2}{\lambda_1^2 + \nu^2}, \ldots\right)M\right] = \\
&= -(I^2 + J^2)\frac{\partial}{\partial \xi}(M^T N M).
\end{aligned}
\tag{2.1.26}
$$

Comparing eqs. (2.1.25) and (2.1.26) one can find a way of destroying instability caused by forces $\varepsilon^2 P_2$ and keeping a stable equilibrium in the first (potential) approximation. For the bodies like rings, plane plate etc. [104, 40] the dependence of forces εP_1 and $\varepsilon^2 P_2$ on frequency is qualitatively the same as in the case of a single non-zero first member of the diagonal matrices N and M. In this case the first members N_{11} and M_{11} of the diagonal matrices N and M are respectively monotonically increasing and decreasing functions of ν.

Let an equilibrium of the body be stable in the potential approximation for some value of ν and be unstable in the higher approximation because of force $\varepsilon^2 P_2$. Let us assume that the field-generating-contour conducts a basic current of frequency ν and an additional current of a lower frequency ν_1. The additional force $\varepsilon \Delta P_1(\nu_1)$ is considerably smaller than the basic force $\varepsilon P_1(\nu)$ and if ν_1 is properly chosen the new equilibrium position will be stable in the first approximation and close to the original one. The additional dissipation will be greater than the original one, can have another sign and stabilize the equilibrium.

In the case of superconductivity, one can prove the potentiality of the mean electromagnetic forces in the first approximation even for the system with several superconducting bodies and any number of the independent prescribed currents creating the field. Instead of eq. (2.1.4) we have

$$\dot{q} = \varepsilon A^{-1}(q)r,$$

$$\dot{r} = \varepsilon \left[-\frac{1}{2} r^{\mathrm{T}} \frac{\partial A^{-1}}{\partial q} r + \sum_s J_s \left(\frac{\partial M_s}{\partial q} \right)^{\mathrm{T}} i + \frac{1}{2} i^{\mathrm{T}} \frac{\partial L}{\partial q} i + Q \right],$$

$$(Li) + \left(\sum_s J_s M_s \right)^{\cdot} = 0 \qquad (2.1.27)$$

Here i denotes the infinite-dimensional vector-column constructed arbitrarily from the vectors of eddy currents in the separate bodies, $L(q)$ is the corresponding matrix of the induction coefficients (in the case a few interactive bodies it depends on their coordinates) and index s marks the prescribed currents creating the field.

For the sake of simplicity we consider the case when the external field is switched on after the bodies have reached the superconducting state. The induction fluxes through the conditional contours in the bodies are equal to zero and system (2.1.27) has a countable number of the integrals which are similar to the cyclic integrals

$$Li + \sum_s J_s M_s = 0. \qquad (2.1.28)$$

Using eq. (2.1.28) we eliminate the currents

$$i = -\sum_s L^{-1} M_s J_s \qquad (2.1.29)$$

from the equations of motion in eq. (2.1.26) and arrive at the standard form of the system. Averaging this system we obtain the expression for the mean electromagnetic force in the first approximation

$$P_1(\xi) = \left\langle \frac{1}{2} i^{\mathrm{T}} \frac{\partial L}{\partial \xi} i + \sum_s J_s \left(\frac{\partial M_s}{\partial \xi} \right)^{\mathrm{T}} i \right\rangle =$$

$$= \left\langle \frac{1}{2} i^{\mathrm{T}} \frac{\partial L}{\partial \xi} i - i^{\mathrm{T}} \frac{\partial L}{\partial \xi} i - \left(\frac{\partial i}{\partial \xi} \right)^{\mathrm{T}} Li \right\rangle =$$

$$= -\left\langle \frac{1}{2} i^{\mathrm{T}} \frac{\partial L}{\partial \xi} i + \left(\frac{\partial i}{\partial \xi} \right)^{\mathrm{T}} Li \right\rangle = -\frac{\partial}{\partial \xi} \langle W_0 \rangle. \qquad (2.1.30)$$

This proves the potentiality of the mean electromagnetic forces in the first approximation. As before, the potential is the mean value of energy of the field of the whirling currents (in this case only on bodies' surface).

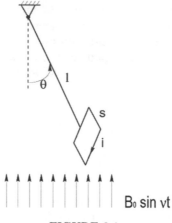

FIGURE 2.1.

In particular, the conclusion about the potentiality of the mean electro-magnetic forces in the first approximation allows us to simplify the analysis of the problem of existence and stability of quasi-static motions. In the first approximation these motions correspond to the values of ξ satisfying the equation $P_1(\xi) + \langle Q_1 \rangle = 0$. Let us consider the case in which the mechanical system consists only of a rigid body having a fixed point and forces Q are a sum of the gravity force (or other potential forces) and the frictional forces. In the general case the potential $\langle W_0 \rangle$ depends on three generalized coordi-nates (angles). Both $\langle W_0 \rangle$ and the potential for the "potential component" $\langle Q_1 \rangle$ are the periodic functions of generalized coordinates. Therefore, gen-erally speaking, the total potential has a minimum, that is, the rigid body in a rapidly changing magnetic field has, in the first approximation in ε, a position of the quasi-equilibrium. The bodies with different types of sym-metry (i.e. when $\langle W_0 \rangle$ does not depend on all generalized coordinates) are an exception.

2.2   Pendulum in the alternating magnetic field

As an example of the electromechanical system in the alternating magnetic field we consider a pendulum, the role of a rigid body being played by a closed contour of current rigidly connected to the suspension by a rigid bar, see Fig. 2.1.

Let us assume that the pendulum is placed in the alternating homoge-neous magnetic field whose frequency ν is much higher than the natural frequency of small oscillations of the pendulum. We designate the angle of deviation of the pendulum from the vertical axis by θ and assume that $\theta = 0$ corresponds to the lower position of the pendulum. The expressions for the kinetic energy T, the magnetic field energy W and the potential energy Π have the form

$$T = \frac{1}{2}I\dot{\theta}^2, \quad W = \frac{1}{2}Li^2 + B_0 S \sin \nu t \sin \theta \, i,$$
$$\Pi = mgl(1 - \cos \theta), \tag{2.2.1}$$

where I denotes the moment of inertia of the contour about the axis passing through the suspension point, m is contour's mass, l is bar's length; L is the coefficient of self-induction of the contour of current i and B_0 is the amplitude of external field.

The Lagrange–Maxwell equations for the considered electromechanical system are

$$I\ddot{\theta} - B_0 S \sin \nu t \cos \theta \, i + mgl \sin \theta = 0,$$
$$L\dot{i} + B_0 S \sin \nu t \cos \theta \, \dot{\theta} + B_0 S \nu \cos \nu t \sin \theta + Ri = 0. \tag{2.2.2}$$

We assume that the frequency of magnetic field is much higher than the natural frequency of the pendulum $k = \sqrt{mgl/I} \ll \nu$ and enter a small parameter $\varepsilon^2 = k^2/\nu^2$. Introducing a dimensionless (fast) time $\tau = \nu t$ and a dimensionless current $i_u = i/i_*$ ($i_* = B_0 S/L$ denotes a basis value of the current) we can write eq. (2.2.2) in terms of the dimensionless variables

$$\ddot{\theta} - \varepsilon^2 \gamma \sin \tau \cos \theta i_u + \varepsilon^2 \sin \theta = 0,$$
$$i_u + \sin \tau \cos \theta \, \dot{\theta} + \cos \tau \sin \theta + ri_u = 0. \tag{2.2.3}$$

In these equations the dimensionless parameters $\gamma = \dfrac{(B_0 S)^2}{Lmgl}$ and $r = \dfrac{R}{L\nu}$ are introduced. In what follows the dimensionless variables use the denotations of dimensional variables. Intending to study the slow motions which are close to free oscillations of the pendulum we rewrite the system (2.2.3) as

$$\dot{\theta} = \varepsilon \omega,$$
$$\dot{\omega} = \varepsilon \gamma \sin \tau \cos \theta \, i - \varepsilon \sin \theta,$$
$$\dot{i} + \sin \tau \cos \theta \, \dot{\theta} + \cos \tau \sin \theta + ri = 0. \tag{2.2.4}$$

System (2.2.4) is a quasi-linear system with a single non-critical fast variable. As shown in Section 2.1 the first approximation of the asymptotic approach describes a conservative system. Electromagnetic forces in the first approximation have a potential which is a mean energy of the eddy currents calculated in the same approximation. Conservatism can be lost in the second approximation to the electromagnetic forces. As shown in Section 2.1, if the field frequency exceeds the value equal to the inverse of the time constant of the conducting contour, that is $\nu > R/L$, the second approximation of electromagnetic forces can be of destabilizing character. Therefore, we derive the averaged equations for slow motions in the second approximation. Assuming $\omega, \theta = \text{const}$ we obtain the second approximation for the current

$$i = -\frac{\sin \theta}{1 + r^2}(\sin \tau + r \cos \tau) + \varepsilon \frac{\omega r \cos \theta}{(1 + r^2)^2}((1 - r^2) \sin \tau + 2r \cos \tau). \quad (2.2.5)$$

Substituting this expression into the first two equations of system (2.2.4) and averaging over the fast time τ we obtain the second order autonomous differential equation

$$\ddot{\theta} - \varepsilon\alpha \cos^2 \theta \, \dot{\theta} + (\beta \cos \theta + 1) \sin \theta = 0, \quad (2.2.6)$$

where

$$\alpha = \frac{\gamma r(1 - r^2)}{2(1 + r^2)^2}, \quad \beta = \frac{\gamma}{2(1 + r^2)}$$

and a dot denotes a differentiation with respect to slow time $t' = kt$. The term $\beta \cos \theta \sin \theta$ describes the potential electromagnetic forces in the first approximation whilst the term $-\varepsilon\alpha \cos^2 \theta \, \dot{\theta}$ stands for "formally dissipative" forces which according to Section 2.1 are of the destabilizing character for $r < 1$. In what follows we consider only the case $r < 1$ and in order to make the approach applicable for analysis of self-oscillations we add a dissipative term in eq. (2.2.6) describing a external viscous damping. As a result the equation of slow oscillations of the pendulum takes the form

$$\ddot{\theta} + \varepsilon(n - \alpha \cos^2 \theta)\dot{\theta} + (\beta \cos \theta + 1) \sin \theta = 0. \quad (2.2.7)$$

Depending on the value of β the pendulum can have either two or four equilibrium positions. If $\beta \leq 1$ there are two equilibrium position $\theta = 0$ and $\theta = \pi$, the lower being stable and the upper being unstable. In the case of $\beta > 1$ there appear two additional saddle equilibrium positions $\theta = \pi \pm \arccos(1/\beta)$. If $n > \alpha$ both upper and lower equilibria are stable, i.e. the unstable upper position becomes stable due to the oscillating electromagnetic forces. For $n = \alpha$ the stable focus in the equilibrium positions $\theta = 0, \pi$ passes to a complex focus of the first order. When n decreases, these equilibrium positions become unstable giving rise to a soft birth of the limit cycles which means originating self-oscillations with frequencies $\sqrt{1 + \beta}$ and $\sqrt{1 - \beta}$ near the lower and upper equilibrium positions respectively.

A further reduce in damping results in a collapse of self-oscillations. The limit cycle near the upper equilibrium position collapses because it first merges with the separatrix from the saddle $\pi - \arccos(1/\beta)$ to saddle $\pi + \arccos(1/\beta)$ and then, in just the same way, it merges with the "own" separatrix near the lower equilibrium. The boundaries of the region of parameters causing the collapse of limit cycles can be found by using

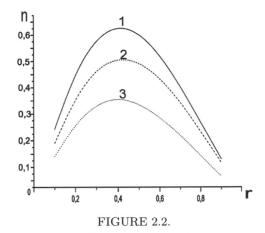

FIGURE 2.2.

the method of small parameter for the quasi-conservative system (2.2.7). Merging the limit cycle with the separatrix of the corresponding conservative systems corresponds to a zero of Pontryagin's function [10]

$$\Psi(h, \phi_1, \phi_2) = 2 \int_{\phi_1}^{\phi_2} (n - \alpha \cos^2 \theta) \sqrt{\beta \cos^2 \theta + 2 \cos \theta + 2hd\theta}, \qquad (2.2.8)$$

where ϕ_1, ϕ_2 denote the coordinates of saddle points connected by the corresponding separatrix, $h = 1/2\beta$ denotes the energy level corresponding to the motion along the separatrix. Thus, for determining $n = n_1(\alpha, \beta)$ for which the limit cycle around $\theta = \pi$ disappears we obtain the relationship

$$\Psi\left(\frac{1}{2\beta}, \pi - \arccos\frac{1}{\beta}, \pi + \arccos\frac{1}{\beta}\right) = 0. \qquad (2.2.9)$$

By analogy we find the expression for determining the parameters $n = n_2(\alpha, \beta)$ for which the cycle around $\theta = 0$ disappears

$$\Psi\left(\frac{1}{2\beta}, -\pi + \arccos\frac{1}{\beta}, \pi - \arccos\frac{1}{\beta}\right) = 0. \qquad (2.2.10)$$

Transforming relationships (2.2.9) and (2.2.10) to parameters γ and r we obtain the implicit dependences $n_1(r, \gamma)$ and $n_2(r, \gamma)$. All bifurcation curves $n(r)$ are shown in Fig. 2.2 for $\gamma = 5$.

In this figure the boundaries of stability of the equilibrium positions, existence of the limit cycles around $\theta = \pi$ and $\theta = 0$ are marked by 1,2 and 3 respectively.

In addition to slow oscillations, the electromechanical system exhibits also fast rotatory motions. These can be investigated by change of variables

$\psi = \theta - \Omega \nu t = \theta - \Omega \tau$ in eq. (2.2.3)

$$\ddot{\psi} - \varepsilon \gamma \cos(\psi + \Omega \tau) \sin \tau i + \varepsilon^2 \sin(\psi + \Omega \tau) + \varepsilon^2 n(\dot{\psi} + \Omega) = 0,$$
$$i^{\cdot} + \cos(\psi + \Omega \tau) \sin \tau (\dot{\psi} + \Omega) + \sin(\psi + \Omega \tau) \cos \tau + ri = 0. \quad (2.2.11)$$

Here in contrast to eq. (2.2.3) the mechanical dissipation is already taken into account. The physical meaning of ψ is the deviation of the trajectory of motion of the pendulum from the rotation with a constant angular speed $\Omega \nu$. Assuming that the rate of change of ψ is small, eq. (2.2.11) can be rewritten in the following form

$$\dot{\psi} = \varepsilon \omega,$$
$$\dot{\omega} = \varepsilon \gamma i \cos(\psi + \Omega \tau) \sin \tau - \varepsilon \sin(\psi + \Omega \tau) - \varepsilon n(\dot{\psi} + \Omega),$$
$$i^{\cdot} + (\dot{\psi} + \Omega) \cos(\psi + \Omega \tau) \sin \tau + \sin(\psi + \Omega \tau) \cos \tau + ri = 0. \quad (2.2.12)$$

Similar to the case of slow oscillations system (2.2.12) contains two slow variables and one non-critical fast variable. Using the previous procedure we find the first approximation for the current

$$i_0 = -\frac{1-\Omega}{2((1-\Omega)^2 + r^2)} \left[((1-\Omega) \sin \psi - r \cos \psi) \sin(1-\Omega)\tau + \right.$$
$$+ (r \sin \psi + (1-\Omega) \cos \psi) \cos(1-\Omega)\tau \left] - \right.$$
$$- \frac{1+\Omega}{2((1+\Omega)^2 + r^2)} \left[((1+\Omega) \sin \psi + r \cos \psi) \sin(1+\Omega)\tau + \right.$$
$$+ r (\sin \psi - (1+\Omega) \cos \psi) \cos(1+\Omega)\tau \left]. \right. \quad (2.2.13)$$

Inserting eq. (2.2.13) into eq. (2.2.12) and averaging the obtained expressions over time we arrive at the equation for the first approximation to ψ

$$\ddot{\psi} - \varepsilon^2 \frac{\gamma r}{8} \left(\frac{1-\Omega}{(1-\Omega)^2 + r^2} - \frac{1+\Omega}{(1+\Omega)^2 + r^2} \right) + \varepsilon^2 n \Omega = 0. \quad (2.2.14)$$

Obviously, the solutions of system (2.2.12) which are close to stationary rotations describe the solution $\psi = $ const of the averaged equation. This solution exists if the following condition

$$\frac{\gamma r}{8} \left(\frac{1-\Omega}{(1-\Omega)^2 + r^2} - \frac{1+\Omega}{(1+\Omega)^2 + r^2} \right) - n \Omega = 0 \quad (2.2.15)$$

holds. This equation provides one with the dependence of the frequency of rotation of the pendulum on the system parameters γ, r, n. Obviously, Ω can not exceed the unity since eq. (2.2.15) has no solution for any r, γ and n if $\Omega > 1$. Consequently, the frequency of stationary rotation of the pendulum can not exceed the frequency of oscillation of the magnetic field.

For analysis of stability of stationary rotations one must consider the second approximation as it adds an additional term proportional to $\dot{\psi}$ in eq. (2.2.14). As it will be shown below in this case the stability of rotations can be judged from qualitative reasoning.

The algebraic equation (2.2.15) can be rewritten in the form of a polynomial of fifth degree for Ω without free term. Hence it has a root $\Omega = 0$ for any parameters. This particular solution makes no sense since the very procedure of deriving eq. (2.2.14) is violated for $\Omega = 0$. In fact, this root corresponds to slow oscillations obtained above.

In order to determine the possible frequencies of stationary rotation of the pendulum we obtain the biquadratic equation with the roots

$$\Omega_{1,2}^2 = \frac{-(r\gamma - 8n + 8nr^2) \pm \sqrt{r^2\gamma^2 - 256n^2r^2}}{8n}. \qquad (2.2.16)$$

One can see from this expression that $\Omega \to 0$ if $n \to \dfrac{\gamma r(1 - r^2)}{4(1 + r^2)^2}$. It follows from eq. (2.2.16) that if $\gamma/16 < n$ and

$$\frac{\gamma r(1 - r^2)}{4(1 + r^2)^2} < n < \frac{\gamma r}{8(1 - r^2)} \qquad \left(r > \sqrt{2} - 1\right),$$

then no rotation exists. If $\dfrac{\gamma r(1 - r^2)}{4(1 + r^2)^2} > n$ there are two rotations with frequencies equal in value but opposite in sign which corresponds to rotations in opposite directions. For $\dfrac{\gamma r(1 - r^2)}{4(1 + r^2)^2} > n$ slow motion of the pendulum is a monotonic motion from the equilibrium position. No self-oscillations appear and thus the obtained rotations must be stable. When the condition

$$\frac{\gamma r(1 - r^2)}{4(1 + r^2)^2} < n < \frac{\gamma}{16}r < \sqrt{2} - 1$$

holds true the system assumes four rotations. In the phase space the trajectories of these rotations are symmetric about axis $\theta = 0$. In the region of these parameters, slow motions of the system are stable, i.e. there exist limit cycles or stable equilibrium positions and all slow motions tend to them. Thus the rotations whose angular velocities are smaller (greater) in absolute value will be unstable (stable). The regions of parameters in the plane r, n with the qualitatively different types of motions are shown in Fig. 2.3.

Numeral analysis of integral (2.2.8) shows that the values $n = n_2(r, \gamma)$ describing the slow outgoing motions (i.e. the limit cycle near the equilibrium position $\theta = 0$ disappears) are close to $n = \dfrac{\gamma r(1 - r^2)}{4(1 + r^2)^2}$. Thus, the transition from four rotations to two rotations is realized by merging the

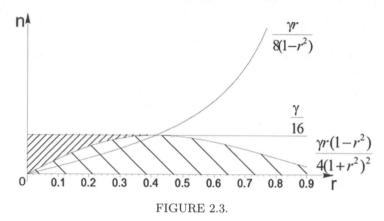

FIGURE 2.3.

unstable rotation and the limit cycle near $\theta = 0$ with the corresponding separatrix.

We can conclude that, depending on the relationship between the parameters and the initial conditions, the pendulum with a single conducting contour tends either to one of the equilibrium positions $\theta = 0, \pi$ or to the limit cycles near these equilibrium positions (that corresponds to self-oscillations) or to rotations with a frequency smaller than the frequency of external magnetic field.

2.3   Magnetic suspension in the field of alternating currents

It is shown in this section that the action of fast oscillating electromagnetic forces can be used for creating a new passive magnetic suspension of inductor type. In a simple case this suspension is the system of two rings with a joint vertical axis. One of the rings is fixed whereas the other is free. The rings conduct fast-oscillating alternating currents and this results in some stable equilibrium positions. Such suspensions can serve as an alternative to the superconducting suspension in a constant magnetic field known as the "magnetic potential well" [64].

A simple model of the contactless passive magnetic suspension can be presented in the form of a system of two aligned thin closed conducting rings (one above the other) in the gravity field, see Fig. 2.4. One of the rings is fixed and conducts an alternating current of amplitude i_* and frequency ω. The second ring is free and a voltage of the same frequency $U_c \cos \omega t + U_s \sin \omega t$ is applied to it. Let us assume that the geometrical sizes (the radius a and the cross-sectional diameter $2b$), the electric and magnetic parameters (inductance L and resistance R) of both rings are coincident. Let us first consider the case in which the free ring can move only vertically,

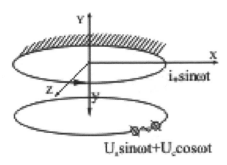

FIGURE 2.4.

i.e. parallel to the upper ring. Let y denote the distance between the rings, i be the current in the free ring and $L_e(y)$ be the coefficient of mutual induction of the rings. The Lagrange-Maxwell equations for the free ring have the form

$$m\ddot{y} - ii_* \sin \omega t \frac{dL_e}{dy} \perp mg = 0,$$

$$L\dot{i} + Ri + L_e i_* \omega \cos \omega t + i_* \sin \omega t \frac{dL_e}{dy} \dot{y} = U_g \sin \omega t + U_s \cos \omega t. \quad (2.3.1)$$

In the first equation, the sign $+ (-)$ in front of the gravity force corresponds to the case when the free ring is above (below) the fixed one, that is, axis y is directed upward (downward).

The coefficient of mutual induction of two conductors modelled by the linear currents is determined by the formula

$$L_e = \frac{\mu_0}{4\pi} \oint \oint \frac{d\mathbf{l}_a \cdot d\mathbf{l}_b}{R_*} = \frac{\mu_0}{4\pi} \oint \oint \frac{a^2 \cos(\phi - \psi) d\phi d\psi}{\sqrt{2a^2(1 - \cos(\phi - \psi)) + y^2}}. \quad (2.3.2)$$

Here $d\mathbf{l}_a$ and $d\mathbf{l}_b$ designate the elementary vectors directed along the tangent to the contours of the conductors, R_* is the distance between the corresponding elements.

Evaluating the double integral along the length of the rings we obtain

$$L_e(y) = \frac{\mu_0 a}{k} \left[(2 - k^2) K(k) - 2E(k) \right], \qquad k^2 = \frac{4a^2}{4a^2 + y^2}, \quad (2.3.3)$$

where $K(k)$ and $E(k)$ denote the complete elliptic Legendre integrals of the first and second kind respectively.

Introducing the dimensionless time $\tau = \omega t$ and the small parameter $\varepsilon^2 = \frac{g}{a\omega^2}$ ($\varepsilon \ll 1$) in eq. (2.3.1) we obtain the equations for the current and motion of the free ring in terms of the dimensionless variables

$$i_u + ri_u + L_{eu} \cos \tau + \frac{dL_{eu}}{dy_u} \sin \tau \dot{y}_u = u_s \sin \tau + u_c \cos \tau \,,$$

$$\ddot{y}_u - i_u \gamma \varepsilon^2 \sin \tau \frac{dL_{eu}}{dy_u} \pm \varepsilon^2 = 0 \,, \qquad (2.3.4)$$

where

$$\gamma = \frac{Li_*^2}{mga}, u_s = \frac{U_s}{L\omega i_*}, u_c = \frac{U_s}{L\omega i_*}, r = \frac{R}{L\omega}, i_u = \frac{i}{i_*}, y_u = \frac{y}{a}, L_{eu} = \frac{L_e}{L}.$$

In what follows we omit the subscript u (which denotes the dimensionless quantities).

The solution of the problem is carried out by a modified method of averaging for quasi-linear systems, see Appendix A. Assuming $y = y_0 = $ const in eq. (2.3.4) we obtain that the first approximation for the current satisfies the equation

$$\dot{i}_0 + ri_0 = u_s \sin \tau + (u_c - L_e) \cos \tau \,. \qquad (2.3.5)$$

Hence, the first approximation to the stationary current in the free ring is given by

$$i_0 = \frac{1}{1+r^2} \left[(u_c - L_e + ru_s) \sin \tau + ((u_c - L_e)r - u_s) \cos \tau \right] \,. \qquad (2.3.6)$$

Inserting eq. (2.3.6) into the first equation of system (2.3.4) and averaging over τ we obtain the first approximation to the averaged equation of motion for the free ring

$$\ddot{y} - \varepsilon^2 \gamma \langle i \sin \tau \rangle \frac{dL_e}{dy} \pm \varepsilon^2 = 0 \,, \qquad (2.3.7)$$

where the denotation $\langle f(t) \rangle$ means the averaging over the period, that is,

$$\langle f(t) \rangle = \frac{1}{T} \int_T f(t) dt.$$

The mean electromagnetic force acting on the free ring is equal to

$$P_W = \gamma \langle i \sin \tau \rangle \frac{dL_e}{dy} = \frac{\gamma}{2(1+r^2)} (u_c + ru_s - L_e) \frac{dL_e}{dy} \,. \qquad (2.3.8)$$

It follows from eq. (2.3.8) that in the first approximation the electromagnetic force has the following potential

$$\Pi_W = -\frac{\gamma}{2(1+r^2)} \left[(u_c + ru_s)L_e - \frac{L_e^2}{2} \right] \,. \qquad (2.3.9)$$

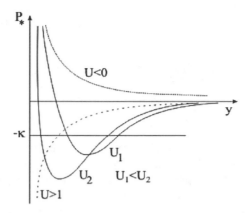

FIGURE 2.5.

In the equilibrium position the following relationship must hold

$$\frac{d\Pi_W}{dL_e}\frac{dL_e}{dy} \pm 1 = 0 \quad \Rightarrow \quad (U - L_e)\frac{dL_e}{dy} \pm \kappa = 0. \qquad (2.3.10)$$

Here parameter $U = u_c + ru_s$ characterizes the voltage in the free ring and parameter $\kappa = \dfrac{2(1 + r^2)}{\gamma}$ is proportional to the ring weight.

As the derivative $\dfrac{dL_e}{dy} < 0$ for any y, for existence of the equilibrium position it is necessary that $\dfrac{d\Pi_W}{dL_e} > 0$ (i.e. $U - L_e < 0$) in the case of the upper free ring. The necessary and sufficient conditions for stability of this position are as follows

$$\frac{d^2\Pi_W}{dy^2} = \frac{\partial^2\Pi_W}{\partial L_e^2}\left(\frac{\partial L_e}{\partial y}\right)^2 + \frac{\partial\Pi_W}{\partial L_e}\frac{\partial^2 L_e}{\partial y^2} > 0. \qquad (2.3.11)$$

This inequality holds for any y and parameters U and κ inasmuch as $\dfrac{\partial^2\Pi_W}{\partial L_e^2} = \dfrac{1}{\kappa} > 0$ and $\dfrac{\partial^2 L_e}{\partial y^2} > 0$ for any y. Thus, if equilibrium position above the fixed ring exists it is always stable with respect to y. Additionally, it follows from eq. (2.3.10) and inequality (2.3.11) that a stable equilibrium position of the upper free ring exists even in the case $U = 0$, i.e. if only the fixed ring conducts a current.

Let us consider the case when the free ring is under the fixed one. For existence of the equilibrium position the condition $\dfrac{\partial\Pi_W}{\partial L_e} < 0$ must hold, that is $U - L_e > 0$. The stability of this position is determined by inequality

(2.3.11) whose explicit form has a sufficiently complex structure

$$\frac{\mu_0 a}{L}\left[(2-k^2)E-2(1-k^2)K\right]^2 -$$
$$- (U-L_e)k^3\left[(1-k^2)K-(1-2k^2)E\right] > 0. \quad (2.3.12)$$

However, the existence and stability of the equilibrium positions of the free ring can be judged by means of the graphic approach. Figure 2.5 displays the graphs of the dependence $P_*(y) = 2P_W(1+r^2)/\gamma$ (dimensionless electromagnetic force) for several values of parameter U. For $0 < U < 1$ the points of intersection of the graph and the line $P_* = -\kappa$ determine two positions of equilibrium, the first (closer to the fixed ring) being stable and the second (distant) being unstable. One can see from the graphs that the closer U to 1, the greater weight can be stably suspended.

If $U < 0$, the average magnetic force is positive (i.e. is directed downward) and consequently, there is no equilibrium position under the fixed ring. If $U > 1$, the magnetic force is negative (is directed upwards) and increases monotonically with the growth of y therefore there exists only one equilibrium position which is unstable.

The characteristic phase trajectories obtained by means of the numeral integration of the averaged equation (2.3.7) are shown in Fig. 2.6 in the case of two equilibrium positions.

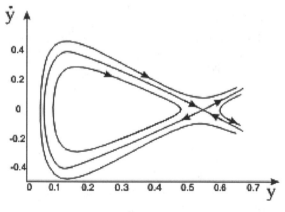

FIGURE 2.6.

The system of the first approximation (2.3.7) is conservative, thus in order to judge the actual stability of the equilibrium it is necessary to construct the second approximation to the current.

We assume the slow motion along axis y and consider the system

$$\dot{y} = \varepsilon x \,,$$

$$\dot{x} = \varepsilon \left(\frac{\gamma i}{2(1 + r^2)} \frac{\partial L_e}{\partial y} \sin \tau + 1 \right) \,,$$

$$\ddot{i} + ri + L_e \cos \tau + \frac{dL_e}{dy} \sin \tau \varepsilon x = u_s \sin \tau + u_c \cos \tau \,. \qquad (2.3.13)$$

Applying the averaging method to eq. (2.3.13) (Appendix A) we obtain

$$i_1 = \frac{r(1 - r^2)}{(1 + r^2)^2} \frac{dL_e}{dy} \sin \tau + \frac{2r^2}{(1 + r^2)^2} \frac{dL_e}{dy} \cos \tau \,. \qquad (2.3.14)$$

The second approximation to the equation of motion for the free ring has the form

$$\ddot{y} + \varepsilon^2 \frac{\gamma r(1 - r^2)}{2(1 + r^2)^2} \left(\frac{dL_e}{dy} \right)^2 \dot{y} - \varepsilon^2 \left((U - L_e) \frac{\gamma}{2(1 + r^2)} \frac{dI_{'c}}{dy} + 1 \right) = 0 \,. \qquad (2.3.15)$$

It follows from eq. (2.3.15) that for $r < 1$ the considered system has a dissipative element. In this case the equilibrium position becomes a stable focus (Fig. 2.7a). There is no stable suspension of the free ring if $r > 1$, that is, both equilibrium positions are unstable, Fig. 2.7b.

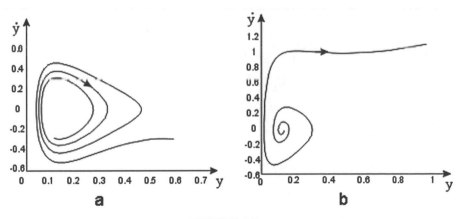

FIGURE 2.7.

Let us now consider the case of free motions of the suspended ring. It has six degrees of freedom described in terms of the coordinates ρ, α_*, y of the center of the free ring in the cylindrical system of coordinate with the origin at the center of the fixed ring (Fig. 2.8) and Euler's angles θ, ψ, ϕ for the orientation of the rings. The kinetic and potential energies and the

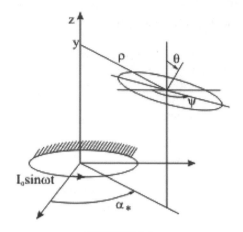

FIGURE 2.8.

energy of magnetic field have the form

$$T = \frac{1}{2}\left(m\dot{\rho}^2 + m\rho^2\dot{\alpha}_*^2 + m\dot{y}^2 + I_1\left(\dot{\theta}^2 + \dot{\psi}^2\sin^2\theta\right) + I_2\left(\dot{\phi} + \dot{\psi}\cos\theta^2\right)\right),$$

$$\Pi = \pm mgy, \quad W = \frac{1}{2}Li^2 + iL_e(\rho,\theta,y,\alpha_* - \psi)I_0\sin\omega t, \quad (2.3.16)$$

where I_1 and I_2 denote the moments of inertia about the vertical and horizontal axes of symmetry respectively.

The coefficient of mutual induction describing the mutual orientation of the rings is determined by the integral

$$L_e = \frac{\mu_0 a}{\pi}\int_0^{2\pi}\frac{1}{k\rho_*^{3/2}}\left[\left(1 - \frac{k^2}{2}\right)K(k) - E(k)\right](\cos\theta +$$

$$+ \rho(\sin\alpha\sin(\alpha_* - \psi) + \cos\alpha\cos\theta\cos(\alpha_* - \psi)))d\alpha. \quad (2.3.17)$$

Here

$$k^2 = \frac{4\rho_*}{(1+\rho_*)^2 + z^2}, \quad z = \sin\alpha\sin\theta + \frac{y}{a},$$

$$\rho_* = 1 - \sin^2\alpha\sin^2\theta + \frac{\rho}{a}\left(\frac{\rho}{a} + 2\cos\alpha\cos(\alpha_* - \psi) + \right.$$

$$+ 2\sin\alpha\cos\theta\sin(\alpha_* - \psi))^{1/2}. \quad (2.3.18)$$

The first approximation of the vector of dimensionless electromagnetic forces acting on the free ring is given by the following formula $P_W = \gamma\left\langle\sin\tau\,(\partial L_e/\partial\xi)^T i_0\right\rangle$, where $\xi = \left[\frac{\rho}{a}\theta, \frac{y}{a}, \psi, \phi, \alpha_*\right]^T$. Substituting expression (2.3.6) for the current i_0 we obtain

$$P_W = \frac{\gamma}{2(1+r^2)}(U - L_e)\left(\frac{\partial L_e}{\partial\xi}\right)^T. \quad (2.3.19)$$

These forces also have the potential

$$\Pi_W = \Pi_W(\rho, \theta, y, \alpha_* - \psi) = -\frac{\gamma}{2(1+r^2)}\left(UL_e - \frac{L_e^2}{2}\right). \qquad (2.3.20)$$

Since the combined potential energy $\Pi_\Sigma = \Pi + mga\Pi_W$ depends only on the difference of angles α_* and ψ it is reasonable to enter a new variable $\sigma = \alpha_* - \psi$. The system of Lagrange equations that govern motions of the free ring in the field of the fixed ring conducting the prescribed alternating current has the form

$$I_2(\dot\phi + \dot\psi\cos\theta) = -\frac{\partial\Pi_\Sigma}{\partial\phi} = 0,$$

$$m\ddot\rho - m\rho(\dot\sigma + \dot\psi)^2 = -\frac{\partial\Pi_\Sigma}{\partial\rho},$$

$$m\rho^2(\ddot\sigma + \ddot\psi) + 2m(\dot\sigma + \dot\psi)\rho\dot\rho = -\frac{\partial\Pi_\Sigma}{\partial\sigma},$$

$$m\ddot y = -\frac{\partial\Pi_\Sigma}{\partial y},$$

$$I_1\ddot\theta - I_1\dot\psi^2\sin\theta\cos\theta + I_2(\dot\phi + \dot\psi\cos\theta)\dot\psi\sin\theta = -\frac{\partial\Pi_\Sigma}{\partial\theta},$$

$$(m\rho^2(\dot\sigma + \dot\psi) + I_1\dot\psi\sin^2\theta + I_2(\dot\phi + \dot\psi\cos\theta)\cos\theta) = -\frac{\partial\Pi_\Sigma}{\partial\psi} = 0, \quad (2.3.21)$$

These equations have two cyclic variables: ϕ and ψ. Entering two corresponding generalized momenta $p_1 = I_2(\dot\phi + \dot\psi\cos\theta)$ and $p_2 = m\rho^2(\dot\sigma + \dot\psi) + I_1\dot\psi\sin^2\theta + I_2(\dot\phi + \dot\psi\cos\theta)\cos\theta$ we obtain the dimensionless potential energy in the Routh form

$$\Pi_R = -\frac{\gamma}{2(1+r^2)}\left[UL_e - \frac{L_e^2(\rho,\theta,y,\sigma)}{2}\right] \pm y - \frac{1}{2}\frac{(r_1 - r_2\cos\theta)^2}{(ma^2\rho^2 + I_1\sin^2\theta)mga}, \qquad (2.3.22)$$

where y and ρ are the dimensionless sizes attributed to the ring radii. The equilibrium positions are determined from the condition $\dfrac{\partial\Pi_R}{\partial\xi} = 0$. Reasoning from symmetry of the problem one can conclude that the positions of equilibrium are possible only on axis z (Fig. 2.8) and hence this condition can be rewritten in the form $\dfrac{\partial\Pi_R}{\partial y}\bigg|_{\rho=0,\theta=0,\sigma=0} = 0$. As shown above, this condition determines two positions of equilibrium under the fixed ring if $0 < U < 1$ and one equilibrium position above the fixed ring if $U < 1$. Let us find the parametric boundaries of stability of the equilibrium positions with respect to all coordinates.

A stable equilibrium position is given by a minimum of the potential energy which is characterized by the positive definiteness of the matrix of second derivatives of the energy in Routh's form which holds under the condition

$$\frac{d\Pi_R^2}{d\xi^2} = \frac{\partial \Pi_R^2}{\partial L_e^2}\left(\frac{\partial L_e}{\partial \xi}\right)^2 + \frac{\partial \Pi_R}{\partial L_e}\frac{\partial^2 L_e}{\partial \xi^2} > 0\,.$$

It is necessary to determine the values $\dfrac{\partial L_e}{\partial \xi}$ and $\dfrac{\partial^2 L_e}{\partial \xi^2}$ at the investigated equilibrium positions. Using expression (2.3.17) we find

$$\left.\frac{\partial L_e}{\partial \rho}\right|_0 = \left.\frac{\partial L_e}{\partial \theta}\right|_0 = \left.\frac{\partial L_e}{\partial \sigma}\right|_0 = 0,$$

$$\left.\frac{\partial^2 L_e}{\partial \sigma^2}\right|_0 = \left.\frac{\partial^2 L_e}{\partial \sigma \partial \rho}\right|_0 = \left.\frac{\partial^2 L_e}{\partial \sigma \partial \theta}\right|_0 = \left.\frac{\partial^2 L_e}{\partial \sigma \partial y}\right|_0 = \left.\frac{\partial^2 L_e}{\partial \rho \partial y}\right|_0 = \left.\frac{\partial^2 L_e}{\partial \theta \partial y}\right|_0 = 0,$$

$$\left.\frac{\partial L_e}{\partial y}\right|_0 = -\frac{\lambda}{2(1-k^2)^{1/2}}[(2-k^2)E - 2(1-k^2)K],$$

$$\left.\frac{\partial^2 L_e}{\partial \rho^2}\right|_0 = \frac{\lambda k^3}{8(1-k^2)}[(1-2k^2)E - (1-k^2)K],$$

$$\left.\frac{\partial^2 L_e}{\partial \rho \partial \theta}\right|_0 = \frac{\lambda \sin \sigma}{8(1-k^2)^{1/2}}[(4-k^2-2k^4)E + (-4+3k^2+k^4)K],$$

$$\left.\frac{\partial^2 L_e}{\partial y^2}\right|_0 = \frac{\lambda k^3}{4(1-k^2)}[(1-k^2)K - (1-2k^2)E],$$

$$\left.\frac{\partial^2 L_e}{\partial \theta^2}\right|_0 = \frac{\lambda k}{8(1-k^2)}[(2-3k^2+2k^4)E + (-2+3k^2-k^4)K]. \quad (2.3.23)$$

Here $\lambda = \mu_0 a/L$ $k^2|_0 = 4/(4+y^2)$ and the denotation $|_0$ means that the function is taken at $\rho = 0, \theta = 0, \sigma = 0$.

Let us apply Routh's theorem on stability for $p_1 = p_2 = 0$. Taking into account eq. (2.3.23) we can set the second derivatives of the modified potential energy at zero generalized momenta in the form

$$\left.\frac{\partial^2 \Pi_R}{\partial \rho^2}\right|_0 = \left.\frac{\partial \Pi_R}{\partial L_e}\frac{\partial^2 L_e}{\partial \rho^2}\right|_0, \qquad \left.\frac{\partial^2 \Pi_R}{\partial \rho \partial \theta}\right|_0 = \left.\frac{\partial \Pi_R}{\partial L_e}\frac{\partial^2 L_e}{\partial \rho \partial \theta}\right|_0, \qquad (2.3.24)$$

$$\left.\frac{\partial^2 \Pi_R}{\partial \theta^2}\right|_0 = \left.\frac{\partial \Pi_R}{\partial L_e}\frac{\partial^2 L_e}{\partial \theta^2}\right|_0, \qquad \left.\frac{\partial^2 \Pi_R}{\partial y^2}\right|_0 = \left.\frac{\partial^2 \Pi_R}{\partial L_e^2}\left(\frac{\partial L_e}{\partial y}\right)^2\right|_0 + \left.\frac{\partial \Pi_R}{\partial L_e}\frac{\partial^2 L_e}{\partial y^2}\right|_0.$$

Since the equilibrium positions are "indifferent" with respect to σ the following conditions

$$1.\ \left.\frac{\partial^2 \Pi_R}{\partial \rho^2}\right|_0 > 0, \quad 2.\ \left.\frac{\partial^2 \Pi_R}{\partial \rho^2}\frac{\partial^2 \Pi_R}{\partial \theta^2} - \left(\frac{\partial^2 \Pi_R}{\partial \rho \partial \theta}\right)^2\right|_0 > 0, \quad 3.\ \left.\frac{\partial^2 \Pi_R}{\partial y^2}\right|_0 > 0.$$

$$(2.3.25)$$

are necessary and sufficient for stability of the suspended ring.

As already shown above, the third condition states that the equilibrium position above the fixed ring is stable with respect to y for all parameters ensuring $U < 1$ and stability of the equilibrium position under the fixed ring is achieved only for $0 < U < 1$.

As $\dfrac{\partial L_e}{\partial \rho} = 0$ the first condition in (2.3.25) can be rewritten as follows $\dfrac{\partial \Pi_R}{\partial L_e} \dfrac{\partial^2 L_e}{\partial \rho^2}\bigg|_0 > 0$. For identical rings $\dfrac{\partial^2 L_e}{\partial \rho^2}\bigg|_0 < 0$ for all values of y. Since $\dfrac{\partial \Pi_R}{\partial L_e} > 0$ in the equilibrium position above the fixed ring, the first condition does not hold for all parameters and hence the global stability is not possible.

Contrary to the previous case, $\dfrac{\partial \Pi_R}{\partial L_e} < 0$ in the lower equilibrium position, hence it is stable with respect to y, and in turn it is stable "with respect to ρ" for all parameters.

The second condition in eq. (2.3.25) is to be rewritten in the following form $\dfrac{\partial \Pi_R}{\partial L_e} \dfrac{\partial^2 L_e}{\partial \rho^2} \dfrac{\partial \Pi_R}{\partial L_e} \dfrac{\partial^2 L_\theta}{\partial \theta^2} - \left(\dfrac{\partial \Pi_R}{\partial L_e} \dfrac{\partial^2 L_{\prime_0}}{\partial \rho \partial \theta} \right)^2 > 0$, and consequently, the condition $F_{lim}(y_*) = \dfrac{\partial^2 L_e}{\partial \rho^2} \dfrac{\partial^2 L_e}{\partial \theta^2} - \left(\dfrac{\partial^2 L_e}{\partial \rho \partial \theta} \right)^2 > 0$ must hold. This inequality does not depend explicitly on parameters U and κ however it depends on y_* which is the coordinate of equilibrium position. This dependence is displayed in Fig. 2.9. As calculations show the critical value of y determining the zone of stability "with respect to θ" is $y_{lim} = 1.9029$. For $y_* > y_{lim}$ the equilibrium is stable. Figure 2.10 shows the parametric boundaries for the existing stable position of equilibrium of the suspended ring, namely the boundary for existing equilibrium position $\max\left[(U - L_e)\dfrac{\partial L_e}{\partial y} \right] = \kappa$ and the condition $y_* = y_{lim}$ are marked as 1 and 2 respectively.

Let us consider a numerical example. Let the radius of each ring be $a = 10$cm and the diameter of the cross-section be $b = 1$mm. We take the parameters ensuring the stability of the maximum weight with respect to all coordinates, i.e. $U = 0.02$ and $\kappa = 5 \cdot 10^{-5}$. Let us determine the strength of current needed for hanging a weight of 10g. Assuming that parameter r is close to 1 we obtain the strength of current $I \approx 10$kA. One can see from these calculations that a great alternating current is required for a stable suspending of such a small weight.

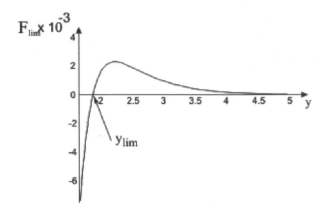

FIGURE 2.9.

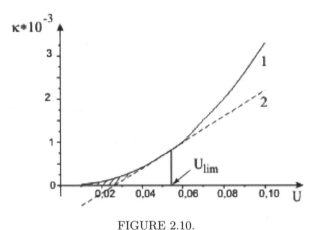

FIGURE 2.10.

Since the physical instability "with respect to θ" is related to the substantial heterogeneity of magnetic field near the fixed rings one can try to decrease the influence of this heterogeneity by considering a system in which the fixed ring has a greater radius than that of the suspended ring. Let us designate the ratio of rings' radii as $\eta = a_2/a_1 > 1$ where a_1 and a_2 denote radius of the free and fixed ring respectively. In this case the coefficient of mutual induction has the form

$$L_e = \frac{\mu_0 a_1 \sqrt{\eta}}{\pi} \int_0^{2\pi} \frac{1}{k\rho_*^{3/2}} \left[(1 - \frac{k^2}{2})K(k) - E(k) \right]$$

$$\times (\cos\theta + \rho(\sin\alpha\sin\sigma + \cos\alpha\cos\theta\cos\sigma))d\alpha, \quad (2.3.26)$$

where $k^2 = \dfrac{4\eta\rho_*}{(\eta + \rho_*)^2 + z^2}$ (all dimensional coordinates are related to radius a_1). The derivatives of L_e are given by

$$\left.\frac{\partial L_e}{\partial y}\right|_0 = -\frac{\lambda(1 - k^2)^{3/2}}{2\eta}[(2 - k^2)E - 2(1 - k^2)K],$$

$$\left.\frac{\partial^2 L_e}{\partial \rho^2}\right|_0 = \frac{\lambda k^3}{8\sqrt{\eta}}\left[(1 - k^2)K - (1 + k^2)E + \right.$$

$$\left. + \frac{(1 + \eta)^2}{4\eta}\left[(1 - k^2)(2 - k^2)(E - K) + k^2(1 + k^2)E\right]\right],$$

$$\left.\frac{\partial^2 L_e}{\partial \rho \partial \theta}\right|_0 = \frac{\lambda yk\sin\sigma}{32\sqrt{\eta}}\left[(8 - 16k^2 + 7k^4 + k^6)K - (8 - 12k^2 + 2k^6)E - \right.$$

$$\left. \frac{k^2}{\eta}[2(1 - k^2 + k^4)E - (2 - 3k^2 + k^4)K]\right],$$

$$\left.\frac{\partial^2 L_e}{\partial y^2}\right|_0 = \frac{\lambda k^3}{4\sqrt{\eta}}\left[(1 + k^2)E - (1 - k^2)K - \right. \tag{2.3.27}$$

$$\left. - \frac{(1 + \eta)^2}{4\eta}\left[(1 - k^2)(2 - k^2)(E - K) + k^2(1 + k^2)E\right]\right],$$

$$\left.\frac{\partial^2 L_e}{\partial \theta^2}\right|_0 = \frac{\lambda\sqrt{\eta}k^3}{8}\left[(1 - k^2)E - \frac{y^2}{4\eta^2}\left[2(1 - k^2 + k^4)E - (2 - 3k^2 + k^4)K\right]\right]$$

where $\lambda = \mu_0 a_1/(1 - k^2)^2$.

Figure 2.11 displays the dependences L_e on ρ, θ and y for different values of η. One can see that $\left.\dfrac{\partial L_e}{\partial y}\right|_0 = 0$ for any y (in contrast to the case of identical rings in which $\left.\dfrac{\partial L_e}{\partial y}\right|_0 = -\infty$). Using this property, one can construct the qualitative dependences of the dimensionless electromagnetic force $(P_* = P_W\kappa)$ on y for different U (Fig. 2.12). The curve 1 corresponds to the case $U < 0$, thus there is no equilibrium positions whilst the curve 2 describes the case $0 < U < L_{e*}$ $\left(L_{e*} = \left.\dfrac{L_e}{L_1}\right|_0\right)$. The curve 3 $(U > L_{e*})$ corresponds to the case of two equilibrium positions under the fixed ring, the position close to (far from) the fixed ring being stable (unstable).

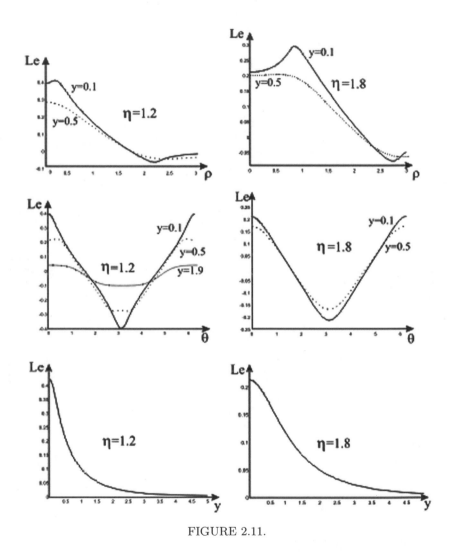

FIGURE 2.11.

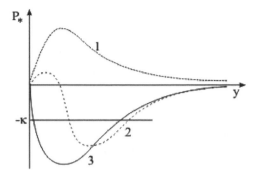

FIGURE 2.12.

The first condition of stability in eq. (2.3.25) holds for $\dfrac{\partial^2 L_e}{\partial \rho^2} < 0$. It determines the critical value of $y_\rho(U, \kappa, \eta)$. For $y_* > y_\rho$ the lower equilibrium position is stable "with respect to ρ" and unstable for $y_* < y_\rho$. The second condition determines another critical value y_θ, which corresponds to the boundary of stability "with respect to θ". The numeral analysis shows that $y_\theta > y_\rho$ for any η from the interval $1 \leq \eta \leq 2$.

The boundaries of existence of the stable position are shown in Fig. 2.13 for different values of η. One can see than the greater η, the broader the parametric region of stability, however the suspended weight becomes smaller for the same value of U. In addition to this, the stability region increases insignificantly as η grows and exceeds some critical value.

A further increase in the difference in radii of the rings does not improve the situation because the maximum electromagnetic force (attainable in the area of stability) exhibits an essentially nonlinear dependence on η and begins to decrease for $\eta > 1.8$. In other words, a smaller weight can be stably suspended for larger values of η, see Fig. 2.14.

Let us consider a numerical example. Let the suspended ring have the above geometrical sizes and the ratio of radii be equal to $\eta = 1.8$. We take the parameters which are needed for a stable suspension of the ring, that is $U = 0.3$ and $\kappa \approx 0.02$. The strength of current which is necessary for holding a weight of $10g$ is equal to $I \approx 1kA$ and the voltage in the free ring is $U_c \sim U_s \approx 0.15V$. Thus an increase in radius of the fixed ring leads to a tenfold decrease in the strength of current nevertheless the current remains too large. Since an equilibrium position (stable with respect to y) exists for large values of U and different radii of the rings one can stably hang any weight with a single degree of freedom on a shorter distance by means of a smaller current. For example, to fix a weight of $10g$ in the distance of $6cm$ from the fixed ring the required strength of current should be about $100A$ at voltage $5V$ or $20A$ at voltage $10V$.

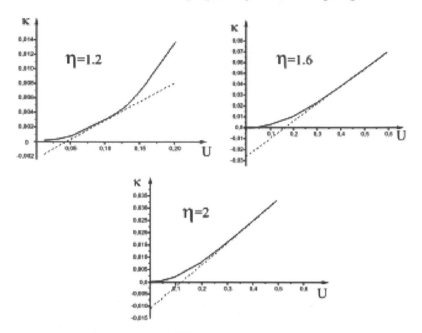

FIGURE 2.13.

In contrast to the case of two identical rings, the derivative $\dfrac{\partial^2 L_e}{\partial \rho^2}$ changes its sign for different values of y, hence for $y_* < y_{\rho g}$ the equilibrium position above the fixed ring is stable with respect to ρ. However the second condition of stability with respect to θ in eq. (2.3.25) holds only for $y_* > y_{\theta g}$, therefore a total stability of the overhead equilibrium position is not possible ($y_{\theta g} > y_{\rho g}$).

A practical realization of a ring suspension with a prescribed alternating current is rather difficult. In practice, an alternating voltage in the fixed ring is simpler to achieve.

Let us consider two identical rings (one fixed and one free) with the given voltages $U_1 \sin \omega t + V_1 \cos \omega t$ and $U_2 \sin \omega t + V_2 \cos \omega t$ respectively. The Lagrange-Maxwell equations for the free ring with a single vertical degree of freedom take the form

$$m\ddot{y} - i_1 i_2 \frac{dL_e}{dy} \pm mg = 0,$$

$$L\ddot{i_1} + R i_1 + L_e \ddot{i_2} + \frac{dL_e}{dy} \dot{y} t i_2 = u_1 \sin \omega t + v_1 \cos \omega t,$$

$$L\ddot{i_2} + R i_2 + L_e \ddot{i_1} + \frac{dL_e}{dy} \dot{y} t i_1 = u_2 \sin \omega t + v_2 \cos \omega t. \qquad (2.3.28)$$

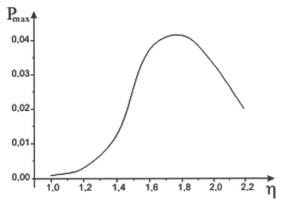

FIGURE 2.14.

Entering the dimensionless time $\tau = \omega t$ and the small parameter $\varepsilon = g/(a\omega^2)$ we obtain a dimensionless form of equations (2.3.28). Then applying the above method of averaging we obtain in the first approximation

$$i_{01} = c_1 \sin \tau + d_1 \cos \tau, \quad i_{02} = c_2 \sin \tau + d_2 \cos \tau, \qquad (2.3.29)$$

where the coefficients are as follows

$$s_1 = \frac{L_e^3 v_2 + L_e^2(ru_1 - v_1) + L_e(v_2(r^2 - 1) - 2u_2 r) + (ru_1 + v_1)(1 + r^2)}{\Delta},$$

$$s_2 = \frac{L_e^3 v_1 + L_e^2(ru_2 - v_2) + L_e(v_1(r^2 - 1) - 2u_1 r) + (ru_2 + v_2)(1 + r^2)}{\Delta},$$

$$d_1 = \frac{-L_e^3 u_2 + L_e^2(rv_1 + u_1) + L_e(u_2(1 - r^2) - 2v_2 r) + (rv_1 - u_1)(1 + r^2)}{\Delta},$$

$$d_2 = \frac{-L_e^3 u_1 + L_e^2(rv_2 + u_2) + L_e(u_1(1 - r^2) - 2v_1 r) + (rv_2 - u_2)(1 + r^2)}{\Delta},$$

$$\Delta = (L_e^2 + r^2 - 1)^2 + 4r^2. \qquad (2.3.30)$$

The denotation for all parameters in eq. (2.3.30) coincides with the previous ones except for i_* which denotes a basic current.

Using the obtained expressions for the currents we can determine the average electromagnetic force acting on the free ring:

$$P = \gamma \left\langle i_1 i_2 \frac{dL_e}{dy} \right\rangle = \frac{\gamma}{2} \frac{dL_e}{dy} [s_1 s_2 + d_1 d_2] =$$

$$= \frac{\gamma}{2\Delta} \left(pL_e^2 - sL_e + r(1 + r^2) \right) \frac{dL_e}{dy}, \qquad (2.3.31)$$

where $p = u_1 u_2 + v_1 v_2$ and $s = u_1^2 + u_2^2 + v_1^2 + v_2^2$. Depending on the relationship between the parameters p, s and r this force can have a qualitatively

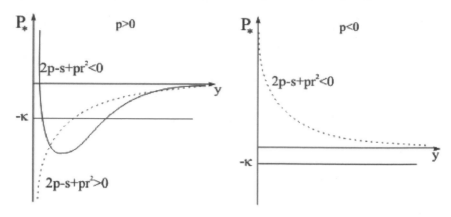

FIGURE 2.15.

different character, see Fig. 2.15. The positions of equilibrium are given by

$$-(pL_e^2 - sL_e + r(1+r^2))\frac{1}{\Delta}\frac{dL_e}{dy} \pm \kappa = 0, \quad \kappa = \frac{2mga}{Li_*^2}. \qquad (2.3.32)$$

The necessary condition for the equilibrium position of the free ring above the fixed ring is $pL_e^2 - sL_e + p(1 + r^2) < 0$ which limits the value of the voltage. This is possible when the following conditions hold true: $s^2 - 4p^2(1 + r^2) > 0$ (the discriminant is positive, i.e. L_e is real-valued) and $0 < \dfrac{s - \sqrt{s^2 - 4r^2(1+r^2)}}{2p} < 1$. These two conditions reduce to a single condition, namely $\dfrac{s}{2p} > 1 + \dfrac{r^2}{2}$ for $p > 0$ and $\dfrac{s}{2p} < 1 + \dfrac{r^2}{2}$ for $p < 0$. The first inequality (for $p > 0$) is a necessary condition for existence of two positions of equilibrium in the case when the suspended ring is beneath. If this condition is not satisfied the suspended ring has only one equilibrium position which is unstable. The dependence of κ/p on s/p is shown in Fig. 2.16 and determines the boundary for the equilibrium position under the fixed ring, this position being stable with respect to y.

For analysis of the total stability of the suspended ring (i.e. with respect to all coordinates) one can make use of eq. (2.3.21) with the only difference that the potential of electromagnetic forces Π_W has a more difficult character of the dependence on L_e

$$\Pi_W = -\frac{\gamma}{8r}\left((2r - s)\arctan\frac{L_e - 1}{r} + (2r + s)\arctan\frac{L_e + 1}{r}\right). \qquad (2.3.33)$$

Since in the stable equilibrium position $pL_e^2 - sL_e + p(1 + r^2) > 0$ and $\dfrac{\partial^2 L_e}{\partial\rho^2} < 0$ (by virtue of the property of the coefficient of mutual induction

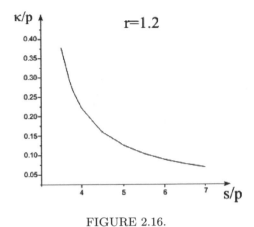

FIGURE 2.16.

of identical rings) this position is stable with respect to ρ for all parameters. The second condition in eq. (2.3.25) determines the critical value y_{lim} which, with the help of eq. (2.3.32), yields the dependence κ/p on s/p describing the border of total stability (in Fig. 2.17 the region of stability is marked by a shading).

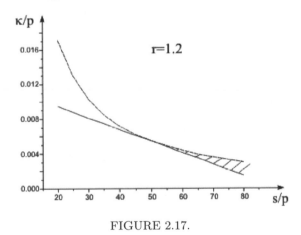

FIGURE 2.17.

The computations show that, in spite of small voltages needed for a stable suspension of the weight, the currents in the rings are too great. However similar to the case of rings with the prescribed current, elimination of the rotational degree of freedom leads to a considerable decrease in the strength of current and makes such a construction feasible. Additionally, an increase of the electromagnetic power can be achieved if a system of electromagnetic rings in series is used.

3
Dynamics of synchronous electric machines

3.1 The idealized model of synchronous machine

The basics of the mathematical model of synchronous machine are well known (see, for example [126]). When the transient processes are under consideration we can make the following assumptions:

1. The magnetic permeability of the armature core is assumed to be unbounded, i.e. saturation of the magnetic conductor and magnetic hysteresis are not taken into account. This condition allows us to use a linear approximation for computing electromagnetic field and the principle of superposition under the joint action of magnetomotive forces of the windings of the electric machine.

2. Only the first harmonic with the spatial period equal to the double pole division is taken into account when the spatial distribution of the fields of self-induction of windings of the rotor and stator is considered.

3. A real damper winding (for the explicit-pole machines) or the rotor body acting as a damper winding (for implicit-pole machines) is replaced by two equivalent damper contours located in longitudinal (contour t) and transversal (contour k) axes of the machine. Longitudinal axis of the rotor d passes through the mid-pint of the rotor pole and has the direction coinciding with the direction of the magnetic field of the excitation winding. The direction of the current in the longitudinal damper contour is positive if the direction of magnetic flux due to this current is coincident with the positive direction of axis d. The transverse axis of rotor q is located between the neighbouring magnetic poles of rotor and forms the angle of 90 electric

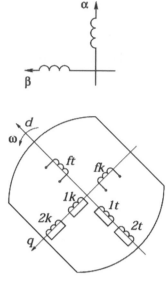

FIGURE 3.1.

degree to axis d. The positive direction of axis q is that for which this axis is behind axis d. The positive direction of current in contour k coincides with that in contour t.

The phase windings of the stator (anchor) are spools distributed over the stator circumferences and connected to each other by means of the star-connection or delta connection. The magnetic axes of the three-phase system of spools a, b, c are shifted to each other on 120 electric degree[1]. The location of the contours of the idealized model of synchronous machine is shown in Fig. 3.1.

The self-inductions of the stator phases are the following functions of the electric angle of rotation of the rotor

$$L_a = L + L_2 \cos 2\theta,$$

$$L_b = L + L_2 \cos \left(2\theta + \frac{2\pi}{3} \right),$$

$$L_s = L + L_2 \cos \left(2\theta - \frac{2\pi}{3} \right). \tag{3.1.1}$$

Here and in what follows the second harmonic is observed only for the explicit-pole rotors.

----

[1] Without loss of generality, in what follows we consider the model of implicit-pole synchronous machine with a three-phase stator winding.

The mutual inductances of the stator windings are given by

$$M_{ab} = M \cos \frac{2\pi}{3} + L_2 \cos \left(2\theta - \frac{2\pi}{3} \right),$$

$$M_{ac} = M \cos \frac{2\pi}{3} + L_2 \cos \left(2\theta - \frac{2\pi}{3} \right),$$

$$M_{bc} = M \cos \frac{2\pi}{3} + L_2 \cos 2\theta. \tag{3.1.2}$$

The mutual inductance of the excitation winding and the phase windings of stator are as follows

$$M_{af} = M_f \cos \theta,$$

$$M_{bf} = M_f \cos \left(\theta - \frac{2\pi}{3} \right),$$

$$M_{cf} = M_f \cos \left(\theta + \frac{2\pi}{3} \right), \tag{3.1.3}$$

where M_f denotes the mutual inductance of these windings when their magnetic axes coincide.

In a similar manner we can express the mutual inductances of the damper contours and the phase windings of stator for the longitudinal damper contours

$$M_{at} = M_t \cos \theta, \quad M_{bt} = M_t \cos \left(\theta - \frac{2\pi}{3} \right), \quad M_{ct} = M_t \cos \left(\theta + \frac{2\pi}{3} \right) \tag{3.1.4}$$

and the transversal damper contours

$$M_{ak} = -M_t \sin \theta, \quad M_{bk} = -M_t \sin \left(\theta - \frac{2\pi}{3} \right), \quad M_{ck} = -M_t \sin \left(\theta + \frac{2\pi}{3} \right), \tag{3.1.5}$$

respectively.

The coefficients of self-induction and mutual induction of the rotor contours do not depend on the rotor rotation and are given by the constant matrix

$$L_r = \begin{pmatrix} L_f & M_{ft} & 0 \\ M_{ft} & L_t & 0 \\ 0 & 0 & L_k \end{pmatrix}. \tag{3.1.6}$$

3.2 Equation of the synchronous machine in the case of a powerful network

The problem of reducing equations of the synchronous machine in the case of a powerful network is considered in most books on transient processes in synchronous machines (see for example [119]) and is of a crucial importance. Earlier, this problem was solved in the following way. The Park-Gorev equations were written down and the flux linkages Ψ_d, Ψ_q corresponding to the stationary regime $\Psi_d = u\cos\delta, \Psi_q = -u\sin\delta$ (u is the voltage amplitude in network and δ is the phase shift between the angle of rotation of the rotor and the network voltage) were substituted in the equations for the rotor circuit and the equation for rotation. After this substitution the equations for the stator circuit were no longer considered. The ways of improving the simplified equations were also offered, in particular by means of taking into account the resistance of stator circuits, [129], however no mathematical substantiation for these procedures was given. The region of applicability of the simplified equations and their accuracy were not estimated except for an indication that the simplified equations were not applicable for switching stator circuits [119].

A natural mathematical procedure for derivation of the simplified equations for synchronous electric machines is provided by asymptotic methods of the theory of nonlinear vibration. Halanay's article [38] seems to be the first work in this direction. In this paper the assumption of smallness of some parameters corresponding to the case of high-power synchronous machines was made. However this paper contains no thorough analysis of the rate of transient processes, in particular, slow processes in the stator circuits were not obtained. Nevertheless, using asymptotic methods Halanay derived the simplified equations which are coinciding with those obtained earlier. New conclusions about applicability of the simplified equations and accuracy of the solution are not contained in [38].

A complete mathematical substantiation of the simplified equations for synchronous machine in the case of a high-power network and a qualitative analysis were given in [55], [109] by means of the asymptotic methods of separation of motion. In contrast to [38], paper [55] takes into account an additional small parameter characterising small dissipation between the excitation winding and the damper contour in the longitudinal axis, which is typical for turbogenerators. This allowed one to decrease the number of simplified equations by one. In addition to this, the hidden slow variables in the stator circuits were obtained due to transition to the variables in axes "$\alpha, \beta, 0$". A new important conclusion about applicability of the simplified equations for analysis of commutation of circuits stator was made in [55].

Let us consider an implicit-pole synchronous machine connected to a powerful network with a given voltage by using a circuit with a zero wire. Basically, other connections can be analysed by analogy.

We begin with the standard model of synchronous machine with the damper contours t and k (the location of these contours of the idealized model of synchronous machine is shown in Fig. 3.1).

The Lagrange-Maxwell equations for synchronous machine connected to the powerful network have the form

$$\frac{d}{dt}\left[Li_a + M\cos\frac{2\pi}{3}(i_b + i_c) + (M_f i_f + M_t i_t)\cos\vartheta - M_k i_k \sin\vartheta\right]$$
$$+ R_a i_a + R_0 i_0 = U\cos\Omega_0 t, \quad (a,b,c),$$

$$\frac{d}{dt}\left\{M_f\left[i_a\cos\vartheta + i_b\cos\left(\vartheta - \frac{2\pi}{3}\right) + i_c\cos\left(\vartheta + \frac{2\pi}{3}\right)\right] + L_f i_f + M_{ft} i_t\right\}$$
$$+ R_f i_f = E_f,$$

$$\frac{d}{dt}\left\{M_t\left[i_a\cos\vartheta + i_b\cos\left(\vartheta - \frac{2\pi}{3}\right) + i_c\cos\left(\vartheta + \frac{2\pi}{3}\right)\right] + L_t i_t + M_{ft} i_f\right\}$$
$$+ R_t i_t = 0,$$

$$\frac{d}{dt}\left\{-M_k\left[i_a\sin\vartheta + i_b\sin\left(\vartheta - \frac{2\pi}{3}\right) + i_c\sin\left(\vartheta + \frac{2\pi}{3}\right)\right] + L_k i_k\right\}$$
$$+ R_k i_k = 0,$$

$$J\frac{d\Omega}{dt} = -\left\{(M_f i_f + M_t i_t)\left[i_a\sin\vartheta + i_b\sin\left(\vartheta - \frac{2\pi}{3}\right) + i_c\sin\left(\vartheta + \frac{2\pi}{3}\right)\right]\right.$$
$$\left.+ M_k i_k\left[i_a\cos\vartheta + i_b\cos\left(\vartheta - \frac{2\pi}{3}\right) + i_c\cos\left(\vartheta + \frac{2\pi}{3}\right)\right]\right\} + M_m. \quad (3.2.1)$$

Here ϑ denotes the angle of the rotor rotation between axis a and the winding axis f, $\Omega - d\vartheta/dt$ is the angular frequency of the rotor rotation, Ω_0 is the network frequency, i_a, i_b, i_c are currents in the stator phases, i_f, i_t, i_k are currents in the excitation winding and the damper contours respectively, L, L_f, L_t, L_k denote the coefficients of self-induction of the stator phases, contour of excitation and damper contours respectively, $M\cos\frac{2\pi}{3}$ are the coefficients of mutual inductions of the stator contours, M_f, M_t, M_k are the "amplitudes" of the coefficients of mutual induction between the stator contours and the excitation contour and the damper contours respectively, M_{ft} are the coefficients of mutual induction between contour excitations and the damper contour t, E_f is the excitation voltage, R_a and R_0 are active resistances of the stator windings together with the corresponding parts of connecting wires and the resistance of zero wire respectively, R_f, R_t, R_k are active resistances of the excitation contour and the damper contours respectively, J is the moment of inertia of the rotor. The external torque applied to the rotor is denoted by M_m. If $M_m > 0$ then it is the torque of a turbine or other engine and the machine works as a generator and if $M_m < 0$ then it is the load torque and the machine acts as an engine. Generally speaking the rotor torque M_m can depend on time and the angular velocity of rotor ϑ. Here and in what follows, the symbol (a, b, c) in eq. (3.2.1) implies that the equation needs to be completed by two similar equations for phases b and c with the help of replacements $\vartheta \to \vartheta - 2\pi/3$ and $\vartheta \to \vartheta + 2\pi/3$.

Generally speaking, the system of equations (3.2.1) is nonlinear. Its "electric" part has the coefficients which are periodic functions of the angle of rotor rotation. The electric subsystem is described by linear equations with time-periodic coefficients in some cases in which the change in the angular velocity of rotor is negligible, for example, stationary rotation or short-time short circuit. In this case the electric subsystem is transformed to equations with constant coefficients for a rather narrow class of transient regimes with the symmetric load of phases. A linear transformation of currents (or flux linkages) resulting in equations with constant coefficients was offered by Park [94]. This transformation is given by

$$i_d = \frac{2}{3}\left[i_a \cos\vartheta + i_b \cos\left(\vartheta - \frac{2\pi}{3}\right) + i_c \cos\left(\vartheta + \frac{2\pi}{3}\right)\right],$$

$$i_q = -\frac{2}{3}\left[i_a \sin\vartheta + i_b \sin(\vartheta - \frac{2\pi}{3}) + i_c \sin(\vartheta + \frac{2\pi}{3})\right],$$

$$i_0 = \frac{1}{3}(i_a + i_b + i_c). \tag{3.2.2}$$

In terms of axes d and q the equations for stator circuits take the form

$$\dot{\Psi}_d - \omega\Psi_q + R_a i_d = U\cos\delta,$$

$$\dot{\Psi}_q + \omega\Psi_q + R_a i_q = -U\sin\delta. \tag{3.2.3}$$

In what follows, while considering equations of the synchronous machine working on the powerful network we also use transformation to axes α and β

$$i_\alpha = \frac{2}{3}\left[i_a - \frac{1}{2}(i_b + i_c)\right],$$

$$i_\beta = \frac{1}{\sqrt{3}}(i_b - i_c),$$

$$i_0 = \frac{1}{3}(i_a + i_b + i_c), \tag{3.2.4}$$

see [119]. It is worth mentioning the relation between the variables in axes α, β and d, q

$$i_d = i_\alpha \cos\vartheta + i_\beta \sin\vartheta,$$

$$i_q = i_\alpha \sin\vartheta - i_\beta \cos\vartheta \tag{3.2.5}$$

As it will be shown below, the variables in axes α, β, turn out to be comfortable for separating slow processes in the stator circuits when the machine is connected to a network of unbounded power.

The equations for electromechanical processes in synchronous machines are difficult to solve not only because of the nonlinearity and high order. Another difficulty is that they describe transient processes with essentially different rates. The system of contours of the idealized model of synchronous machine is known to have essentially different time constants. This fact is reflected in the theory of electric machines by means of introducing two typical time constants of the rotor contours in the longitudinal axis. A small time constant (subtransient one) is caused by a small dissipation between the damper contour t and the excitation winding f, whereas the transient time constant determined by the equivalent inductance and the resistance of the excitation winding has a substantially higher value. The time constants of the damper contours and the mechanical time constant determined by inertia of the rotor of synchronous machine are also essentially different.

This obstacle remains covert when any standard systems of the relative units is used. For this reason, the further exposition is based on another method of introducing the relative units enabling explicit separation of small parameters in the equations for synchronous machine. This strategy determines the mathematical structure of nonlinear equations for synchronous machine and provides one with a possibility of the asymptotic simplification with the help of the methods of nonlinear mechanics.

First we introduce non-dimensional currents and flux linkages. The procedure is more transparent in terms of axes d and q. We use the well-known relationship between the currents and flux linkages in axes d

$$\Psi_d = L_d i_d + M_f i_f + M_t i_t,$$
$$\Psi_f = \frac{3}{2} M_f i_d + L_f i_f + M_{ft} i_t,$$
$$\Psi_t = \frac{3}{2} M_t i_d + M_{ft} i_f + L_t i_t. \tag{3.2.6}$$

Let us divide each of these relations by the base value of the flux linkage determined in terms of the base inductance and currents

$$\Psi_{d*} = L_{ad} i_{d*}, \Psi_{f*} = L_{af} i_{f*}, \Psi_{t*} = L_{at} i_{t*}. \tag{3.2.7}$$

The base current of stator is equal to the nominal current. The other five base values are taken from the conditions of symmetry of the non-dimensional matrix of inductance and the requirement that all non-diagonal elements are equal to unity. It is easy to notice that these requirements are reduced to five independent relations

$$\frac{3}{2} L_{ad} i_{d*}^2 = L_{af} i_{f*}^2 = L_{at} i_{t*}^2,$$
$$\frac{M_f i_{f*}}{L_{ad} i_{d*}} = \frac{M_t i_{t*}}{L_{ad} i_{d*}} = \frac{M_{ft} i_{t*}}{L_{af} i_{f*}} = 1 \tag{3.2.8}$$

We obtain from these relations

$$L_{ad} = \frac{3}{2}\frac{M_f M_t}{M_{ft}}, \quad L_{af} = \frac{M_f M_{ft}}{M_t}, \quad L_{at} = \frac{M_t M_{ft}}{M_f},$$

$$i_{f*} = \frac{L_{ad}}{M_f}i_{d*}, \quad i_{t*} = \frac{L_{ad}}{M_t}i_{d*} . \qquad (3.2.9)$$

Similar transformations for the flux linkages in axis q result in the following expressions for the corresponding base values [2]

$$L_{aq} = \frac{3}{2}\frac{M_k^2}{L_k}, i_{k*} = \frac{L_{aq}}{M_k}i_{d*} \qquad (3.2.10)$$

The final expressions for the flux linkages in the introduced system of relative units are as follows

$$\begin{aligned}
\Psi_d &= (1+\sigma_d)i_d + i_f + i_t, \\
\Psi_f &= i_d + (1+\varepsilon_r\sigma_f)i_f + i_t, \\
\Psi_t &= i_d + i_f + (1+\varepsilon_r\sigma_t)i_t, \\
\Psi_q &= (1+\sigma_q)i_q + i_k, \\
\Psi_k &= i_q + i_k,
\end{aligned} \qquad (3.2.11)$$

where

$$1+\sigma_d = \frac{L_d}{L_{ad}}, \quad 1+\sigma_q = \frac{L_q}{L_{aq}}, \quad 1+\varepsilon_r\sigma_f = \frac{L_f}{L_{af}}, \quad 1+\varepsilon_r\sigma_t = \frac{L_t}{L_{at}}.$$

It is also possible to pass on to the introduced system of relative units from any known system of relative units, say from the system x_{ad} or from the system of equal magnetomotive forces (mmf), [119]. Parameters $\varepsilon_r\sigma_f$ and $\varepsilon_r\sigma_t$ characterize the dissipation between contours t and f. Subtracting expression for Ψ_t from expression for Ψ_f in equations (2.2.6) we obtain

$$\varepsilon_r\Psi_r = \Psi_f - \Psi_t = \varepsilon_r(\sigma_f i_f - \sigma_t i_t),$$

where the quantity Ψ_r is proportional to the flux linkage of the dissipation between the excitation winding and the damper contour in the longitudinal axis.

Parameter ε_r is small because this dissipation is small as compared with the main flux. However parameter ε_r is not small for hydrogenerators and some types of turbogenerators, in this case the asymptotic transformation of equations of synchronous machine is carried out in [125].

Introducing the basis values by eqs. (3.2.9) and (3.2.10) determines the procedure of obtaining the non-dimensional equations for synchronous machine. So the equations for the transient processes in the case of machines

---

[2]This choice of the base values is more convenient for application of asimptotic methods. It was first suggested in [126].

working on a network with a given voltage in terms of variables $\alpha, \beta, 0$ are written in the form

$$[(1 + \sigma_d)i_\alpha + (i_f + i_t)\cos\vartheta - i_k\sin\vartheta]^\bullet + \varepsilon_n\nu_a i_\alpha = (u\cos\vartheta_0)^\cdot,$$
$$[(1 + \sigma_d)i_\beta + (i_f + i_t)\sin\vartheta + i_k\cos\vartheta]^\bullet + \varepsilon_n\nu_a i_\beta = (u\sin\vartheta_0)^\cdot$$
$$[i_\alpha\cos\vartheta + i_\beta\sin\vartheta + i_f(1 + \varepsilon_r\sigma_f) + i_t]^\bullet + \varepsilon_f\nu_f i_f = \varepsilon_f e_f,$$
$$[i_\alpha\cos\vartheta + i_\beta\sin\vartheta + i_f + i_t(1 + \varepsilon_r\sigma_t)]^\bullet + \nu_t i_t = 0,$$
$$[i_\beta\cos\vartheta - i_\alpha\sin\vartheta + i_k]^\bullet + \nu_k i_k = 0,$$
$$\ddot{\vartheta} = -\frac{\varepsilon_\omega}{\varepsilon}[(i_f + i_t)(i_\alpha\sin\vartheta - i_\beta\cos\vartheta) + i_k(i_\alpha\cos\vartheta + i_\beta\sin\vartheta)] + \frac{\varepsilon_\omega}{\varepsilon}m,$$
$$\dot{\vartheta}_0 = \frac{1}{\varepsilon}. \quad (3.2.12)$$

Here a dot means the derivative with respect to the non-dimensional time $\tau = t/T_k$ where $T_k = L_k/R_k$ denotes the time constant of the damper contour in axis k. The other parameters are introduced by the relations

$$\nu_n \quad - \quad \frac{R_a T_k}{L_{ad}}, \quad \nu_t = \frac{R_t T_k}{L_{at}}, \quad \nu_k = 1, \quad u = \frac{U}{L_{ad}i_*\omega_0},$$
$$\varepsilon_f\nu_f \quad = \quad \frac{R_f T_k}{l_{af}}, \quad \varepsilon_f e_f = \frac{E_f T_k}{L_{af}i_{d*}}, \quad \varepsilon_\omega = \frac{3}{2}\frac{L_{ad}i_{d*}^2 T_k}{J\omega_0}. \quad (3.2.13)$$

The small parameter $\varepsilon = 1/\omega_0 T_k$, where ω_0 denotes the synchronous frequency, is given by the ratio of the period of network voltage to the time constant of the damper contour. This parameter is small for all synchronous machines and is the main small parameter of the problem (for turbogenerators, a characteristic value of ε is 0.02). An analysis of the technical requirements to synchronous machines indicates that parameters ε_f and ε_ω are small (characteristic values are about 0.1). Indeed, by virtue of definition $\varepsilon_f = T_k/T_b$, where T_b is the time constant of the excitation winding, thus $T_b \gg T_k$ and ε_f is a small parameter. By analogy, $\varepsilon_\omega = T_m/T_k$, where T_m is the mechanical time constant, therefore $\varepsilon_\omega \ll 1$. In what follows, this fact is used for simplifying equations for electromechanical processes in the case of autonomous work of synchronous machine on the load. The value of parameter ε_n is small due to a relative smallness of active resistances of the stator circuits relative to the inductive resistance.

Before we proceed to derivation of the asymptotically simplified equations for synchronous machine let us explain a necessity of using axes $\alpha, \beta, 0$ for separation of slow processes in the stator circuits. As follows from equations for the stator circuits in axes d, q (3.2.3), in the transient regime variables Ψ_d, Ψ_q have a fast oscillating component. In the case of the small sliding these equations are quasi-linear, therefore transition to the axes α, β (3.2.4) is somehow equivalent to the Van-der-Pol replacement in the theory of nonlinear vibrations.

Using equations in the dimensionless variables (3.2.12) in axes α, β we introduce dimensionless flux linkages related to the dimensionless currents in the following manner

$$
\begin{aligned}
\Psi_\alpha &= (1 + \sigma_d)i_\alpha + (i_f + i_t)\cos\vartheta - i_k\sin\vartheta - u\cos\vartheta_0, \\
\Psi_\beta &= (1 + \sigma_d)i_\beta + (i_f + i_t)\sin\vartheta + i_k\cos\vartheta - u\sin\vartheta_0, \\
\Psi_f &= i_\alpha\cos\vartheta + i_\beta\sin\vartheta + (1 + \varepsilon_r\sigma_f)i_f + i_t, \\
\Psi_k &= i_\beta\cos\vartheta - i_\alpha\sin\vartheta + i_k, \\
\Psi_r &= \sigma_f i_f - \sigma_t i_t.
\end{aligned}
\tag{3.2.14}
$$

Variables Ψ_f, Ψ_k are dimensionless flux linkages of the excitation contour and the damper contour in the transverse axis, Ψ_r is a difference of the flux linkage of the excitation contour and damper contour in the longitudinal axis, attributed to the value of small parameter ε_r, variables Ψ_α, Ψ_β differ from the stator flux linkage in axes α, β by additional integrals of the network voltage attributed to these axes.

Let us introduce a new variable $\delta = \vartheta - \vartheta_0$ which is the angle of the rotor rotation "relative to the network". Expressing currents from eq. (3.2.14) in terms of the introduced flux linkages we arrive at the equations

$$
\begin{aligned}
\frac{d\Psi_\alpha}{d\vartheta_0} &= -\frac{\varepsilon\varepsilon_n\nu_n}{\sigma_d(\sigma_d + \kappa)}[\sigma_d + \kappa\sin^2(\vartheta_0 + \delta)]\Psi_\alpha - \kappa\sin(\vartheta_0 + \delta)\cos(\vartheta_0 + \delta)\Psi_\beta \\
&\quad - \sigma_d(1 - \kappa)\cos(\vartheta_0 + \delta)\Psi_f + (\sigma_d + \kappa)\sin(\vartheta_0 + \delta)\Psi_k \\
&\quad + \kappa\frac{\sigma_d}{\sigma_t}\cos(\vartheta_0 + \delta)\Psi_r + \sigma_d u\cos\vartheta_0 + \kappa u\sin\delta\sin(\vartheta_0 + \delta),
\end{aligned}
$$

$$
\begin{aligned}
\frac{d\Psi_\beta}{d\vartheta_0} &= -\frac{\varepsilon\varepsilon_n\nu_n}{\sigma_d(\sigma_d + \kappa)} - \kappa\sin(\vartheta_0 + \delta)\cos(\vartheta_0 + \delta)\Psi_\alpha + \\
&\quad + [\sigma_d + \kappa\cos^2(\vartheta_0 + \delta)]\Psi_\beta - \sigma_d(1 - \kappa)\sin(\vartheta_0 + \delta)\Psi_f - \\
&\quad - (\sigma_d + \kappa)\cos(\vartheta_0 + \delta)\Psi_k + \kappa\frac{\sigma_d}{\sigma_t}\sin(\vartheta_0 + \delta)\Psi_r + \sigma_d u\sin\vartheta_0 - \\
&\quad - \kappa u\sin\delta\cos(\vartheta_0 + \delta),
\end{aligned}
$$

$$
\begin{aligned}
\frac{d\Psi_f}{d\vartheta_0} &= -\varepsilon\varepsilon_f\nu_f\frac{1 - \kappa}{\sigma_f + \sigma_t} - \frac{\sigma_t}{\sigma_d + \kappa}[\cos(\vartheta_0 + \delta)\Psi_\alpha + \sin(\vartheta_0 + \delta)\Psi_\beta] + \\
&\quad + \frac{1 + \sigma_d}{\sigma_d + \kappa}(\sigma_t\Psi_f + \Psi_r) - \frac{\sigma_t}{\sigma_d + \kappa}u\cos\delta + \varepsilon\varepsilon_f e_f,
\end{aligned}
$$

$$
\frac{d\Psi_k}{d\vartheta_0} = -\frac{\varepsilon\nu_k}{\sigma_d}[\sin(\vartheta_0 + \delta)\Psi_\alpha - \cos(\vartheta_0 + \delta)\Psi_\beta + (1 + \sigma_d)\Psi_k + u\sin\delta],
$$

$$\frac{d\Psi_r}{d\vartheta_0} = -\frac{\varepsilon}{\varepsilon_r}\frac{1-\kappa}{\sigma_f+\sigma_t}\frac{\varepsilon_f\nu_f\sigma_t-\nu_t\sigma_f}{\sigma_d+\kappa}[-\cos(\vartheta_0+\delta)\Psi_\alpha-\sin(\vartheta_0+\delta)\Psi_\beta+$$

$$(1+\sigma_d)\Psi_f+\frac{\varepsilon_f\nu_f\sigma_d}{\sigma_d+\kappa}+\frac{\nu_t}{\sigma_t}\left(\frac{\sigma_t+\kappa\sigma_f}{1-\kappa}+\frac{\kappa\sigma_f}{\sigma_d+\kappa}\right)\Psi_r-$$

$$\frac{\varepsilon_f\nu_f\sigma_t-\nu_t\sigma_f}{\sigma_d+\kappa}u\cos\delta+\frac{\varepsilon\varepsilon_fe_f}{\varepsilon_r}, \tag{3.2.15}$$

$$\frac{ds}{d\vartheta_0} = -\frac{\varepsilon\varepsilon_\omega}{\sigma_d+\kappa}[\sin(\vartheta_0+\delta)\Psi_\alpha-\cos(\vartheta_0+\delta)\Psi_\beta+u\sin\delta][(1-\kappa)\Psi_f-$$

$$-\frac{\kappa}{\sigma_t(\sigma_d+\kappa)}\Psi_r\Big] + (1+\kappa/\sigma_d)\left[\cos(\vartheta_0+\delta)\Psi_\alpha+\sin(\vartheta_0+\delta)\Psi_\beta+\right.$$

$$+u\cos\delta\big]\Psi_k+\frac{\kappa}{\sigma_d(\sigma_d+\kappa)}[\sin(\vartheta_0+\delta)\Psi_\alpha-\cos(\vartheta_0+\delta)\Psi_\beta+u\sin\delta]\times$$

$$\times[\cos(\vartheta_0+\delta)\Psi_\alpha+\sin(\vartheta_0+\delta)\Psi_\beta+u\cos\delta]+\varepsilon\varepsilon_\omega m,$$

$$\frac{d\delta}{d\vartheta_0} = s.$$

The equation with ι_0 is separated from the others and is not considered here. In eq. (3.2.15) the following denotation is introduced

$$\kappa = \frac{\varepsilon_r\sigma_t\sigma_f}{\sigma_f+\sigma_t(1+\varepsilon_r\sigma_f)}.$$

As one can see, system (3.2.15) contains the variables with essentially different time-rates. Variable Ψ_r is a fast one and its derivative is proportional to the ratio of small parameters $\varepsilon/\varepsilon_r$. The derivatives of variables $\Psi_\alpha, \Psi_\beta, \Psi_f, \Psi_k$ are proportional to ε and these variables should be considered as slow (relative to Ψ_r). Variable δ is also formally fast (the rate of change is of the order of unity). However in the following we consider only the motions under which sliding s is small and δ changes slowly. This corresponds to analysis of particular solutions of the system with two fast phases in the case of principle resonance. It is possible to seek different motions of such sort having the order of $s = O(\varepsilon^h)$ during the whole motion. Accuracy of the motion calculation by means of the averaging method and the time interval ensuring the assumed order of smallness s depend on h. The maximum interval of order $1/\sqrt{\varepsilon\varepsilon_\omega}$ and the least error in the first approximation of order $\sqrt{\varepsilon\varepsilon_\omega}$ are achieved for motions with

$$s = \sqrt{\frac{\varepsilon\varepsilon_\omega}{\sigma_d}}S, \quad S = O(1). \tag{3.2.16}$$

These motions are considered in what follows. Accordingly, the initial sliding $s(0)$ is also of the same order. Substituting relations (3.2.16) into eq. (3.2.15) allows us to rewrite the last two equations in (3.2.15) in the form

$$\frac{dS}{d\vartheta_0} = \sqrt{\frac{\varepsilon\varepsilon_\omega}{\sigma_d}}(\ldots), \quad \frac{d\delta}{d\vartheta_0} = \sqrt{\frac{\varepsilon\varepsilon_\omega}{\sigma_d}}S. \tag{3.2.17}$$

Thus, system (3.2.15) along with replacement (3.2.17) is a system with many slow variables, one non-critical fast variable Ψ_r and one fast phase ϑ_0. The method asymptotic integration of such systems was offered by V.M.Volosov [128]. According to [128], in order to construct a first approximation it is necessary to integrate the equation with $d\Psi_r/d\vartheta_0$, assuming the other variables be independent of ϑ_0 and substitute the obtained result $\Psi_r = \Psi_r(\vartheta_0, \Psi_\alpha, \Psi_\beta \ldots, \delta)$ into the other equations. Only the terms of lower order in ε_r are needed to be retained in the right hand sides of equations, i.e. to put $\kappa = 0$ in eq. (3.2.15). The result is

$$
\Psi_r = C_r \exp\left[-\frac{(\varepsilon_f \nu_f + \nu_t)\varepsilon}{(\sigma_f + \sigma_t)\varepsilon_r}\vartheta_0\right] + \frac{\varepsilon^2(\varepsilon_f \nu_f \sigma_t - \nu_t \sigma_f)(\varepsilon_f \nu_f + \nu_t)}{\sigma_d[\varepsilon^2(\varepsilon_f \nu_f + \nu_t)^2 + \varepsilon_r^2(\sigma_f + \sigma_t)^2]} \times
$$

$$
\times \left\{\cos(\vartheta_0 + \delta)\Psi_\alpha + \sin(\vartheta_0 + \delta)\Psi_\beta + \frac{\varepsilon_r(\sigma_f + \sigma_t)}{\varepsilon(\varepsilon_f \nu_f + \nu_t)} \left[\sin(\vartheta_0 + \delta)\Psi_\alpha - \right.\right.
$$

$$
- \cos(\vartheta_0 + \delta)\Psi_\beta]\} - \frac{\nu_f \sigma_t - \nu_t \sigma_f}{\sigma_d(\nu_f + \nu_t)}[(1 + \sigma_d)\Psi_f - u\cos\delta] + \frac{\sigma_f + \sigma_t}{\varepsilon_f \nu_f + \nu_t}e_f.
$$

$$(3.2.18)$$

Here C_r denotes an integration constant determined from the initial conditions.

Let us substitute eq. (3.2.18) into the other equations of system (3.2.15), take into account eq. (3.2.17) and average the right hand sides of these equations with respect to explicit time ϑ_0. As a result we obtain the equations for slow transient processes

$$
\frac{d\Psi_\alpha}{d\vartheta_0} = -\varepsilon\frac{\nu_n}{\sigma_d}\Psi_\alpha, \qquad \frac{d\Psi_\beta}{d\vartheta_0} = -\varepsilon\frac{\nu_n}{\sigma_d}\Psi_\beta,
$$

$$
\frac{d\Psi_f}{d\vartheta_0} = -\varepsilon\frac{\varepsilon_f \nu_f \nu_t}{\sigma_d(\nu_f + \nu_t)}[(1 + \sigma_d)\Psi_f - u\cos\delta - \sigma_d e_f],
$$

$$
\frac{d\Psi_k}{d\vartheta_0} = -\varepsilon\frac{\nu_k}{\sigma_d}[(1 + \sigma_d)\Psi_k + u\sin\delta],
$$

$$
\frac{dS}{d\vartheta_0} = -\sqrt{\frac{\varepsilon\varepsilon_\omega}{\sigma_d}}[u(\Psi_f \sin\delta + \Psi_k \cos\delta) - \sigma_d m],
$$

$$
\frac{d\delta}{d\vartheta_0} = \sqrt{\frac{\varepsilon\varepsilon_\omega}{\sigma_d}}S. \qquad (3.2.19)
$$

The equations for the stator contours (the first two equations of system (3.2.19)) are separated from "the rotors equations" and are easily integrated. The solutions are slowly attenuating exponential functions. In contrast to the solution obtained above, the equation for the damper contour in longitudinal axis is eliminated in eq. (3.2.19), however this equation should be considered in the case of a hydrogenerator.

Generally speaking, the solutions of equations (3.2.19) approximate the exact solutions of eq. (3.2.15) in the time interval ϑ_0 of the order of $\sqrt{\sigma_d/\varepsilon\varepsilon_\omega}$

with the error of the order of $\sqrt{\varepsilon\varepsilon_w/\sigma_d}$. Nonetheless a qualitative accordance of the solutions is ensured as time tends to infinity. This aspect in discussed in detail in Section 3.5. As follows from the derivation, equations (3.2.19) can be also used in the case of commutation of the stator circuits.

The accuracy of solutions of eq. (3.2.19) relies on the values of parameters ε_w, σ_d. For high-power turbogenerators, coefficients $\varepsilon\nu_k/\sigma_d$ and $\sqrt{\varepsilon\varepsilon_w/\sigma_d}$ are approximately equal, although formally speaking, the first contains the first degree of the main small parameter ε and the second one has a square root of ε. For example, a machine taken for the further numeral work has the following values

$$\varepsilon = 0.0134,\ \varepsilon_f\nu_f = 0.075\nu_k = 1,\ \sigma_d = 0.35,\ \sigma_f = 1,\ \sigma_t = 0.5,\ \varepsilon_w = 0.04,$$

$$\varepsilon_r = 0.03,\ \frac{\varepsilon\nu_k}{\sigma_d} = 0.038,\ \sqrt{\frac{\varepsilon\varepsilon_w}{\sigma_d}} = 0.039,\ \kappa = 0.001,\ u = 1,\ \varepsilon_n\nu_n = 0.023.$$

In this case all variables are obtained from eq. (3.2.19) with the error of order ε/σ_d.

However the case of large ε_w is possible when the difference between these coefficients is considerable since one coefficient is proportional to ε and the other is proportional to $\sqrt{\varepsilon}$. Then it is necessary to improve the accuracy of calculation of sliding by ensuring the same accuracy as in the case of Ψ_k. In this case, improving accuracy is achieved by calculation of the second approximation in $\sqrt{\varepsilon}$.

Let us ignore the equation containing $d\Psi_r/d\vartheta_0$ in eq. (3.2.15), retain only terms of order ε in the right hand sides of the other equations and insert Ψ_r from eq. (3.2.18). Then we replace the last two equations in eq. (3.2.15) by eq. (3.2.17). We arrive at the system

$$\frac{dx}{d\tau} = \varepsilon X(h,\delta,\tau),$$

$$\frac{dS}{d\tau} = \sqrt{\varepsilon}\Omega(h,\delta,\tau),$$

$$\frac{d\delta}{d\tau} = \sqrt{\varepsilon}\sqrt{\frac{\varepsilon_w}{\sigma_d}}S. \tag{3.2.20}$$

Here $x = (\Psi_\alpha, \Psi_\beta, \Psi_f, \Psi_k)^{\mathrm{T}}$ and multiplier $\sqrt{\varepsilon_w/\sigma_d}$ is introduced in Ω since it does not affect the comparative smallness of the right hand sides.

After truncation of the terms proportional to κ, variable Ψ_r does not appear in the last two equations of system (3.2.15). Therefore when constructing equations (3.2.20) and calculating the second approximation in $\sqrt{\varepsilon}$ (the terms of the order higher than ε are truncated) it is sufficient to take expression (3.2.18) for Ψ_r.

Generally speaking, in the second approximation it is necessary to find the change of variables

$$x = \xi + \sqrt{\varepsilon} u_{x1}(\xi, \zeta, \gamma, \tau),$$
$$S = \zeta + \sqrt{\varepsilon} u_{s1}(\xi, \zeta, \gamma, \tau),$$
$$\delta = \gamma + \sqrt{\varepsilon} u_{\delta 1}(\xi, \zeta, \gamma, \tau). \tag{3.2.21}$$

However it is obvious that $u_{x1} = 0$. The terms of order $\sqrt{\varepsilon}$ in the equation with $d\xi/d\tau$ are also equal to zero. Therefore the equations of second approximation are

$$\frac{d\xi}{d\tau} = \varepsilon \Xi_2(\xi, \zeta, \gamma),$$
$$\frac{d\zeta}{d\tau} = \sqrt{\varepsilon} Z_1(\xi, \zeta, \gamma) + \varepsilon Z_2(\xi, \zeta, \gamma),$$
$$\frac{d\gamma}{d\tau} = \sqrt{\varepsilon} \Gamma_1(\xi, \zeta, \gamma) + \varepsilon \Gamma_2(\xi, \zeta, \gamma). \tag{3.2.22}$$

Functions Ξ_2, Z_1 in eq. (3.2.22) are the same as those in eq. (3.2.19) and $\Gamma_1 = \sqrt{\varepsilon_\omega / \sigma_d} \zeta$. According to the general idea of the averaging method the functions $u_{s1}, u_{\delta 1}$ are governed by the equations

$$Z_1 + \frac{\partial u_{s1}}{\partial \tau} = \Omega,$$
$$\Gamma_1 + \frac{\partial u_{\delta 1}}{\partial \tau} = \sqrt{\frac{\varepsilon_\omega}{\sigma_d}} \zeta. \tag{3.2.23}$$

It is known that u_{s1} and $u_{\delta 1}$ are determined up to an arbitrary function of ξ, ζ, γ. At this place, it is convenient to put $u_{\delta 1} = 0$ and $\langle u_{s1} \rangle = 0$ where u_{s1} is given by the quadrature

$$u_{s1} = \int (\Omega - Z_1) d\tau.$$

Then

$$Z_2 = -\left\langle \frac{\partial u_{s1}}{\partial \gamma} \Gamma_1 \right\rangle = 0, \qquad \Gamma_2 = \langle u_{s1} \rangle = 0. \tag{3.2.24}$$

The result is as follows. In the second approximation, the averaged equations (3.2.22) and the variable appearing x and δ have the form of the first approximation. Hence there is no need to introduce new denotations for ξ and γ. Only S gains a correction proportional to $\sqrt{\varepsilon}$

$$S = \zeta + \sqrt{\frac{\varepsilon \varepsilon_\omega}{\sigma_d}} [\cos(\vartheta_0 + \delta)\Psi_\alpha - \sin(\vartheta_0 + \delta)\Psi_\beta]\Psi_f -$$
$$- [\sin(\vartheta_0 + \delta)\Psi_\alpha - \cos(\vartheta_0 + \delta)\Psi_\beta]\Psi_k. \tag{3.2.25}$$

Therefore the equations of the second approximation has the form of eq. (3.2.19), with S being replaced by ζ. Having solved these equations and obtained S from eq. (3.2.25) we obtain all variables with the error of order of ε in the time interval τ of the order of $1/\sqrt{\varepsilon}$.

FIGURE 3.2.

3.3 Systems with electromagnetic vibration absorbers

Before we proceed to analysis of the asymptotically simplified equations of synchronous machine let us study some general properties of the electromechanical systems with magneto-electric extinguishers. Let us notice that the asymptotically simplified equations of the synchronous machine working on a powerful network obtained in the previous section are also valid for these systems.

The magneto-electric extinguisher of vibrations contains one or several short-circuited conducting contours or massive conducting bodies moving in a constant magnetic field. The conductors are also the elements of the oscillating system. When the conductors move in the magnetic field, the mechanical forces appear due to the vortex currents and counteract vibrations of the conductors and the whole system. This sort of damping is referred to as magneto-electric.

A shortcoming of the magneto-electric extinguishers is that it is necessary to ensure the currents needed for the constant magnetic field. Nevertheless these extinguishers are used inasmuch as they possess a number of advantages in comparison with the viscous friction dampers. In particular, the damping properties of the magneto-electric extinguisher can be changed with a relative ease by changing intensity of the constant magnetic field. One of the magneto-electric extinguishers is shown in Fig. 3.2.

A similar method of extinguishing of vibrations is widely used in electric machines (generators, engines etc.), for example, for damping of swinging rotor when the synchronous machine starts to work on a network or the network load is changing. For the implicit-pole synchronous electric machine, the vibration extinguisher is the rotor body itself in which the vortex currents appear, provided that the angular velocity of rotation does not coincide with the frequency of rotation of the magnetic field created by the stator winding. In contrast to the magneto-electric extinguishers with a constant magnetic field, the rotor motion occurs in the magnetic field which is variable both in time and space. Using asymptotic methods of the motion separation the equations of swinging the rotor of synchronous machine can be transformed such that they coincide with the equations of motion of the pendulum with magneto-electric extinguishers. This is due to the fact that the averaged equations of the rotor swinging have the structure of Routh's equations however they are autonomous and have a considerably low order.

Let $q = (q_1..., q_n)$ and $i = (i_1..., i_m)$ denote the vector of generalized coordinates of oscillating systems and the vector of currents in the extinguisher contours respectively. The number of the extinguishers attached to the oscillating system, the number of contours in each extinguisher and numbering of the extinguishers and currents is not important. If the extinguishers contain some volumetric conductors (massive conducting bodies) m is equal to infinity. A discrete form of the equations of electromechanics for volumetric conductors is suggested in [90]. The constraints in the oscillating system are considered to be holonomic and stationary. The expressions for the kinetic energy T and the energy of magnetic field W are as follows

$$T = \frac{1}{2} \dot{q}^{\mathrm{T}} A \dot{q},$$

$$W = W_1 + W_2 + W_3 = \frac{1}{2} i^{\mathrm{T}} L i + J^{\mathrm{T}} M i + \frac{1}{2} J^{\mathrm{T}} L_0 J. \qquad (3.3.1)$$

Here $A(q)$ denotes a symmetric positive definite matrix of the inertia coefficients, $J = (J_1, .., .J_r)$ is the column of currents creating a constant magnetic field in the extinguishers, L is the symmetric positive definite matrix of coefficients of self-induction and mutual induction of the contours of currents $i_1, ..., i_m$ and $M(q)$ is the rectangular matrix of the dimension $p \times m$ of the coefficients of mutual induction between the contours of currents i and J. The mutual location of the conducting contours in the extinguisher does not change under vibration and the contours in different extinguishers are not linked magnetically. This corresponds to the condition $L = \text{const}$. Currents J are assumed to be constant and prescribed, i.e. we do not take into account the influence of emf induced in the current contour J by current i. It is possible if the coefficients of self-induction or resistance of the contours with currents J are sufficiently large and this contour is connected to the high-power voltage source, that corresponds to standard technical conditions. In addition to this, matrix L_0 of the coefficients of self-induction and mutual induction of contours with currents J is considered to be constant. The latter implies that term W_3 in the expression for energy of the magnetic field is not important. The generalized mechanical forces are considered as potential ones whereas the electric forces are the voltage drops on the active resistances of the conducting contours.

Designating the potential energy of mechanical systems by Π and the positive definite matrix of active resistances of contours with currents $i_1 ..., i_m$ by $R = \text{const}$ we can write down the Lagrange-Maxwell equations (1.2.14) in the vectorial form

$$L\ddot{i} + \frac{\partial M^{\mathrm{T}}}{\partial q} J\dot{q} + Ri = 0,$$

$$(A\dot{q})^{\cdot} - \frac{1}{2}\dot{q}^{\mathrm{T}}\frac{\partial A}{\partial q}\dot{q} - J^{\mathrm{T}}\frac{\partial M}{\partial q}i + \frac{\partial \Pi}{\partial q} = 0. \qquad (3.3.2)$$

Here the terms $\partial M^{\mathrm{T}}/\partial q\, J\dot{q}$ and $-J^{\mathrm{T}}\partial M/\partial q\, i$ describe respectively the emf of the motion and ponderomotive forces. Inasmuch as the currents are similar to the generalized velocities and energy of the magnetic field is analogous to the kinetic energy, these terms due to the term W_2 in the expression for W are linear in the currents and similar to the expressions for gyroscopic forces. In eq. (3.3.2) the considered terms relate the subsystem describing mechanical motions with the subsystem describing the currents. Only the term Ri in the whole system corresponds to the dissipation.

Let us consider the influence of magneto-electric extinguishers on stability of equilibrium of the mechanical oscillating systems. The results of the present section obtained for the systems with magneto-electric extinguishers are similar to the theorems by Thomson, Tait and Chetaev on the influence of viscous damping on the equilibrium stability [85]. Considering stability of equilibrium with respect to variables $q, \dot{q}, i$ in the system with extinguishers, we assume that the stability or instability of equilibrium in undamped system can be judged by terms of the second order in the series expansion of Π in the vicinity of the equilibrium. It is possible to attain a greater generality if we apply Krasovskiy's theorem however in this case it is difficult to prove the forthcoming equalities which are specific for the systems under consideration.

Let us assume that $q = q_0, i = 0$ in an equilibrium position. We introduce the perturbation vector $u = q - q_0$ and express variables $q, \dot{q}$ in terms of $u, \dot{u}$ in eq. (3.3.2). Substituting q and i into eq. (3.3.2) we obtain equations of the perturbed motion. The following power relationship

$$\left[\frac{1}{2}\dot{u}^{\mathrm{T}}A\,\dot{u} + W_1 + \Pi(q_0 + u)\right]^{\cdot} = -i^{\mathrm{T}}Ri \qquad (3.3.3)$$

is valid. If we retain the terms in the equations for the perturbed motion, which are linear in the unknown variables, we arrive at the equations in variations (the first approximation equations). Keeping the denotations of the perturbed motion for the unknown variables in these equations we have

$$L\ddot{i} + Ri + \Gamma\dot{u} = 0,$$

$$A_0\ddot{u} + Cu - \Gamma^{\mathrm{T}}i = 0, \qquad (3.3.4)$$

where

$$A_0 = A(q_0) , \quad C = \left(\frac{\partial^2 \Pi}{\partial q^{\mathrm{T}} \partial q} \right)_{q=q_0} , \quad \Gamma = \left(\frac{\partial M^{\mathrm{T}}}{\partial q} \right)_{q=q_0} J. \qquad (3.3.5)$$

The following relationship

$$(\Delta T + \Delta W + \Delta \Pi)^{\cdot} = -\Delta \Phi \qquad (3.3.6)$$

holds for system (3.3.3) where

$$\Delta T = \frac{1}{2} \dot{u}^{\mathrm{T}} A_0 \dot{u}, \quad \Delta W = \frac{1}{2} i^{\mathrm{T}} L i,$$

$$\Delta \Pi = \frac{1}{2} u^{\mathrm{T}} C u, \quad \Delta \Phi = i^{\mathrm{T}} R i. \qquad (3.3.7)$$

1. Let the equilibrium in the undamped system be stable and none of the particular solutions $w(t)$ of the system

$$A_0 \ddot{w} + C w = 0 \qquad (3.3.8)$$

satisfies simultaneously the system of m equalities

$$\Gamma \dot{w} \equiv 0. \qquad (3.3.9)$$

Then the equilibrium in the system with extinguishers is asymptotically stable.

First we will show that it is stable. As far as the stability of equilibrium in the undamped system is determined by the terms of second order in the series expansion of Π, matrix C is positive definite and there exist such $\delta_1, \ldots, \delta_n$ that $\Pi > 0$ for $|u_r| < \delta_r, u_r \neq 0,$ $r = 1, \ldots, n$. For such u_r and any $\dot{u}_r, i_k$ the function in the square brackets in the left hand side of eq. (3.3.3) is positive definite while the function in the right hand side is negative. Then Lyapunov's theorem ensures stability of equilibrium.

While analysing stability of equilibrium the complete set of linearly independent particular solutions of equations in variations is composed of the solutions of three types: (i) attenuating, (ii) describing harmonic vibrations and (iii) a set of constant values. The latter does not appear when matrices R and C are positive definite. The oscillating solutions do not satisfy eq. (3.3.6) if $i_j \neq 0$ for at least one j. When all $i_j \equiv 0$ the solutions do not exist since the solutions of system (3.3.8) do not satisfy eq. (3.3.9). Therefore all solutions of system (3.3.4) are attenuating, i.e. the solution is asymptotically stable.

On the other side, if equalities (3.3.9) hold for a solution of system (3.3.8) the equations in variations (3.3.4) assume a particular solution for which all $i_j \equiv 0$ and $u_r(t)$ describe this particular solution of system (3.3.8). Then the stability is not asymptotic.

Equalities (3.3.9) mean that there can exist such vibrations of the undamped system that no emf motions are induced in the extinguisher contours. Usually it presents no problem to find whether this is the case. Another form of equalities (3.3.9) is more convenient in more difficult cases. Let $u^{(\beta)}$ designate the vector of normal modes of system (3.3.8). For asymptotic stability none of the modes must satisfy the following m equalities

$$\Gamma u^{(\beta)} = 0. \tag{3.3.10}$$

For the systems with viscous damping there exists a general (i.e. independent of the normal modes of the undamped system) condition of asymptotic stability of equilibrium which implies the positive definiteness of matrix of the damping coefficients. There is no analogous condition for the systems with the magneto-electric extinguishers. The above assertion is analogous to the following statement for the systems with viscous damping: let the matrix of the damping coefficients be non-negative and the normal modes of the undamped system do not coincide with the eigenvectors of this matrix corresponding to zero eigenvalues. Then the stability of equilibrium is asymptotic.

2. Let the equilibrium of the undamped system be unstable. Then the equilibrium in the system with extinguishers is unstable, too. Inasmuch as instability is determined by the terms of second order in expansion of Π, it is possible to choose such $u_{10}, \ldots, u_{n0}$ that $\Delta\Pi(u_{10}, \ldots, u_{n0}) < 0$. Let us consider the solution of system (3.3.4) with the initial conditions $u(t_0) = u_0, \dot{u}(t_0) = 0, i(t_0) - 0$. This solution can not attenuate (when all $u_r \to 0, \dot{u}_r \to 0, i_j \to 0$ or $i_j \equiv 0$), since it follows from eq. (3.3.6) that $\Delta T + \Delta W + \Delta\Pi \leq \Delta\Pi(u_0) < 0$ whereas for the attenuating solution $\Delta T + \Delta W + \Delta\Pi \to 0$.

Let us assume that not all $i_j \equiv 0$ in the considered solution of system (3.3.4). Then $u_r(t)$, $r = 1, \ldots, n$ do not satisfy the equalities

$$\sum_{r=1}^{n} \Gamma_{kr} \dot{u}_r \equiv 0, \tag{3.3.11}$$

otherwise $\dot{u}_r$ would be absent in the first m equations (3.3.4) and it would follow from $i_j(t_0) = 0$, $j = 1, \ldots, m$ that all $i_j \equiv 0$. Hence if $u_r(t)$ describe the undamped vibrations or superposition of the undamped vibrations and attenuating motions, then at least one $i_j(t)$ contains an undamped component. Such a set $u_r(t)$, $i_j(t)$ can not be the solution of system (3.3.4) because it does not satisfy relation (3.3.6). Therefore, the considered solution is unbounded and the equilibrium is unstable.

If all $i_j \equiv 0$ in this solution then $u_r(t)$ is changing in accordance with one of the particular solutions of system (3.3.8). But any of the solution satisfying the condition $\Delta\Pi(t_0) < 0$ is unbounded. Consequently, for the above initial conditions the solution of system (3.3.4) is unbounded regardless of $i_j \equiv 0$, and the equilibrium of the system with extinguishers is unstable.

3. Let us consider the case when one group of equations in variations contains terms describing "pure mechanical" gyroscopic forces and has the form

$$A\ddot{u} + G\dot{u} + Cu - \Gamma^{\mathrm{T}}i = 0,$$
$$G = -G^{\mathrm{T}}. \tag{3.3.12}$$

The other set of equations in variations is coincident with the first m equations in (3.3.4).

Such equations can appear if the constraints in the oscillating system are non-stationary, but the expression for kinetic energy does not explicitly contain time. Coefficients C_{rs} are obtained from the difference $\Pi - T_0$ where T_0 is the term in the expression for kinetic energy which does not depend on the generalised velocities. Equations (3.3.12) appear also in the case in which the mechanical system contains cyclic coordinates and one analyses the equilibrium in a positional subsystem with fixed cyclic momenta. In this case Routh equations which can contain gyroscopic terms appear in eq. (3.3.12) instead of the second groups of Lagrange equations.

3. Let the equilibrium of the linear system, described by the variational equations for the system without extinguishers

$$A_0\ddot{w} + G\dot{w} + Cw = 0 \tag{3.3.13}$$

have a temporal stability. Besides, none of the particular solutions of eq. (3.3.13) is assumed to satisfy m equalities of eq. (3.3.9). Then the equilibrium in the system with extinguishers is unstable.

Let us take the same initial conditions for the solution of equations in variations for the system with extinguishers as in the case of unstable equilibrium of the undamped system. Under temporal stability the inequality $\Delta\Pi < 0$ is possible and relationship (3.3.6) holds true. Not all $i_j \equiv 0$ in the considered solution because $u_r(t)$ coincide with the solutions of system (3.3.13) at $i \equiv 0$ and the latter do not satisfy equalities (3.3.9) which are necessary for $i \equiv 0$. Instability in the system with extinguishers is proved in the same manner as the instability of equilibrium of undamped systems, except for the case $i \equiv 0$.

However if equalities (3.3.9) are met, then all particular solutions of system (3.3.13) for which $\Delta\Pi < 0$ satisfy these equalities. Then the extinguishers do not destroy temporal stability, though they do not extinguish vibration. Neither magneto-electric extinguishers nor viscous damping dampers are able to eliminate vibrations of the gyroscopically stabilised system without destroying stability.

3.4   Optimal parameters of the magneto-electric extinguisher of small vibrations of a single-degree-of-freedom system

Let us consider the problem of optimal choice of parameters of the extinguisher with a single contour current aimed at eliminating small vibration of a single-degree-of-freedom system. In this case the equations for small vibration are obtained from the first approximation equations (3.3.4) with $m = 1, n = 1$

$$Li\ddot{} + \Gamma\dot{u} + Ri = 0,$$
$$A\ddot{u} - \Gamma i + Cu = 0. \tag{3.4.1}$$

Here A, Γ, C, L, R are scalars, $C > 0$ since the equilibrium of the undamped system must be stable, the sign of Γ is not important and the other coefficients are positive due to Section 3.3. Let all parameters but R be prescribed. Clearly, for very small values of R or for $R - \infty$ (open circuit contour) the extinguisher is inefficient. The same statement is valid if $\Gamma - 0$ or all parameters are prescribed except for Γ. For sufficiently large values of $|\Gamma|$ the extinguisher is also inefficient for the reason that a viscous damper with a very large damping coefficient is inefficient. Therefore the problem of the optimum choice of Γ (or R) for prescribed values of other parameters has a solution. We assume that the optimum values of parameters correspond to the maximum stability, i.e. to the maximum absolute value of the real part of that root of the characteristic equation which is the closest to the imaginary axis. This is a frequently applied optimum criterion (see for example [19]).

Supposing that variables u, i in eq. (3.4.1) are proportional to $\exp(\lambda t)$ we obtain the characteristic equation

$$A L\lambda^3 + A R\lambda^2 + (C\Gamma + \Gamma^2)\lambda + R C = 0. \tag{3.4.2}$$

As shown in Section 3.3 the roots of eq. (3.4.2) are either negative or have negative real parts. It is easy to prove by applying Hurwitz's stability criterion.

Assuming the parameters of oscillating system to be prescribed, we introduce a new unknown parameter $\mu = \lambda/\Omega$, $\Omega^2 = C/A$ into eq. (3.4.2) and write down the equation for μ

$$\mu^3 + \rho\mu^2 + \gamma\mu + \rho = 0. \tag{3.4.3}$$

It contains two independent parameters

$$\rho = \frac{R}{L\Omega}, \quad \gamma = 1 + \frac{\Gamma^2}{CL}.$$

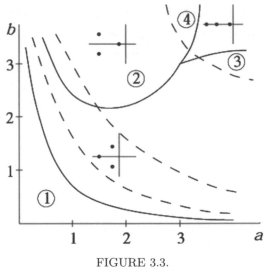

FIGURE 3.3.

Introducing $\tilde{\mu} = \mu/\rho^{1/3}$ in eq. (3.4.3) results in the canonical form of equation for $\tilde{\mu}$

$$\tilde{\mu}^3 + a\tilde{\mu}^2 + b\tilde{\mu} + 1 = 0, \tag{3.4.4}$$

where $a = \rho^{2/3}$, $b = \gamma/\rho^{2/3}$. The condition of asymptotic stability of equilibrium in the form of Hurwitz's criterion is satisfied for any $\rho > 0$, $\gamma > 1$

$$ab = \gamma > 1, \ a > 0, \ b > 0.$$

A typical root locus of eq. (3.4.4) is displayed in Fig. 3.3. The curve 1 is the hyperbola $ab = 1$, the curve 2 corresponds to equal real parts of the roots of equation (3.4.4) and consequently eq. (3.4.3), the curves 3 and 4 correspond to the condition of existence of the double order of the root. These curves divide the plane of parameters on the regions with a different location of the roots of eq. (3.4.3), see Fig. 3.3.

First we consider the problem of the optimum choice of both parameters γ and ρ (by the way, the previous analysis does not ensure that this problem has a solution). When two parameters are sought, three coefficients of equation (3.4.3) must be related by a relationship which is independent of these parameters. This relationship is obvious, namely the second coefficient must be equal to the free term. This yields the following relation between the roots $\mu_1 + \mu_2 + \mu_3 = \mu_1\mu_2\mu_3$. Consequently, parameters μ_1, μ_2, μ_3 must satisfy this relation and the optimum condition which results in the expressions for the coefficients in eq. (3.4.3). This choice should be made for two cases: (i) when the roots are real-valued and (ii) when one root is real-valued and two roots are complex-conjugated. Having analysed these cases one should choose the best variant.

Let all roots be real-valued. Let us express one root in terms of the others $\mu_3 = (\mu_1 + \mu_2)/(\mu_1\mu_2 - 1)$. The derivatives $\partial\mu_3/\partial\mu_1$ and $\partial\mu_3/\partial\mu_2$ are negative wherever they exist. Hence for the case of the real-valued roots the optimum solution corresponds the triple root $\mu_1 = \mu_2 = \mu_3 = -\sqrt{3}$ for the values of parameters $\rho = 3\sqrt{3}$, $\gamma = \mu_1\mu_2 + \mu_2\mu_3 + \mu_3\mu_1 = 9$. Indeed, assuming that μ_1 and μ_2 are smaller than $-\sqrt{3}$ we obtain $\mu_3 > -\sqrt{3}$.

Let μ_3 be a real-valued root and $\mu_{1,2} = \alpha \pm i\beta$ be two complex-conjugated roots. Expressing μ_3 in terms of α, β yields $\mu_3 = 2\alpha/(\alpha^2 + \beta^2 - 1)$. The stability for numbers $\alpha, \mu_3(\alpha, \beta = 0)$ is not worse than that for numbers $\alpha, \mu_3(\alpha, \beta)$ at any $\beta \neq 0$. Therefore the optimum solution should be sought for $\beta = 0$ that results in already considered case of three real-valued roots. However, it is possible to use the relation $\mu_3 = 2\alpha/(\alpha^2 - 1)$ where $\alpha < -1$. The derivative $d\mu_3/d\alpha$ is negative that immediately yields $\mu_3 = \alpha = -\sqrt{3}$.

Of interest is also the problem of the optimum choice of γ for fixed values of ρ since the choice of the optimum value of the current creating a constant magnetic field for the prescribed parameters of contour reduces to this problem.

In this case the roots of eq. (3.4.3) satisfy two relations not containing the sought-for parameter γ. If one root, say μ_3, is real-valued and two others are complex-conjugated $\mu_{1,2} = \alpha \pm i\beta$, these relations have the form $2\alpha + \mu_3 = -\rho$, $\mu_3(\alpha^2 + \beta^2) = -\rho$. Let us choose the optimum values α and μ_3 that satisfy only the first relation. The solution of this problem is obvious: $\alpha = \mu_3 - -\rho/3$. If we determine $\beta^2 > 0$ from the second relation for obtained values of α and μ_3 then these values of α, μ_3, β correspond to the optimum solution of the whole problem. The condition $\beta^2 > 0$ yields $\rho < 3\sqrt{3}$. For these values of ρ the optimum value is $\gamma = 2\rho^2/9 + 3$. For $\rho = 3\sqrt{3}$ the optimum solution corresponding to the triple root is the same as in the case of the optimum choice of γ and ρ.

The optimization with respect to γ for a given value of ρ reduces to study of changes in roots' location along the vertical lines $a = \mathrm{const}$ in Fig. 3.3. The case $\rho < 3\sqrt{3}$ corresponds to direct lines left to the direct line $a = 3$. An optimum location of roots is given by curve 2, thus all three roots have the same real part.

Let $\rho > 3\sqrt{3}$. In this case the lines $a = \mathrm{const}$ cross three regions: two regions with two complex-conjugated and one real-valued roots and a region with three real-valued roots. Let us consider first the case of the complex-conjugated roots. From two equalities relating μ_3, α and β^2 we obtain the relationship between β^2 and α

$$\beta^2 = \frac{\rho - \rho\alpha^2 - 2\alpha^3}{\rho + 2\alpha} = \frac{\rho}{\rho + 2\alpha} - \alpha^2. \tag{3.4.5}$$

Given β^2, it is possible to find α from eq. (3.4.5) and then $\mu_3 = -\rho - 2\alpha$. For sufficiently large values $\beta^2 > 1$ eq. (3.4.5) has a unique solution $\alpha = \alpha_1(\beta^2)$ thus $-\rho/2 < \alpha_1 < -\rho/3$, $\mu_3(\alpha_1) > \alpha_1$. This can be easily proved if one analyses the qualitative dependence of the right hand side of eq. (3.4.5) from α and takes into account that $\beta^2 < 0$ for $\alpha = -\rho/3, \rho > 3\sqrt{3}$. When β^2 decreases, eq. (3.4.5) gains two other roots α_2 and α_3, such that $-\rho/3 < \alpha_2 < 0$, $\alpha_3 > 0$, $\mu_3(\alpha_2) < \alpha_2$. When β^2 decreases up to $\beta^2 = 0$ then α_1 increases, and μ_3 decreases such that the value $\beta^2 = 0$ provides one with the better value of the optimum criterion $|\mu_3| = \rho + 2\alpha_1(0)$ than any other value of β^2. When β^2 decreases, α_2 decreases and the best value of the optimum criterion is $|\alpha_2|$ is also reached at $\beta^2 = 0$.

What remains is to compare the values $\mu_3 = -\rho - 2\alpha_1(0)$ and $\alpha_2(0)$. Let us show that $\alpha_2(0) < -\rho - 2\alpha_1$. The considered values of α are the roots of equation $\rho - \rho\alpha^2 - 2\alpha^3 = 0$ and satisfy the relations $\alpha_1\alpha_2 + \alpha_2\alpha_3 + \alpha_3\alpha_1 = 0$, $\alpha_1\alpha_2\alpha_3 = \rho/2$, hence $-\rho - 2\alpha_1 = -\rho/\alpha_1^2$. Using the second relation between the roots we can transform the inequality to the form $\alpha_1 < -2\alpha_3$. Inserting the expression $\alpha_3 = -\alpha_1\alpha_2/(\alpha_1 + \alpha_2)$ from the first relation, after some simple transformations we obtain the obvious inequality $\alpha_1 < \alpha_2$.

Thus, the maximal degree of stability in the case under consideration is equal to $|\alpha_2(\rho)|$ where α_2 is the greater negative root of the equation $Q(\alpha, \rho) = \alpha^3 + (\rho/2)\alpha^2 - \rho/2 = 0$. This root is a multiple root of the initial equation (3.4.3) and consequently is a root of the derivative of the polynomial in the left hand side of eq. (3.4.3). This yields the optimum value which is $\gamma = -3\alpha_2^2 - 2\rho\alpha_2$.

The analysis of the case in which the roots of eq. (3.4.3) are real-valued does not affect the conclusion that obtained degrees of stability and γ are optimal. The optimum location of the roots at $\rho > 3\sqrt{3}$ corresponds to curve 3 in Fig. 3.3.

Let us notice some properties of the dependence α_2 upon ρ. At $\alpha = -1$ we have $Q < 0$ and $\partial Q/\partial\alpha < 0$. For the given location of roots of $Q(\alpha, \rho)$ we conclude that $\alpha_2 < -1$ for all $\rho > 3\sqrt{3}$. It is also obvious that $\alpha_2 \to -1$ as $\rho \to \infty$. Differentiating $Q(\alpha, \rho)$ with respect to ρ and taking into account that α is a function of ρ we obtain that $d\alpha_2/d\rho = 0$ at $\alpha_2 = -1$. Since value $\alpha_2 = -1$ is not achieved, α_2 monotonically increases from $\alpha = -\sqrt{3}$ (at $\rho = 3\sqrt{3}$) as ρ grows.

Thus we proved the following result of the optimisation problem. The best degree of stability achieved for given values ρ by varying γ coincides with the best degree of stability achieved by varying γ and ρ simultaneously.

Having a qualitative dependence $\alpha_2(\rho)$ at disposal, it is easy to analyse the dependence $\gamma(\rho)$, hence $\gamma(\rho)$ is a monotonically increasing function of ρ and $9 \le \gamma < \infty$ at $3\sqrt{3} \le \rho < \infty$.

The third possible problem is an optimum choice of ρ for prescribed values of γ and determining maximal degree of stability as a functions of γ. Without going into detail we indicate two simple cases and demonstrate the final numerical result. Similar to the previous problem we have two relations between the roots which are independent of the sought-for parameter. They are as follows $\mu_1 + \mu_2 + \mu_3 = \mu_1\mu_2\mu_3$, $\mu_1\mu_2 + \mu_2\mu_3 + \mu_3\mu_1 = \gamma$. For $1 \le \gamma \le 9$ the optimum solution corresponds to the case in which $\mu_{1,2} = \alpha \pm i\beta$ and

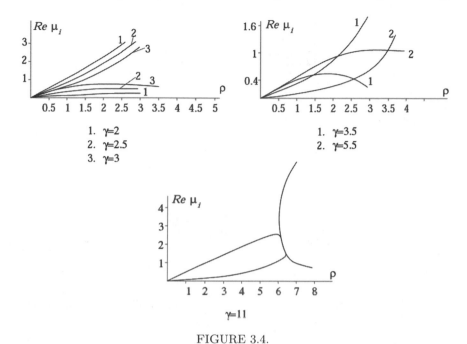

FIGURE 3.4.

μ_3 is real-valued. If $1 \leq \gamma \leq 5$ then $\alpha \geq \mu_3$ and the maximal degree of stability $|\alpha| = (\gamma - 1)/4$ is achieved at $\rho = (\gamma + 1)/2$. If $5 \leq \gamma \leq 9$, then $|\alpha| = |\mu_3| = \sqrt{(\gamma - 3)/2}$ and $\rho = 3|\alpha|$. In the latter case we have the relation between γ and ρ and between ρ and α as in one of the variants of solution of the previous problem $\alpha = -\rho/3, \gamma = 2\rho^2/9 + 3$.

The optimization of damping with respect to ρ for prescribed value of γ corresponds to the study of location of the roots of characteristic equation (3.4.3) in Fig. 3.3 under motion on the hyperbolae $ab = \gamma$. For $1 < \gamma < 3$ the corresponding hyperbolae lie in the region between the curves 1, 2, 3, for $3 \leq \gamma < 9$ and $\gamma > 9$ they cross curve 2 and curves 3, 4 respectively. In the case $1 < \gamma < 5$ the search of the optimum reduces to looking for the analytical maximum of the absolute value of the real part of the complex-conjugated roots (Fig. 3.4). For $5 < \gamma < 9$ the optimum lies on curve 2 and corresponds to equal real parts of the roots of the characteristic equation (Fig. 3.3). For $\gamma > 9$ (Fig. 3.3) the optimum location of the roots corresponds to curve 3. It is interesting to notice that for $\gamma > 5$ the optimum value of ρ corresponds to the non-analytical maximum of dependence of the maximal degree of stability on ρ. The complete dependence of the optimum value ρ and the maximal degree of stability m upon γ are presented in Fig. 3.5.

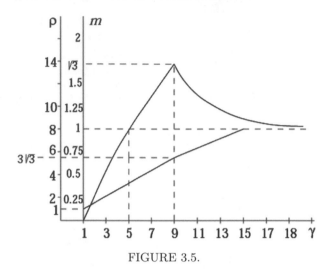

FIGURE 3.5.

3.5   Swinging of the rotor of synchronous machine and motion of the pendulum with magneto-electric extinguishers

The asymptotically simplified equations of the transient processes in the synchronous machine working on the powerful network obtained in Section 3.2, eq. (3.2.19), can be written down in the form

$$\dot{\Psi}_f + \frac{r_f}{l}(\Psi_f - \gamma \cos \delta) = e_f,$$

$$\dot{\Psi}_k + \frac{r_k}{l}(\Psi_k + \gamma \sin \delta) = 0,$$

$$\kappa \ddot{\delta} + \frac{\gamma}{l}(\Psi_f \sin \delta + \Psi_k \cos \delta) = m. \qquad (3.5.1)$$

Here Ψ_f and Ψ_k denote the slow components of flux linkage of the excitation contour and the transversal damper contour respectively, δ is the corner of the rotor swinging between two axes perpendicular to the axis of the rotor rotation, one axis being fixed in the rotor and the other rotating with the synchronous angular velocity. The sets of dimensionless parameters having appeared in eq. (3.2.19) after applying the asymptotic method are denoted in eq. (3.5.1) by l , γ etc.

$$r_f = \sqrt{\frac{\varepsilon \sigma_d}{\varepsilon_\omega}} \frac{\varepsilon_f \nu_f \nu_t}{(\nu_f + \nu_t)}, \quad r_k = \sqrt{\frac{\varepsilon \sigma_d}{\varepsilon_\omega}} \nu_k \kappa = \frac{1}{\sigma_d}, \quad l = \frac{\sigma_d}{1 + \sigma_d}, \quad \gamma = \frac{u}{\sigma_d}.$$

Additionally, another dimensionless "slow" time $\tau = \sqrt{\varepsilon \varepsilon_\omega / \sigma_d} \vartheta_0$ is introduced in eq. (3.5.1).

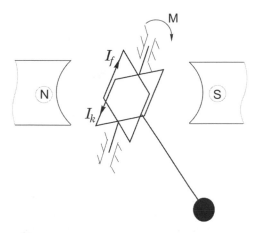

FIGURE 3.6.

Equations (3.5.1) have a structure of Routh's equations in which Ψ_f and Ψ_k are the quasi-cyclic generalized momenta and δ is the positional generalized coordinate. This allows one to make use of Lagrange equations which are more comfortable in this case. Let us introduce the currents (generalized velocities) I_f and I_k by the relations

$$I_f = \frac{1}{l}\left(\Psi_f - \gamma\cos\delta - \frac{le_f}{r_f}\right), \quad I_k = \frac{1}{l}(\Psi_k + \gamma\sin\delta). \qquad (3.5.2)$$

We obtain the equations

$$l\dot{I}_f - \gamma\dot{\delta}\sin\delta + r_f I_f = 0,$$
$$l\dot{I}_k - \gamma\dot{\delta}\cos\delta + r_k I_k = 0,$$
$$\kappa\ddot{\delta} + \gamma(I_f\sin\delta + I_k\cos\delta) + \frac{\gamma e_f}{r_f}\sin\delta = m. \qquad (3.5.3)$$

These equations are a special case of equations (3.3.2) and describe both swinging of the rotor of synchronous machine and the motions of pendulum with the magneto-electric extinguishers subjected to gravity and external moment m with the only difference that for the pendulum the factor γe_f must be replaced by another dimensionless parameter. In this case the extinguishers have a single contour with currents I_f and I_k and these contours are rigidly attached to the pendulum, the contours' planes being mutually perpendicular and the coefficient of mutual induction between the contours being equal to zero. The contours are placed in the homogeneous magnetic field and exhibit angular vibrations or revolve when the pendulum vibrates and rotates, see Fig. 3.6.

In the case in which the inductance l is negligible in eq. (3.5.3) and the resistances are equal to $r_f = r_k = r$, system (3.5.3) reduces to the well-known Trikomi equation [117]

$$\ddot\delta + \beta\dot\delta + \sin\delta = \sin\alpha,$$

where the differentiation is performed with respect to the dimensionless time $\tau = \Omega t$ and $\Omega^2 = \gamma e_f/r\kappa$, $\beta = \gamma^2/r\Omega$, $\sin\alpha = m/\kappa\Omega^2$.

If $\gamma e_f > |m|$, system (3.5.3) has two equilibrium positions

$$\delta = \begin{cases} \delta_* = \arcsin\dfrac{m}{\gamma e_f}, \\ \pi - \delta_* \end{cases}$$

for the zero currents I_k, I_f and sliding $\dot\delta = 0$. For the pendulum without extinguishers the first equilibrium is stable and the second one is unstable. As proved above, the extinguishers strengthen stability, i.e. a stable equilibrium becomes asymptotically stable whereas an unstable equilibrium remains unstable.

A positive torque m corresponds to a synchronous machine acting as a generator, in this case $\delta_* > 0$ and the rotor leads the stator field. A negative torque describes a machine acting as an engine, then $\delta_* < 0$ and the rotor lags behind the rotating field of stator by phase δ_*.

The following power relationship

$$\left[\frac{1}{2}l(I_f^2 + I_k^2) + \frac{1}{2}\kappa\dot\delta^2 + \frac{\gamma e_f}{r_f}(1 - \cos\delta)\right]^{\cdot} = -(r_f I_f^2 + r_k I_k^2) + m\dot\delta \quad (3.5.4)$$

holds true for system (3.5.3).

The properties of solution of equations describing the transient processes in different synchronous machines are discussed in the book by A.H.Gelig, G.A.Leonov and V.Ya.Yakubovich [35]. However [35] neither mentions nor uses the fact that these equations have a Lagrange structure[3], at least for a number of important cases, and the relationships like eq. (3.5.4) are valid. For the above mentioned systems the proof of theorems in [35] can be essentially simplified. For instance, it is easy to show by means of eq. (3.5.4) that system (3.5.3) is dichotomous at $m = $ const or $m = m(\delta)$ where $m(\delta)$ is a periodic function of δ with a non-zero mean value. Dichotomy means that the motions are either unbounded or describe equilibria or tend to an equilibrium (under such a definition the rotations are understood as unbounded motions [35]).

Indeed, for bounded motions the integral of $r_f I_f^2 + r_k I_k^2$ is bounded as $t \to \infty$. Also I_f and I_k are bounded. Consequently $I_f, I_k \to 0$. Taking

---

[3]The conclusion that the structure of the equations for synchronous machines written in the form of Routh's equations does not change after appllying the asymptotic method (generally speaking, this is not necessarily the case) and the transition to equations similar to (3.5.3) were performed by O.M.Lev and K.Sh.Khodgaev for the explicit-pole generator.

derivative with respect to time of the left hand sides of two equations (3.5.3) we can see that I_f and I_k are limited. From here and the fact that I_f, I_k are integrals of $\dot{I}_f, \dot{I}_k$ we obtain that $\dot{I}_f, \dot{I}_k \to 0$. From the same equations (3.5.3) it follows that $\ddot{\delta} \to 0$. One can prove by analogy that $\dddot{\delta}$ is a limited function of t and $\dddot{\delta} \to 0$. The latter equation in (3.5.3) yields that δ either corresponds to an equilibrium or $\delta(t)$ tends to the equilibrium position.

If the mean value $m(\delta)$ is zero (the machines working as a synchronous compensator satisfy this requirement), then system (3.5.3) is globally asymptotically stable, [35].

First we show that currents I_f and I_k in the extinguisher contours are bounded functions of time under any motion.

From the first equation (3.5.3) we obtain the equality relating I_f and δ

$$I_f(t) = I_f(t_0) \exp\left(-r_f(t - t_0)/l\right) + \frac{\gamma}{l} \int_{t_0}^{t} \exp\left(-r_f(t - \theta)/l\right) \frac{d\cos\delta(\theta)}{d\theta} d\theta.$$

$$(3.5.5)$$

In order to remove the derivative of $\cos\delta$ one preforms partial integration in eq. (3.5.5) then it becomes obvious that $I_f(t)$ is a bounded function of time for any solution of systems (3.5.3). The same is valid for I_k. It is also clear that the estimates obtained for $|I_f|$ and $|I_k|$ are the same for all solutions with the same initial conditions for the currents.

Let us set the power relation (3.5.4) in the form

$$\left[\frac{1}{2}l(I_f^2 + I_k^2) + \frac{1}{2}\kappa\dot{\delta}^2 + \frac{\gamma e_f}{r_f}(1 - \cos\delta) - \int_0^\delta m(\xi)d\xi\right]^{\bullet} = -(r_f I_f^2 + r_k I_k^2).$$

$$(3.5.6)$$

It follows from relation (3.5.6) that if $\int_{\delta_0}^{\delta_0+2\pi} m(\xi)d\xi = 0$ then unbounded motions are impossible in the system. Indeed, the function of phase variables in the square brackets is non-negative, its time-derivative is non-positive by virtue of relation (3.5.6), hence the motions are bounded for any initial δ. Repeating the reasoning used for proof of dichotomy we obtain that any motion tends to the equilibrium position, i.e. the system is globally asymptotically stable.

Let the mean value of m be positive. In addition to [35] we will show that system (3.5.3) assumes unbounded solutions with unbounded values $\dot{\delta} \to \infty$. We use the inequality

$$\dot{\delta}(t) \geq \dot{\delta}(t_0) - c(t - t_0), \qquad (3.5.7)$$

which is easy to derive having integrated the third equation in (3.5.3). It is essential that constant $c > 0$ in eq. (3.5.7) does not depend on the initial value $\dot{\delta}(t_0)$. It follows from eq. (3.5.7) that for a sufficiently large value $\dot{\delta}(t_0)$ one can ensure that $\dot{\delta}(t) > s_*$ where s_* is any number in any interval $t_0 \le t \le t_1$. Let us take t_1 such that $\delta(t_1) = \delta_0 + 2\pi$, $\delta_0 = \delta(t_0)$ (by the way, the larger $\dot{\delta}(t_0)$ the smaller difference $t_1 - t_0$) and introduce argument δ in eq. (3.5.3) instead of time t. The first equation in (3.5.3) takes the form

$$\frac{dI_f}{d\delta} = \frac{\gamma}{l}\sin\delta - \frac{r_f}{l}\frac{I_f}{s}, \qquad (3.5.8)$$

where $\dot{\delta} = s$. Integrating the similar equation with I_k yields

$$I_f(\delta) = I_f(t_0) + \frac{\gamma}{l}(\cos\delta_0 - \cos\delta) - \frac{r_f}{l}\int_{\delta_0}^{\delta}\frac{I_f(\xi)}{s(\xi)}d\xi,$$

$$I_k(\delta) = I_k(t_0) + \frac{\gamma}{l}(\sin\delta_0 - \sin\delta) - \frac{r_k}{l}\int_{\delta_0}^{\delta}\frac{I_k(\xi)}{s(\xi)}d\xi. \qquad (3.5.9)$$

Let us substitute eq. (3.5.9) into the equation obtained from the third equation in (3.5.3) after replacing the argument and integrating over δ. The result is the following relationship

$$\frac{1}{2}\kappa[s^2(\delta_0 + 2\pi) - s^2(\delta)] = \int_{\delta_0}^{\delta_0+2\pi} m(\xi)d\xi +$$

$$+ \frac{\gamma r_f}{l}\int_{\delta_0}^{\delta_0+2\pi}\sin\xi\int_{\delta_0}^{\xi}\frac{I_f(\eta)}{s(\eta)}d\eta d\xi + \frac{\gamma r_k}{l}\int_{\delta_0}^{\delta_0+2\pi}\cos\xi\int_{\delta_0}^{\xi}\frac{I_k(\eta)}{s(\eta)}d\eta d\xi. \qquad (3.5.10)$$

Choosing a sufficiently large value $s_0 = \dot{\delta}(t_0)$ we can ensure that the sum of two last terms in eq. (3.5.10) is less than the first term. In this case $s(\delta_0 + 2\pi) > s(\delta_0)$. Repeating this reasoning for the interval $\delta_0 + 2\pi \le \delta \le \delta_0 + 4\pi$ etc. we obtain that $s(\delta)$ and $\dot{\delta}(t)$ are unbounded for the taken initial condition. Interestingly enough, this is possible for the mean value m as small as is wished.

This conclusion is based on the observation that the terms in eq. (3.5.9), that do not have s in the denominator, do not contribute to increase in s within the period of change of δ. The latter are a result of Lagrange's structure of equations (3.5.3) and the gyroscopic terms relating the equation of motion with the equations for currents.

The pendulum under the action of external torque and a moment of damping proportional to the angular velocity can not accomplish such a motion. This indicates that the action of magneto-electric extinguishers under rotatory motions is "weaker" than the forces of viscous damping.

The same conclusion is valid for the pendulum with extinguishers. However dealing with electric machines it is necessary to take into account that equations (3.5.3) are obtained by the asymptotic method and are valid only for sufficiently small sliding. This fact should be taken into account when the results of [35] are applied.

If the mechanical dissipation is taken into account in the equation for rotation (for example, assuming the static characteristic of turbine in vicinity of the synchronous frequency $m(\dot{\delta}) = m_0 - \beta\dot{\delta}$ where $\beta > 0$) then the unbounded growth of $\dot{\delta}$ is impossible. This is easy to show by using the power relation with account of the mechanical damping

$$\left[\frac{1}{2}l(I_f^2 + I_k^2) + \frac{1}{2}\kappa\dot{\delta}^2 + \frac{\gamma e_f}{r_f}(1 - \cos\delta)\right]^{\cdot} =$$

$$= -(r_f\,I_f^2 + r_k\,I_k^2) + m_0\dot{\delta} - \beta\dot{\delta}^2. \quad (3.5.11)$$

Motions unbounded in δ (rotation of the pendulum with increasing angular velocity) are impossible since for sufficiently large values of $\dot{\delta}$ the right hand side of eq. (3.5.11) becomes negative and consequently the non-negative function in the square brackets can no longer increase. But an unbounded growth of this function is possible only under unbounded growth in δ by virtue of boundedness of currents I_f and I_k.

As numeral computations have shown, under certain relations between the external torque m and the damping coefficient β the motions unbounded in phase δ tend to the periodic motion in $\dot{\delta}$ with a constant mean value after the sliding period $\dot{\delta} > 0$ (for $m_0 > 0$). In the case of the synchronous machine it is equivalent to asynchronous motion.

Rotatory motions of the equivalent pendulum are analysed with the help of method of harmonic balance. In order to determine the main independent parameters of the problem we introduce the new time and the current by means of the relations: $lI_*^2 = \kappa/T_*^2$, $\tau' = \tau/T_* = \Omega\tau$ and $1/T_* = \sqrt{\dfrac{\gamma e_f}{\kappa}}$.

Taking into account the necessity of existence for synchronous motion we introduce a non-dimensional moment $\dfrac{m}{\gamma e_f} = \sin\delta_* < 1$. The system of equations (3.5.3) is reset in the following form

$$\dot{I}_f - \gamma\dot{\delta}\sin\delta + \tilde{r}_f I_f = 0,$$
$$\dot{I}_k - \gamma\dot{\delta}\cos\delta + \tilde{r}_k I_k = 0,$$
$$\ddot{\delta} + \gamma(I_f\sin\delta + I_k\cos\delta) + \sin\delta = \sin\delta_*, \quad (3.5.12)$$

where $\tilde{r}_f = r_f T_*/l$ and $\tilde{r}_k = r_k T_*/l$. Introducing the independent variable δ in eq. (3.5.12) yields

$$\frac{dI_f}{d\delta}\omega - \gamma\omega \sin\delta + \tilde{r}_f I_f = 0,$$

$$\frac{dI_k}{d\delta}\omega - \gamma\omega \cos\delta + \tilde{r}_k I_k = 0,$$

$$\frac{d\omega}{d\delta}\omega + \gamma(I_f \sin\delta + I_k \cos\delta) + \sin\delta = \sin\delta_*. \tag{3.5.13}$$

Approximate rotatory solutions of eq. (3.5.13) is sought in the form

$$\omega = \omega_0 + a_s \sin\delta + a_c \cos\delta,$$
$$I_f = I_{f0} + I_{fs} \sin\delta + I_{fc} \cos\delta,$$
$$I_k = I_{k0} + I_{ks} \sin\delta + I_{kc} \cos\delta. \tag{3.5.14}$$

Inserting eq. (3.5.14) into eq. (3.5.13) and balancing the constant and harmonic terms results in the system of linear equations in the expansion coefficients (3.5.14)[4]

$$\omega_0 I_{fs} + \tilde{r}I_{fc} = 0,$$
$$-\omega_0 I_{fc} + \tilde{r}I_{fs} = \gamma\omega_0,$$
$$\tilde{r}I_{f0} - \frac{1}{2}a_s I_{fc} + \frac{1}{2}a_c I_{fs} - \frac{1}{2}\gamma a_s = 0,$$
$$\omega_0 I_{ks} + \tilde{r}I_{kc} = \gamma\omega_0,$$
$$-\omega_0 I_{kc} + \tilde{r}I_{ks} = 0,$$
$$\tilde{r}I_{k0} + \frac{1}{2}a_c I_{ks} - \frac{1}{2}a_s I_{kc} - \frac{1}{2}\gamma a_s = 0,$$
$$\frac{1}{2}\gamma I_{fs} + \frac{1}{2}\gamma I_{kc} = \sin\delta_*,$$
$$a_s\omega_0 + \gamma I_{k0} = 0,$$
$$-a_s\omega_0 + \gamma I_{f0} = -1. \tag{3.5.15}$$

The coefficients of harmonics of the current $I_{fs}, I_{fc}, I_{ks}, I_{kc}$ are determined independently

$$I_{fs} = I_{kc} = \frac{\gamma\omega_0 \tilde{r}}{(\omega_0)^2 + \tilde{r}^2},$$

$$I_{ks} = -I_{fc} = \frac{\gamma\omega_0^2}{(\omega_0)^2 + \tilde{r}^2}. \tag{3.5.16}$$

Substituting relations (3.5.16) in the seventh equation in (3.5.15) we obtain the equation for the mean angular velocity of rotation of the equivalent

---

[4]In what follows, the simplest case $\tilde{r}_f = \tilde{r}_k = \tilde{r}$ is considered.

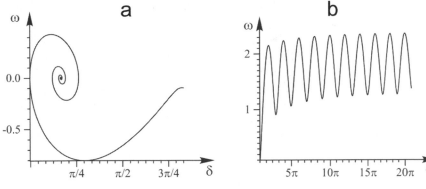

FIGURE 3.7.

pendulum

$$\mu(\omega_0) = \frac{\mu_0 \omega_0}{\omega_0^2 + \lambda^2} = \sin \delta_*, \qquad (3.5.17)$$

where $\mu_0 - \tilde{r}\gamma^2$. Curve $\mu(\omega_0)$ has a maximum equal to $\mu_0/2\lambda$ at frequency $\omega_0 = \lambda$. Thus under the condition $\sin \delta_* < \mu_0/2\lambda$ system (3.5.13) has two rotatory solutions among which the smaller and larger mean angular veloc ity ω_0 corresponds respectively to stable and unstable stationary motion. Taking into account the dissipative term in the equation of rotation (the last equation of system (3.5.13)) the equation for determining the mean angular velocity of the "pendulum" (sliding of synchronous machine) takes the form

$$\mu_1(\omega_0) = \frac{\mu_0 \omega_0}{\omega_0^2 + \lambda^2} + \beta\omega_0 = \sin \delta_*. \qquad (3.5.18)$$

Curve $\mu_1(\omega_0)$ can have a maximum and a minimum provided that

$$\omega_0^{(1,2)\,2} = \frac{\mu_0}{2\beta} - \lambda^2 \mp \frac{\mu_0}{2\beta}\sqrt{1 - \frac{8\beta}{\mu_0}\lambda^2}. \qquad (3.5.19)$$

The region where curve $\mu_1(\omega_0)$ has two extreme values is determined by the condition of non-negative radicand

$$\lambda^2 < \frac{\mu_0}{8\beta}. \qquad (3.5.20)$$

Under this condition in the interval

$$\mu_1(\omega_0^{(2)}) < \sin \delta_* < \mu_1(\omega_0^{(1)}) \qquad (3.5.21)$$

equation (3.5.18) has three solutions, the extreme solution and the middle solution corresponding to stable and unstable rotations respectively.

The main types are motions described by eq. (3.5.3) are shown in Fig. 3.7 for the following values of parameters $l = 0.4, \gamma = 0.65, r_f = r_k = 1, \alpha = \dfrac{\pi}{6}$.

Figure 3.8a demonstrates the effect of dissipation in the equation of rotation. The variant of existence of two stationary rotations in the system (3.5.3) ($\beta = 0.2$) is shown in Fig. 3.8b. The averaged equations of swinging of the generator rotor allows a simpler determination of the region of attraction for the synchronous regime. The phase space of system (3.5.1) is a topological product of the cylindrical surface δ, $\dot{\delta}$ and the plane Ψ_f, Ψ_k. Given initial currents or flux linkages, the construction of region of attraction reduces to determination of the region of initial conditions in angle δ and sliding s which results in the synchronous regime $s = 0, \delta = \delta_*$.

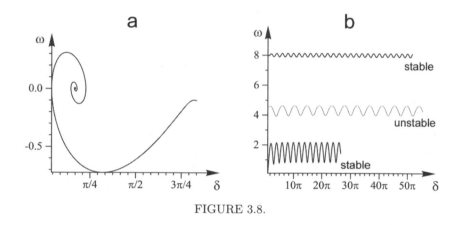

FIGURE 3.8.

A characteristic feature of the region of attraction is non-uniqueness of the initial sliding in a certain range of change of the initial angle, cf. Fig. 3.9.

An important technical problem of determining the limiting load moment of synchronous motor which ensures the synchronism is also reduced to constructing the region of attraction. In fact this problem reduces to the following: given a "new" load moment, it is necessary to establish whether the point of stationary solution under the "old" moment belongs to the region of attraction of the "new" stationary regime. The numeral solution of this problem is also considerably simplified when the averaged equations are used.

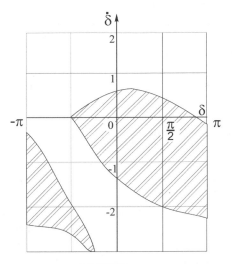

FIGURE 3.9.

3.6   Dynamics of synchronous machine that autonomously works on load

The present section is devoted to the asymptotic transformation and simplification of equations for synchronous machines for any feasible load. Such transformations are very useful for study of dynamics of machines working autonomously on load, and for the brief-action-generators acting by using kinetic energy of the recuperator [36].

When one considers a machine with an arbitrary load, the whole system "machine + load" is seen as consisting of two parts: (i) the machine itself described by certain variables and certain equations and (ii) a load whose description is not yet required.

Let us introduce the following sought-for variables: the currents in the stator windings i_a, i_b, i_c, the currents in the equivalent damper contours i_t, i_k; the excitation current i_f, the angular velocity ω and the angle of rotation ϑ. We also introduce Park's variables related to the phase currents by means of relationships (3.2.2). Variables i_d, i_q, i_0 are used in the following as the intermediate ones although in some cases they are more convenient new unknown variables than i_a, i_b, i_c.

In order to describe the processes in the load circuits it is necessary to introduce the variables $y_1, y_2, \ldots, y_n$ which are the currents in inductances and the charges of condensers (if they exists). In particular, in the problem of generator feeding an inductive accumulator of energy, the unknown variables are currents in the contours of rectifier and the current in the accumulator [127]. The mechanical variables among y_j are also possible, for example the angle of rotation and the angular velocity of rotor of the motor connected to a generator.

Thus, the load can be either pure electrical or electromechanical system fed by the phase voltages U_a, U_b, U_c of the machine and possibly by prescribed voltages of other sources. It is essential that variables $y_1, \ldots, y_n$ can be only found if voltages U_a, U_b, U_c are given as functions of time. To this end, it is necessary to calculate electrical or electromechanical processes in the load under given voltages U_a, U_b, U_c. The latter presents a problem which has no relation to the theory of synchronous machines, and this problem must be solved regardless of the method of studying the processes in the system "machine + load". As it is shown below, analysis of the slow processes requires only computation of quasi-stationary processes in the load.

The closed system of equations describing the machine with load has the following structure. First, it has the equations for the rotor contours of synchronous machine which in terms of the system of relative units (3.2.8)-(3.2.10) are as follows

$$
\begin{aligned}
[i_d + (1 + \varepsilon_r \sigma_f)i_f + i_t]^{\cdot} + \varepsilon_f \nu_f i_f &= \varepsilon_f e_f, \\
[i_d + i_f + (1 + \varepsilon_r \sigma_t)i_t]^{\cdot} + \nu_t i_t &= 0, \\
(i_q + i_k)^{\cdot} + \nu_k i_k &= 0, \\
\dot{\omega} &= \varepsilon[(i_f + i_t)i_q - i_k i_d + m].
\end{aligned} \tag{3.6.1}
$$

We resolve relations (3.2.11) for currents by retaining only the terms without ε_r

$$
\begin{aligned}
i_f &= \frac{\sigma_t}{\sigma_t + \sigma_f}\left(\Psi_f - i_d - \frac{1}{\sigma_t}\Psi_r\right), \\
i_t &= \frac{\sigma_f}{\sigma_t + \sigma_f}\left(\Psi_f - i_d - \frac{1}{\sigma_f}\Psi_r\right), \\
i_k &= \Psi_k - i_q.
\end{aligned} \tag{3.6.2}
$$

Let us substitute eq. (3.6.2) in (3.6.1) and subtract the second equation from the first. The difference is a new equation for $\dot{\Psi}_r$ and for this reason we omit the second equation. As a result we obtain the equations having no terms proportional to ε_r in the right hand sides

$$\varepsilon_r \dot{\Psi}_r = \frac{\nu_t + \varepsilon_f \nu_f}{\sigma_t + \sigma_f} \Psi_r + \frac{\nu_t \sigma_f - \varepsilon_f \nu_f \sigma_t}{\sigma_t + \sigma_f} (\Psi_f - i_d) + \varepsilon_f e_f,$$

$$\dot{\Psi}_k = -\nu_k (\Psi_k - i_q),$$

$$\dot{\Psi}_f = -\frac{\varepsilon_f \nu_f \sigma_t}{\sigma_t + \sigma_f} \left(\Psi_f - i_d + \frac{1}{\sigma_t} \Psi_r \right) + \varepsilon_f e_f,$$

$$\dot{\omega} = \varepsilon_\omega (\Psi_f i_q - \Psi_k i_d + m),$$

$$\varepsilon \dot{\vartheta} = \omega. \tag{3.6.3}$$

A dot in eq. (3.6.3) means differentiation with respect to the dimensionless time $\tau = t/T_k$ where $T_k = L_k/R_k$ denotes the time constant of the damper contour in axis k. Angle ϑ is the angle of rotation of the rotor and ω is the dimensionless angular velocity of rotor equal to $\dot{\vartheta}/\omega_*$ where ω_* is the characteristic (base) frequency. In contrast to eq. (3.6.1) the equations in (3.6.3) allow one to indicate immediately the fast and slow variables which is not possible by resolving eq. (3.6.1) for derivatives of the currents.

Equations (3.6.3) should be completed by the equations for the stator circuits whose particular form depends on the character of load. The phase voltages are determined as derivatives of the flux linkage of the stator windings with a minus sign

$$u_a = -\varepsilon[(1 + \sigma_d)i_a - \mu i_0 + (i_f + i_t)\cos\vartheta - i_k \sin\vartheta]\,\dot{},$$

$$a \to b \to s, \qquad \vartheta \to \vartheta - \frac{2\pi}{3} \to \vartheta + \frac{2\pi}{3}. \tag{3.6.4}$$

Let us now express the phase voltages in the load circuits in terms of the new dimensionless variables. For this aim we substitute eq. (3.6.2) into eq. (3.6.4) and take into account eq. (3.6.3). The result is given by

$$u_a = \omega \Psi_m \cos(\vartheta - \varphi) - \varepsilon \sigma_d (i_a - i_0)\,\dot{},$$

$$u_b = \omega \Psi_m \cos \left(\vartheta - \varphi - \frac{2\pi}{3} \right) - \varepsilon \sigma_d (i_b - i_0)\,\dot{},$$

$$u_c = \omega \Psi_m \cos \left(\vartheta - \varphi + \frac{2\pi}{3} \right) - \varepsilon \sigma_d (i_c - i_0)\,\dot{}. \tag{3.6.5}$$

Here $\Psi_m = \sqrt{\Psi_f^2 + \Psi_k^2}$, $\varphi = \arctan(\Psi_f/\Psi_k)$ and, as before, the terms proportional to $\varepsilon, \varepsilon_r$ are dropped.

The equations in (3.6.3) contain variables with essentially different timerates. The faster variables are variable ϑ and Ψ_r having derivatives of the order $1/\varepsilon$ and $1/\varepsilon_r$. The slow variable Ψ_k has a derivative of order of the unity and there are the slower variables Ψ_f and ω which have small derivatives (of the order of ε_f and ε_ω).

The fast and slow variables can be selected in the equations of the load circuits. Their number depends on the particular type of load. For example, in the case of the linear active-inductive load the currents y_1, y_2, y_3 are fast variables and in the case of machine working on the inductive accumulator there are fast variables (currents in the rectifier circuits) and one slow variable which is the accumulator current, [37].

Let us assume that it is possible to select, among variables y_j, s slow variables $y_1, y_2, \ldots, y_s$ with the velocities of order of unity or $\varepsilon_f, \varepsilon_\omega$ and $n - s$ fast variables $y_{n-s}, \ldots, y_n$ with the velocities of order $1/\varepsilon$.

Using difference in velocities of the variables we apply the asymptotic method of separation of motions [128] to the system "machine+load" and consider only the first approximation without distinguishing among variables Ψ_k, Ψ_f, ω.

Following this method one needs to find the general solution of the fast subsystem under the assumption that the slow variables are constant parameters. The obtained general solution is substituted in the equations for slow processes and the result is averaged over the fast time. The resulting equations are substantially simpler than the initial ones as they are autonomous and have a lower order and less variables.

For the problems of theory of electric machines and considered method of introduction of small parameters the expression for fast variables are functions of time t and fast time t/ε. Firstly, the fast time appears in terms of the fast variables of the phase type: angle ϑ and the phases of the external alternating voltages applied to the load (if these exist). Of crucial interest for the systems with several phases is the resonance case in which the frequencies of the external voltage and the machine rotations are close in the whole interval of the transient process (the problem of switching machine on the powerful network considered in the previous section belongs to this case). Introducing a necessary number of phase differences and considering these as slow variables we arrive at the system with one fast phase. Secondly, the fast time explicitly appears, for example in the case of the linear load, in the exponential dependence describing the fast aperiodic process. For this reason, the averaging needs to be performed over the phase and explicit fast time independently.

Realising the above method we first take into account that only the parameters $\varepsilon_r, \varepsilon$ are small and ignore the distinctions between the variables $\Psi_k, \Psi_f, \omega, y_1, \ldots, y_s$.

Let us consider the problem of determining the fast variables $y_{s+1}, \ldots, y_n$. To this end, quantities $\Psi_m, \varphi, \omega, y_1, \ldots, y_n$ are considered as prescribed parameters. Then in the transient process, $y_{s+1}, \ldots y_n$ are functions of phase ϑ and time t appearing in combinations $t/T_{s+1}, t/T_{s+2}$ etc. where $T_{s+1}, T_{s+2}, \ldots$ are characteristic small constants of the load circuits. Further it is possible to take $T_{s+1} \approx T_{s+2} \approx \ldots \approx \varepsilon$. Thus, in the first approximation, the complete expressions for $y_{s+1}, \ldots, y_n$ contain variables $\Psi_k, \Psi_f, \omega, y_1, \ldots, y_s, \vartheta$ and explicit fast time t/ε.

It is important that variable Ψ_r does not appear in expressions (3.6.5). If the elements of the stator circuits described by terms $\sigma(i_0 - i_a)^{\cdot}, \sigma(i_0 - i_b)^{\cdot}$ and $\sigma(i_0 - i_c)^{\cdot}$ in eq. (3.6.5) are considered as a load, then one needs to compute the load circuits under the symmetric system of voltages

$$\tilde{u}_a = \omega \Psi_m \cos(\vartheta - \varphi),$$

$$\tilde{u}_b = \omega \Psi_m \cos\left(\vartheta - \varphi - \frac{2\pi}{3}\right),$$

$$\tilde{u}_s = \omega \Psi_m \cos\left(\vartheta - \varphi + \frac{2\pi}{3}\right) \tag{3.6.6}$$

and complemented by these elements. It presents no problem for the circuits without zero wire or with the symmetric load. In this case three equal inductance $\varepsilon\sigma$ should be added to the load.

The problem of computing the load circuits under prescribed voltages (3.6.6) is a problem of the theory of circuits. As we show in the following, in order to derive the averaged equations for slow variables it is necessary to determine only the steady-state 2π−periodic electrical or electromechanical vibrations in the load circuits in argument ϑ under the assumption that $\Psi_k, \Psi_f, \omega, y_1, \ldots, y_s$ are constant parameters. For applications of the method of the motion separation the load needs to have the following property: any fast transient processes in the load (i.e. processes subjected to any possible initial conditions) must tends to the same regime which must be 2π−periodic in argument ϑ. This is normally the case in practice.

Thus, the fast variables in load can be considered as given functions of $\vartheta, t/\varepsilon, \Psi_m, \varphi, \omega, y_1, \ldots, y_s$ as well as the phase currents i_a, i_b, i_c and the currents i_d, i_q. It is worth noting that since all fast processes attenuate rapidly the dependence of $y_{s+1}, \ldots, y_n$ and, consequently, i_a, i_b, i_c, i_d, i_q on the explicit fast time is considerable in a small time interval (of the order of ε) after switching or changing of load. The corresponding transient processes attenuate and these variables are coinciding with their steady-state components.

Among the fast variables we must consider the flux linkage Ψ_r determined by the first equation in (3.6.3), the latter being a linear equation with the "forcing factors" i_d, Ψ_f, e_f. The two latter add to Ψ_r only an additional constant component

$$\Psi_{rs} = \frac{\nu_t \sigma_f - \varepsilon_f \nu_f \sigma_t}{\nu_t + \varepsilon_f} \Psi_f + \frac{\sigma_t + \sigma_f}{\nu_t + \varepsilon_f} \varepsilon_f e_f. \tag{3.6.7}$$

Function Ψ_r depends on the fast times $t/T_{s+1}, t/T_{s+2}, \ldots$, in terms of i_d but these components in Ψ_r disappear within a small time after the process has begun. The homogeneous solution

$$C_r \exp\left[-\frac{(\nu_t + \varepsilon_f \nu_f)t}{(\sigma_t + \sigma_f)\varepsilon_r}\right]$$

also disappears as it is function of the fast time t/ε_r. Hence it is sufficient to calculate the steady-state component Ψ_{rp} which is $2\pi-$periodic in ϑ. To this end, we assume that ϑ is a new argument and arrive at the equation

$$\varepsilon_r \frac{d\Psi_{rp}}{d\vartheta} = -\frac{\varepsilon}{\omega}\left[\frac{\nu_t + \varepsilon_f\nu_f}{\sigma_t + \sigma_f}\Psi_{rp} + \frac{\nu_t\sigma_f - \varepsilon_f\nu_f\sigma_t}{\sigma_t + \sigma_f}i_{dp}(\vartheta, \Psi_m, \varphi, \omega, y_1, ..., y_s)\right],$$

(3.6.8)

where i_{dp} is a steady-state component of i_d which is $2\pi-$periodic in ϑ.

Function Ψ_{rp} is now determined as a $2\pi-$periodic solution of eq. (3.6.8) at $\Psi_m, \varphi, \omega, y_1, \ldots, y_s = \text{const}$. Thus after a small time after switching the load the entire function Ψ_r is determined and equal to $\Psi_{rs} + \Psi_{rp}$.

The following step in separation of the non-stationary processes is to substitute the expressions for the fast variables in the remaining slow subsystem and to average the result over phase ϑ and the fast times. It is necessary to take the last three equations among the rotor equations. This leads to the sought-for equations

$$\dot{\Psi}_k = -\nu_k(\Psi_k - \langle i_q\rangle),$$
$$\dot{\Psi}_f = -\frac{\varepsilon_f\nu_f\sigma_t}{\sigma_t + \sigma_f}\left(\Psi_f - \langle i_d\rangle + \frac{1}{\sigma_t}\langle\Psi_r\rangle\right) + \varepsilon_f e_f,$$
$$\dot{\omega} = \varepsilon_\omega(\Psi_f\langle i_q\rangle - \Psi_k\langle i_d\rangle + m).$$

(3.6.9)

Here $\langle i_d\rangle, \langle i_q\rangle, \langle\Psi_r\rangle$ denote mean values of the corresponding variables averaged over phase ϑ and the fast times.

Determination of the mean values reduces to omitting all terms containing the fast times $t/T_{s+1}, t/T_{s+2}\ldots, t/\varepsilon_r$ in the complete expressions for i_d, i_q, Ψ_r and to calculating the constant components. This yields

$$\langle i_d\rangle = \langle i_{dp}\rangle, \langle i_q\rangle = \langle i_{qp}\rangle,$$

(3.6.10)

$$\langle\Psi_r\rangle = \Psi_{rs} + \langle\Psi_{rp}\rangle = \frac{\nu_t\sigma_f - \varepsilon_f\nu_f\sigma_t}{\nu_t + \varepsilon_f\nu_f}(\Psi_f - \langle i_{dp}\rangle) + \frac{\sigma_t + \sigma_f}{\nu_t + \varepsilon_f\nu_f}\varepsilon_f e_f,$$

where i_{dp}, i_{qp} are steady-state components of i_d, i_q which are $2\pi-$periodic in ϑ. As follows from eq. (3.6.10), for deriving the equations for slow transient processes it is necessary to know only the steady-state values of variables i_d, i_q and, consequently, variables $i_a, i_b, i_c, y_{s+1}\ldots, y_n$.

Inserting eq. (3.6.10) into eq. (3.6.9) yields the final form of the slow non-stationary processes of synchronous machine

$$\dot{\Psi}_k = -\nu_k(\Psi_k - \langle i_{qp}\rangle),$$
$$\dot{\Psi}_f = -\frac{\varepsilon_f\nu_f\nu_t}{\nu_t + \varepsilon_f}(\Psi_f - \langle i_{dp}\rangle - e_f),$$
$$\dot{\omega} = \varepsilon_\omega(\Psi_f\langle i_{qp}\rangle - \Psi_k\langle i_{dp}\rangle + m).$$

(3.6.11)

These equations need to be completed by the averaged equations for slow variables in the load circuits $y_1, \ldots, y_s$. The result is a closed system of $s+3$ equations.

When deriving equations (3.6.11) angle ϑ was taken as the fast phase. There exist no other fast phases for load without external alternating voltages. However if there are prescribed voltages and the resonance case is considered then it is reasonable to view phase ϑ_0 of the external voltage as the fast phase. In this case all obtained relations hold true however the fast variables should be considered as functions of ϑ_0 and the averaging should be performed over ϑ_0.

Let us notice that the averaged equations for the rotor (3.6.11) written in terms of the mean values $\langle i_{dp} \rangle$, $\langle i_{qp} \rangle$ have the same form for any load. Therefore, in order to write these equations for any particular load one needs to determine $\langle i_{dp} \rangle$, $\langle i_{qp} \rangle$ as functions of the slow variables and insert them in equations (3.6.11).

Determining the slow variables from eq. (3.6.11) and other s equations for slow nonstationary processes and substituting the result in the expressions for fast variables we obtain the complete solution of the problem in first approximation. Clearly this way is much more simpler than a direct integration of the initial system. The explanation is as follows: firstly, the number of equations for slow processes ($s + 3$) is much smaller than the number of equations in original system ($s + 3$ in place of $n + 5$) and secondly, the equations for slow nonstationary processes contain no rapidly changing variables (in particular angle ϑ). For this reason the qualitative analysis of equations for nonstationary processes is facilitated. The numeral integration is also simplified because no temporal boundary layer is observed in the averaged equations (and consequently we remove the problem of a considerable difference in arguments of fast and slow variables). The latter is a considerable advantage of the asymptotic methods. The fact that parameters $\varepsilon_f, \varepsilon_\omega$ are small against ν_k allows one to perform the further separation of motions as variables Ψ_f, ω are slower in comparison with Ψ_k. Let us assume that among the slow variables there are variables $y_1, \ldots, y_l$ which are slower and whose time-rates are characterized by a small parameter ε_n. Usually one can assume that $\varepsilon_f, \varepsilon_\omega$ and ε_n are values of the same order. Separation of the "slowest" nonstationary processes is performed by analogy to the previous case. First it is necessary to determine the variables $\Psi_k, y_{l+1} \ldots, y_s$ as functions of time t and the "slowest" variables $\Psi_f, \omega, y_1, \ldots, y_l$ from $s - l + 1$ equations, the latter variables being considered time-independent parameters. The obtained expressions for $\Psi_k, y_{l+1}, \ldots, y_s$ are to be substituted in the last two equations in (3.6.11) and the remaining equations for $y_1, \ldots, y_l$ and then the right hand sides of all these equations are to be averaged over time t. This results in the equations of the "slowest" nonstationary processes which are even simpler than equations (3.6.11). When these equations are resolved the variables $\Psi_k, y_{l+1}, \ldots, y_s$ can be found by using the expressions relating these variables with $t, \Psi_f, \omega, y_{l+1}, \ldots, y_s$.

As mentioned above, there are the cases in which the ratio ε_r / ν_t is not small. Then variable Ψ_r can not be considered as fast and the scheme of

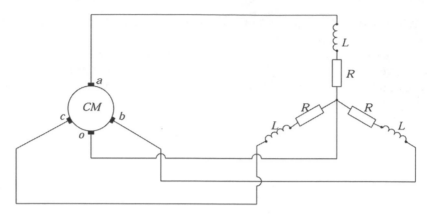

FIGURE 3.10.

separating fast and slow processes differs from the above. In this case the equations for slow processes have another form rather than eq. (3.6.11). However it is possible to separate the slowest processes because even if ε_r/ν_t is equal to unity, the variable Ψ_r is faster than Ψ_f and ω.

When the machine works on a symmetric active-inductive load (Fig. 3.10) the computation of quasi-stationary currents in the stator circuits reduces to determining a particular periodic solution of the system

$$\sigma \dot{i}_a + \nu_n i_a = u_a, \quad a \to b \to c, \tag{3.6.12}$$

where $\sigma = \sigma_d + \sigma_n$. Next, σ_n and ν_n denote respectively the dimensionless inductance and resistance of load (the resistance of load is assumed to have the order of the inductive resistance of the stator circuits) and u_a, u_b, u_c form the system of phase voltages determined by relations (3.6.6). In order to determine the mean currents $\langle i_d \rangle, \langle i_q \rangle$ we introduce axes d, q in equations (3.6.12)

$$\sigma \dot{i}_d - \omega \sigma i_q + \nu_n i_d = \omega \Psi_k,$$
$$\sigma \dot{i}_q + \omega \sigma i_d + \nu_n i_d = -\omega \Psi_f. \tag{3.6.13}$$

Mean currents $\langle i_d \rangle, \langle i_q \rangle$ are determined as the constant particular solution of eq. (3.6.13)

$$\langle i_d \rangle = \frac{\omega \nu_n \Psi_k - \omega^2 \sigma \Psi_f}{\nu_n^2 + \omega^2 \sigma^2},$$

$$\langle i_q \rangle = -\frac{\omega \nu_n \Psi_f + \omega^2 \sigma \Psi_k}{\nu^2 + \omega^2 \sigma^2}. \tag{3.6.14}$$

After substituting the obtained mean currents into eq. (3.6.11) we arrive at the system of equations of the third order

$$\dot{\Psi}_k = -\nu_k[\Delta_1(\omega)\Psi_k + \Delta_2(\omega)\Psi_f],$$

$$\dot{\Psi}_f = -\frac{\varepsilon_f \nu_t}{\nu_t + \varepsilon_f}[\Delta_1(\omega)\Psi_f - \Delta_2(\omega)\Psi_k - e_f/\nu_f],$$

$$\dot{\omega} = -\varepsilon_\omega[\Delta_2(\omega)(\Psi_f^2 + \Psi_k^2) - m], \qquad (3.6.15)$$

where

$$\Delta_1(\omega) = \frac{\sigma^2\omega^2 + \sigma\omega^2 + \nu_n^2}{\sigma^2\omega^2 + \nu_n^2}, \quad \Delta_2(\omega) = \frac{\nu_n\omega}{\sigma^2\omega^2 + \nu_n^2}, \quad \sigma = \sigma_d + \sigma_n.$$

In order to find the stationary solution it is necessary to resolve the system of transcendent equations obtained from eq. (3.6.15) for $\dot{\Psi}_k, \dot{\Psi}_f, \dot{\omega} = 0$. To this end, it is convenient to apply the following method. First, the linear system of first two equations for Ψ_f and Ψ_k with coefficients depending on ω is to be solved. Then the obtained expressions for $\Psi_f(\omega)$ and $\Psi_k(\omega)$ are inserted in the first term of the third equation determining the electromagnetic moment m_e. As a result we obtain a single transcendent equation for frequency ω

$$m_e = \frac{(e_f/\nu_f)^2\Delta_2(\omega)}{\Delta_1^2(\omega) + \Delta_2^2(\omega)} = m. \qquad (3.6.16)$$

It is important that the resistance of load and the frequency appear in the expressions for $\Delta_1(\omega)$ and $\Delta_2(\omega)$ only in the form of the ratio ν_n/ω. It results in that the maximum electromagnetic moment being a function of the frequency of stationary regime is reached at the frequencies proportional to ν_n and is independent of the load resistance. The dependences $m_e(\omega)$ for some values of ν_n are displayed in Fig. 3.11. When the constant external moment $m = $ const is less than the maximum of electromagnetic moment there are two stationary values of frequency, the lower and higher frequency corresponding to stable and unstable stationary regime respectively. If the motor torque depends on frequency ω and monotonically decreases as ω grows, the system assumes either one stable stationary solution or three stationary solutions, the higher and lower frequencies describing stable stationary motion and the middle one corresponding to unstable stationary motion.

Using the fact that, by virtue of eq. (3.6.15), variable Ψ_k is faster than variables Ψ_f and ω one can pursue a further simplification. System (3.6.15) is quasi-linear and Ψ_k is considered as a fast variable.

In the first approximation with respect to small parameters ε_f and ε_ω, one should set $\dfrac{\nu_t}{\nu_t + \varepsilon_f} = 1$. Following the method of separation of motions we find the general solution of the first equation in (3.6.15) by assuming

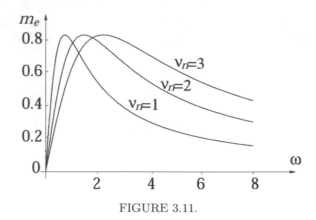

FIGURE 3.11.

$\Psi_f, \omega = $ const. Then we have

$$\Psi_k = C_k \exp\{-\nu_k \Delta_1(\omega)\tau\} - \frac{\Delta_2(\omega)}{\Delta_1(\omega)}\Psi_f. \qquad (3.6.17)$$

Substituting the obtained solution in the slow subsystem of equations in eq. (3.6.15) and averaging the right hand sides over the explicit time (in this case this is equivalent to omitting rapidly attenuating exponential function in eq. (3.6.17) we obtain the following system of the second order

$$\dot{\Psi}_f = -\varepsilon_f \left[\left(\Delta_1(\omega) + \frac{\Delta_2^2(\omega)}{\Delta_1(\omega)} \right) \Psi_f - e_f/\nu_f \right],$$

$$\dot{\omega} = -\varepsilon_\omega \left[\frac{\Delta_2(\omega)}{\Delta_1^2(\omega)} \left(\Delta_1^2(\omega) + \Delta_2^2(\omega) \right) \Psi_f - m(\omega) \right]. \qquad (3.6.18)$$

Reducing equations (3.6.15) to the second-order autonomous system allows one to make use of the method of phase plane for study of dynamics of the synchronous generator with active inductive load. A phase portrait of system (3.6.18) in case of three stationary modes is shown in Fig. 3.12.

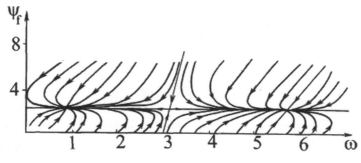

FIGURE 3.12.

Two nodal states of equilibrium corresponding to stationary motions, are separated by the separatrices of the saddle point corresponding to unstable solution. Thus depending upon the initial values of flux linkages and frequency all motions tend to one of the stable stationary regimes. By virtue of the definition [35] the system possesses the point stability as a whole.

When the external torque is independent of frequency, i.e. $m = \text{const}$, there are only two equilibrium positions which are a stable node and a saddle. The domain of attractions of the stable stationary regime is separated from the domain of unbounded motions by separatrices converging to the saddle. In this case system (3.2.6) possesses the property of dichotomy.

3.7 Asymptotic transformation of equations when synchronous machine works on load through rectifier

As explained in the previous section, the method of asymptotic transformation of equations of the synchronous machine working on the active-inductive load can be applied for deriving equations of slow nonstationary processes in the machine working on load through a rectifier. Nevertheless certain small changes in regard of computation of quasi-stationary regime in the stator circuits are needed.

Let us consider Larionov's circuit [95] which is the most widespread circuit of rectification of three-phase current. The valves are considered to be ideal and the active resistances of the stator windings are assumed to be small as compared with their inductive resistances. As before, the load resistance is considered to have the order of the inductive resistance of the stator windings. Similar to the previous case, the currents in the stator phases are fast variables. For closing the system of equations it is necessary to introduce another fast variable which is the current i_n in load. The equation for i_n is given by

$$\sigma_n \dot{i}_n + \nu_n i_n = u_n, \tag{3.7.1}$$

where u_n is the rectified voltage of the load. There is no need to write down the other equation of the stator circuits because only a quasi-stationary solution for the fast subsystem is required for asymptotic transformations.

Construction of this solution reduces to computation of $2\pi-$periodic regime in the rectifier circuits subjected to a three-phase system of voltages at the stator windings given by formulae (3.6.6). While determining the quasi-stationary regime, the amplitude and phase of the stator emf should be considered as constant values as they are functions of low variables. This regime in the rectifier circuits is realized after establishing a regular sequence of switching of valves that takes place during a few rotations of

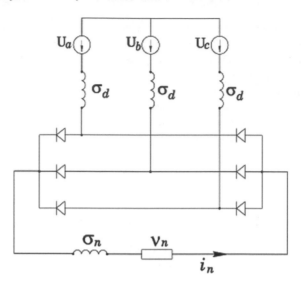

FIGURE 3.13.

the rotor after connecting the load and fading of the aperiodic component of the load current with the characteristic time σ_n/ν_n.

In the case of infinitely large inductance of load and sinusoidal three-phase voltage, the stationary processes in Larionov's circuit are studied for all possible regimes of valves' operations [95] (such circuit is presented in Fig. 3.13).

Let us consider the steady-state regime if the overlap angle γ is smaller than $\pi/3$. This regime is determined by the following relation between the parameters $\sigma\omega \leq \pi\nu_n/9$ which is fulfilled for the majority of the systems under consideration.

In the steady-state regime the phase currents are $2\pi-$periodic functions of ϑ. In the interval $\varphi - \pi/3 \leq \vartheta \leq 2\pi + \varphi - \pi/3$ we have

$$
i_a(\vartheta) = \begin{cases}
i_1(\vartheta) & \varphi - \pi/3 \leq \vartheta \leq \gamma + \varphi - \pi/3, \\
i_n(\vartheta) & \gamma + \varphi - \pi/3 \leq \vartheta \leq \varphi - \pi/3, \\
i_2(\vartheta) & \varphi + \pi/3 \leq \vartheta \leq \gamma + \varphi + \pi/3, \\
0 & \gamma + \varphi + \pi/3 \leq \vartheta \leq \varphi + 2\pi/3, \\
i_a(\vartheta - \pi) & \varphi + 2\pi/3 \leq \vartheta \leq 2\pi + \varphi - \pi/3.
\end{cases} \tag{3.7.2}
$$

The relations for currents i_b, i_c are obtained from eq. (3.7.2) when ϑ is replaced by $\vartheta - 2\pi/3$ and $\vartheta + 2\pi/3$ respectively, and the expressions for i_1, i_2 are shown below.

The load current i_n in the steady-state regime is $\pi/3$-periodic function of ϑ. Let us make use of that the pulsations of the rectified voltage are small at any γ smaller than $\pi/3$ [95]. Neglecting these pulsations we obtain the expressions for i_1, i_2, i_n [95]. In the case of the diode rectifiers these expressions are given by

$$i_1(\vartheta) = i_n + \frac{\sqrt{3}\Psi_m}{2\sigma}[\cos(\vartheta - \varphi + \pi/3) - 1],$$

$$i_2(\vartheta) = \frac{\sqrt{3}\Psi_m}{2\sigma}[1 - \cos(\vartheta - \varphi + \pi/3)],$$

$$i_n = \frac{3\sqrt{3}\omega\Psi_m}{2\pi\nu_n}(1 + \cos\gamma), \qquad \cos\gamma = 1 - \frac{2i_n\sigma}{\sqrt{3}\Psi_m}. \qquad (3.7.3)$$

Values of currents i_a, i_b, i_c, i_n are close to the true values with the accuracy of the order of $\varepsilon, \varepsilon_n$ in large time intervals except for, probably, a small interval after the rectifier has been switched. Within this interval the fast transient process ends and the proper regime of switching of the valves establishes.

In the problem under consideration current i_n as well as currents $i_a, i_b, i_c,$ i_d, i_q are formally considered as fast variables. However i_n differs from currents i_a, i_b, i_c in that its fast component has small pulsations which can be neglected. This allows one to use relations (3.7.3) which provide one with the slow component i_n. The dependence of currents i_d, i_q on the fast phase is substantial, therefore their slow components should be found by averaging. Using expressions (3.7.2) for the mean currents i_d, i_q, we obtain

$$\langle i_d \rangle = S_1(\omega)\Psi_k + S_2(\omega)\Psi_f,$$
$$\langle i_q \rangle = S_2(\omega)\Psi_k - S_1(\omega)\Psi_f, \qquad (3.7.4)$$

where functions $S_1(\omega), S_2(\omega)$ are determined in terms of the overlap angle γ

$$S_1(\omega) = \frac{3\omega}{\pi\nu_n}\left\{\frac{\nu_n}{2\sigma_d\omega}[1 - \cos\gamma(\omega)]^2 + \frac{3}{\pi}[1 + \cos\gamma(\omega)]\cos\gamma(\omega)\right\}$$

$$S_2(\omega) = \frac{3\omega}{\pi\nu_n}\left\{\frac{\nu_n}{2\sigma_d\omega}[2\sin\gamma(\omega) - \gamma(\omega) - \frac{1}{2}\sin 2\gamma(\omega)] - \right.$$
$$\left. -\frac{3}{\pi}[1 + \cos\gamma(\omega)]\sin\gamma(\omega)\right\} \qquad (3.7.5)$$

and angle γ is given by the relation

$$\cos\gamma = \frac{\pi\nu_n - 3\sigma_d\omega}{\pi\nu_n + 3\sigma_d\omega}.$$

Substituting the obtained values of the mean currents in the averaged equations (3.6.11) we arrive at the equations for slow nonstationary processes

$$\dot{\Psi}_k = -\nu_k\{[1 - S_2(\omega)]\Psi_k + S_1(\omega)\Psi_f\},$$

$$\dot{\Psi}_f = -\frac{\varepsilon_f\nu_t}{\nu_t + \varepsilon_f}\{[1 - S_2(\omega)]\Psi_f - S_1(\omega)\Psi_k - e_f\},$$

$$\dot{\omega} = -\varepsilon_\omega[S_1(\omega)(\Psi_f^2 + \Psi_k^2) - m]. \qquad (3.7.6)$$

Using smallness of parameters $\varepsilon_f, \varepsilon_\omega$ which is valid for the high-power turbogenerators, it is possible to carry our the second averaging similar to that of the previous section. The general solution for the fast variable Ψ_k at fixed Ψ_f, ω consists of a rapidly attenuating exponential function and the constant component

$$\Psi_k = C_k \exp\{-\nu_k[1 - S_2(\omega)](t - t_0)\} - \frac{S_1(\omega)}{1 - S_2(\omega)}\Psi_f. \qquad (3.7.7)$$

Averaging Ψ_k, Ψ_k^2 in eq. (3.7.6) reduces to omitting the exponential function in eq. (3.7.7). Then only the constant component appears in eq. (3.7.6) which results in the equations for the slowest processes

$$\dot{\Psi}_f = -\varepsilon_f S_3(\omega)\Psi_f + \varepsilon_f e_f,$$

$$\dot{\omega} = -\varepsilon_\omega \frac{S_3(\omega)S_1(\omega)}{1 - S2(\omega)}\Psi_f^2 + \varepsilon_\omega m, \qquad (3.7.8)$$

where

$$S_3(\omega) = \frac{1}{1 - S_2(\omega)}[1 - 2S_2(\omega) + S_2^2(\omega) + S_1(\omega)].$$

As a result, the study of dynamics of synchronous machine working on load through a rectifier is reduced to construction of the phase portrait of the second-order autonomous system (3.7.8). As numeral computations showed, this phase portrait does not qualitatively differs from the above in the case when the synchronous machine works directly on the active-inductive load.

3.8   Stationary regimes of operation of rectifier working on pure active load

We[5] consider the steady-state regimes of operating three-phase rectifier (Fig. 3.14) that works on the active load R and is fed by the systems of three-phases emf

$$e_a = A\sin(\vartheta - \varphi), \quad e_b = A\sin(\vartheta - \varphi - 2\pi/3),$$

$$e_c = A\sin(\vartheta - \varphi + 2\pi/3). \qquad (3.8.1)$$

The denotations in eq. (3.8.1) are as follows: A and φ denote respectively the amplitude and phases of the feeder voltage, ϑ is the "electric" angle of rotation of the machine rotor relative to the axis of phase a of the stator winding. While determining the steady-state regime parameters A and φ

---

[5]The results of this section are obtained together with A.D. Sablin.

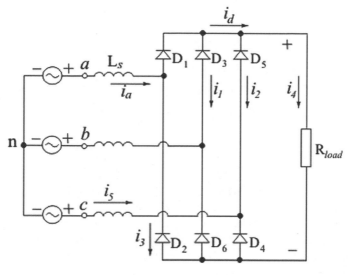

FIGURE 3.14.

are considered to be prescribed and one looks for the mean values $\langle i_d \rangle$, $\langle i_q \rangle$ of currents i_d, i_q and also the mean value $\langle i_n \rangle$ of current i_n in the load and the mean electromagnetic torque $\langle m_e \rangle$. For computation of these quantities it is necessary to find the phase currents i_a, i_b, i_c of the rectifier and the current in load i_n.

In order to compute the processes in the rectifier it is convenient to introduce a new phase variable ϑ_1 in eq. (3.8.1) by means of the relationship

$$\vartheta = \vartheta_1 + \psi + \varphi,$$

where ψ is the phase of voltage e_a at time instant of initiation of the first valve, ϑ_1 is the phase of the feeder voltage relative to the time instant of initiation of the first valve. In contrast to prescribed A and φ, quantity ψ is unknown as a function of the system parameters and must be found under calculation of processes in the rectifier.

After the above replacement the expression for the feeder voltages are set in the form

$$e_a = A\sin(\vartheta_1 + \psi), \quad e_b = A\sin(\vartheta_1 + \psi - 2\pi/3),$$
$$e_c = A\sin(\vartheta_1 + \psi + 2\pi/3). \tag{3.8.2}$$

The mean values $\langle i_d \rangle, \langle i_q \rangle$ of currents i_d, i_q are now given by the relationships

$$\langle i_d \rangle = i_{ds}\cos(\psi + \varphi) - i_{qs}\sin(\psi + \varphi),$$
$$\langle i_q \rangle = i_{ds}\sin(\psi + \varphi) + i_{qs}\cos(\psi + \varphi), \tag{3.8.3}$$

where

$$i_{ds} = \frac{1}{2\pi} \int\limits_0^{2\pi} i_d(\vartheta_1)d\vartheta_1, \quad i_{qs} = \frac{1}{2\pi} \int\limits_0^{2\pi} i_q(\vartheta_1)d\vartheta_1. \qquad (3.8.4)$$

The mean current $\langle i_n \rangle$ in the load and the mean electromagnetic torque
are as follows

$$\langle i_n \rangle = \frac{1}{2\pi} \int\limits_0^{2\pi} i_n(\vartheta_1)d\vartheta_1, \quad \langle m_e \rangle = \frac{r}{2\pi} \int\limits_0^{2\pi} i_n^2(\vartheta_1)d\vartheta_1. \qquad (3.8.5)$$

Thus, to determine the above quantities it is necessary to find the phase
currents in the rectifier in the steady-state regime of operations. The steady-
state regime in the rectifier is 2π−periodic, thus the periodicity interval is
split into six subintervals of the duration of $\pi/3$. Within each subinterval
some valves of the anode group (the valves 2,4,6) and some of the cathode
group (the valves 1,3,5) conduct current, see Fig. 3.13. It is sufficient to de-
termine the phase currents $i_a(\vartheta_1) = i_1, i_b(\vartheta_1) = i_2, i_c(\vartheta_1) = i_3$ only in the
first subinterval $0 \leq \vartheta_1 \leq \pi/3$ because in the subsequent intervals the expression for the phase currents are calculated with the help of the obtained
dependences $i_1(\vartheta_1), i_2(\vartheta_1), i_3(\vartheta_1)$ provided that the order of switching the
valves is known in advance.

Let us go over from parameters R and L_s of the circuits (Fig. 3.14) to
the dimensionless ones in the way explained in Section 3.2.

In the case in which the inductance x in the rectifier circuits is absent,
i.e. $x = 0$, the order of switching the valves is determined by a simple
rule: from the anode group (valves with the numbers 2,4,6 in Fig. 3.14)
the valve conducts current whose anode is supplied by the greater positive
voltage and among the cathode group the valve conducts current whose
anode is supplied by the negative voltage of greater absolute value. When
the inductance is absent, the first and sixth valves conduct in the interval
$0 \leq \vartheta_1 \leq \pi/3$, the first and second valves conduct in the interval $\pi/3 \leq
\vartheta_1 \leq 2\pi/3$ etc. In the case of non-zero inductance x, for example, at the
time instant of initiation of the first valve the current through the valve 5
does not vanish, therefore during the time interval $0 \leq \vartheta_1 \leq \gamma$ (γ is the
overlap angle) two valves (1 and 5) of the cathode group and one valve
(6) of the anode group conduct current. Depending on the relationship
between R of the active load and inductance x in the feeder circuits there
are different combinations of parameters γ and ψ.

The simplest case is that in which the overlap angle γ is equal to $\pi/3$,
i.e. in any time interval only three valves conduct current: either one of the
anode group and two of the cathode group or vice versa two of the anode
group and one of the cathode group. So in the first interval $0 \leq \vartheta_1 \leq \pi/3$
the valves 1,6,5 conduct, in the second interval the valves 1,6,2 conduct, in
the third interval valves 1,2,3 conduct etc.

In the first interval the processes in the system are described by the equations

$$x\frac{di_a}{d\vartheta_1} - x\frac{di_s}{d\vartheta_1} = e_a - e_c, \quad x\frac{di_a}{d\vartheta_1} + x\frac{di_n}{d\vartheta_1} + ri_n = e_a - e_b,$$

$$i_n = i_a + i_c, \qquad 0 \le \vartheta_1 \le \pi/3. \tag{3.8.6}$$

Because at $\vartheta_1 = 0$ the first valve initiates $i_a = 0$ and at $\vartheta_1 = \pi/3$ the fifth valve switches off, then $i_c(\pi/3) = 0$. Because the regime is periodic then $i_n(0) = i_n(\pi/3) = i_{n0}$ where i_{n0} is a constant which is determined during the calculation.

The solution of eq. (3.8.6) satisfying the above conditions has the form

$$i_a = i_1 = \frac{A}{2x\sqrt{1+\nu^2}} \left[\frac{3}{2}\nu(1 - \cos\vartheta_1) + \left(2 - \frac{\sqrt{3}}{2}\nu\right)\sin\vartheta_1 \right],$$

$$i_b = i_2 = -i_n = -\frac{A}{x\sqrt{1+\nu^2}}\sin(\vartheta_1 + \pi/3), \tag{3.8.7}$$

$$i_c - i_3 - \frac{A}{2x\sqrt{1+\nu^2}} \left[\left(\frac{3}{2}\nu + \sqrt{3}\right)\cos\vartheta_1 - \left(1 - \frac{\sqrt{3}}{2}\right)\sin\vartheta_1 - \frac{3\nu}{2} \right],$$

where

$$\nu = \frac{2r}{3x}.$$

In addition, we have the relation determining the angle of initiation ψ and an auxiliary value i_{n0} which is the value of current in the load at $\vartheta_1 = 0$

$$\tan\psi = \frac{1}{\nu}, \quad i_{n0} = \frac{\sqrt{3}A}{2x}\sin\psi. \tag{3.8.8}$$

The expressions for the phase currents in the subsequent intervals $i_{ak}(\vartheta_1)$, $i_{bk}(\vartheta_1)$ and $i_{ck}(\vartheta_1)$, with k denoting the interval number, are obtained from the case $k = 1$ by a cyclic permutation of indexes with the simultaneous change of the sign

$$i_{a1} = i_1(\vartheta_1), \; i_{b1} = i_2(\vartheta_1), \; i_{c1} = i_3(\vartheta_1),$$
$$i_{a2} = -i_2(\vartheta_1 + \pi/3), \; i_{b2} = -i_3(\vartheta_1 + \pi/3), \; i_{c2} = -i_1(\vartheta_1 + \pi/3),$$
$$i_{a3} = i_3(\vartheta_1 + 2\pi/3), \; i_{b3} = i_1(\vartheta_1 + 2\pi/3), \; i_{c3} = -i_2(\vartheta_1 + 2\pi/3),$$
$$\dots \tag{3.8.9}$$

Taking into account eq. (3.8.6) yields

$$i_{ds} = -\left(\frac{1}{2} - \frac{3\sqrt{3}}{4\pi}\right)\frac{\nu A}{x\sqrt{1+\nu^2}}, i_{qs} = \frac{A}{x\sqrt{1+\nu^2}}, \langle i_n \rangle = \frac{3}{\pi}\frac{A}{x\sqrt{1+\nu^2}}. \tag{3.8.10}$$

In addition to this, we determine the mean moment $\langle m_e \rangle$ of the electromagnetic forces

$$\langle m_e \rangle = \left(\frac{1}{2} + \frac{3\sqrt{3}}{4\pi} \right) \frac{A^2}{x^2(1+\nu^2)}. \tag{3.8.11}$$

Expressions (3.8.7) and (3.8.9) completely determine the regime of valves' operations when the overlap angle γ is exactly $\pi/3$ however we have not yet established under what conditions imposed on parameters r, x this regime is feasible. To obtain these conditions we consider the second regime when the overlap angle γ is smaller than $\pi/3$, i.e. in the part $0 \le \vartheta_1 \le \gamma$ of the interval $0 \le \vartheta_1 \le \pi/3$ three valves conduct current and in the other part $\gamma \le \vartheta_1 \le \pi/3$ only two valves conduct current. The equations for processes in the rectifier are written in the form

$$x\frac{di_a}{d\vartheta_1} - x\frac{di_c}{d\vartheta_1} = e_a - e_c, \quad x\frac{di_a}{d\vartheta_1} + x\frac{di_n}{d\vartheta_1} + ri_n = e_a - e_b,$$

$$i_n = i_a + i_c \quad 0 \le \vartheta_1 \le \gamma,$$

$$2x\frac{di_a}{d\vartheta_1} + ri_a = e_a - e_b, \quad i_b = -i_a, i_c = 0,$$

$$i_n = i_a \quad \gamma \le \vartheta_1 \le \pi/3. \tag{3.8.12}$$

The solution $i_a = i_{a1}(\vartheta_1), i_b = i_{b1}(\vartheta_1), i_c = i_{c1}(\vartheta_1)$ at $0 \le \vartheta_1 \le \gamma$ and $i_a = i_{a2}(\vartheta_1), i_b = i_{b2}(\vartheta_1), i_c = i_{c2}(\vartheta_1)$ at $\gamma \le \vartheta_1 \le \pi/3$ is sought under the conditions: $i_{a1}(0) = 0$ at $\vartheta_1 = 0$ the first valve initiates; $i_{c1}(\gamma) = 0$ at $\vartheta_1 = \gamma$ the fifth valve switches off; at $\vartheta_1 = \gamma$ the current in load is continuous, then $i_{a1}(\gamma) + i_{c1}(\gamma) = i_{a2}(\gamma)$. The current in load is periodic then $i_{c1}(0) = i_{n0} = i_{a1}(\pi/3)$, at $\vartheta_1 = \pi/3$ when the second valve initiates and valves 1 and 6 are functioning, the voltage at the second valve 2 must vanish. The latter condition relates the value of current i_{n0} in the load at $\vartheta_1 = 0$ (also at $\vartheta_1 = \pi/3$) to the value of the initiation angle ψ and the system parameters

$$i_{n0} = \frac{3A}{r} \sin\psi. \tag{3.8.13}$$

Relations (3.8.8) and (3.8.12) determine the conditions of realization of the regime of operation of the rectifier with the overlap angle $\gamma = \pi/3$. This regime is realized at $0 \le \nu \le 4/\sqrt{3}$. The remaining conditions result in the system of two transcendent equations relating the value of the overlap angle γ and initiation angle ψ with the parameter $\nu = 2r/3x$ characterizing the load. Dependences $\gamma = \gamma(\nu), \psi = \psi(\nu)$ were determined numerally with the further approximation by polynomials of fourth order. The dependences $\gamma(\nu), \psi(\nu)$ for $4/\sqrt{3} \le \nu \le 5$ are shown in Fig. 3.15. It should be noted that for a sufficiently accurate calculation of the phase currents of the rectifier one needs a high accuracy of approximation of dependences $\gamma(\nu)$ and $\psi(\nu)$.

Similar to the regime $\nu < 4\sqrt{3}$ the phase currents of the rectifier are proportional to amplitude A of the feeder voltage. For the regime $\nu > 4/\sqrt{3}$

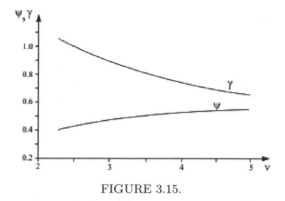

FIGURE 3.15.

and the amplitude $A = 1$ the expressions have the following form

$$0 \le \vartheta_1 \le \gamma, \quad i_a = i_1(\vartheta_1), \quad i_b = i_2(\vartheta_1), \quad i_c = i_3(\vartheta_1),$$

$$i_1 = \left\{ \frac{1}{\nu x} \sin \psi - \frac{1}{2x(1+\nu^2)} [\nu \sin(\psi + \pi/3) - \cos(\psi + \pi/3)] \right\} \exp(-\nu\vartheta_1)$$

$$+ \frac{1}{2x(1+\nu^2)} [\nu \sin(\vartheta_1 + \psi + \pi/3) - \cos(\vartheta_1 + \psi + \pi/3)] +$$

$$+ \frac{\sqrt{3}}{2x} [\cos(\psi - \pi/6) - \cos(\vartheta_1 + \psi - \pi/6)] - \frac{1}{\nu x} \sin \psi,$$

$$i_2 = -\left\{ \frac{2}{\nu x} \sin \psi - \frac{1}{x(1+\nu^2)} [\nu \sin(\psi + \pi/3) - \cos(\psi + \pi/3)] \right\} \exp(-\nu\vartheta_1)$$

$$- \frac{1}{2x(1+\nu^2)} [\nu \sin(\vartheta_1 + \psi + \pi/3) - \cos(\vartheta_1 + \psi + \pi/3)],$$

$$i_3 = \left\{ \frac{1}{\nu x} \sin \psi - \frac{1}{2x(1+\nu^2)} [\nu \sin(\psi + \pi/3) - \cos(\psi + \pi/3)] \right\} \exp(-\nu\vartheta_1)$$

$$+ \frac{1}{2x(1+\nu^2)} [\nu \sin(\vartheta_1 + \psi + \pi/3) - \cos(\vartheta_1 + \psi + \pi/3)] +$$

$$+ \frac{1}{\nu x} \sin \psi - \frac{\sqrt{3}}{2} [\cos(\psi - \pi/6) - \cos(\vartheta_1 + \psi - \pi/6)],$$

$$\gamma \le \vartheta_1 \le \pi/3$$

$$i_1 = \left[\frac{2}{x\nu} \sin \psi - \frac{\sqrt{3}}{2x(1 + (\frac{3}{4}\nu)^2)} (\frac{3}{4}\nu \cos \psi + \sin \psi) \right] \exp(-\frac{3}{4}\nu(\vartheta_1 - \pi/3))$$

$$+ \frac{\sqrt{3}}{2x(1 + (\frac{3}{4}\nu)^2)} \left[\frac{3}{4} \sin(\vartheta_1 + \psi + \pi/6) - \cos(\vartheta_1 + \psi + \pi/6) \right],$$

$$i_2(\vartheta_1) = -i_1(\vartheta_1), \quad i_3(\vartheta_1) = 0 \qquad (3.8.14)$$

The expressions for the phase currents $i_a(\vartheta_1), i_b(\vartheta_1), i_c(\vartheta_1)$ in the subsequent intervals $(k-1)\pi/3 \le \vartheta_1 \le k\pi/3 k = 2, \ldots, 6$ are obtained by a cyclic permutation of indexes with the change of sign (similar to for the regime with $\nu \leq 4/\sqrt{3}$).

The calculated regimes engulf all whole range of the regimes of operations of the rectifier from the short circuit to the idling. The formulae derived in the present section allow one to determine all quantities in the steady-state regimes of rectifier provided that the amplitude A of the emf in the circuits is known.

3.9   Control of the load current

In the case of quasi-stationary regimes with a fixed load $r = \text{const}$, current $\langle i_n \rangle$ in the load is determined by frequency ω of the rotor rotation and the excitation voltage e_f. The frequency of the rotor rotation depends on the angular velocity of the generator drive and can vary within certain limits, thus the current in the load is also changing. Generally speaking, in order to maintain a certain current $\langle i_n \rangle$ or to guarantee a prescribed change in the current under the variable frequency of rotation there are two possibilities: (i) to change the excitation voltage to ensure the prescribed current in the load in spite of change in the frequency of rotation or (ii) to change the angle of initiation of valves ψ. For the regime $\nu < 4/\sqrt{3}$ a prescribed value of $\langle i_n \rangle$ can be achieved only by the excitation voltage. Indeed, for the above regime the angle ψ is determined by the relation

$$\tan \psi = 1/\nu \tag{3.9.1}$$

and it is impossible to change the initiation angle without changing frequency ω and load r because this regime no longer exists when condition (3.9.1) is violated. Thus, the only possibility to ensure the required value $\langle i_n \rangle$ for the regime at $\nu < 4/\sqrt{3}$ is changing the excitation voltage. One can derive the formula relating current $\langle i_n \rangle$ with the excitation voltage and the frequency of rotation ω with the help of relations of the previous section. Hence for $\nu < 4/\sqrt{3}$ we obtain for machines with the three-phase winding that quantities $\langle i_n \rangle, e_f$ and ω are related by the formula

$$\langle i_n \rangle = e_f F(\nu, \sigma), \tag{3.9.2}$$

$$F(\nu, \sigma) = \frac{3}{\pi} \frac{\Delta_1}{\Delta}, \quad \nu = \frac{2r}{3(1-\sigma)\omega},$$

$$\Delta_1 = \frac{1}{(1-\sigma)\sqrt{1+\nu^2}} \left\{ \left[1 + \frac{\sigma}{1-\sigma} \frac{1+\kappa\nu^2}{1+\nu^2} \right]^2 + \left[\frac{\sigma}{1-\sigma} \frac{\nu(1-\kappa)}{1+\kappa^2} \right]^2 \right\}^{1/2},$$

$$\Delta = 1 + 2\frac{\sigma}{1-\sigma} \frac{1+\kappa\nu^2}{1+\nu^2} + \frac{\sigma^2}{(1-\sigma)^2} \frac{1+\kappa^2\nu^2}{1+\nu^2}, \quad \kappa = \frac{1}{2} - \frac{3\sqrt{3}}{4\pi}. \tag{3.9.3}$$

The relations (3.9.3) allow one to find the excitation voltage needed for the prescribed current in the load under variable frequency of rotation.

In the case of $\nu > 4/\sqrt{3}$ the load current can also be controlled by change in the angle of valves' initiation ψ. The angle of delay in initiation ψ is considered to be variable, that is $\psi = \psi_0 + \beta$, where ψ_0 is the angle of initiation if the artificial delay is absent and β is the angle of delay initiation, this angle playing role of the control signal. Equations determining the steady-state regime of rectifier for $\nu > 4/\sqrt{3}$ are obtained in Section 3.7 and have the form

$$\left\{ i_{n0} + \frac{A}{x(1+\nu^2)} \left[\nu \sin\left(\psi + \frac{\pi}{3}\right) - \cos\left(\psi + \frac{\pi}{3}\right) \right] \right\} \exp(-\nu\gamma) +$$

$$\frac{A}{x(1+\nu^2)} \left[\nu \sin\left(\gamma + \psi + \frac{\pi}{3}\right) - \cos\left(\gamma + \psi + \frac{\pi}{3}\right) \right] =$$

$$\frac{A\sqrt{3}}{x} \left[\cos\left(\psi - \frac{\pi}{6}\right) - \cos\left(\gamma + \psi - \frac{\pi}{6}\right) \right] - i_{n0}, \tag{3.9.4}$$

$$\left\{ i_{n0} - \frac{A\sqrt{3}}{2x} \frac{1}{1+(3/4\nu)^2} \left(\frac{3}{4}\nu \cos\psi + \sin\psi \right) \right\} \exp\left(-\frac{3}{4}\nu\left(\gamma - \frac{\pi}{3}\right) \right) +$$

$$+ \frac{A\sqrt{3}}{2x} \frac{1}{1+(3/4\nu)^2} \left[\frac{3}{4}\nu \sin\left(\gamma + \psi + \frac{\pi}{6}\right) - \cos\left(\gamma + \psi + \frac{\pi}{6}\right) \right] =$$

$$= \frac{A\sqrt{3}}{x} \left[\cos\left(\psi - \frac{\pi}{6}\right) - \cos\left(\gamma + \psi - \frac{\pi}{6}\right) \right] - i_{n0}, \tag{3.9.5}$$

$$i_{n0} = \frac{3A}{r} \sin\psi. \tag{3.9.6}$$

Relationship (3.9.6) expresses that fact that the valve voltage (say valve 2) must be equal to zero when this valve is connected to two conducting valves (say 1 and 6), otherwise the initiation is impossible. In the case of the artificial increase in the initiation angle the latter condition is replaced by the inequality

$$i_{n0} < \frac{3A}{r} \sin\psi, \tag{3.9.7}$$

that describes the situation that the connected valve initiates when its voltage is already positive. To solve the system of equations (3.9.4-3.9.6) we put

$$i_{n0} = \frac{3A}{r} \sin\psi_0, \quad \psi = \psi_0 + \beta, \quad \beta > 0. \tag{3.9.8}$$

Inserting i_{n0} and ψ from eq. (3.9.8) into relations (3.9.4) and (3.9.5) we obtain two equations determining the dependences $\gamma = \gamma(\nu, \beta)$ and $\psi = \psi(\nu, \beta)$ on parameter ν and the angle of delay in initiation β. The further solution of the problems is constructed by analogy to Section 3.7. Introducing angle β as a control parameter allows one to reduce the mean current in the load $\langle i_n \rangle$ against its value at $\beta = 0$.

3.10 Work of generator on counter-electromotive force via rectifier

This section deals with the case in which the circuit of load consists of the active resistance r in series with the receiver of energy having the counter-electromotive force E, Fig. 3.16. This circuit can serve as an idealized model for work of a generator on a battery.

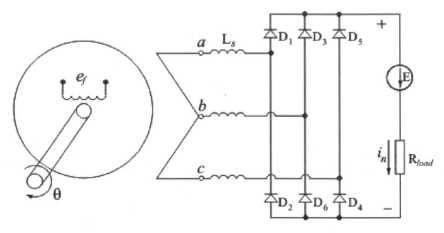

FIGURE 3.16.

As above, while analysing the steady-state regime we consider the interval $0 \leq \vartheta \leq \pi/3$ between the moment of initiation of valves 1 and 2, where ϑ denotes the "electric" angle of rotation of the machine rotor relative to the moment of initiation of valve 1.

When the circuit contains an energy receiver with the counter-electromotive force, in the regime of idling the voltage u at the energy receiver is maximal and equal to $u = A\sqrt{3}$ where A is the amplitude of voltage at the rectifier input. The duration of passage of the anode current is thus equal to zero. As the load increases, voltage u decreases and duration λ of passage of the anode current increases, $0 < \lambda < \pi/3$. At the small load the rectified current has discontinuous character, i.e. in the interval $0 \leq \vartheta \leq \lambda$ the current in the load is not zero whereas in the interval $\lambda \leq \vartheta \leq \pi/3$ of the main interval the current in the load is equal to zero. The increase in duration of the current passage results in increasing number of simultaneously initiating valves and the current in load becomes continuous. Depending upon the parameters, the system can function from idling to short circuit in three different regimes which differ from each other by the number of simultaneously initiating valves. In the first regime with discontinuous current in the load two valves initiate simultaneously, for example, the valves 1 and 6 for $0 \leq \vartheta \leq \lambda$ and for $\lambda \leq \vartheta \leq \pi/3$ all valves do not conduct current. Further, for $\pi/3 \leq \vartheta \leq \pi/3 + \lambda$ the valves 1 and 2 conduct etc. There are also two regimes (second and third regimes) with the continuous current in the load. In the second regime, for $0 \leq \vartheta \leq \gamma$ three valves (1,5 and 6) conduct current and for $\gamma \leq \vartheta \leq \pi/3$ valve 5 does not initiate and the current is conducted by valves 1 and 6. In this regime the overlap angle γ changes from zero to $\pi/3$. As the current in load increases the system passes to the third regime when in the interval $0 \leq \vartheta \leq \pi/3$ three valves initiate simultaneously and the overlap angle is $\gamma = \pi/3$.

If the system parameters are arbitrary then it is not known in advance what regime is realized. Therefore for computation of the steady-state processes in the rectifier it is expedient to choose inverse strategy. Namely, one assumes that a certain regime exists and then finds the proper parameters. The aim of such strategy is to determine the value of counter-electromotive force and the corresponding value of current proper in the load for all possible regimes of work of the rectifier.

Let us describe a method of calculation on the example of second regime provided that the overlap angle γ lies in the range $0 \leq \gamma \leq \pi/3$. First the processes in the rectifier are calculated for the unit amplitude of voltage $A = 1$ at the rectifier input. These relations rely on parameter $q = E/A\sqrt{3}$ characterizing the ratio of the counter-electromotive force to the amplitude of feeder voltage. The value of parameter q is a function of γ and is determined from the condition of existence of the considered regime. The mean current in the load $\langle i_n \rangle$ and the mean quasi-currents $\langle i_d \rangle, \langle i_q \rangle$ are simultaneously determined by the method described above. These mean values are proportional to the excitation voltage

$$\langle i_n \rangle = i_n e_f, \quad \langle i_d \rangle = i_d e_f, \quad \langle i_q \rangle = i_q e_f. \tag{3.10.1}$$

Next, in terms of the obtained values of i_d, i_q we calculate the voltage amplitude a at the rectifier input for the unit excitation voltage $e_f = 1$

$$a = \omega\sqrt{(1 + \sigma i_d)^2 + \sigma^2 i_q}. \tag{3.10.2}$$

Finally, the relative value of the counter-electromotive force e is determined: $e = E/e_f$ where $e = \sqrt{3}aq$. This value causes the relative current $i_n = i_n a$ of the load . The dependence $i_n = i_n(e)$ for $r = 1, \sigma = 0.5, \omega = 1$ is displayed in Fig. 3.17.

3.11 Control of excitation of brief action turbogenerators

The equations for slow processes in synchronous machine working on the active-inductive load derived in Section 3.6 can be used for computation of the transient regime when the machine works as a generator of brief

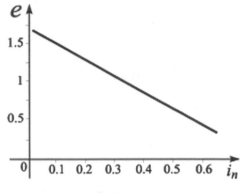

FIGURE 3.17.

action. Turbogenerators of brief action operate in the impulsive regime (of duration of the order of few seconds) due to the kinetic energy of rotating flywheel attached mechanically to the rotor of the generator.

The principle of work of the turbogenerators of brief action is the following. First, the flywheel together with the generator rotor spins by a relative low-power motor up to a nominal angular velocity, the circuit being open and the excitation being absent. After that the excitation is provided and when the proper voltage at the stator is established at no load, then the load is thrown on. The generating regime begins and the stored kinetic energy is transformed into electric one, partly dissipates at the active resistance of the load and partly accumulates in the inductive elements. When the load in the rotor contours is thrown on, the electromagnetic transient processes change the values of flux in the gap. These processes are characterized by several time constants, the largest being the time constants of the damper contour in the transversal axis and the excitation contour. They are of the order of duration of the operation process. For this reason the stator voltage and the power are changing while the machine operates. In addition to this, changing these values also takes place due to reduction of the angular frequency of the rotor rotation.

If any change in power at the constant excitation voltage due to reducing angular velocity is inadmissible from the practical standpoint, then the control of the operating generator is required for maintaining the angular velocity. If there are no additional feeders for maintaining the prescribed power in the load then two methods are applied: (i) introducing controllers directly in the stator circuits or control of the excitation voltages [36]. The second way seems to be more rational and requires calculation of the program of control of the excitation voltage. The method based on application of the simplified equations of transient processes in the synchronous machine is suggested in what follows.

A number of works, e.g. [98], are devoted to study of dynamics of controlled generators. However in [98] the damper contours are not taken into account and additionally the dependence of the stator flux linkage on time and angular frequency are actually guessed right (generally speaking they are a part of the solution of problem). Therefore the method [98] applied only to the active-inductive load can not be viewed sufficiently general and substantiated. There exist also the works with the results of experimental investigations of controlled regimes, e.g. [75].

The working process in the generators of brief action is an aggregate of transient processes of different duration. Immediately after the load is connected the fast transient processes in the stator circuits begin and take time comparable with the period of rotations of the rotor of synchronous machine. In this time interval the very notion of the mean power and the control problem has not sense, inasmuch as these processes can not be considered as quasi-stationary with the slowly varying parameters. When the fast aperiodic processes have be decayed all characteristics of the regime (including mean power) are governed by the equations of slow nonstationary processes (3.6.9) applicable for any practically relevant load.

The equations of slow nonstationary processes obtained above for the case of machine operating into the active-inductive load (3.6.15) (the motor torque in them must vanish, $m = 0$) admits a simpler way of calculation of change in angular velocity, power, phase voltage, current in load and other variables describing the transient process. Using these equations one determines the required law of changing the excitation voltage needed for maintaining the prescribed power. The control strategy is suggested for the machines operating into rectifier with active-inductive [56] and pure inductive [127] load, as well as for the plasmatron of alternating current [58].

Let us consider the problem of maintaining a prescribed power of generator for any autonomous load described by eq. (3.6.9) (for the problems of maintaining prescribed values of other quantities like voltage, current and so on, see below). The generator power is changing in the following way. Immediately after the load is connected there begin fast transient processes lasting for short time of the order of period of the rotor revolution. In practice, it is impossible to maintain the prescribed power by means of the excitation voltage within this time interval. After attenuating fast processes all characteristics of the regime, including the power, are determined by slow processes described by equations (3.6.9). In order to provide a prescribed change in power $N(t)$ in this time interval it is necessary to change the excitation voltage $e_f(t)$ such that the relation

$$N(t) = \langle U_a i_{ap} + U_b i_{bp} + U_c i_{bc}\rangle = 3/2\omega(\Psi_k\langle i_{dp}\rangle - \Psi_k\langle i_{qp}\rangle) \qquad (3.11.1)$$

holds. Here currents i_{ap}, i_{bp}, i_{cp} are stationary periodic currents in the stator circuits. Let $t = 0$ correspond to the time instant when the load is applied. Within the small time interval of attenuation of fast aperiodic processes the slow variables are changing in small values of order of ε, and in the first approximation it is possible to assume that they keep their values at $t = 0$. This allows one to state the problem of maintaining power $N(t)$ at all times beginning at $t = 0$. However it is necessary to recall that the problem has physical meaning only within a sort-time interval after the load is connected.

Let us consider equations (3.6.9). In accordance with eq. (3.11.1) the equation of rotation is separated from the others

$$\dot{\omega} = \frac{2}{3}\frac{N(t)}{\omega} + \varepsilon_\omega M. \tag{3.11.2}$$

In many cases equations (3.6.9) completely describe slow nonstationary processes in the system "machine+load", [56, 127, 58]. However there are problems requiring additional slow variables y_j for complete description of processes in the load (for example, in [58] such a variable is the current of inductive storage). In these cases eq. (3.6.9) should be completed by averaged equations with $\dot{y}_j$ [58].

The quantities $\langle i_{dp}\rangle, \langle i_{qp}\rangle$ in eq. (3.6.9) are functions of variables Ψ_k, Ψ_f, ω, y_j and determined by the form of a particular load. In order to find these functions it is necessary to calculate first the steady-state processes in the circuits of load subjected to three-phase symmetric system of phase voltages (3.6.6). While calculating the load circuits the variables $\Psi_k, \Psi_f, \omega, y_j$ are considered as time-independent parameters, therefore the very calculation presents a standard problem of the theory of electric circuits with prescribed system of emf.

When the steady-state 2π−periodic processes in the load circuits are calculated, then with the help of eq. (3.2.2) relating the phase currents i_a, i_b, i_c and angle of the rotor rotation ϑ to variables i_d and i_q, the steady-state values $\langle i_{dp}\rangle, \langle i_{qp}\rangle$ are calculated, which in turn determine the form of equations for slow nonstationary processes.

Let us now proceed to relationship (3.11.1) allowing one to express one of the slow variables in terms of the others. It is convenient to express Ψ_f such that the equations for the remaining variables do not contain e_f

$$\Psi_f = \Psi_f[\Psi_k, \omega, y_j, N(t)]. \tag{3.11.3}$$

Let us insert eq. (3.11.3) and the solution of equation (3.11.2) in the first equation (3.6.9) and the averaged equations of the load circuits. Then we arrive at the closed system of differential equations for Ψ_k and y_j. Let us assume that this system and eq. (3.11.2) are solved under prescribed initial conditions. Then we can determine Ψ_f as a function of time from eq. (3.11.3). Utilising the second equation in (3.6.9) we obtain e_f

$$e_f = \frac{\nu_t + \varepsilon_f}{\varepsilon_f \nu_t}\dot{\Psi}_f + \Psi_f - \langle i_{dp}\rangle. \tag{3.11.4}$$

Usually the above equations for Ψ_k, y_j have to be integrated numerally. If eq. (3.11.4) is used and $\dot{\Psi}_f$ is differentiated numerally then this can lead to the loss of accuracy. For this reason it is better to use the expression for derivative $\dot{\Psi}_f$ obtained by differentiating eq. (3.11.3)

$$\dot{\Psi}_f = \frac{\partial \Psi_f}{\partial \Psi_k}\dot{\Psi}_k + \sum_j \frac{\partial \Psi_f}{\partial u_j}\dot{u}_j + \frac{\partial \Psi_f}{\partial \omega}\dot{\omega} + \frac{\partial \Psi_f}{\partial N}\dot{N}. \tag{3.11.5}$$

Inserting eq. (3.11.5) into (3.11.4) and taking into account eqs. (3.6.9) and (3.11.2) yields

$$e_f = \frac{\nu_t + \varepsilon_f}{\varepsilon_f \nu_t}\left[\nu_k \frac{\partial \Psi_f}{\partial \Psi_k}\left(\langle i_{qp}\rangle - \Psi_k\right) + \sum_j \frac{\partial \Psi_f}{\partial u_j}\dot{u}_j + \right.$$

$$\left. +\varepsilon_\omega \frac{\partial \Psi_f}{\partial \omega}\left(M - \frac{2}{3}\frac{N}{\omega}\right) + \frac{\partial \Psi_f}{\partial N}\dot{N}\right] + \Psi_f - \langle i_{dp}\rangle. \tag{3.11.6}$$

When applying eq. (3.11.6) to particular problems the derivative $\dot{u}_j$ should be replaced by the right hand sides of the corresponding averaged equations.

An important question is the value of $e_f(0)$. Before the load is exerted the value $e_f = e_{f-}$ should guarantee the value of $\Psi_f(0)$ determined by relation (3.11.3)

$$\Psi_f(0) = \Psi_f[\Psi_k(0)\omega(0), u_j(0), N(0)]. \tag{3.11.7}$$

At no load it is possible to obtain this value $\Psi_f(0)$ by setting $e_{f-} = \Psi_f(0)$. The value $e_f(0)$ immediately after switching the load is calculated from eq. (3.11.6) at $t = 0$. Generally speaking $e_f \neq e_f(0)$. This means that at the moment of switching the load the excitation voltage needs to be changed stepwise from e_f to $e_f(0)$. In practice e_f can be changed within the time of decaying fast transient processes.

As an example we demonstrate derivation of equations of slow nonstationary processes and determination of the law of program control by voltage excitations for the case of the symmetric linear load characterized by certain amplitude-frequency $A(\omega)$ and phase-frequency $\Phi(\omega)$ and voltage-current characteristics. The steady-state components i_{ap}, i_{bp}, i_{cp} of currents in phases of this load satisfy the relationship

$$i_{ap} = \omega \Psi_m A(\omega) \cos[\vartheta - \varphi + \Phi(\omega)], \qquad a \to b \to c. \tag{3.11.8}$$

For any symmetric load the mean values $\langle i_{dp}\rangle, \langle i_{qp}\rangle$ are known to be equal to the coefficients at $\cos \vartheta, -\sin \vartheta$ respectively in the expansion of function $i_{ap}(\vartheta)$ in Fourier series. Therefore

$$\langle i_{dp}\rangle = \omega A(\omega)[\Psi_k \cos \Phi(\omega) + \Psi_f \sin \Phi(\omega)],$$
$$\langle i_{qp}\rangle = \omega A(\omega)[\Psi_k \sin \Phi(\omega) - \Psi_f \cos \Phi(\omega)]. \tag{3.11.9}$$

Substituting eq. (3.11.9) into (3.6.9) we obtain the sought-for equations of slow nonstationary processes

$$\dot{\Psi}_k = -\nu_k[\Delta_1(\omega)\Psi_k + \Delta_2(\omega)\Psi_f],$$

$$\dot{\Psi}_f = -\frac{\varepsilon_f \nu_t}{\varepsilon_f + \nu_t}[\Delta_1(\omega)\Psi_f - \Delta_2(\omega)\psi_k - e_f],$$

$$\dot{\omega} = -\varepsilon_\omega[\Delta_2(\omega)(\Psi_f^2 + \Psi_k^2) - M], \qquad (3.11.10)$$

where

$$\Delta_1(\omega) = 1 - \omega A(\omega)\sin\Phi(\omega),$$

$$\Delta_2(\omega) = \omega A(\omega)\cos\Phi(\omega). \qquad (3.11.11)$$

Following the above explanation we insert relations (3.11.9) into eq. (3.11.1) and express the flux linkage Ψ_f as a function of Ψ_k, ω and N

$$\Psi_f[\Psi_k, \omega, N(t)] = \left[\frac{2N(t)}{3\omega\Delta_2(\omega)} - \Psi_k^2\right]^{1/2}. \qquad (3.11.12)$$

Differentiating (3.11.12) we obtain the derivative appearing in eq. (3.11.6)

$$\frac{\partial\Psi_f}{\partial\Psi_k} = -\frac{\Psi_k}{\Psi_f}, \quad \frac{\partial\Psi_f}{\partial N} = \frac{1}{3\omega\Delta_2(\omega)\Psi_f},$$

$$\frac{\partial\Psi_f}{\partial\omega} = -\frac{N}{3\omega^2\Delta_2(\omega)^2\Psi_f}\Delta_2(\omega) + \omega\frac{d\Delta_2(\omega)}{d\omega}. \qquad (3.11.13)$$

Inserting eq. (3.11.13) into eq. (3.11.6) and taking into account eqs. (3.11.2) and (3.11.10) results in the expression for e_f

$$e_f(\Psi_k, \omega, t) = \left[\frac{\nu_k(\nu_t + \varepsilon_f)}{\nu_t\varepsilon_f} - 1\right]\Delta_2(\omega)\Psi_k + \Delta_1(\omega)\left[\frac{2N(t)}{3\omega\Delta_2(\omega)} - \psi_k^2\right]^{1/2} +$$

$$\frac{\nu_t + \varepsilon_f}{\nu_t\varepsilon_f}\frac{2N(t)}{3\omega\Delta_2(\omega)} - \Psi_k^{2-1/2}\{\nu_k\Delta_1(\omega)\Psi_k^2 -$$

$$\frac{\varepsilon_\omega N(t)}{3\omega^2\Delta_2(\omega)^2}\left[\Delta_2(\omega) + \omega\frac{d\Delta_2(\omega)}{d\omega}\right]\left[M - \frac{2N(t)}{3\omega}\right] + \frac{\dot{N}(t)}{3\omega\Delta_2(\omega)}\}. \qquad (3.11.14)$$

Equation (3.11.14) needs to be completed by eqs. (3.11.12), (3.11.2) and the equation for Ψ_k which, in this case, has the following form

$$\dot{\Psi}_k = -\nu_k\left\{\Delta_1(\omega)\Psi_k + \Delta_2(\omega)\left[\frac{2N(t)}{3\omega\Delta_2(\omega)} - \psi_k^2\right]^{1/2}\right\}. \qquad (3.11.15)$$

Solving consequently equations (3.11.2),(3.11.15) and substituting the solutions obtained into eq. (3.11.14) we find the required program control

$e_f(t)$. Before the load is switched the excitation voltage e_{f-} is given by the formula

$$e_{f-} = \Psi_f(0) = \left[\frac{2N(0)}{3\omega(0)\Delta_2(\omega(0))} - \Psi_k(0)^2\right]^{1/2}. \qquad (3.11.16)$$

In the important case in which the generator of brief action operates due to the energy of flywheel and the rotor, i.e. $M = 0$, the variables in eq. (3.11.2) are separated and the angular frequency is the following function of time

$$\omega(t) = \left[\omega(0)^2 - \frac{4}{3}\varepsilon_\omega \int_0^t N(t)dt\right]^{1/2}. \qquad (3.11.17)$$

Let us now take into account that parameters ε_f and ε_ω are small. As a rule the smallness of parameters $\varepsilon_f, \varepsilon_\omega$ is typical for high-power turbogenerators and corresponds to large (as compared with the unity of dimensionless time) values of the time constant of contour excitations and the mechanical time constant. Then among the slow transient processes it is possible to select the processes with small (of order of $\varepsilon_f, \varepsilon_\omega$) velocities, such processes are further referred to as the slowest ones. The equations for the slowest processes are obtained from equations (3.6.9) and the averaged equations for y_j by means of another asymptotic separation of motions which is carried out similar to the separation of fast and slow processes.

At small $\varepsilon_f, \varepsilon_\omega$ it is also possible to state the problem of maintaining the slowest component of the power in the controlled generators of brief action. The prescribed value N must be a rather slowly changing function such that it can be considered as a function of slow time $\tau = \varepsilon_f t$. The way of solving this problem does not change with the only difference that now the solutions of equations of the slowest processes are to be substituted into all relationships and the steady-state solutions Ψ_{kc}, u_{kc}, obtained under the assumption that the other variables are time-independent parameters, must be substituted into the variables whose derivatives are not small (Ψ_k and part of variables y_j).

For instance, in the case of given amplitude-frequency and phase-frequency characteristics of the load we have from the first equation in (3.11.10)

$$\Psi_{kc} = -\frac{\Delta_2(\omega)}{\Delta_1(\omega)}\Psi_f. \qquad (3.11.18)$$

Let us insert eq. (3.11.18) in eq. (3.11.12) and express the flux linkage Ψ_f. The result is

$$\Psi_f = \Psi_f(\omega, N(\tau)) = \Delta_1(\omega)\left\{\frac{2N(\tau)}{3\omega\Delta_2(\omega)[\Delta_1(\omega)^2 + \Delta_2(\omega)^2]}\right\}^{1/2}. \qquad (3.11.19)$$

Substituting eqs. (3.11.18) and (3.11.19) into eq. (3.11.14) and assuming $\nu_t/(\nu_t + \varepsilon_f) = 1$ we arrive at the expression for control e_f as a function of slow time τ

$$e_f(\tau) = \left[\frac{2N(\tau)[\Delta_1(\omega(\tau))^2 + \Delta_2(\omega(\tau))^2]}{3\omega(\tau)\Delta_2(\omega(\tau))} \right]^{1/2} +$$

$$+ \frac{1}{\varepsilon_f \Psi_f(\omega(\tau), N(\tau))} \left\{ \varepsilon_f \frac{dN(\tau)}{d\tau} \frac{1}{3\omega(\tau)\Delta_2(\omega(\tau))} - \right. \tag{3.11.20}$$

$$\left. - \frac{\varepsilon_\omega N(\tau)}{3\omega(\tau)^2 \Delta_2(\omega(\tau))^2} \Delta_2(\omega(\tau)) + \omega(\tau) \frac{d\Delta_2(\omega(\tau))}{d\omega} \left[M - \frac{2N(\tau)}{3\omega(\tau)} \right] \right\}.$$

Here $\omega(\tau)$ is the solution of the equation for the slowest nonstationary processes

$$\frac{d\omega}{d\tau} = -\frac{2}{3} \left[\frac{\varepsilon_\omega}{\varepsilon_f} \frac{N(\tau)}{\omega} - M \right]. \tag{3.11.21}$$

The obtained law of control of the excitation voltage (3.11.20) is a more smooth function of time than (3.11.14) and more preferable for technical realization. However the control $e_f(\tau)$ ensures a prescribed value only of the slowest component of power which can substantially differ from the prescribed power $\langle N_r \rangle$ in the interval of attenuation of slow transient processes. If considerable deviation of the mean power is inadmissible then in the case of small $\varepsilon_f, \varepsilon_\omega$ the control (3.11.14) should be applied. This gives rise to the following peculiarity. One can see from eq. (3.11.3) that in general Ψ_f is a function of time t rather than slow time τ, because expression (3.11.3) contains variables with non-small derivatives. Therefore the derivative $\dot{\Psi}_f$ has also a value of the order of unity. It follows from eq. (3.11.4) that e_f is a non-small quantity of the order of $1/\varepsilon_f$, at least until slow transient processes die out. Because of this feature the problem of maintaining the power at small $\varepsilon_f, \varepsilon_\omega$ requires a special consideration. For the case of active-inductive load such a consideration is carried out in [56] where it is shown that large value of e_f is merely a formality. In practice function Ψ_f in the interval of attenuation of slow processes slightly changes and the value of e_f at the beginning of process is not greater than that in the end.

Let us consider the problem of maintaining the prescribed amplitude $\omega\Psi_m$ of voltages U_a, U_b, U_c. Let us assume that there is no need to introduce variables y_j for describing slow processes in the load. Then the mean value of the power is a function of amplitude $\omega\Psi_m$ and the angular velocity ω. By virtue of eqs. (3.6.9) and (3.11.1) the equation of rotation has the form

$$\dot{\omega} = -\frac{2}{3} \varepsilon_\omega \frac{N(\omega\Psi_m, \omega)}{\omega} + \varepsilon_\omega M. \tag{3.11.22}$$

Inasmuch as $\omega\Psi_m = U(t)$ is a prescribed function of time, then as well as before, ω is determined regardless of the other variables and the general

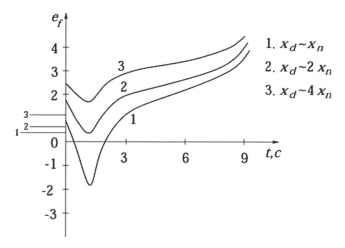

FIGURE 3.18.

strategy of solution is coincident with that of the problem of maintaining prescribed power.

The problem of maintaining a prescribed amplitude of the phase currents

$$I_m(\omega\Psi_m,\omega) = \left(\langle i_{dp}\rangle^2 + \langle i_{qp}\rangle^2\right)^{1/2} \qquad (3.11.23)$$

is solved by analogy. Power N should be presented as a function of $\omega\Psi_m$ and I_m, and the flux linkage Ψ_f should to be determined from the equation

$$I_m\left(\omega(\Psi_f^2 + \Psi_k^2)^{1/2},\omega\right) = I(t), \qquad (3.11.24)$$

where $I(t)$ is a given law of change in the amplitude of currents.

Provided that variables y_j are present the mean power is a function of y_j and $\omega\Psi_m,\omega$ (or I_m,ω). Then the equation for ω is not separated and is to be solved together with the other equations of slow nonstationary processes.

The result of numeral computation of the excitation voltage of a turbogenerator operating into active-inductive load is presented in Fig. 3.18. Here x_n denotes the dimensionless inductance of the load in system x_{ad}. The law of change of the excitation voltage is characterised by a pit in the first seconds of the transient process, caused by the fact that the flux linkage in the damper contour in the transverse axis Ψ_k grows faster than the flux linkage Ψ_f of the excitation winding decreases. At the constant excitation voltage this corresponds to the peak of power in the beginning of operating condition.

A similar problem of control of the excitation voltage for maintaining direct current in the load when the synchronous machine operates into active-inductive load through rectifier in the regime of generator of brief action is considered in [56].

3.12   Dynamics of two synchronous generators operating in parallel into a common active-inductive load

The problem of the parallel work of synchronous generators into a common active-inductive load that considered in this section is a generalization of the above problem of the autonomous work of one generator into an active-inductive load and study of the process of synchronisation of two generators with the inverse-parallel connection [54].

Such facilities can serve as autonomous sources of power supply for different electrophysical devices and use in both steady-state and transient regimes, for example, as a generator of brief action. Using two generators in place of one is crucial, in particular, in the space-system engineering in which for vanishing the total angular moment of orbital station at least two generators are needed, whose rotors are rotating in the opposite directions.

The mathematical background of solution is provided by the method of averaging the systems with many fast variables that belongs to the asymptotic methods of theory of nonlinear vibrations. Among the fast variables there are two phases which are the angles of rotation of the rotor of each machine. We consider the main resonance, i.e. the motions of the system when the difference in the angular velocities of rotation of both generators is small. The objective of study is derivation of the simplified equations and determination of possible regimes, in particular, analysis of electromechanical processes under synchronizations of both machines, study of the process of energy transmission and reducing the angular velocity under joint operation of the generators in the regime of brief action.

The obtained results are easily generalised on the case, when generators operate into load via rectifier rather than directly. Direct numeral integration of the initial equations presents an extraordinary challenging problem as one needs to describe operation of the valve system under non-sinusoidal voltages at the input and to sort out all possible combinations of the opened and locked valves. This problem is acknowledged to be difficult by other authors and utilising the analog processors is offered for avoiding numerical solution of the rectifier equations [25]. These and other numerical difficulties caused by "rigidity" of the initial systems of equations, completely disappear when the asymptotic methods are applied.

Let us consider two parallel synchronous generators operating into a three-phase active-inductive load. The stator windings of both generators are assumed to be connected by star-star connection with zero wire (Fig. 3.19). We assume the standard generator schematization of the theory of

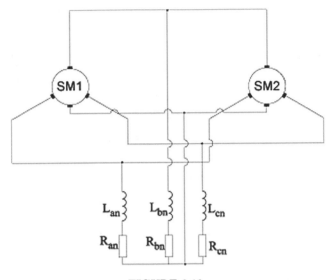

FIGURE 3.19.

electric machines generator, in particular, it is considered that three windings are located on the rotor: the excitation winding f and two damper windings t and k, the axes of t and k being respectively parallel and perpendicular to the axis of f winding. The higher harmonics and saturation of the magnetic circuit are neglected which is typical for study of nonstationary processes.

The system of contours of the electric circuit of the system is chosen such that the branch with the load belongs to one of the contours of the three-phase system only. This is attributed to that the active resistance of load is considered as a quantity of the order of the inductive resistance of the stator circuits. Therefore Kirchhoff equations for the contours with load describe formally fast processes while the slow processes and the corresponding slow variables are "hidden". For separating hidden variables it is necessary to transform the equations of circuits which is equivalent to introduction of the contours including only the stator circuits of two machines.

Let us first write down the Lagrange-Maxwell equations for the system of machines in terms of dimensional variables. Equations of the contours with the like phases of both generators connected by the zero wire, equations of the rotor contours and equations of the rotor rotation are given by

$$\frac{d}{dt}\left[L_1 i_{a1} + M_1 \cos\frac{2\pi}{3}(i_{b1} + i_{c1}) + (M_{f1}i_{f1} + M_{t1}i_{t1})\cos\vartheta_1 - \right.$$

$$\left. -M_{k1}i_{k1}\sin\vartheta_1\right] - \frac{d}{dt}\left[L_2 i_{a2} + M_2\cos\frac{2\pi}{3}(i_{b2} + i_{c2}) + \right.$$

$$\left. + (M_{f2}i_{f2} + M_{t2}i_{t2})\cos\vartheta_2 - M_{k2}i_{k2}\sin\vartheta_2\right] +$$

$$+R_{a1}i_{a1} - R_{a2}i_{a2} + R_{01}i_{01} - R_{02}i_{02} = 0, \quad (a, b, c),$$

$$\frac{d}{dt}\left\{M_{fi}\left[i_{a1}\cos\vartheta_i + i_{bi}\cos\left(\vartheta_i - \frac{2\pi}{3}\right) + i_{ci}\cos\left(\vartheta_i + \frac{2\pi}{3}\right)\right] + \right.$$

$$\left. +L_{fi}i_{fi} + M_{fti}i_{ti}\right\} + R_{fi}i_{fi} = E_{fi},$$

$$\frac{d}{dt}\left\{M_{ti}\left[i_{ai}\cos\vartheta_i + i_{bi}\cos\left(\vartheta_i - \frac{2\pi}{3}\right) + i_{ci}\cos\left(\vartheta_i + \frac{2\pi}{3}\right)\right] + \right.$$

$$\left. +L_{ti}i_{ti} + M_{fti}i_{fi}\right\} + R_{ti}i_{ti} = 0,$$

$$\frac{d}{dt}\left\{-M_{ki}\left[i_{ai}\sin\vartheta_i + i_{bi}\sin\left(\vartheta_i - \frac{2\pi}{3}\right) + i_{ci}\sin\left(\vartheta_i + \frac{2\pi}{3}\right)\right] + \right.$$

$$\left. +L_{ki}i_{ki}\right\} + R_{ki}i_{ki} = 0, \quad i = 1, 2,$$

$$J_i\frac{d\Omega_i}{dt} = -\left\{(M_{fi}i_{fi} + M_{ti}i_{ti})\left[i_{ai}\sin\vartheta_i + i_{bi}\sin\left(\vartheta_i - \frac{2\pi}{3}\right) + \right.\right.$$

$$\left. +i_{ci}\sin\left(\vartheta_i + \frac{2\pi}{3}\right)\right] + M_{ki}i_{ki}\left[i_{ai}\cos\vartheta_i + \right.$$

$$+i_{bi}\cos\left(\vartheta_i - \frac{2\pi}{3}\right) + i_{ci}\cos\left(\vartheta_i + \frac{2\pi}{3}\right)\right]\right\} + M_{mi},$$

$$\frac{d\vartheta_i}{dt} = \Omega_i \quad i = 1, 2. \tag{3.12.1}$$

The contours with the load are those with the phase windings of stator of the first machine, connected in series with the corresponding branches of the load. The corresponding equations have the form

$$\frac{d}{dt}\left[L_1 i_{a1} + M_1\cos\frac{2\pi}{3}(i_{b1} + i_{c1}) + (M_{f1}i_{f1} + M_{t1}i_{t1})\cos\vartheta_1 - \right.$$

$$\left. -M_{k1}i_{k1}\sin\vartheta_1\right] + \frac{d}{dt}(L_{an}i_{an}) + R_{an}i_{an} + R_{0n}i_{0n} + R_{01}i_{01} = 0, \quad (a, b, c)$$

$$i_{0i} = i_{ai} + i_{bi} + i_{ci}, \quad i = 1, 2$$

$$i_{0n} = i_{an} + i_{bn} + i_{cn}. \tag{3.12.2}$$

Here ϑ_i denotes the angle of rotation of the rotor of $i-th$ machine ($i = 1, 2$) measured from axis a to the axis of winding f in the direction of rotation, $\Omega_i = d\vartheta_i/dt$ is the angular frequency of rotation of the corresponding rotor; i_{ai}, i_{bi}, i_{ci} are currents in the stator phases, i_{fi}, i_{ti}, i_{ki} are respectively the currents in the excitation windings and the damper windings,

$L_i, L_{fi}, L_{ti}, L_{ki}$ are the coefficients of self-induction in the stator phases, in the excitation contour and in the damper contours, $M_i \cos \frac{2\pi}{3}$ is the coefficient of mutual induction of the stator contours, M_{fi}, M_{ti}, M_{ki} are the "amplitude" coefficients of the mutual induction between stator contours and respectively the excitation contour and the damper contours, M_{fti} is the coefficient of mutual induction between the excitation contour and the damper contour t, E_{fi} is the excitation voltages, R_i, R_{0i} are the active resistances of the stator windings together with the corresponding parts of connecting wires and the corresponding resistances of parts of zero wire, R_{fi}, R_{ti}, R_{ki} are active resistances of the excitation contours and the damper contours, J_i is the moment of inertia of the rotor; M_{mi} is the motor torque, R_{an}, R_{bn}, R_{cn} and L_{an}, L_{bn}, L_{cn} are active resistances and inductances of the load, respectively. The symbols a, b, c in eq. (3.12.1) and in what follows mean that the equation should be completed by two similar equations for phases b and c.

We also have equations relating currents in the load circuits i_a, i_b, i_c and the currents in the phases of stator windings

$$i_{a1} + i_{a2} = i_{an} \ (a, b, c). \tag{3.12.3}$$

Next we consider the operation of both generators into symmetric load and assume $R_{an} = R_{bn} = R_{cn}$, $L_{an} = L_{bn} = L_{cn}$.

In what follows we make use of equations in terms of the dimensionless variables introduced by the relations $t_u = t/t_*$, $\omega = \Omega/\Omega_*$, $i_{aiu} = i_{ai}/i_{ai*}$, $i_{fiu} = i_{fi}/i_{fi*}$ etc. where an asterisk designates the basis (characteristic) values. Let us introduce the nominal current of stator of the first machine i_* and the nominal (basis) synchronous frequency Ω_*. Let us put $i_{ai*} = i_{bi*} = i_{ci*} = i_*$. Similar to Section 3.2 the other basic values are taken as follows

$$i_{fi*} = \frac{L_{adi}}{M_{fi}} i_*, i_{ti*} = \frac{L_{adi}}{M_{ti}} i_*, i_{ki*} = \frac{L_{adi}}{M_{ki}} i_*, t_* = \frac{L_{k1}}{R_{k1}}, \tag{3.12.4}$$

where

$$L_{adi} = \frac{3}{2} \frac{M_{fi} M_{ti}}{M_{fti}}, \ L_{aqi} = \frac{3}{2} \frac{M_{ki}^2}{L_{ki}}, \ L_{afi} = \frac{M_{fi} M_{fti}}{M_{ti}}, \ L_{ati} = \frac{M_{ti} M_{fti}}{M_{fi}}. \tag{3.12.5}$$

For many synchronous machines the equality $L_{adi} = L_{aqi}$ holds with accuracy of few percents. Only this case is considered in the following. If $L_{adi} \neq L_{aqi}$ then there appear the same difficulties as in the case of considerable dissipation between contours t and f. These cases require additional consideration.

Under the above choice of basic values the "non-dimensional" equations have the form (the subscript u of dimensionless variables is omitted)

$$[(1 + \sigma_{d1})i_{a1} - \mu_1 i_{01} + (i_{f1} + i_{t1}) \cos\vartheta_1 - i_{k1} \sin\vartheta_1]^{\cdot} -$$
$$- \lambda[(1 + \sigma_{d2})i_{a2} - \mu_2 i_{02} + (i_{f2} + i_{t2}) \cos\vartheta_2 - i_{k2} \sin\vartheta_2]^{\cdot} +$$
$$+ \varepsilon_n \nu_{a1} i_{a1} - \varepsilon_n \nu_{a2} i_{a2} + \varepsilon_n \nu_0 i_0 = 0 \ (a, b, s),$$
$$[i_{di} + (i_{fi} + i_{ti}) + \varepsilon_r \sigma_{fi} i_{fi}]^{\cdot} + \varepsilon_f \nu_{fi} i_{fi} = \varepsilon_f e_{fi},$$
$$[i_{di} + (i_{fi} + i_{ti}) + \varepsilon_r \sigma_{ti} i_{ti}]^{\cdot} + \nu_{ti} i_{ti} = 0,$$
$$[i_{qi} + i_{ki}]^{\cdot} + \nu_{ki} i_{ki} = 0,$$
$$\omega_i^{\cdot} = -\varepsilon_{\omega i}[i_{di} i_{ki} - i_{qi}(i_{fi} + i_{ti})] + \varepsilon_{\omega i} m_i, \quad i = 1, 2,$$
$$[(1 + \sigma_{d1})i_{a1} - \mu_1 i_{01} + (i_{f1} + i_{t1}) \cos\vartheta_1 - i_{k1} \sin\vartheta_1]^{\cdot} +$$
$$+ x_n i_{an}^{\cdot} + \frac{\nu_{an}}{\varepsilon} i_{an} + \varepsilon \nu_0 i_{0n} + \varepsilon \nu_{01} i_{01} = 0 \qquad (a, b, c). \qquad (3.12.6)$$

Here

$$1 + \sigma_{di} = \frac{2L_i + M_i}{2L_{adi}}, \ \lambda = \frac{L_{ad2}}{L_{ad1}}, \ \mu_i = \frac{M_i}{2L_{adi}},$$

$$\varepsilon_n \nu_{\alpha i} = \frac{R_{ai} t_*}{L_{ad1}}, \ \varepsilon_n \nu_0 = \frac{R_0 t_*}{L_{adi}}, \ \varepsilon_f e_{fi} = \frac{E_{fi} t_*}{L_{afi} i_*}, \ \varepsilon_f \nu_{fi} = \frac{R_{fi} t_*}{L_{afi}},$$

$$\nu_{ti} = \frac{R_{ti} t_*}{L_{ati}}, \ \nu_{ki} = \frac{R_{ki} t_*}{L_{ki}}, \ \varepsilon_{\omega i} = \frac{3}{2} \frac{L_{adi} i_*^2 t_*}{J_i \Omega_*},$$

$$\varepsilon = \frac{1}{\omega_0 t_*}, \ 1 + \varepsilon_r \sigma_{fi} = \frac{L_{fi}}{L_{afi}}, \ 1 + \varepsilon_r \sigma_{ti} = \frac{L_{ti}}{L_{ati}},$$

$$h_n = \frac{L_n}{L_{ad}}, \ \frac{\nu_{an}}{\varepsilon} = \frac{R_n t_*}{L_{ad}}. \qquad (3.12.7)$$

Here we introduced the following denotations for the currents in d and q axes [119]

$$i_{di} = \frac{2}{3}\left[i_{ai}\cos\vartheta_i + i_{bi}\cos\left(\vartheta_i - \frac{2\pi}{3}\right) + i_{ci}\cos\left(\vartheta_i + \frac{2\pi}{3}\right)\right],$$

$$i_{qi} = -\frac{2}{3}\left[i_{ai}\sin\vartheta_i + i_{bi}\sin\left(\vartheta_i - \frac{2\pi}{3}\right) + i_{ci}\sin\left(\vartheta_i + \frac{2\pi}{3}\right)\right], \quad (3.12.8)$$

i_{di}, i_{qi} being considered as functions of i_{ai}, i_{bi}, i_{ci} rather than additional unknown variables.

Similar to the case of one machine the obtained dimensionless equations contain small parameters. This fact allows one to use the asymptotic methods of theory of nonlinear vibrations. As above, the basic small parameter is $\varepsilon = 1/\Omega_* t_*$ which is proportional to the ratio of the period of rotation to the time constant of contour k and small for all synchronous machines (for high-powerful generators the characteristic value $\varepsilon \simeq 0.02$). Another small parameter is ε_r characterising the leakage flux between the contour of the excitation winding and the damper contour t. The value of ε_r is small for most implicit-pole machines (the characteristic value is 0.05).

The analysis of technical requirements to synchronous machines shows that parameters $\varepsilon_f, \varepsilon_\omega$ are small (characteristic value is 0.1), however this fact is not further used for simplifications of equations. The value of parameter ε_n is not essential for application of the asymptotic method. The other dimensionless parameters in the above equations (including parameter σ_d characterising the leakage between the rotor and stator windings) are considerably larger than $\varepsilon, \varepsilon_r$ and considered as having the order of unity.

Instead of currents $i_{fi}, i_{ti}, i_{ki}, i = 1, 2$ we introduce the new variables $\Psi_{fi}, \Psi_{ri}, \Psi_{ki}, i = 1, 2$ in equations (3.12.6) by means of the following relationships

$$\Psi_{fi} = i_{di} + (i_{fi} + i_{ti}) + \varepsilon_r \sigma_{fi} i_{fi},$$
$$\varepsilon_r \Psi_{ri} = \varepsilon_r(\sigma_{fi} i_{fi} - \sigma_{ti} i_{ti}),$$
$$\Psi_{ki} = i_{qi} + i_{ki}. \tag{3.12.9}$$

The values Ψ_{fi}, Ψ_{ki} are dimensionless flux linkages in the rotor contours and Ψ_{ri} is proportional to leakage fluxes between the damper contour in axis t and the contour of the excitation winding. The dimensionless flux linkages of the stator contours are described by relations, see eq. (3.2.14)

$$\Psi_{ai} = (1 + \sigma_{di})i_{ai} + (i_{fi} + i_{ti})\cos\vartheta_i - i_{ki}\sin\vartheta_i, \ i = 1, 2 \ (a, b, c). \tag{3.12.10}$$

The stator flux linkages in axes α, β are as follows

$$\Psi_{\alpha i} = (1 + \sigma_{di})i_{\alpha i} + (i_{fi} + i_{ti})\cos\vartheta_i - i_{ki}\sin\vartheta_i,$$
$$\Psi_{\beta i} = (1 + \sigma_{di})i_{\beta i} + (i_{fi} + i_{ti})\sin\vartheta_i + i_{ki}\cos\vartheta_i. \tag{3.12.11}$$

In order to obtain eq. (3.12.11) we need to transform eq. (3.12.10) to axes $\alpha, \beta, 0$. The transformation $\alpha, \beta, 0$ of currents i_{ai}, i_{bi}, i_{ci} is determined by relations [119]

$$i_{\alpha i} = \frac{2}{3}[i_{ai} - \frac{1}{2}(i_{bi} + i_{ci})],$$

$$i_{\beta i} = \frac{1}{\sqrt{3}}(i_{bi} - i_{ci}),$$

$$i_{0i} = \frac{1}{3}(i_{ai} + i_{bi} + i_{ci}) \tag{3.12.12}$$

and the similar transformation takes place for the flux linkages. It is also necessary to relate the currents in systems α, β and d, q

$$i_{di} = i_{\alpha i}\cos\vartheta_i + i_{\beta i}\sin\vartheta_i,$$
$$i_{qi} = i_{\beta i}\cos\vartheta_i - i_{\alpha i}\sin\vartheta_i, \qquad i = 1, 2.$$

Let us introduce two new unknown variables

$$\Psi_\alpha = \Psi_{\alpha 1} - \lambda \Psi_{\alpha 2},$$
$$\Psi_\beta = \Psi_{\beta 1} - \lambda \Psi_{\beta 2}, \qquad (3.12.13)$$

and perform the transformation $\alpha, \beta, 0$ of the first three equations in (3.12.6) (from these equations only one equation for a phase is written down). After "0−transformation" the equations for i_{0i}, i_{0n}

$$(1 + \sigma d - \mu 1)i_{01} - \lambda(1 + \sigma_d - \mu_2)i_{02} + \varepsilon_n \nu_{a1} i_{01} - \varepsilon_n \nu_{a2} i_{02} + \frac{\nu_n}{\varepsilon} i_{0n} = 0,$$

$$i_{01} + i_{02} = i_{0n} \qquad (3.12.14)$$

are separated from the others. The remaining two equations obtained after $\alpha, \beta-$transformations have the form

$$\dot\Psi_\alpha + \varepsilon_n \nu_{a1} i_{\alpha 1} + \varepsilon_n \nu_{a2} i_{\alpha 2} = 0,$$
$$\dot\Psi_\beta + \varepsilon_n \nu_{a1} i_{\beta 1} + \varepsilon_n \nu_{a2} i_{\beta 2} = 0. \qquad (3.12.15)$$

Let us now express the currents i_{fi}, i_{ti}, i_{ki} in terms of the flux linkages and currents i_{di}, i_{qi}

$$i_{fi} = \frac{\Psi_{ri} + \sigma_{ti}(\Psi_{fi} - i_{di})}{\sigma_{fi} + \sigma_{ti} + \varepsilon_r \sigma_{fi} \sigma_{ti}},$$

$$i_{ti} = \frac{\sigma_{fi}(\Psi_{fi} - i_{di}) - (1 + \varepsilon_r \sigma_{fi})\Psi_{ri}}{\sigma_{fi} + \sigma_{ti} + \varepsilon_r \sigma_{fi} \sigma_{ti}},$$

$$i_{ki} = \Psi_{ki} - i_{qi}. \qquad (3.12.16)$$

Substituting expressions (3.12.16) into the "rotor" equations of system (3.12.6) yields the equations

$$\dot\Psi_{fi} + \varepsilon_f \nu_{fi} \frac{\sigma_{ti}(\Psi_{fi} - i_{di}) + \Psi_{ri}}{\sigma_{fi} + \sigma_{ti} + \varepsilon_r \sigma_{ti} \sigma_{fi}} = \varepsilon_f e_{fi}, \qquad \varepsilon_r \dot\Psi_{ri} +$$

$$+\varepsilon_f \nu_{fi} \frac{\sigma_{ti}(\Psi_{fi} - i_{di}) + \Psi_{ri}}{\sigma_{fi} + \sigma_{ti} + \varepsilon_r \sigma_{fi} \sigma_{ti}} - \nu_{ti} \frac{\sigma_{fi}(\Psi_{fi} - i_{di}) - \Psi_{ri}(1 + \varepsilon_r \sigma_{fi})}{\sigma_{fi} + \sigma_{ti} + \varepsilon_r \sigma_{fi} \sigma_{ti}} = \varepsilon_f e_{fi},$$

$$\dot\Psi_{ki} + \nu_{ki}(\Psi_{ki} - i_{qi}) = 0,$$

$$\dot\omega_i = -\varepsilon_{wi} \left\{ i_{di}(\Psi_{ki} - i_{qi}) - i_{qi} \frac{(\sigma_{ti} + \sigma_{fi})(\Psi_{fi} - i_{di}) - \varepsilon_r \sigma_{fi} \Psi_{ri}}{\sigma_{ti} + \sigma_{fi} + \varepsilon_r \sigma_{ti} \sigma_{fi}} + m_i \right\},$$

$$\vartheta_i = \frac{\omega_i}{\varepsilon}, \quad i = 1, 2. \qquad (3.12.17)$$

Equations of the contours with the load are not presented.

For further analysis we need the expressions for the flux linkage of the stator in terms of flux linkages Ψ_{ri}, Ψ_{ki} instead of currents i_{fi}, i_{ti}, i_{ki}

$$\Psi_{\alpha i} = (1 + \sigma_{di})i_{\alpha i} + \frac{(\sigma_{ti} + \sigma_{fi})(\Psi_{fi} - i_{di}) - \varepsilon_r \sigma_{fi}\Psi_{ri}}{\sigma_{ti} + \sigma_{fi} + \varepsilon_r \sigma_{ti}\sigma_{fi}} \cos\vartheta_i +$$
$$- (\Psi_{ki} - i_{qi})\sin\vartheta_i,$$
$$\Psi_{\beta i} = (1 + \sigma_{di})i_{\beta i} + \frac{(\sigma_{ti} + \sigma_{fi})(\Psi_{fi} - i_{di}) - \varepsilon_r \sigma_{fi}\Psi_{ri}}{\sigma_{ti} + \sigma_{fi} + \varepsilon_r \sigma_{ti}\sigma_{fi}} \sin\vartheta_i -$$
$$+ (\Psi_{ki} - i_{qi})\cos\vartheta_i. \quad (3.12.18)$$

Expressions for Ψ_{ai}, Ψ_{bi}, Ψ_{ci} differ from those in eq. (3.12.18) only in currents i_{ai}, i_{bi}, i_{ci} instead of $i_{\alpha i}$, $i_{\beta i}$.

After the replacement of all variables the unknown variables are now the flux linkages Ψ_α, Ψ_β, Ψ_{fi}, Ψ_{ri}, Ψ_{ki}, angles $\vartheta_1\vartheta_2$, angular velocities ω_1, ω_2 and the phase currents of stator of one of the machines. However in place of these currents it is more comfortable to use currents i_{an}, i_{bn}, i_{cn} in the load circuits which are related to the phase currents by eq. (3.12.3).

In order to obtain the system for the above variables we need to express them in terms of currents $i_{\alpha i}$, $i_{\beta i}$.

Let us obtain the expressions for currents $i_{\alpha i}$, $i_{\beta i}$. Neglecting the terms proportional to ε_r in eq. (3.12.18) yields

$$(\sigma_{d1} + \lambda\sigma_{d2})i_{\alpha 1} = \Psi_\alpha + \lambda\sigma_{d2}i_{an} - (\Psi_{f1}\cos\vartheta_1 - \Psi_{k1}\sin\vartheta_1) +$$
$$+\lambda(\Psi_{f2}\cos\vartheta_2 - \Psi_{k2}\sin\vartheta_2),$$
$$(\sigma_{d1} + \lambda\sigma_{d2})i_{\alpha 2} = -\Psi_\alpha + \sigma_{d1}i_{an} + (\Psi_{f1}\cos\vartheta_1 - \Psi_{k1}\sin\vartheta_1) -$$
$$-\lambda(\Psi_{f2}\cos\vartheta_2 - \Psi_{k2}\sin\vartheta_2),$$
$$(\sigma_{d1} + \lambda\sigma_{d2})i_{\beta 1} = \Psi_\beta + \lambda\sigma_{d2}i_{\beta n} - (\Psi_{f1}\sin\vartheta_1 + \Psi_{k1}\cos\vartheta_1) +$$
$$+\lambda(\Psi_{f2}\sin\vartheta_2 + \Psi_{k2}\cos\vartheta_2),$$
$$(\sigma_{d1} + \lambda\sigma_{d2})i_{\beta 2} = -\Psi_\beta + \sigma_{d1}i_{\beta n} + (\Psi_{f1}\sin\vartheta_1 + \Psi_{k1}\cos\vartheta_1) - \quad (3.12.19)$$
$$-\lambda(\Psi_{f2}\sin\vartheta_2 + \Psi_{k2}\cos\vartheta_2).$$

Currents $i_{\alpha n}$, $i_{\beta n}$ in this equations are obtained by α, β–transformation of currents i_{an}, i_{bn}, i_{cn}.

Let us now apply $\alpha, \beta, 0$–transformation of the equations for the first machine and the load. In the case of the symmetric three-phase load the equation for "zero" currents i_{01}, i_{0n} is separated and given by

$$(1 + \sigma_{d1} - \mu_1)i_{0n} + \varepsilon_n\nu_{a1}i_{01} + \frac{\nu_n}{\varepsilon}i_{0n} = 0. \quad (3.12.20)$$

The other two equations (in axes α, β) take the form

$$[(1+\sigma_{d1})i_{\alpha 1} + (i_{f1}+i_{t1})\cos\vartheta_1 - i_{k1}\sin\vartheta_1]^{\cdot} + x_n i_{\alpha n} + \frac{\nu_n}{\varepsilon}i_{\alpha n} = 0,$$

$$[(1+\sigma_{d1})i_{\beta 1} + (i_{f1}+i_{t1})\sin\vartheta_1 + i_{k1}\cos\vartheta_1]^{\cdot} + x_n i_{\beta n} + \frac{\nu_n}{\varepsilon}i_{\beta n} = 0.$$

$$(3.12.21)$$

Substituting expressions for currents i_{f1}, i_{t1} in terms of the flux linkages (3.12.16) and relations (3.12.19) for currents $i_{\alpha 1}, i_{\beta 1}$ and neglecting the small elements of the order of $\varepsilon_r, \varepsilon_n$ we arrive at the equations

$$\frac{\lambda\sigma_{d1}\sigma_{d2}}{\sigma_{d1}+\lambda\sigma_{d2}}i_{\alpha n}^{\cdot} + \frac{\lambda\sigma_{d2}}{\sigma_{d1}+\lambda\sigma_{d2}}(\Psi_{f1}\cos\vartheta_1 - \Psi_{k1}\sin\vartheta_1)^{\cdot} +$$

$$+\frac{\lambda\sigma_{d1}}{\sigma_{d1}+\lambda\sigma_{d2}}(\Psi_{f2}\cos\vartheta_2 - \Psi_{k2}\sin\vartheta_2)^{\cdot} + x_n i_{\alpha n} + \frac{\nu_n}{\varepsilon}i_{\alpha n} = 0,$$

$$\frac{\lambda\sigma_{d1}\sigma_{d2}}{\sigma_{d1}+\lambda\sigma_{d2}}i_{\beta n}^{\cdot} + \frac{\lambda\sigma_{d2}}{\sigma_{d1}+\lambda\sigma_{d2}}(\Psi_{f1}\sin\vartheta_1 + \Psi_{k1}\cos\vartheta_1)^{\cdot} +$$

$$+\frac{\lambda\sigma_{d1}}{\sigma_{d1}+\lambda\sigma_{d2}}(\Psi_{f2}\sin\vartheta_2 + \Psi_{k2}\cos\vartheta_2)^{\cdot} + x_n i_{\beta n} + \frac{\nu_n}{\varepsilon}i_{\beta n} = 0. \quad (3.12.22)$$

The derivatives of the sums in brackets in eq. (3.12.22), express the emf in the phase windings of stators in axes α, β

$$E_{\alpha i} = -(\Psi_{fi}\cos\vartheta_i - \Psi_{ki}\sin\vartheta_i)^{\cdot},$$

$$E_{\beta i} = -(\Psi_{ki}\cos\vartheta_i + \Psi_{fi}\sin\vartheta_i)^{\cdot}. \quad (3.12.23)$$

The sum of the first three terms in eq. (3.12.22) determines the joint voltage drop across two parallel branches of phase windings including inductance σ_{di} and emf $E_{ai}(a,b,s)$ transformed to axes α, β. Indeed, by virtue of Kirchhoff relations for two parallel branches we have the condition of equal voltages

$$\sigma_{d1}i_{a1}^{\cdot} + E_{a1} = \lambda\sigma_{d2}i_{a2}^{\cdot} + \lambda E_{a2}, \quad (a,b,c)$$

and the obvious relation for the currents

$$i_{a1} + i_{a2} = i_{an} \quad (a,b,c).$$

Now we can express i_{a1}, i_{a2} in terms of i_{an} and obtain the expression for voltage $\sigma_{d1}i_{ai}^{\cdot} + E_{ai}$ in terms of i_{an}. This expression provides us with parameters of the branch which is equivalent to the circuit of two parallel branches with a common load with respect to current and voltage

$$E_a = \frac{\lambda\sigma_{d1}E_{a2} + \lambda\sigma_{d2}E_{a1}}{\sigma_{d1}+\lambda\sigma_{d2}}, \quad (a,b,c),$$

$$\sigma_d = \frac{\lambda\sigma_{d1}\sigma_{d2}}{\sigma_{d1}+\lambda\sigma_{d2}}. \quad (3.12.24)$$

One can see from eqs. (3.12.17) and (3.12.22) that the sought-for variables are divided into two groups: slow variables having derivatives of the order of unity and fast variables with derivatives proportional to $1/\varepsilon$ and $1/\varepsilon_r$. The fast variables are flux linkage Ψ_r, angles ϑ_1, ϑ_2 and the load currents i_{an}, i_{bn}, i_{cn}. According to technical meaning of the problem only those processes are of interest under which the frequencies of rotation of both machines are close (in the theory of asymptotic methods such motions are called "main resonance" or "resonance of order 1:1"). Considering these motions we introduce new variables δ and S by the relationships $\delta = \vartheta_2 - \vartheta_1$, $\sqrt{\varepsilon \varepsilon_{\omega 1}} S = \omega_2 - \omega_1$ (S is proportional to sliding). Instead of two last equations in (3.12.17) we have

$$\dot{\delta} = \sqrt{\frac{\varepsilon_{\omega 1}}{\varepsilon}} S, \tag{3.12.25}$$

$$\dot{S} = \sqrt{\frac{\varepsilon_{\omega 1}}{\varepsilon}} \left\{ i_{d1}(\Psi_{k1} - i_{q1}) - i_{q1} \frac{(\sigma_{t1} + \sigma_{f1})(\Psi_{f1} - i_{d1}) - \varepsilon_r \sigma_{f1} \Psi_{r1}}{\sigma_{f1} + \sigma_{t1} + \varepsilon_r \sigma_{f1}\sigma_{t1}} + m_1 \right\}$$
$$- \sqrt{\frac{\varepsilon_{\omega 2}}{\varepsilon}} \left\{ i_{d2}(\Psi_{k2} - i_{q2}) - i_{q2} \frac{(\sigma_{t2} + \sigma_{f2})(\Psi_{f2} - i_{d2}) - \varepsilon_r \sigma_{f2} \Psi_{r2}}{\sigma_{f2} + \sigma_{t2} + \varepsilon_r \sigma_{f2}\sigma_{t2}} + m_2 \right\}.$$

Thus, we arrived at the system with one fast phase ϑ_1 and parameter $\varepsilon_{\omega 1}/\varepsilon$. The values δ and S should be attributed to slow variables (their "velocities" are of the order of $\varepsilon_{\omega 1}/\varepsilon$ rather than $1/\varepsilon$). The fast variables are ϑ_1, Ψ_r and currents i_a, i_b, i_c, these currents having the frequency equal to the frequency of rotation. The slow variables are the remaining flux linkages and frequency ω_1.

Let us now apply the asymptotic method of separation of the system motion with many fast variables [128] to the obtained equations. Let us consider the first approximation. The calculation scheme in this approximation is as follows. From the system of equations we select a "fast" subsystem, i.e. the equations with derivatives of the fast variables. This subsystem needs to be integrated, the slow variables being considered as time-independent values. It is sufficient to find a particular solution which is a periodic function of ϑ_1. The obtained solutions need to be substituted into the "slow" subsystem and the right hand sides of the obtained equations are to be averaged over ϑ_1. It results in the equations for slow nonstationary processes.

The first stage which is determining the fast variables under constant slow variables includes calculation of currents $i_{\alpha n}, i_{\beta n}$. As follows from eq. (3.12.22), for $\Psi_{fi}, \Psi_{ki} = \text{const}$ this problem reduces to computation of vibrations in three-phase linear circuit with resistances and inductance subjected to harmonic emf. By introducing argument ϑ_1 the equations for this circuit transformed to the axes α, β have the form

$$\omega_1 \sigma \frac{di_\alpha}{d\vartheta_1} + \nu_n i_\alpha = \gamma_1 \omega_1 \Psi_{m1} \sin(\vartheta_1 + \varphi_1) + \gamma_2 \omega_2 \Psi_{m2} \sin(\vartheta_1 + \delta + \varphi_2),$$

$$\omega_1 \sigma \frac{di_\beta}{d\vartheta_1} + \nu_n i_\beta = -\gamma_1 \omega_1 \Psi_{m1} \cos(\vartheta_1 + \varphi_1) - \gamma_2 \omega_2 \Psi_{m2} \sin(\vartheta_1 + \delta + \varphi_2),$$

$$(3.12.26)$$

where

$$\Psi_{mi} = \sqrt{\Psi_{ki}^2 + \Psi_{fi}^2}, \ \tan\varphi_i = \frac{\Psi_{ki}}{\Psi_{fi}},$$

$$\gamma_1 = \frac{\lambda \sigma_{d2}}{\sigma_{d1} + \lambda \sigma_{d2}}, \ \gamma_2 = \frac{\lambda \sigma_{d1}}{\sigma_{d1} + \lambda \sigma_{d2}}, \ \sigma = \sigma_d + x_n.$$

In order to derive the averaged equations in terms of the slow variables it is necessary to find the solution of this problem which is $2\pi-$periodic in ϑ_1. By virtue of linearity of equations (3.12.26) this solution is unique and asymptotically stable. The exponentially attenuating terms do not affect the averaged slow equations.

The periodic solution for currents $i_{\alpha n}, i_{\beta n}$ have the form

$$i_{\alpha p} = \gamma_1 \Psi_{m1}[S_1(\omega_1)\sin(\vartheta_1 + \varphi_1) - S_2(\omega_1)\cos(\vartheta_1 + \varphi_1)]+$$
$$+ \gamma_2 \Psi_{m2}\frac{\omega_2}{\omega_1}[S_1(\omega_1)\sin(\vartheta_1 + \delta + \varphi_2) - S_2(\omega_1)\cos(\vartheta_1 + \delta + \varphi_2)],$$

$$i_{\beta r} = -\gamma_1 \Psi_{m1}[S_1(\omega_1)\cos(\vartheta_1 + \varphi_1) + S_2(\omega_1)\sin(\vartheta_1 + \varphi_1)]-$$
$$- \gamma_2 \Psi_{m2}\frac{\omega_2}{\omega_1}[S_1(\omega_1)\cos(\vartheta_1 + \delta + \varphi_2) + S_2(\omega_1)\sin(\vartheta_1 + \delta + \varphi_2)],$$

$$(3.12.27)$$

where

$$S_1(\omega) = \frac{\nu_n \omega}{\omega^2 \sigma^2 + \nu_n^2}, \ S_2(\omega) = \frac{\sigma \omega^2}{\omega^2 \sigma^2 + \nu_n^2}, \ \omega_2 = \omega_1 + \sqrt{\varepsilon \varepsilon_\omega} S.$$

In the first approximation the fast variables i_{an}, i_{bn}, i_{cn} can be considered as known functions of ϑ_1 and the slow variables. Then by virtue of relations (3.12.19) the currents in the stator phases of both machines i_{ai}, i_{bi}, i_{ci}, and consequently currents i_{di}, i_{qi} should be viewed as known functions of phase ϑ_1 and the slow variables

$$i_{d1} = \frac{1}{\sigma_{d1} + \lambda \sigma_{d2}}(\Psi_\alpha \cos\vartheta_1 + \Psi_\beta \sin\vartheta_1) + \frac{\lambda \sigma_{d2}}{\sigma_{d1} + \lambda \sigma_{d2}}(i_{\alpha n}\cos\vartheta_1 +$$

$$+ i_{\beta n}\sin\vartheta_1) - \frac{1}{\sigma_{d1} + \lambda \sigma_{d2}}\Psi_{f1} + \frac{\lambda}{\sigma_{d1} + \lambda \sigma_{d2}}(\Psi_{f2}\cos\delta - \Psi_{k2}\sin\delta),$$

$$i_{q1} = \frac{\lambda \sigma_{d2}}{\sigma_{d1} + \lambda \sigma_{d2}}(\Psi_\beta \cos\vartheta_1 - \Psi_\alpha \sin\vartheta_1) + \frac{\lambda \sigma_{d2}}{\sigma_{d1} + \lambda \sigma_{d2}}(i_{\beta n}\cos\vartheta_1 -$$

$$- i_{\alpha n}\sin\vartheta_1) - \frac{1}{\sigma_{d1} + \lambda \sigma_{d2}}\Psi_{k1} + \frac{\lambda}{\sigma_{d1} + \lambda \sigma_{d2}}(\Psi_{f2}\sin\delta + \Psi_{k2}\cos\delta),$$

$$i_{d2} = -\frac{1}{\sigma_{d1} + \lambda\sigma_{d2}}(\Psi_\alpha \cos\vartheta_2 + \Psi_\beta \sin\vartheta_2) + \frac{\sigma_{d1}}{\sigma_{d1} + \lambda\sigma_{d2}}(i_{\alpha n}\cos\vartheta_2 +$$

$$+i_{\beta n}\sin\vartheta_2) + \frac{\lambda}{\sigma_{d1} + \lambda\sigma_{d2}}\Psi_{f2} + \frac{1}{\sigma_{d1} + \lambda\sigma_{d2}}(\Psi_{f1}\cos\delta + \Psi_{k1}\sin\delta),$$

$$i_{q2} = -\frac{1}{\sigma_{d1} + \lambda\sigma_{d2}}(\Psi_\beta \cos\vartheta_2 - \Psi_\alpha \sin\vartheta_2) + \frac{\sigma_{d1}}{\sigma_{d1} + \lambda\sigma_{d2}}(i_{\beta n}\cos\vartheta_2 -$$

$$-i_{\alpha n}\sin\vartheta_2) - \frac{\lambda}{\sigma_{d1} + \lambda\sigma_{d2}}\Psi_{k2} - \frac{1}{\sigma_{d1} + \lambda\sigma_{d2}}(\Psi_{f1}\sin\delta - \Psi_{k1}\cos\delta).$$

$$(3.12.28)$$

Now we need to consider the flux linkage Ψ_{ri} among the fast variables. Following the averaging method it is necessary to find the general solution of the third and fourth equations in (3.12.17) at $\Psi_{fi} = \text{const}$ by substituting the periodic solution for currents i_{di}, i_{qi} obtained from the first approximation to calculation of equivalent circuits. Under the "frozen" slow variables and prescribed $2\pi-$periodic in ϑ_1 currents i_{di}, i_{qi} determined by substituting the steady-state solutions of equations (3.12.17) $i_{\alpha p}, i_{\beta r}$ into eq. (3.12.19) the equations for Ψ_{ri} have the structure of linear differential equations with the $2\pi-$periodic right hand side

$$\omega_1 \frac{\varepsilon_r}{\varepsilon} \frac{d\Psi_{ri}}{d\vartheta_1} + \frac{\nu_{ti} + \varepsilon_f \nu_{fi}}{\sigma_{li} + \sigma_{fi}}\Psi_{ri} = \frac{\nu_{ti}\sigma_{fi} - \varepsilon_{fi}\nu_{fi}\sigma_{ti}}{\sigma_{li} + \sigma_{fi}}(\Psi_{fi} - i_{di}) + \varepsilon_{fi}e_{fi}.$$

$$(3.12.29)$$

There is no need to calculate the exponentially attenuating homogeneous solution as it disappears under averaging equations (3.12.17). When small terms of the first order only are considered, the slow equations of system (3.12.17) are linear in variables $\Psi_{ri}, i_{di}, i_{qi}$, that is, the averaged equations requires only the mean values of the fast variables

$$\langle i_{d1}\rangle = \frac{\lambda\sigma_{d2}}{\sigma_{d1} + \lambda\sigma_{d2}}\langle i_{\alpha p}\cos\vartheta_1 + i_{\beta r}\sin\vartheta_1\rangle - \frac{1}{\sigma_{d1} + \lambda\sigma_{d2}}\Psi_{f1} +$$

$$+ \frac{\lambda}{\sigma_{d1} + \lambda\sigma_{d2}}(\Psi_{f2}\cos\delta - \Psi_{k2}\sin\delta),$$

$$\langle i_{q1}\rangle = \frac{\lambda\sigma_{d2}}{\sigma_{d1} + \lambda\sigma_{d2}}\langle i_{\beta p}\cos\vartheta_1 - i_{\alpha r}\sin\vartheta_1\rangle - \frac{1}{\sigma_{d1} + \lambda\sigma_{d2}}\Psi_{k1} +$$

$$+ \frac{\lambda}{\sigma_{d1} + \lambda\sigma_{d2}}(\Psi_{f2}\sin\delta + \Psi_{k2}\cos\delta),$$

$$\langle i_{d2} \rangle = \frac{\sigma_{d1}}{\sigma_{d1} + \lambda \sigma_{d2}} \langle i_{\alpha p} \cos \vartheta_2 + i_{\beta r} \sin \vartheta_2 \rangle - \frac{\lambda}{\sigma_{d1} + \lambda \sigma_{d2}} \Psi_{f2} +$$

$$+ \frac{1}{\sigma_{d1} + \lambda \sigma_{d2}} (\Psi_{f1} \cos \delta + \Psi_{k1} \sin \delta),$$

$$\langle i_{q2} \rangle = \frac{\sigma_{d1}}{\sigma_{d1} + \lambda \sigma_{d2}} \langle i_{\beta p} \cos \vartheta_2 - i_{\alpha r} \sin \vartheta_2 \rangle - \frac{1}{\sigma_{d1} + \lambda \sigma_{d2}} \Psi_{k2} -$$

$$- \frac{1}{\sigma_{d1} + \lambda \sigma_{d2}} (\Psi_{f1} \sin \delta - \Psi_{k1} \cos \delta), \qquad (3.12.30)$$

$$\langle \Psi_{ri} \rangle = \frac{\nu_{ti} \sigma_{fi}}{\nu_{ti} + \varepsilon_{fi} \nu_{fi}} (\Psi_{fi} - \langle i_{di} \rangle) + \frac{\sigma_{ti} + \sigma_{fi}}{\nu_{ti} + \varepsilon_{fi} \nu_{fi}} \varepsilon_{fi} e_{fi}.$$

For determining the above mean currents one needs to average $\langle i_{\alpha p} \cos \vartheta_i + i_{\beta p} \sin \vartheta_i \rangle$, $\langle i_{\beta p} \cos \vartheta_i - i_{\alpha p} \sin \vartheta_i \rangle$ over the period of fast phase ϑ_1 which is easy to perform after substituting $i_{\alpha r}, i_{\beta p}$ from eq. (3.12.27)

$$\langle i_{\alpha p} \cos \vartheta_1 + i_{\beta p} \sin \vartheta_1 \rangle = \gamma_1 [S_1(\omega_1) \Psi_{k1} - S_2(\omega_1) \Psi_{f1}] +$$

$$+ \gamma_2 \frac{\omega_2}{\omega_1} [S_1(\omega_1)(\Psi_{k2} \cos \delta + \Psi_{f2} \sin \delta) - S_2(\omega_1)(\Psi_{f2} \cos \delta - \Psi_{k2} \sin \delta)],$$

$$\langle i_{\beta p} \cos \vartheta_1 - i_{\alpha p} \sin \vartheta_1 \rangle = -\gamma_1 [S_1(\omega_1) \Psi_{f1} + S_2(\omega_1) \Psi_{k1}] -$$

$$- \gamma_2 \frac{\omega_2}{\omega_1} [S_1(\omega_1)(\Psi_{f2} \cos \delta - \Psi_{k2} \sin \delta) + S_2(\omega_1)(\Psi_{k2} \cos \delta + \Psi_{f2} \sin \delta)],$$

$$\langle i_{\alpha p} \cos \vartheta_2 + i_{\beta p} \sin \vartheta_2 \rangle = \gamma_1 [S_1(\omega_1)(\Psi_{k1} \cos \delta - \Psi_{f1} \sin \delta) -$$

$$- S_2(\omega_1)(\Psi_{f1} \cos \delta + \Psi_{k1} \sin \delta)] + \gamma_2 \frac{\omega_2}{\omega_1} [S_1(\omega_1) \Psi_{k2} - S_2(\omega_1) \Psi_{f2}],$$

$$\langle i_{\beta r} \cos \vartheta_2 - i_{\alpha p} \sin \vartheta_2 \rangle = -\gamma_1 [S_1(\omega_1)(\Psi_{f1} \cos \delta + \Psi_{k1} \sin \delta) +$$

$$+ S_2(\omega_1)(\Psi_{k1} \cos \delta - \Psi_{f1} \sin \delta)] - \gamma_2 \frac{\omega_2}{\omega_1} [S_1(\omega_1) \Psi_{f2} + S_2(\omega_1) \Psi_{k2}].$$

$$(3.12.31)$$

Now we can write the equations for slow nonstationary processes which remain after separation of the stator equations (3.12.15)

$$\dot{\Psi}_{fi} + \varepsilon_{fi} \nu_{fi} (\Psi_{fi} - \langle i_{di} \rangle) - \varepsilon_{fi} e_{fi} = 0,$$

$$\dot{\Psi}_{ki} + \nu_{ki} (\Psi_{ki} - \langle i_{qi} \rangle) = 0, \ i = 1, 2,$$

$$\dot{\omega}_1 = -\varepsilon_{\omega 1} \{ \Psi_{k1} \langle i_{d1} \rangle - \Psi_{f1} \langle i_{q1} \rangle \} + \varepsilon_{\omega 1} m_1,$$

$$\dot{S} = \sqrt{\frac{\varepsilon_{\omega 1}}{\varepsilon}} \{ \Psi_{k1} \langle i_{d1} \rangle - \Psi_{f1} \langle i_{q1} \rangle \} - \sqrt{\frac{\varepsilon_{\omega 1}}{\varepsilon}} m_1 -$$

$$- \sqrt{\frac{\varepsilon_{\omega 2}}{\varepsilon}} \{ \Psi_{k2} \langle i_{d2} \rangle - \Psi_{f2} \langle i_{q2} \rangle \} + \sqrt{\frac{\varepsilon_{\omega 2}}{\varepsilon}} m_2,$$

$$\dot{\delta} = \sqrt{\frac{\varepsilon_{\omega 1}}{\varepsilon}} S, \qquad (3.1.2.32)$$

where the mean values $\langle i_{di}\rangle, \langle i_{qi}\rangle$ are determined by relations (3.12.30), (3.12.31) and are functions of slow variables.

While studying the possible motions admitted by the averaged equations (3.12.32) we limit our consideration to the most important case of the identical generators. We will consider two cases: (i) the torques of the drive motors are identical and (ii) they differ in value.

The analysis of motions in the first case is actually reduced to the earlier problem of a single generator operating into active-inductive load. As follows from system (3.12.12) $w_1^{(s)} = w_2^{(s)}$ in the steady-state regime, thus by virtue of symmetry of the system at $m_1 = m_2$ the like variables for different machines are equal in pairs: $\Psi_{f1}^{(s)} = \Psi_{f2}^{(s)}, \Psi_{k1}^{(s)} = \Psi_{k2}^{(s)}$ and the error angle $\delta^{(s)} = 0$. In this case the steady-state flux linkages are given by the formulae

$$
\Psi_{fi}^{(s)} = \frac{\left[1 + \frac{1}{2}S_2(\omega^{(s)})\right] e_f/\nu_f}{\left[1 + \frac{1}{2}S_2(\omega^{(s)})\right]^2 + \frac{1}{4}S_1^2(\omega^{(s)})},
$$

$$
\Psi_{ki}^{(s)} = -\frac{\frac{1}{2}S_1(\omega^{(s)})c_f/\nu_f}{\left[1 + \frac{1}{2}S_2(\omega^{(s)})\right]^2 + \frac{1}{4}S_1^2(\omega^{(s)})}, \qquad (3.12.33)
$$

that within the coefficient of $1/2$ in front of S_1, S_2 coincides with the expressions of flux linkages given by eq. (3.7.6) for a single generator.

Calculation of steady-state frequency $\omega^{(s)}$ reduces to solving the algebraic equation

$$
\frac{\frac{1}{2}S_1(\omega^{(s)})(e_f/\nu_f)^2}{\left[1 + \frac{1}{2}S_2(\omega^{(s)})\right]^2 + \frac{1}{4}S_1^2(\omega^{(s)})} = m(\omega^{(s)}). \qquad (3.12.34)
$$

Dependence of the left hand side of eq. (3.12.34) on frequency $\omega^{(s)}$ has the same properties that the expression of electromagnetic moment (3.6.16) for one generator. The same can be said about the character of the singular points of equations (3.12.32) determined by the values of steady-state frequency obtained from eq. (3.12.34).

In the considered problem one fails to reduce the averaged equations to the system the second order, which is in contrast to the case of one generator. Numeral integration of system (3.12.32) for the initial conditions corresponding to machine starting to operate into load from the regime of no load allows one to highlight the following regularities:

1. Provided that the torques are equal to each other $(m_1(\omega) = m_2(\omega))$ then the synchronisation of generators occurs under any motions, that is, their angular velocity become coincident.

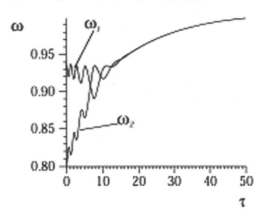

FIGURE 3.20.

2. Under the initial angular velocities which are smaller than the steady-state frequency corresponding to the saddle point of equilibrium, the angular velocities of both machines become equal to the steady-state value of frequency given by eq. (3.12.34) and corresponding to the stable nodal point of equilibrium (Fig. 3.20).

3. If $m_1 = m_2 = $ const and one of the (or both) initial velocities $w_i(0)$ is larger than the "saddle" frequency then $w_i, i = 1, 2$ increases unboundedly, the difference in velocities of both machines vanishing.

4. In the case of $m_i = m_{i0} - \beta_i w_i$, where $m_{10} = m_{20}$, the unbounded motions are impossible. The motions with the initial angular velocities exceeding the "saddle" frequency (one or both) are established in the point corresponding to the second stable nodal point of equilibrium.

While analysing motions in the case of $m_1 \neq m_2$ we limit our consideration to the most important problem, namely study of the transient process caused by disconnection of one of the motors during a steady-state regime.

Numeral solution of this problem, shown in Fig. 3.21 was carried out as follows. For the initial conditions corresponding to the steady-state values of variables in the case of $m_1 = m_2$, the Cauchy problem was solved for the case when one of moments m_i vanishes. The numeral solution demonstrated that after a transient process a new steady-state regime of synchronously operating generator was established, its stationary frequency being lower than the initial one. Such regime of operation of generators with one disconnected motor is known in the electrical engineering and referred to as the regime of "electric shaft".

In conclusion we notice that this method can also be applied, with small modification of calculation of fast variables, to investigation of dynamics of the system of two generators operating into the active-inductive load via rectifier. In this case the calculation of currents in the load reduces to determining the steady-state, 2π—periodic in phase ϑ_1, regime in the circuit consisting of equivalent harmonic emf (3.6.5) operating into load

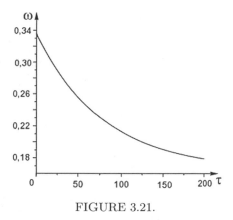

FIGURE 3.21.

via a three-phase rectifier (e.g. of Larionov's type [56]). The solution of this problem is known (see, for example [95]). When the currents of the rectifier input are determined as functions of slow variables and the fast phase ϑ_1 the further analysis is performed similar to the case of machine operating directly into load.

3.13   Equation for electromechanical processes in inductor machines

The inductor machine is an electric machine whose magnetic induction at any point of the operating gap is changing only in value while its direction is not changing. The anchor winding and the excitation winding are placed on the stator and the change in magnetic flux coupled with the anchor winding is due to the periodic change in magnetic conductivity of the air-gap under rotation of the tooth rotor. The inductor machines are extensively used as generators of current of higher frequency (400 Hz and higher) and autonomous sources of power supply in radar and gyroscopic devices, for induction heating etc.

Despite the extensive literature on the theory of inductor machines (see, for example [3, 4, 24]) the equations of transient processes in the inductor machines were derived only for special sorts of the anchor windings with a large number of slots [28]. In this case the coefficients of self-induction and mutual induction allows one to apply Park's transformation [119]. The equations of electromechanical processes take the form of the equations of "standard" synchronous electric machines.

The present section suggests derivation, asymptotic transformation and qualitative analysis of equations of electromechanical processes in the inductor generators and motors in the cases in which these equations are not reduced to the Park-Gorev equations which are well-known in the theory

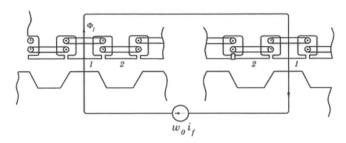

FIGURE 3.22.

of synchronous electric machines. In the following we suggest a method
of asymptotic simplification of equations of electromechanical processes in
the single-phase and three-phase inductor generators and motors with different types of the anchor windings that allows one to leave fast transient
processes out of consideration. As a result we obtain the equations which
are substantially more simple and appropriate for qualitative and numeral
analysis.

The obtained results can be used for computation of nonstationary processes under load-on and load-off, when the machine operates as a generator
of brief action with variable angular velocity etc. In particular, calculation
of nonstationary processes are needed for design of inductor generators for
power supply of space vehicles and electrophysical plants.

Let us consider a machine with a balanced anchor winding operating as
a generator into active-inductive load. (The winding is called balanced if
the mutual inductances between the anchor windings and the excitation
has no constant component [33].) The winding is assumed to consist of
series spools embracing all stator teeth. The stator tooth corresponds to
a half-period of the rotor tooth, and the subtractive polarity is used for
neighbouring spools in pack and the spools in different packs identically
located relative to the rotor teeth, see Fig. 3.22.

For the balanced windings of other types (see for example [33]) the equations in the dimensionless variables have the form of the above winding. The
only difference is the equations relating dimensional and non-dimensional
values. The single-pack machines with opened stator slots and machines
with the opposite poles are considered in Section 3.15.

The magnetic permeability of steel is considered as being infinitely large
and the influence of vortex currents and leakage currents is not taken into
account. The equations can be refined by taking into account these factors,
in particular the vortex currents can be taken into account by introducing
an additional equivalent contour [28].

The form of teeth and cavities of the rotor and hence the dependence of magnetic conductance on the angle of the rotor rotation are not specified. In what follows we consider several problems of choosing these parameters.

Calculation of magnetic conductance presents a separate problem which has been adequately explored (see for example [3]). In the following we assume that the dependence of conductance on the angle of the rotor rotation has been determined before we proceed to solving the equations of electromechanical processes for a particular machine.

The properties of equations of the machine do not change if the inductances are found from calculation of the field without using the concept of magnetic conductivity. It is even possible to keep denotations in the equations provided that we formally introduce the conductivity by dividing the inductance by the appropriate constant values.

Let us construct the expressions for the flux linkages Ψ and Ψ_f of the anchor contours and the excitation contour. Let i and i_f designate the currents in these contours and let w and w_0 denote respectively the numbers of turns of an anchor spool and the excitation windings. There is no need to construct the complete system of Kirchhoff equations for the magnet circuit. It is enough to consider the contour with gaps 1 and 1' in different packs as an independent circuit contour (Fig. 3.22). In this picture the stator packs are conditionally unfolded in a plane and the path of magnetic thread is shown in the contour under consideration. The excitation winding is located on the stator between the generator packs, its axis being coincident with the rotor axis. This winding causes mmf of excitations $w_0 i_f$ in this contour that is also shown in Fig. 3.22. The magnetic fluxes through gaps 1 and 1' are identical, therefore flux Φ_1 through gap 1 is

$$\Phi_1 = (2wi + w_0 i_f)\frac{g_1}{2} = \left(wi + \frac{1}{2}w_0 i_f\right)g_1, \qquad (3.13.1)$$

where $2wi + w_0 i_f$ is the total mmf in the contour and g_1 denotes the magnetic conductivity of gap 1.

When determining flux Φ_2 through gap 2 it is necessary to take into account that the neighbouring anchor spools have subtractive polarity. Designating the conductivity of this gap by g_2 we obtain

$$\Phi_2 = \left(\frac{1}{2}w_0 i_f - wi\right)g_2. \qquad (3.13.2)$$

The fluxes through the other gaps under the anchor spools are equal cither Φ_1 or Φ_2. Now it is possible to find the flux linkages of the anchor contours and the excitation winding

$$\Psi = 2nw(\Phi_1 - \Phi_2) + Li, \quad \Psi_f = nw_0(\Phi_1 + \Phi_2), \qquad (3.13.3)$$

where n denotes the number of indents in one pack of the rotor and L is the load inductance. The sign in front of Φ_2 in the expression for Ψ is explained by the subtractive polarity of the neighbouring anchor spools. Inserting eqs. (3.13.1) and (3.13.2) into eq. (3.13.3) yields

$$\Psi = [2nw^2(g_1 + g_2) + L]i + nww_0(g_1 - g_2)i_f,$$

$$\Psi_f = nww_0(g_1 - g_2)i + \frac{1}{2}nw_0^2(g_1 + g_2)i_f. \tag{3.13.4}$$

The coefficients in front of currents in eq. (3.13.4) are equal to the coefficients of self-induction and mutual induction of the contours of currents i and i_f.

The conductivities g_1, g_2 are changing under rotation of the rotor and are $2\pi-$periodic functions of angle φ related to the angle of rotor rotation ϑ by formula $\varphi = n\vartheta$. Let us indicate some properties of functions $g_1(\varphi)$ and $g_2(\varphi)$. These properties are correct regardless of the method of calculation of the field and magnetic conductivity (see for example [3]). The important properties of $g_1(\varphi)$ and $g_2(\varphi)$ become obvious for the simple method of calculation of conductivity for the in-plane field in gaps and the following expressions for $g_1(\varphi)$ and $g_2(\varphi)$

$$g_1(\varphi) = \mu_0 bR_s \int\limits_{-\varphi}^{\pi-\varphi} \frac{d\psi}{n\Delta(\psi)}, \quad g_2(\varphi) = \mu_0 bR_s \int\limits_{\pi-\varphi}^{2\pi-\varphi} \frac{d\psi}{n\Delta(\psi)}. \tag{3.13.5}$$

Here b is the length of the package along the machine axis, R_c is the radius of boring of the stator, μ_0 is the magnetic permeability of air, $\Delta(\psi)$ is the gap between the stator and the rotor. Angle ψ is measured from the mobile axis fixed in the rotor, one tooth corresponding to period of 2π in ψ. It is assumed that at $\varphi = 0$ the mobile axis is coincident with the immobile one passing the left edge of slot 1. Under rotation of the rotor the angle between axes is equal to φ.

Magnetic conductivities given by eq. (3.13.5) satisfy the relations

$$g_2(\varphi) = g_1(\varphi - \pi),$$

$$g_1(\varphi) + g_2(\varphi) = \mu_0 bR_s \int\limits_{-\varphi}^{2\pi-\varphi} \frac{d\psi}{n\Delta(\psi)} = g_0 = \text{const}. \tag{3.13.6}$$

Functions $g_1(\varphi)$ and $g_2(\varphi)$ can be set as follows $g_1(\varphi) = 1/2\,(g_0 + g(\varphi))$, $g_2(\varphi) = 1/2\,(g_0 - g(\varphi))$ where $g(\varphi)$ denotes a function having only odd harmonics in the Fourier decomposition.

It follows from eq. (3.13.4) and relation $g_1 - g_2 = g$ that the mutual induction between the anchor winding and the excitation winding has no constant component, i.e. the anchor winding is balanced, indeed .

Having obtained the induction coefficients it is possible to write down the expression for the energy of magnetic field

$$W = \frac{1}{2}(2nw^2 g_0 + L)i^2 + nww_0 g\,i\,i_f + \frac{1}{4}nw_0^2 g_0 i_f. \tag{3.13.7}$$

The equations for electromechanical processes for the machine with a load are the Lagrange-Maxwell equations which have the form

$$[(2nw^2g_0 + L)i + nww_0gi_f]^{\cdot} + Ri = 0,$$

$$(nww_0gi + \frac{1}{2}nw_0^2g_0i_f)^{\cdot} + R_fi_f = E_f,$$

$$J\dot{\Omega} = n^2ww_0\frac{dg}{d\varphi}ii_f + M,$$

$$\dot{\vartheta} = \Omega, \tag{3.13.8}$$

where Ω is the frequency of the rotor rotation, R is the active resistance of the load together with the anchor winding, R_f is the resistance of the excitations winding, E_f is the excitation voltage, J is the moment of inertia of the rotor and M is the torque at the generator shaft.

Instead of initial currents i, i_f we introduce the dimensionless currents $i_u = i/i_*$ and $i_{fu} = i_f/i_{f*}$ where i_* is the basic value of the anchor current and $i_{f*} = (2w/w_0)i_*$. Let Ψ_u, Ψ_{fu} designate the dimensionless flux linkages related to dimensional quantities by the relations $\Psi_u = \Psi/\Psi_*, \Psi_{fu} = \Psi_f/\Psi_{f*}, \Psi_* = 2nw^2g_0i_*, \Psi_{f*} = nww_0g_0i_*$. In place of eq. (3.13.4) we obtain the relationship between the dimensionless flux linkages and currents

$$\Psi_u = (1 + l)i_u + g_ui_{fu}, \quad \Psi_{fu} = g_ui_u + i_{fu}, \tag{3.13.9}$$

where $l = L/(2nw^2g_0)$ and $g_u = g/g_0$. Let us also introduce the dimensionless time $t_u = n\Omega_*t$ and the dimensionless frequency of rotation $\omega_u = \Omega/\Omega_*$ where Ω_* denotes the basic frequency of rotation. Let us write down the dimensionless equations of electromechanical processes omitting subscript u at the dimensionless values, a dot denoting now the differentiation with respect to dimensionless time

$$[(1 + l)i + gi_f]^{\cdot} + ri = 0,$$

$$(gi + i_f)^{\cdot} + \varepsilon_fr_fi_f = \varepsilon_fe_f,$$

$$\dot{\omega} = \varepsilon_\omega\frac{dg}{d\varphi}ii_f + \varepsilon_\omega m,$$

$$\dot{\varphi} = \omega, \tag{3.13.10}$$

where

$$r = R/(2n^2w^2g_0\Omega_*), \quad \varepsilon_fr_f = 2R_f/(n^2w_0^2g_0\Omega_*),$$

$$\varepsilon_fe_f = 2E_f/(n^2w_0^2g_0i_{f*}\Omega_*), \quad \varepsilon_\omega m = M/(nJ\Omega_*^2),$$

$$\varepsilon_\omega = 2nw^2g_0i_*^2/(J\Omega_*^2). \tag{3.13.11}$$

Often there is no need to determine angles φ and ϑ. In this case it is more comfortable to view angle φ as an argument. Designating a derivative with respect to φ by a prime we come to the equations

$$[(1 + l)i + gi_f]' + \frac{r}{\omega}i = 0,$$

$$(gi + i_f)' + \frac{\varepsilon r_f}{\omega}i_f = \frac{\varepsilon e_f}{\omega},$$

$$\omega' = \frac{\varepsilon_\omega}{\omega}g'ii_f + \frac{\varepsilon_\omega}{\omega}m. \tag{3.13.12}$$

In the following we consider the most important case when the active resistance of load approximately equals to the inductive resistance of the anchor winding at frequency $n\Omega_*$. The active power consumed by the load is of the order of the nominal power of machine whereas the value of r in eq. (3.13.10) is of the order of unity.

The values of $\varepsilon_f r_f, \varepsilon_f e_f$ and ε_ω are small in many cases. There are various reasons for explaining this. In particular, the smallness of value $\varepsilon_f e_f$ follows from the expression $\varepsilon_f r_f/r = 4(R_f/R)(w/w_0)^2$ obtained from eq. (3.13.11).

The values R_f and R are of the same order and R_f can be even larger than R, [3]. But the number of turns w_0 in the excitation winding is always much larger than the number of turns w in one spool of the anchor winding (in practice $w_0/w = 20 - 100$). It explains the smallness of $\varepsilon_f r_f$ in comparison with r, the latter having the order of unity for the adopted method of introducing the dimensionless values. Value ε_ω is small as the mechanical time constant $T_m = J\Omega_*/(2n^2w^2g_0i_*^2)$ equal to the ratio of basic angular momentum of the revolved parts to the basic electromagnetic torque is much larger than the time constant of the anchor circuit. This meets the natural technical requirement that the irregularity in the rotor rotation due to time-dependent electromagnetic torque should be small. For standard synchronous machines the smallness of parameters in the equations for electromechanical processes is discussed in Section 3.2, and that reasoning is also applicable to the machines under consideration.

The smallness of parameters ε_f and ε_ω allows applying asymptotic methods of nonlinear mechanics to simplify equations (3.13.10) or (3.13.12). According to these methods it is necessary to select in these equations all unknown variables with small derivatives (i.e. the slow variables). Let us view the flux linkage Ψ_f as a new unknown variable and eliminate i_f. Then we express the current from eq. (3.13.9)

$$i_f = \Psi_f - gi \tag{3.13.13}$$

and substitute eq. (3.13.13) into eq. (3.13.12). Then we arrive at the equations

$$[(1 + l - g^2)i + g\Psi_f]' + \frac{r}{\omega}i = 0,$$

$$\Psi_f' + \frac{\varepsilon_f r_f}{\omega}(\Psi_f - gi) = \frac{\varepsilon_f e_f}{\omega},$$

$$\omega' = \frac{\varepsilon_\omega}{\omega}g'i(\Psi_f - gi) + \frac{\varepsilon_\omega}{\omega}m. \tag{3.13.14}$$

Let us consider the first approximation to solutions of equations (3.13.14), i.e. the approximate solutions containing only non-small terms and no terms proportional to $\varepsilon_f, \varepsilon_\omega$ and their degrees. Among the solutions of eq. (3.13.14) the steady-state solution is of special interest. In the first approximation this solution is as follows: $\Psi_f = \text{const}, \omega = \text{const}$ and i is the 2π-periodic solution of the equation

$$[(1 + l - g^2)i]' + \frac{r}{\omega}i = -g'\Psi_f. \tag{3.13.15}$$

This solution contains ω, Ψ_f as parameters and it can be presented in the form $i(\varphi, \omega, \Psi_f) = \Psi_f j(\varphi, \omega)$.

Let us replace the argument φ in eq. (3.13.15) by $\varphi - \pi$. Taking into account that $g(\varphi - \pi) = -g(\varphi)$ we arrive at the following equation

$$[(1 + l - g^2)i(\varphi - \pi)]' + \frac{r}{\omega}i(\varphi - \pi) = g'\Psi_f. \tag{3.13.16}$$

Addition of eq. (3.13.15) and eq. (3.13.16) yields

$$[(1 + l - g^2)(i(\varphi) + i(\varphi - \pi))]' + \frac{r}{w}[i(\varphi) + i(\varphi - \pi)] = 0. \tag{3.13.17}$$

We look for a $2\pi-$periodic solution of the latter equation however eq. (3.13.17) is a linear homogeneous differential equation of the first order in the sum $i(\varphi) + i(\varphi - \pi)$. It is easy to demonstrate by finding, for example, the exact solution of eq. (3.13.17) that the only asymptotically stable periodic solution is $i(\varphi) + i(\varphi - \pi) \equiv 0$. In doing so we proved that under a steady-state regime the current $i(\varphi)$ contains no constant component and even harmonics of φ.

Let us consider nonstationary solutions of system (3.13.14). In accordance with the asymptotic methods of motion separation, in the first approximation these solutions are given as follows. At first it is necessary to find the general solution of equation (3.13.15) under the assumption $\Psi_f = \text{const}, \omega = \text{const}$ (similar to the case of steady-state regime). We obtain function $i(\varphi, C, \omega, \Psi_f)$ which is no longer periodic in φ and depends on the arbitrary constant C and ω, Ψ_f considered as parameters. Then this result should be substituted in the last two equations (3.13.14), whereupon all terms of these equations, but Ψ_f' and ω', are to be replaced by their values averaged over φ. By the mean value of a non-periodic function $F(\varphi)$ we understand the following expression

$$\langle F(\varphi) \rangle = \lim_{T \to \infty} \frac{1}{T} \int\limits_{\varphi_0}^{\varphi_0 + T} F(\varphi)d\varphi.$$

A part of the general solution of equation (3.13.15) that contains the integration constant and the exponentially attenuating terms does not contribute at the averaging. Therefore for determining Ψ_f and ω in the first approximation it is sufficient to put the expression $i = \Psi_f j(\varphi, \omega)$ corresponding to the steady-state solution into the last equations in (3.13.14). Thus the terms in eq. (3.13.14) to be averaged are $2\pi-$periodic functions of φ and their mean values should be calculated as the mean value over the period. As a result we obtain equations for slow nonstationary processes

$$\Psi'_f = \varepsilon_f \frac{e_f}{\omega} - \frac{\varepsilon_f r_f}{\omega} \Psi_f (1 - \langle gj \rangle),$$

$$\omega' = \frac{\varepsilon_\omega}{\omega} \Psi_f^2 \langle g' j (1 - gj) \rangle + \varepsilon_\omega \frac{m}{\omega}. \tag{3.13.18}$$

These equations are essentially simpler than the initial equations (3.13.14) since the argument φ does not appear explicitly and they contain one unknown variable less than eq. (3.13.14).

When the machine starts to operate into load or an additional load is connected etc. the complete nonstationary process consists of two stages. At first, there happens an exponential change in the anchor current (with the characteristic time constant $1/n\Omega_*$) unless it achieves the steady-state solution $\Psi_{f0} j(\varphi, \omega_0)$ corresponding to the initial values of the excitation flux Ψ_{f0} and the angular velocity ω_0. In the first approximation Ψ_f and ω do not change within this time. Then they change slowly (with the characteristic time constant $1/(\varepsilon_f r_f n\Omega_*)$ or $1/(\varepsilon_\omega n\Omega_*)$) in accordance with equations for slow nonstationary processes. In the first approximation the anchor current is given by the relation $i = \Psi_f j(\varphi, \omega)$, i.e. it changes as in the steady-state regime with the slowly changing parameters.

When variables i, Ψ_f are determined it is possible to find the excitation current i_f with the help of eq. (3.13.13). One can see from eq. (3.13.13) that even in the steady-state regime the quantity i_f has a variable component which is not small in the sense that its value is not proportional to $\varepsilon_f, \varepsilon_\omega$ (however this fact is known, see [28, 33]).

The calculation of the fast exponentially attenuating transient process is usually out of interest as this process decays faster than one revolution of the rotor. For this reason, the analysis of dynamics of the considered machines for given forms of teeth and conductivities $g_1(\varphi)$ and $g_2(\varphi)$ requires solving two problems. The first is determination of the periodic current i from eq. (3.13.15) in the form $i = \Psi_f j(\varphi, \omega)$ as functions of angle φ, the values Ψ_f and ω appearing as parameters. The second problem is determination of the mean values in the right hand sides of equations (3.13.18) and integration of these equations, that allows one to find Ψ_f and ω.

There are various methods of calculating a periodic solution of equation (3.13.15). Foremost it is integrated by quadrature. Determining the constant in the general solution from the condition of periodicity yields

$$
i(\varphi) = -\frac{\Psi_f}{1 + l - g^2(\varphi)} \left[\frac{1}{\mu(2\pi) - 1} \int\limits_0^{2\pi} g'(\psi)\mu(\psi - \varphi)d\psi + \right.
$$

$$
\left. + \int\limits_0^{\varphi} g'(\psi)\mu(\psi - \varphi)d\psi \right] , \quad (3.13.19)
$$

where

$$
\mu(\varphi) = \exp \left[\frac{r}{\omega} \int\limits_0^{\varphi} \frac{d\eta}{1 + l - g^2(\eta)} \right] . \quad (3.13.20)
$$

Another method of obtaining the periodic solution of eq. (3.13.15) is determination of the coefficients of a few first harmonics of Fourier series for current i. Given $g(\varphi)$, it reduces to the resolving a system of linear algebraic equations. The difficulty is that the coefficients should be found as functions of ω. There are another approaches to solving equation (3.13.15) based on that the variable component in the expression $1 + l - g^2$ weakly affects the current. Let us point out such a method in the first approximation. The current i is represented by the sum $i_0 + \Delta i$ where i_0 and Δi are determined as periodic solutions of the linear differential equations with constant coefficients

$$
(1 + l - \langle g^2 \rangle)i_0' + \frac{r}{\omega}i_0 = -g'\Psi_f,
$$

$$
(1 + l - \langle g^2 \rangle)\Delta i' + \frac{r}{\omega}\Delta i = \Delta g^2 i_0, \quad (3.13.21)
$$

where $\Delta g^2 = g^2 - \langle g^2 \rangle$ is the variable part of g^2.

Up to now we have discussed the solution of eq. (3.13.15). However for obtaining the steady-state regime it is necessary to proceed to equations (3.13.18) and put $\Psi_f' = 0, \omega' = 0$. This results in the system of two transcendent equations for Ψ_f, ω.

In order to find the periodic solution of equation (3.13.15) we use expansion in Fourier series. In the simple case we assume that the main contribution to decomposition of magnetic permeability is due to the first harmonic, that is $g(\varphi) = a_1 \cos(\varphi)$. In this case, searching for $j(\varphi)$ in the form

$$
j = \sum\limits_{k=1}^{\infty} j_{2k-1,c} \cos(2k - 1)\varphi + j_{2k-1,s} \sin(2k - 1)\varphi \quad (3.13.22)
$$

we obtain the system of linear algebraic equations for the coefficients of harmonics

$$-\left(1+l-\frac{3}{4}a_1^2\right)j_{1,c} + \frac{r}{\omega}j_{1,s} + \frac{1}{4}a_1^2 j_{3,s} = a_1,$$

$$\frac{r}{\omega}j_{1,s} + \left(1+l-\frac{1}{4}a_1^2\right)j_{1,s} - \frac{1}{4}a_1^2 j_{3,s} = 0,$$

$$\frac{3}{4}a_1^2 j_{1,c} - 3\left(1+l-\frac{1}{2}a_1^2\right)j_{3,c} + \frac{r}{\omega}j_{3,c} + \frac{3}{4}a_1^2 j_{5,c} = 0,$$

$$-\frac{3}{4}a_1^2 j_{1,s} + 3\left(1+l-\frac{1}{2}a_1^2\right)j_{3,s} + \frac{r}{\omega}j_{3,s} - \frac{3}{4}a_1^2 j_{5,s} = 0,$$

$$\frac{5}{4}a_1^2 j_{3,c} - 5\left(1+l-\frac{1}{2}a_1^2\right)j_{5,c} + \frac{r}{\omega}j_{5,c} + \frac{5}{4}a_1^2 j_{7,c} = 0,$$

$$-\frac{5}{4}a_1^2 j_{3,s} + 5\left(1+l-\frac{1}{2}a_1^2\right)j_{5,s} + \frac{r}{\omega}j_{5,s} - \frac{5}{4}a_1^2 j_{7,s} = 0,$$

$$\dots \tag{3.13.23}$$

The solution of this system has shown that the coefficients of harmonics rapidly decrease with the growth of number. Besides, the calculation of coefficients of the first harmonics can be carried out separately. The approximate expressions for $j_{1,c}, j_{1,s}$ are as follows

$$j_{1,c} = -\frac{\left(1+l-\frac{1}{4}a_1^2\right)a_1}{\Delta}, \quad j_{1,s} = \frac{\frac{r}{\omega}a_1}{\Delta}, \tag{3.13.24}$$

where $\Delta = (1+l-\frac{1}{4}a_1^2)(1+l-\frac{3}{4}a_1^2) + (r/\omega)^2$.

Taking into account the only the first harmonics of current i we can set the averaged equations (3.13.18) in the form

$$\Psi_f' = -\frac{\varepsilon_f r_f}{\omega}\Psi_f\left(1 - \frac{1}{2}a_1 j_{1,c}\right) + \frac{\varepsilon_f e_f}{\omega},$$

$$\omega' = \frac{\varepsilon_\omega}{\omega}\Psi_f^2\left(-\frac{a_1}{2}j_{1,s} + \frac{a_1^2}{4}j_{1,s}j_{1,c}\right) + \frac{\varepsilon_\omega m}{\omega}. \tag{3.13.25}$$

These equations allow one to determine the static characteristic of the generator which is the dependence of mean stationary electromagnetic moment on the angular velocity. To this end, we obtain the dependence of the steady-state flux linkage of the excitation winding Ψ_f upon frequency ω from the first equation of systems (3.13.25) at $\Psi_f' = 0$ and substitute it into the second equation. The dimensionless electromagnetic moment is determined by the first term in the right hand side of the second equation in (3.13.25). Designating this moment divided by ε_ω as m_e we obtain

$$m_e = -\frac{e_f^2}{r_f^2}\frac{a_1}{2}j_{1,s}\frac{1}{1-\frac{1}{2}a_1 j_{1,c}}. \tag{3.13.26}$$

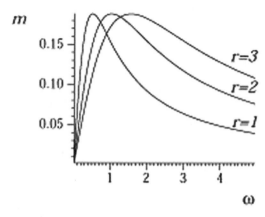

FIGURE 3.23.

Dependence of m_e on ω for several values of the load resistance has a maximum (Fig. 3.23), and the value of this maximum is independent of r since the moment depends only on the ratio r/ω

The static characteristic allows us to define the moment of the motor $m = -m_e$ for which the frequency of rotation in the steady-state regime has a prescribed value. The dependence $m_e(\omega)$ shows that two steady-state regime exist if the value of moment is less than the maximum of the electromagnetic moment. The lower (higher) frequency corresponds to stable (unstable) steady-state regime. In the phase plane Ψ_f, ω, cf. eq. (3.13.25), the lower and higher frequencies correspond to a stable note and a saddle respectively. Hence system (3.13.25) is dichotomic at $m = $ const, that is, all bounded solutions tend to the stable position of equilibrium. The region of bounded motions is separated from the region of the unbounded solutions by separatrices converging to the saddle point.

Let the equations of rotation have a dissipative term, say the static characteristic of the motor is given as $m(\omega) = m_0 - \beta\omega$. In accordance with Fig. 3.23 the system has either one stable position of equilibrium (node) or three ones: two stable nodes separated by a saddle. In the latter case unbounded motions are impossible and the regions of attraction of the stable equilibrium positions are separated by the converging separatrices of the saddle. Thus the structure of the phase plane of averaged equations of the inductor generator autonomously operating into active-inductive load is qualitative similar to the phase portrait for the averaged equations of standard synchronous machine operating into active-inductive load (Fig. 3.12).

As another example of use of the obtained equations we present the results of computation of the transient process in the single-phase inductor machine operating as a generator of brief action. While the circuit is open the rotor of this machine along with the flywheel is spined up by an auxiliary motor. Then the motor stops and the load circuit is closed. Because of the

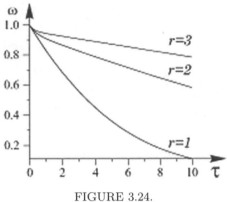

FIGURE 3.24.

power consumptions in the load the kinetic energy of the rotor and flywheel decreases which causes a decrease in the frequency of rotation. Figure 3.24 shows the dependences of the frequency of rotation upon time for several values of the load resistance obtained by numeral integration of equations (3.13.18). The qualitative character of these dependences is coincident with that for the three-phase synchronous generators of brief action [126].

3.14   Some inverse problems of theory of inductor machines

Let us consider some inverse problems of the theory of inductor machines. They are the problems of determining the magnetic conductance and the form of the rotor teeth ensuring the desired properties of the machine. For instance, it is the concern of the analysis (carried out by many authors, see, e.g. [3]) to seek the form of the teeth such that emf is close to sinusoidal at no-load. However even if emf is sinusoidal ($g'\Psi_f \sim \sin\varphi$ in eq. (3.13.15)) the load current in not sinusoidal as the coefficient of i in the first term of eq. (3.13.15) depends on φ. For this reason a more interesting problem is to find the conditions under which current i is harmonic in the steady-state regime. Let us find the required dependence $g(\varphi)$.

Let us put $i = I\cos\varphi$ in eq. (3.13.15) and integrate both sides of the equation

$$(1 + l - g^2)I\cos\varphi + \frac{r}{\omega}I\sin\varphi = -g\Psi_f + C. \qquad (3.14.1)$$

Replacing φ in eq. (3.14.1) by $\varphi - \pi$ and adding the result with eq. (3.14.1) yields $C = 0$. Resolving now this square equation for g we obtain the root

$$g(\varphi) = \frac{1}{2\cos\varphi}\left(\frac{\Psi_f}{I} - \sqrt{\left(\frac{\Psi_f}{I}\right)^2 + 2(1+l) + \frac{2r}{\omega}\sin 2\varphi + 2(1+l)\cos 2\varphi}\right)$$

$$(3.14.2)$$

which is a continuous function of φ. Given values Ψ_f and ω in eq. (3.13.15) and $g(\varphi)$ due to eq. (3.14.2), we obtain the steady-state solution of the latter equation as a harmonic function of amplitude I.

However the required dependence $g(\varphi)$ can be found not for any value of I. The radicand in eq. (3.14.2) is required to be non-negative for any φ. By finding the minimum value of sum of the terms with $\sin 2\varphi$ and $\cos 2\varphi$ we obtain the inequality for the possible values of the current amplitude I

$$I^2 \leq \omega^2\Psi_f^2\frac{1+l}{2r^2}\left(1 + \sqrt{1 + \frac{r^2}{\omega^2(1+l)^2}}\right). \qquad (3.14.3)$$

Now it is necessary to take into account that in the steady-state regime Ψ_f and ω must satisfy the above equations obtained from eq. (3.10.18) at $\Psi_f' = 0, \omega' = 0$. The expression for $g(\varphi)$ from eq. (3.14.2) should be inserted into these equations. If value I is given then Ψ_f and ω can be obtained from the above equations. It is also possible to take the values of I, Ψ_f, ω corresponding, for example, the nominal regime, and to define the excitation voltage $\varepsilon_f e_f$ and the torque $\varepsilon_\omega m$ from the equations. Inequality (3.14.3) should be proved while carrying out the calculation. Another example of the inverse problem is the problem on determining $g(\varphi)$ from the condition of maximum active power transmitted to the load This problem makes sense when non-sinusoidal current in the load is admitted otherwise the matter reduces to inequality (3.14.3).

Another sort of the inverse problems is determining the form of teeth from the obtained dependence $g(\varphi)$. The solution of this problem is not unique which is evident, for example, from expressions (3.10.5). The determined expressions $g_1(\varphi), g_2(\varphi)$ are independent of the even harmonics in the decomposition of the integrand. Consequently these harmonics can be are chosen arbitrarily when $\Delta(\psi)$ is determined in terms of given g_1 and g_2.

3.15 Single-pack and unlike-pole machines, machines with open slot

Let us consider a single-pack machine [24] with the same winding and under the same assumptions as the two-pack one. The single-pack machine has an additional gap between the casing and the rotor bush. Let the contours of magnetic circuit including the gaps under the stator teeth and the additional gap be taken as independent contours. Let Φ_1 and Φ_2 designate the fluxes through the gaps between the neighbouring teeth of stator, then the flux through the additional gap is $n(\Phi_1 + \Phi_2)$ where n is the number of the rotor teeth. By analogy with eq. (3.13.1) and (3.13.2) we obtain

$$\frac{1}{g_1}\Phi_1 + \frac{n}{g_s}(\Phi_1 + \Phi_2) = w_0 i_f + wi,$$

$$\frac{1}{g_2}\Phi_2 + \frac{n}{g_s}(\Phi_1 + \Phi_2) = w_0 i_f - wi. \tag{3.15.1}$$

Here and in what follows we use the above denotation and g_s is magnetic conductivity of the additional gap. From eq. (3.15.1) we have

$$\Phi_1 = \frac{g_1(g_s + 2ng_2)}{g_s + ng_0}wi + \frac{g_s g_1}{g_s + ng_0}w_0 i_f,$$

$$\Phi_2 = -\frac{g_2(g_s + 2ng_1)}{g_s + ng_0}wi + \frac{g_s g_2}{g_s + ng_0}w_0 i_f. \tag{3.15.2}$$

Here, as above, $g_0 = g_1 + g_2 = \text{const}$. The flux linkage of the contours currents i and i_f are

$$\Psi = \left[\frac{nw^2}{g_s + ng_0}(4ng_1 g_2 + g_s g_0) + L\right]i + \frac{nww_0 g_s}{g_s + ng_0}(g_1 - g_2)i_f,$$

$$\Psi_f = \frac{nww_0 g_s}{g_s + ng_0}(g_1 - g_2)i + \frac{nw^2 g_s g_0}{g_s + ng_0}i_f. \tag{3.15.3}$$

One can see from eq. (3.15.3) that the dependence of the coefficient of mutual induction of the contours currents i and i_f of φ does not contain a constant term, i.e. the considered winding is balanced. In contrast to the two-pack machines the self-inductance of the anchor circuit of the single-pack machine depends on φ. As shown in the following, this fact has only a slight effect on the final equations.

The derivation of equations of the electromechanical processes is similar to the above one. Therefore we write down the equations for the dimensionless variables i, Ψ_f, ω

$$[1 + \alpha + l - (1 + \alpha)g^2]i' + \frac{r}{\omega}i = -(g\Psi_f)',$$

$$\Psi_f' + \frac{\varepsilon_f r_f}{\omega}(\Psi_f - gi) = \frac{\varepsilon_f e_f}{\omega},$$

$$\omega' = \frac{\varepsilon \omega}{\omega}g'i[\Psi_f - (1 + \alpha)gi] + \frac{\varepsilon \omega}{\omega}m. \tag{3.15.4}$$

φ being the argument. Here

$$r = R\frac{g_s + ng_0}{n^2w^2g_sg_0\Omega_*}, \quad \alpha = \frac{ng_0}{g_s},$$

$$\varepsilon_f r_f = R\frac{g_s + ng_0}{n^2w^2g_sg_0\Omega_*},$$

$$\varepsilon_f e_f = E_f\frac{g_s + ng_0}{n^2w^2g_s\Omega_*i_{f*}},$$

$$\varepsilon_w = \frac{nw^2g_sg_0i_*^2}{J\Omega_*^2}, \quad \varepsilon_w m = \frac{M}{nJ\Omega_*^2}. \tag{3.15.5}$$

The basic values of the excitation current and the flux linkage are determined as follows

$$i_{f*} = wi_*/w_0, \quad \Psi_* = nw^2g_sg_0i_*/(g_s + ng_0), \quad \Psi_{f*} = w\Psi_*/w_0. \tag{3.15.6}$$

Equations (3.15.4) differ from the corresponding equations (3.13.12) for the two-pack machine in that the first and third equations (3.15.5) have $1 + \alpha$ instead of unity in eq. (3.13.12).

Let us also construct the equations for the two-pack machine with the open slot of stator [28]. The magnetic flux through the slot is not taken into account, otherwise we use the above assumptions. In this case the expressions for the flux linkage have the form of eq. (3.12.4) however the dependences $g_1(\varphi), g_2(\varphi)$ have other properties. These properties can be exposed by means of the simple approximate method of calculations of $g_1(\varphi), g_2(\varphi)$ applied for deriving expressions (3.12.5). Instead of eq. (3.12.5) we have

$$g_1(\varphi) = \mu_0 bR_c \int\limits_{-\varphi}^{\pi-\gamma-\varphi} \frac{d\psi}{\Delta(\psi)}, \quad g_2(\varphi) = \mu_0 bR_c \int\limits_{\pi-\varphi}^{2\pi-\gamma-\varphi} \frac{d\psi}{\Delta(\psi)}, \tag{3.15.7}$$

where γ denotes the angular width of the open slot. By analogy with the above reasoning we obtain that $g_2(\varphi) = g_1(\varphi-\pi)$ and the difference $g_1 - g_2$ does not contain a constant component and even harmonics. The winding under consideration is thus balanced for the open slot. However the sum $g_0(\varphi) = g_1 + g_2$ is no more constant and can be represented by a Fourier expansion in terms of even harmonics.

The equations of electromechanical processes in terms of the dimensionless unknown variables i, Ψ_f, w with argument φ are given by

$$\left(l + g_0 - \frac{g^2}{g_0}\right)\ddot{i} + \frac{r}{w}i = -\frac{g}{g_0}\Psi'_f,$$

$$\Psi'_f + \frac{\varepsilon_f r_f}{wg_0}(\Psi_f - gi) = \frac{\varepsilon_f e_f}{w},$$

$$w' = \frac{\varepsilon_w}{w}\frac{g'}{g_0}i(\Psi_f - gi) + \frac{1}{2}g'_0 i^2 + \frac{1}{2}\frac{g'_0}{g_0^2}(\Psi_f - gi)^2 + \frac{\varepsilon_w m}{w}. \tag{3.15.8}$$

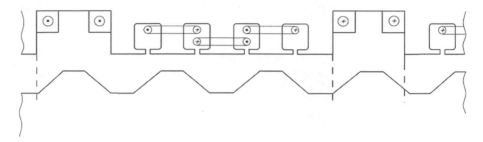

FIGURE 3.25.

The dimensionless unknown variables and parameters in eq. (3.15.8) are determined by analogy with the above approach with the only difference that instead of $g_0 = $ const it is necessary to use $\langle g_0 \rangle$, i.e. the value of g_0 averaged over the period of φ. The dimensionless conductivities g_0, g in eq. (3.15.8) are equal to the ratios of dimensional values $g_0/\langle g_0 \rangle$, $g/\langle g_0 \rangle$ respectively.

The above results about the qualitative properties of transient processes and application of asymptotic method hold true for equations (3.15.4) and (3.15.8). In particular, assuming $i = \Psi_f j(\varphi, \omega)$ we can write down the equations of slow nonstationary processes, similar to equations (3.13.18). The above formal substantiation of that current i has only odd harmonics in the steady-state regime remains valid.

Let us also consider the unlike-pole machine with the radial excitation (see, for example [3]). It is typical for such machines that the spools of anchor and excitations are laid in the stator slots. It is assumed that spools of the excitation winding are connected in series, the neighbouring sides of two excitation spools are laid in the large slot of the stator, one slot of the stator fits one slot of the rotor and an integer number of the rotor periods lie between by two sides of a spool (Fig. 3.25). It is also assumed that the anchor winding is laid in the half-closed slots and consists of series spools embracing the stator teeth, a half of the slot partition of the rotor being placed within one stator tooth. The neighbouring spools of excitation and anchor as well as the anchor spools placed within the neighbouring excitation spools and symmetrically relative to the rotor teeth have subtractive polarity (see Fig. 3.25, where the current directions are shown).

When constructing the equations we make the assumptions for the machine with the axial excitation. The magnetic fluxes through the large slots are not taken into account. Let us consider two contours of the magnetic circuit including gaps between the identically located teeth of the anchor within the neighbouring excitation spools (Fig. 3.25). Using the above denotations we obtain

$$\Phi_1 = g_1(wi + w_0 i_f), \quad \Phi_2 = g_2(w_0 i_f - wi). \tag{3.15.9}$$

The flux linkage of the anchor winding is

$$\Psi = nw(\Phi_1 - \Phi_2) + Li = [nw^2(g_1 + g_2) + L]i + nww_0(g_1 - g_2)i_f, \quad (3.15.10)$$

where n is the number of the stator teeth embraced by the anchor winding.

While constructing expression for the flux linkage of the excitation winding, one should take into account the fluxes through the stator teeth which are not embraced by the anchor windings

$$\Psi_f = nw_0(\Phi_1 + \Phi_2) + n_f w_0^2(g_1 + g_2)i_f =$$
$$= nww_0(g_1 - g_2)i + (n + n_f)w_0^2(g_1 + g_2)i_f, \quad (3.15.11)$$

where n_f is the number of the excitation spools, the total number of the stator teeth being equal to $n + 2n_f$.

The basic values of the excitation current and the flux linkage are given by

$$i_{f*} = wi_*/w_0, \ \Psi_* = nw^2 g_0 i_*, \ \Psi_{f*} = w\Psi_*/w_0 .$$

The equations similar to eqs. (3.13.18), (3.15.4) and (3.15.8) take the form

$$\left(1 + l - \frac{g^2}{1+\sigma}\right)' + \frac{r}{\omega}i = -\frac{g\Psi_f}{1+\sigma}',$$

$$\Psi_f' + \frac{c_f r_f}{\omega(1+\sigma)}(\Psi_f - gi) = \frac{\varepsilon_f e_f}{\omega},$$

$$\omega' = \frac{\varepsilon_\omega}{\omega(1+\sigma)}g'i(\Psi_f - gi) + \frac{\varepsilon_\omega m}{\omega}, \quad (\sigma = n_f/n). \quad (3.15.12)$$

The dimensionless parameters in this equation are determined by formulae different from those in eq. (3.13.11) only in factor 2. The equations of unlike-pole machines with other balanced windings of the anchor have the same form but with another expressions for the parameters. The properties of the nonstationary processes described by equations (3.15.12) and the procedure of applying the asymptotic method remain unchanged.

3.16   Equation of electromechanical processes in three-phase unlike-pole machine

In comparison with the single-phase inductor generators, application of the asymptotic method of separation of motions to the equations of three-phase machines has certain peculiarities caused by a large number of fast variables. As we show in the following, these variables are independent currents in the anchor winding. Interestingly enough, determination of the fast variables in the first approximation is related to solving a differential-

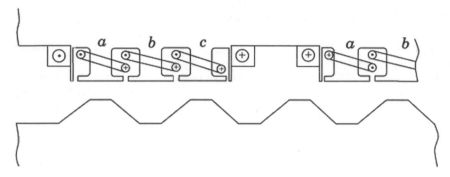

FIGURE 3.26.

difference equation with periodic coefficients. Equation of this type seems to appear first time in the theory of electric machines.

We consider different variants of unlike-pole machines with the teeth winding of anchor [3] operating as generator into symmetric active-inductive load.

The involute of the teeth area of the machine is shown in Fig. 3.26. The three spools of the anchor located within one tooth partition of the rotor are shifted in space onto a $1/3$ of the slot partition of the rotor and form a spool group. The spools identically located within the neighbouring spool groups are connected in series, have subtractive polarity and form a phase. The slots of the stator are half-closed. While constructing the Lagrange-Maxwell equations the magnetic resistance of steel and the vortex currents are not taken into account.

Let i_f, i_a, i_b, i_c designate respectively the currents in the excitation winding and the phase windings, with w_0, w denoting the number of turns. Let the contour currents i_a and i_b be considered as independent, then $i_c = -(i_a + i_b)$. Constructing Kirchhoff's equations for the magnetic circuit it is possible to show that the magnetic potentials of the rotor and the stator steel are identical. Therefore the magnetic fluxes through the stator teeth $\Phi_a \Phi_b, \Phi_c$ are

$$
\begin{aligned}
\Phi_a &= w i_a g_a + w_0 i_f g_a, \\
\Phi_b &= w i_b g_b + w_0 i_f g_b, \\
\Phi_c &= -w(i_a + i_b) g_c + w_0 i_f g_c.
\end{aligned}
\tag{3.16.1}
$$

Here $g_a(\varphi), g_b(\varphi), g_c(\varphi)$ are the magnetic conductivity of the gaps under the corresponding tooth of the stator and $\varphi = z\vartheta$, ϑ and z denoting respectively the angle of rotation of the rotor and the number of the rotor teeth.

When Φ_a, Φ_b, Φ_c are obtained it is possible to find the flux linkage of the contours a, b, f

$$\Psi_a = \frac{1}{2}zw(\Phi_a - \Phi_c) + L_n i_a + L_n(i_a + i_b) =$$
$$= \left[\frac{1}{2}zw^2(g_a + g_c) + 2L_n\right]i_a + \left(\frac{1}{2}zw_c^2 + L_n\right)i_b + \frac{1}{2}zww_0(g_a - g_c)i_f,$$
$$\Psi_b = \frac{1}{2}zw(\Phi_b - \Phi_c) + L_n i_b + L_n(i_a + i_b) =$$
$$= \left(\frac{1}{2}zw^2 g_c + L_n\right)i_a + \left[\frac{1}{2}zw^2(g_b + g_c) + 2L_n\right]i_b + \frac{1}{2}zww_0(g_b - g_c)i_f,$$
$$\Psi_f = \frac{1}{2}zw_0(\Phi_a + \Phi_b + \Phi_c) =$$
$$= \frac{1}{2}zww_0(g_a - g_c)i_a + \frac{1}{2}zww_0(g_b - g_c)i_b + \frac{1}{2}zw_0^2 g_0 i_f. \qquad (3.16.2)$$

Here $g_0 = g_a + g_b + g_c$ denotes the conductivity of the gap under the spool group within one period of the rotor teeth and L_n is the load inductance. The value of g_0 does not depend on φ.

It is worth noting the basic properties of the magnetic conductances $g_a(\varphi), g_b(\varphi), g_c(\varphi)$. Firstly, the following relations for shift $g_b(\varphi) = g_a(\varphi - \frac{2\pi}{3}), g_c(\varphi) = g_a(\varphi + \frac{2\pi}{3})$ hold. Secondly, they contain no harmonics with the numbers divisible by 3 because their sum is equal to a constant.

The coefficients of currents in the expressions for the flux linkage (3.16.2) are the coefficients of self-induction and mutual induction of the considered contours. Designating them by $L_a, L_b, L_f, M_{ab}, M_{af}, M_{bf}$ we can write the expression for energy of magnetic field

$$W = \frac{1}{2}(L i_a^2 + L_b^2 + L_f^2) + M_{ab} i_a i_b + M_{af} i_a i_f + M_{bf} i_b i_f. \qquad (3.16.3)$$

Now it is possible to construct the equation of the rotor rotation

$$J\ddot{\vartheta} = \frac{\partial W}{\partial \vartheta} + M, \qquad (3.16.4)$$

where M denotes the external torque applied to the rotor when the machine operates as a generator. The expression for the electromagnetic moment $\frac{\partial W}{\partial \vartheta}$ is as follows

$$\frac{\partial W}{\partial \vartheta} = \frac{1}{4}z^2 w[w(g_a' + g_c')i_a^2 + w(g_b' + g_c')i_b^2 + 2wg_c' i_a i_b +$$
$$+ 2w_0(g_a' - g_c')i_a i_f + 2w_0(g_b' - g_c')i_b i_f], \qquad (3.16.5)$$

where a prime denotes the differentiation with respect to angle φ. In the case of a rectangular cross-section of the rotor tooth the derivatives g_a', g_b', g_c'

are discontinuous functions of φ (however it is essential for the further analysis).

The determinant Δ composed of the induction coefficients at $L_n = 0$ is as follows

$$\Delta = 9 g_a g_b g_c. \tag{3.16.6}$$

This means that even for a pure active load and negligible dissipative fluxes it is not possible to express one current in terms of the other two currents.

In order to select a small parameters we introduce the dimensionless variables i_{au}, i_{bu}, i_{fu}

$$i_{au} = i_a/i_*, \quad i_{bu} = i_b/i_*, \quad i_{fu} = i_f/i_{f*}. \tag{3.16.7}$$

Here i_* is a characteristic value of the phase current, that is $i_{f*} = (w/w_0)i_*$. The dimensionless flux linkages are designated as $\Psi_{au}, \Psi_{bu}, \Psi_{fu}$

$$\Psi_{au} = \Psi_a / \frac{1}{2} z w^2 g_0 i_*, \quad \Psi_{bu} = \Psi_b / \frac{1}{2} z w^2 g_0 i_*,$$

$$\Psi_{fu} = \Psi_f / \frac{1}{2} z w w_0 g_0 i_*. \tag{3.16.8}$$

Instead of eq. (3.16.2) we obtain the expressions for the dimensionless flux linkages in terms of the dimensionless currents

$$\Psi_{au} = (g_{au} + g_{cu} + 2l)i_{au} + (g_{cu} + l)i_{bu} + (g_{au} - g_{cu})i_{fu},$$

$$\Psi_{bu} = (g_{cu} + l_n)i_{au} + (g_{bu} + g_{cu} + 2l_n)i_{bu} + (g_{bu} - g_{cu})i_{fu},$$

$$\Psi_{fu} = (g_{au} - g_{cu})i_{au} + (g_{bu} - g_{cu})i_{bu} + i_{fu}. \tag{3.16.9}$$

Here $g_{au} = g_a/g_0, g_{bu} = g_b/g_0, g_{cu} = g_c/g_0$ are the dimensionless magnetic conductivities and $l_n = L_n/z w^2 g_0$ is the dimensionless inductance of the load.

Let us introduce the dimensionless time $\tau = z \Omega_* t$ and the dimensionless angular velocity $\omega = \vartheta/\Omega_*$, where Ω_* is the characteristic angular velocity of the rotor rotation. Let us write down the dimensionless Lagrange-Maxwell equations, omit the subscript u at the dimensionless variables and coefficients and denoting the differentiation with respect to dimensionless time by a prime

$$[(g_a + g_c + 2l_n)i_a + (g_c + l_n)i_b + (g_a - g_c)i_f] + r(2i_a + i_b) = 0,$$

$$[(g_c + l_n) + (g_b + g_c + 2l_n)i_b + (g_b - g_c)i_f] + r(2i_b + i_a) = 0,$$

$$[(g_a - g_c)i_a + (g_b - g_c)i_b + i_f] + \varepsilon_f r_f i_f = \varepsilon_f e_f,$$

$$\dot{\omega} = \varepsilon_\omega \left[\frac{1}{2}(g_a' + g_c')i_a^2 + \frac{1}{2}(g_b' + g_c')i_b^2 + g_c' i_a i_b + (g_a' - g_c')i_a i_f + \right.$$

$$\left. + (g_b' - g_c')i_b i_f \right] + \varepsilon_\omega m, \qquad \dot{\varphi} = \omega. \tag{3.16.10}$$

Here $r = R/(z^2 w^2 g_0 \Omega_*)$, $\varepsilon_f r_f = R_f/(z^2 w_0^2 g_0 \Omega_*)$, $\varepsilon_f e_f = E_f/(z^2 w_0^2 g_0 \Omega_* i_*)$, $\varepsilon_w = z w^2 g_0 i_*^2/J\Omega_*^2$, $\varepsilon_w m = M/Jz\Omega_*^2$, R and R_f denote respectively the active resistance of the load and the excitation circuit, E_f is the excitation voltage. The reason for introducing coefficients ε_f and ε_w is explained further.

We consider the case in which the active resistance of the load is approximately equal to the characteristic inductive resistance of the anchor windings. Then the power consumed by the load is of the order of the nominal power of machine. In this case r in eq. (3.16.10) is of the order of unity. The values $\varepsilon_f r_f, \varepsilon_w$ and $\varepsilon_f e_f$ are essentially smaller than unity, and the small parameters $\varepsilon_f, \varepsilon_w$ are introduced just to stress the smallness of these values. The parameters ε_f and ε_w are small as the characteristic time constant of the excitation circuit and mechanical time constant are much greater than the characteristic time constant of the anchor circuits. The smallness of these parameters in the dimensionless Lagrange-Maxwell equations is explained in detail in Section 3.13 for the single-phase inductor machine, and this reasoning is also applicable to the three-phase machines under consideration.

The smallness of parameters ε_f and ε_w enables using the asymptotic methods of nonlinear mechanics for simplification of equations (3.16.10). It is worth considering the flux linkage Ψ_f as a new unknown variable by excluding the excitation current i_f from the equations. Let us express current i_f from eq. (3.16.9) in terms of Ψ_f, i_a, i_b

$$i_f = \Psi_f - (g_a - g_c)i_a - (g_b - g_c)i_b. \tag{3.16.11}$$

Inserting eq. (3.16.11) into eq. (3.16.10) and considering angle φ as the argument (instead of time) we arrive at the equations

$$\frac{d}{d\varphi}[g_a + g_c + 2l_n - (g_a - g_c)^2]i_a + [g_c + l_n - (g_a - g_c)(g_b - g_c)]i_b +$$

$$+(g_a - g_c)\Psi_f + \frac{r}{\omega}(2i_a + i_b) = 0,$$

$$\frac{d}{d\varphi}[g_c + l_n - (g_a - g_c)(g_b - g_c)]i_a + [g_b + g_c + 2l_n - (g_b - g_c)^2]i_b +$$

$$+(g_b - g_c)\Psi_f + \frac{r}{\omega}(2i_b + i_a) = 0,$$

$$\frac{d}{d\varphi}\Psi_f + \frac{\varepsilon_f r_f}{\omega}[\Psi_f - (g_a - g_c)i_a - (g_b - g_c)i_b] = \frac{\varepsilon_f e_f}{\omega},$$

$$\frac{d\omega}{d\varphi} = \frac{\varepsilon_w}{\omega}\left[\frac{1}{2}(g_a' + g_c') - (g_a' - g_c')(g_a - g_c)\right]i_a^2 + \left[\frac{1}{2}(g_b' + g_c') -\right.$$

$$-(g_b' - g_c')(g_b - g_c)]i_b^2 + [g_c' - (g_a' - g_c')(g_b - g_c) - (g_b' - g_c')(g_a - g_c)]i_a i_b +$$

$$+(g_a' - g_c')i_a\Psi_f + (g_b' - g_c')i_b\Psi_f + \frac{\varepsilon_w}{\omega}m. \tag{3.16.12}$$

There are two groups of variables in eq. (3.16.12): fast variables i_a, i_b and slow variables Ψ_f and ω. According to the averaging method [128] one needs to resolve two first equations for i_a and i_b under the assumption that Ψ_f and ω are constant parameters. Then these solution should be substituted into the other two equations (3.16.12), the result being averaged over φ. The obtained equations for Ψ_f and ω are the sought equations of slow nonstationary processes.

As for the solutions $i_a(\varphi, \Psi_f, \omega)$, $i_b(\varphi, \Psi_f, \omega)$ of two first equations in (3.16.12) at Ψ_f, $\omega = $ const one can determine only their steady-state part not taking into account rapidly attenuating functions of φ, which do not contribute to the general solution under averaging. The steady-state solutions for i_a, i_b are 2π-periodic functions of φ which are proportional to Ψ_f.

Let us prove that the steady-state solutions i_a, i_b are related as follows $i_b(\varphi) = i_a(\varphi - \frac{2\pi}{3})$. We write down three equations: the first equation for the fast subsystem and two equations obtained from the first one by replacing φ by $\varphi - \frac{2\pi}{3}$ and $\varphi + \frac{2\pi}{3}$. Taking into account the properties of magnetic conductances g_a, g_b, g_c and the assumed equalities $i_a(\varphi) = i_b(\varphi + \frac{2\pi}{3})$, $i_b(\varphi - \frac{2\pi}{3}) = i_a(\varphi + \frac{2\pi}{3})$ we have

$$[g_a + g_c + 2l_n - (g_a - g_c)^2]i_a(\varphi) +$$

$$+[g_c + l_n - (g_a - g_c)(g_b - g_c)]i_a\left(\varphi - \frac{2\pi}{3}\right) +$$

$$+(g_a - g_c)\Psi'_f + \frac{r}{\omega}\left(2i_a(\varphi) + i_a\left(\varphi - \frac{2\pi}{3}\right)\right) = 0,$$

$$[g_b + g_a + 2l_n - (g_b - g_a)^2]i_a\left(\varphi - \frac{2\pi}{3}\right) + [g_a + l_n - (g_b - g_a)(g_c - g_a)] \times$$

$$\times i_a\left(\varphi + \frac{2\pi}{3}\right) + (g_b - g_a)\Psi'_f + \frac{r}{\omega}\left(2i_a\left(\varphi - \frac{2\pi}{3}\right) + i_a\left(\varphi + \frac{2\pi}{3}\right)\right) = 0,$$

$$[g_c + g_b + 2l_n - (g_c - g_b)^2]i_a\left(\varphi + \frac{2\pi}{3}\right) +$$

$$+[g_b + l_n - (g_c - g_b)(g_a - g_b)]i_a(\varphi) +$$

$$+(g_c - g_b)\Psi'_f + \frac{r}{\omega}\left(2i_a\left(\varphi + \frac{2\pi}{3}\right) + i_a(\varphi)\right) = 0.$$

One obtains the linear homogeneous differential equation of the first order for the sum $i_a(\varphi) + i_a\left(\varphi - \frac{2\pi}{3}\right) + i_a\left(\varphi + \frac{2\pi}{3}\right) = i_0(\varphi)$ by summing these equations

$$[g_a + g_b + g_c + 3l_n - g_a^2 - g_b^2 - g_c^2 + g_a g_c + g_b g_a + g_c g_b]i_0(\varphi)' +$$

$$+ 3\frac{r}{\omega}i_0(\varphi) = 0. \quad (3.16.13)$$

This equation has the only periodic solution $i_0 \equiv 0$.

Now we replace argument φ in the first equation of fast subsystem by $\varphi + \frac{2\pi}{3}$. Under the assumption that $i_b(\varphi) = i_a(\varphi - \frac{2\pi}{3})$ and with the proved relation $i_0 \equiv 0$ we obtain

$$
-[g_a + g_b + 2l_n - (g_c - g_b)^2]i_a\left(\varphi - \frac{2\pi}{3}\right) -
$$
$$
-[g_c + l_n - (g_b - g_c)(g_a - g_c)]i_a(\varphi) -
$$
$$
-(g_c - g_b)\Psi'_f - \frac{r}{\omega}\left[2i_a\left(\varphi - \frac{2\pi}{3}\right) + i_a(\varphi)\right] = 0.
$$

This equation coincides with the second equation in (3.16.12) if one assumes $i_b(\varphi) = i_a(\varphi - \frac{2\pi}{3})$.

Thus, we proved the following. Let a 2π-periodic solution $i_a(\varphi)$ of the differential-difference equation

$$
\left\{[g_a + g_c + 2l_n - (g_a - g_c)^2]i_a(\varphi) + \right.
$$
$$
\left. +[g_c + l_n - (g_a - g_c)(g_b - g_c)]i_a\left(\varphi - \frac{2\pi}{3}\right)\right\}' +
$$
$$
+\frac{r}{\omega}\left[2i_a(\varphi) + i_a\left(\varphi - \frac{2\pi}{3}\right)\right] = (y_c - y_a)'\Psi_f \qquad (3.16.14)
$$

be found for $\Psi_f, \omega = $ const. The equation obtained from eq. (3.16.14) by replacing φ by $\varphi + \frac{2\pi}{3}$ is equivalent to eq. (3.16.14) and has the solution $i_a\left(\varphi + \frac{2\pi}{3}\right)$. However under condition $i_0 \equiv 0$ this "shifted" equation coincides with the second equation (3.16.12) if $i_b(\varphi) = i_a\left(\varphi - \frac{2\pi}{3}\right)$. Consequently $i_a(\varphi)$ and $i_b(\varphi) = i_a\left(\varphi - \frac{2\pi}{3}\right)$ are 2π-periodic solution of the fast subsystem.

In order to obtain this solution one can use, for example, the method of harmonic balance. It is obvious that the Fourier expansion of the steady-state currents i_a, i_b does not have harmonics with numbers divisible by 3 because of the sum $i_a(\varphi) + i_a\left(\varphi - \frac{2\pi}{3}\right) + i_a\left(\varphi + \frac{2\pi}{3}\right) = 0$.

One should substitute the obtained steady-state currents $i_a(\varphi), i_b(\varphi)$ into the slow subsystem and average the right hand sides over φ. Then the right hand sides are functions only of Ψ_f and ω

$$\Psi'_f = -\frac{\varepsilon_f r_f}{\omega}\langle\Psi_f - (g_a - g_c)i_a - (g_b - g_c)i_b\rangle + \frac{\varepsilon_f e_f}{\omega},$$

$$\omega' = \frac{\varepsilon_\omega}{\omega}\langle[\frac{1}{2}(g'_a + g'_c) - (g'_a - g'_c)(g_a - g_c)]i_a^2 + \qquad (3.16.15)$$

$$+[\frac{1}{2}(g'_b + g'_c) - (g'_b - g'_c)(g_b - g_c)]i_b^2 + [g'_c - (g'_a - g'_c)(g_b - g_c) -$$

$$-(g'_b - g'_c)(g_a - g_c)]i_a i_b + (g'_a - g'_c)i_a \Psi_f + (g'_b - g'_c)i_b \Psi_f\rangle + \frac{\varepsilon_\omega}{\omega}m,$$

where $\langle f(\varphi)\rangle = \frac{1}{2\pi}\int\limits_{T}^{T+2\pi} f(\varphi)d\varphi.$

The obtained equations enables calculating both steady-state regime of operation and a nonstationary regime for the machine operating as a generator of brief action, change in the load etc.

Equation (3.16.14) is a linear differential-difference equation with periodic coefficients and plays an important role. Presumably, such an equation is studied in the theory of electric machines for the first time.

As an example of use of the simplified equations (3.16.15) we present calculation of the static characteristic of the three-phase generator (that is, the dependence of electromagnetic moment on the angular velocity of steady-state regime) and calculation of the transient process under a discontinuous change in the load resistance.

In order to determine the steady-state solution of equation (3.16.14) we use the Fourier expansion of current i_a, that is, we seek $i_a(\varphi)$ in the form of a series without harmonics with numbers divisible by 3

$$i_a(\varphi) = \sum_{\substack{k=1\\k\neq3}}^{\infty}(i_{kc}\cos k\varphi + i_{ks}\sin k\varphi). \qquad (3.16.16)$$

To simplify the derivation we present the magnetic conductivity as a sum of only the mean component and the first harmonics of angle, i.e. $g_a(\varphi) = 1/3 + a_1\cos\varphi$ (constant term equals $1/3$ due to the equality $g_a + g_b + g_c = 1$). The coefficients of the current harmonics are the solution of the following system of linear algebraic equations

$$-\left[\sqrt{3}\left(1 + l_n - \frac{3}{2}a_1^2\right) - \frac{r}{\omega}\right]i_{1c} + \left(1 + l_n + \sqrt{3}\frac{r}{\omega}\right)i_{1s} -$$

$$-\frac{a_1}{2}\left(\sqrt{3}i_{2c} - i_{2s}\right) = \sqrt{3}a_1\Psi_f,$$

$$\left[\sqrt{3}(1 + l_n) - \frac{r}{\omega}\right]i_{1s} + \left(1 + l_n - \frac{3}{2}a_1^2 + \sqrt{3}\frac{r}{\omega}\right)i_{1c} +$$

$$+\frac{a_1}{2}\left(i_{2c} + \sqrt{3}i_{2s}\right) = -a_1\Psi_f,$$

$$- a_1 \left(\sqrt{3}i_{1c} + i_{1s} \right) - \left[2 \left(1 + l_n - \frac{3}{4}a_1^2 \right) \sqrt{3} + \frac{r}{\omega} \right] i_{2c} -$$

$$- \left[2 \left(1 + l_n - \frac{3}{4}a_1^2 \right) - \sqrt{3}\frac{r}{\omega} \right] i_{2s} + \frac{3}{2}a_1^2 \left(\sqrt{3}i_{4c} + i_{4s} \right) = 0,$$

$$- a_1 \left(i_{1c} - \sqrt{3}i_{1s} \right) - \left[2 \left(1 + l_n - \frac{3}{4}a_1^2 \right) - \sqrt{3}\frac{r}{\omega} \right] i_{2c} +$$

$$+ \left[2\sqrt{3} \left(1 + l_n - \frac{3}{4}a_1^2 \right) + \frac{r}{\omega} \right] i_{2s} + \frac{3}{2}a_1^2 \left(i_{4c} - \sqrt{3}i_{4s} \right) = 0,$$

$$\dots \quad (3.16.17)$$

Let us perform the reduction of this system similar to the case of single-phase machine. The approximate values of coefficients of the first harmonic $i_a(\varphi)$ are equal to $i_{c1} = \Psi_f j_{1c}$, $i_{s1} = \Psi_f j_{1s}$ where

$$j_{1c} = \frac{-a_1 (1 + l_n)}{d}, \quad j_{1s} = \frac{-a_1 r/\omega}{d},$$

$$d - \left(1 + l_n - \frac{3}{2}u_1^2 \right)(1 + l_n) + (r/\omega)^2. \quad (3.16.18)$$

Taking into account only the first harmonic of the current we can write the averaged equations in the form

$$\Psi_f' = -\frac{\varepsilon_f r_f}{\omega}\Psi_f \left(1 - \frac{3}{2}a_1 j_{1c} \right) + \frac{\varepsilon_f e_f}{\omega},$$

$$\omega' = \frac{\varepsilon_\omega}{\omega}\Psi_f^2 \left[-\frac{3}{2}a_1 j_{1s} + \frac{9}{8}a_1 j_{1s} j_{1c} - \frac{3\sqrt{3}}{8}a_1^2 (j_{1s}^2 + j_{1c}^2) \right] + \frac{\varepsilon_\omega}{\omega}m.$$

$$(3.16.19)$$

The obtained equations allow one to determine the dependence of the mean electromagnetic moment on the angular velocity of the steady-state regime. To this end, one needs to obtain the steady-state value of flux linkage of the excitation winding $\Psi_f(\omega)$ from the first equation at $\Psi_f' = 0$ and substitute it in the second one. The dimensionless electromagnetic moment is equal to the first term in the right hand side of the second equation in eq. (3.16.19). By analogy with the calculation of the steady-state regime of "standard" synchronous machines and the single-phase inductor machines, the dependence of the static electromagnetic moment on frequency of the steady-state regime has a maximum whose value does not depend on the load resistance. The further reasonings in regard of the number of the steady-state regimes, stability and types of the phase portrait of system (3.16.19) are completely similar to the single-phase inductor machines, see Section 3.11.

Along with the above study of the steady-state regime the calculation of the transient behaviour under discontinuous change of the load resistance

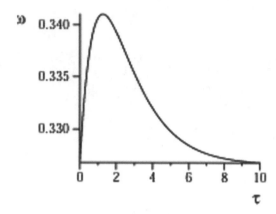

FIGURE 3.27.

was performed. In order to keep the frequency of the generator, the motor torque is changed such that its new value corresponds to the steady-state regime of the same frequency but a new load resistance. The dependence of the angular velocity on time in the transient behavior is displayed in Fig. 3.27 for the case in which the dimensionless resistance is changing from $r = 1$ to $r = 3$.

3.17 Feature of transient processes for star connection with zero wire

In this case the electric circuit of the anchor has three independent contour currents i_a, i_b, i_c. The flux linkages of all contours are found by analogy with the previous case

$$
\begin{aligned}
\Psi_a &= (zw^2 g_a + L_n)i_a + zww_0 g_a i_f, \\
\Psi_b &= (zw^2 g_b + L_n)i_b + zww_0 g_b i_f, \\
\Psi_c &= (zw^2 g_c + L_n)i_c + zww_0 g_c i_f, \\
\Psi_f &= zww_0 g_a i_a + zww_0 g_b i_b + zww_0 g_c i_c + zw_0^2 g_0 i_f.
\end{aligned}
\tag{3.17.1}
$$

The expression for energy of the magnetic field is given by

$$
W = \frac{1}{2}(zw^2 g_a + L_n)i_a^2 + \frac{1}{2}(zw^2 g_b + L_n)i_b^2 + \frac{1}{2}(zw^2 g_c + L_n)i_c^2 +
$$
$$
+ \frac{1}{2}zw_0^2 g_0 i_f + zww_0 g_a i_a i_f + zww_0 g_b i_b i_f + zww_0 g_c i_c i_f.
\tag{3.17.2}
$$

For $L_n = 0$ we obtain the relationship

$$\Psi_f = \frac{w}{w_0}(\Psi_a + \Psi_b + \Psi_c).$$ (3.17.3)

It follows from this equation that for a pure active load ($L_n = 0$) and negligible dissipation fluxes, four electric contours of the machine are fully magnetically coupled and one current can be expressed with the help of the Lagrange-Maxwell equations in terms of the other currents. At first glance there are substantial differences from the case of circuit without zero wire. Let us show, that this is not the case. We will also show that for $l_n = 0$ and $l_n \neq 0$ the final form of equations does not change.

Let us introduce the dimensionless variables (similar to the above) and take into account the dissipation fluxes in the expressions for the flux linkages

$$\Psi_a - (g_a + l_n)i_a + g_a i_f + \varepsilon_r \gamma i_a,$$
$$\Psi_b = (g_b + l_n)i_b + g_b i_f + \varepsilon_r \gamma i_b,$$
$$\Psi_c - (g_c + l_n)i_c + g_c i_f + \varepsilon_r \gamma i_c,$$
$$\Psi_f = g_a i_a + g_b i_b + g_c i_c + i_f + \varepsilon_r \gamma_f i_f.$$ (3.17.4)

Here we take the simplest expressions for $\varepsilon_r \gamma i_a, \varepsilon_r \gamma i_b$ etc. for the dissipation fluxes. A particular form of these expressions is not so important, it is only important that the dissipation exists and is small, i.e. ε_r is a small value.

Having relations (3.17.2) and (3.17.4) at disposal it is possible to construct the Lagrange-Maxwell equations in terms of dimensionless variables

$$[(g_a + l_n + \varepsilon_r \gamma)i_a + g_a i_f]^{\cdot} + r i_a = 0,$$
$$[(g_b + l_n + \varepsilon_r \gamma)i_b + g_b i_f]^{\cdot} + r i_b = 0,$$
$$[(g_c + l_n + \varepsilon_r \gamma)i_c + g_c i_f]^{\cdot} + r i_c = 0,$$
$$[g_a i_a + g_b i_b + g_c i_c + (1 + \varepsilon_r \gamma_f)i_f]^{\cdot} + \varepsilon_f r_f i_f = \varepsilon_f e_f,$$
$$\dot{\omega} = \varepsilon_\omega \left[\frac{1}{2}g_a' i_a^2 + \frac{1}{2}g_b' i_b^2 + \frac{1}{2}g_c' i_c^2 + g_a' i_a i_f + g_b' i_b i_f + g_c' i_c i_f \right] + \varepsilon_\omega m,$$
$$\dot{\varphi} = \omega.$$ (3.17.5)

Let us introduce the new variables $i_0 = i_a + i_b + i_c$ and Ψ_f instead of i_c and i_f

$$i_c = i_0 - i_a - i_b,$$
$$i_f = \Psi_f - (g_a - g_c)i_a - (g_b - g_c)i_b - g_c i_0.$$ (3.17.6)

Let us add the first three equations in (3.17.5) and subtract the fourth equation. Replacing i_c, i_f according to eq. (3.17.6) yields

$$[(g_a + l_n - g_a^2 + g_a g_c)i_a - g_a(g_b - g_c)i_b - g_a g_c i_0 + g_a \Psi_f]^{\cdot} + r i_a = 0,$$

$$[(g_b + l_n - g_b^2 + g_b g_c)i_b - g_b(g_a - g_c)i_b - g_b g_c i_0 + g_b \Psi_f]^{\cdot} + r i_b = 0,$$

$$[l_n + \gamma \varepsilon_r + \varepsilon_r \gamma_f g_c)i_0 + \varepsilon_r \gamma_f(g_a - g_c)i_a + \varepsilon_r \gamma_f(g_b - g_c)i_b - \varepsilon_r \gamma_f \Psi_f]^{\cdot} +$$

$$+ r i_0 - \varepsilon_f r_f[\Psi_f - (g_a - g_c)i_a - (g_b - g_c)i_b - g_c i_0] = \varepsilon_f e_f,$$

$$\dot{\omega} = \varepsilon_\omega \left\{ \left[\frac{1}{2}(g_a' + g_c') - (g_a' - g_c')(g_a - g_c) \right] i_a^2 + \left[\frac{1}{2}(g_b' + g_c') - \right. \right.$$

$$(g_b' - g_c')(g_b - g_c)] i_b^2 + [g_c' - (g_a' - g_c')(g_b - g_c) - (g_b' - g_c')(g_a - g_c)]i_a i_b +$$

$$(g_a' - g_c')i_a \Psi_f + (g_b' - g_c')i_b \Psi_f - [g_c' + (g_a' - g_c')g_c + g_c'(g_a - g_c)]i_a i_0 -$$

$$[g_c' + (g_b' - g_c')g_c + g_c'(g_b - g_c)]i_b i_0 + g_c' i_0 \Psi_f - g_c' g_c i_0^2 \} + \varepsilon_\omega m,$$

$$\dot{\varphi} = \omega. \tag{3.17.7}$$

In the first two equations in (3.17.7) we omit the small terms proportional to ε_r since they are insignificant. Vice versa, the small terms in the third equation are required for analysis of the case of $l_n = 0$ or small l_n. In this case the unknown variables in the Lagrange-Maxwell equations are split in three groups: slow variables Ψ_f and ω with the derivatives of the order of $\varepsilon_f, \varepsilon_\omega$; fast variables i_a, i_b, φ with the derivatives of the order of unity and one unknown variable i_0 which is ultrafast and has a derivative of the order of $1/\varepsilon_r$. For example, the transient process at the start of machine has respectively three stages. First, in the dimensionless time of order of ε_r (in the first approximation) total current i_0 vanishes. Then, in the time of order of unity the fast processes in the anchor circuits decay and the anchor currents approaches a quasi-stationary regime. Within this time Ψ_f and ω do not change in the first approximation. Finally, in accordance with the equations for slow nonstationary processes the machine approaches the steady-state regime in the time of order of $\max(1/\varepsilon_f, 1/\varepsilon_\omega)$.

When l_n is considerable, the ultrafast process does not exist and all three variables i_a, i_b, i_0 approach a quasi-stationary regime in the time of the order of unity. In the first approximation $i_0 = 0$ in this regime.

Now we show that the equations of the quasi-stationary regime for the machine with zero wire coincide, in the first approximation, with those in eq. (3.16.12). Putting $i_0 = 0$ in the first two equations in (3.17.7) we multiply the first equation by two and add with the second one. Taking into account that $g_a + g_b + g_c = 1$ yields the following expressions for the coefficients in front of i_a and i_b

$$2g_a + 2l_n - 2g_a^2 + 2g_a g_c + g_b g_c - g_a g_b = g_a + g_c + 2l_n - (g_a - g_c)^2,$$

$$-2ga(g_b - g_c) + g_b + l_n - g_b^2 + g_b g_c = g_c + l_n - (g_a - g_c)(g_b - g_c).$$

The second equation from eq. (3.16.12) is obtained summing the first equation in (3.17.7) with the doubled second one. The coefficients of Ψ_f

are obtained in the similar manner, for example

$$[(2g_a + g_b)\Psi_f]' = [(1 + g_a - g_c)\Psi_f]' = (g_a - g_c)'\Psi_f.$$

The other equations from (3.16.12) for the slow variables Ψ_f, ω are obtained from eq. (3.17.7) by substituting $i_0 = 0$.

Thus, if one is not interested in rapidly attenuating "initial" processes then the machines with and without zero wire are described by the same equations. For "standard" synchronous machines this can be proved by Park's transformation, "$\alpha, \beta, 0$" transformation etc. and is correct for any values of parameters. In the present case the proof is more difficult and substantially relies on smallness of parameters $\varepsilon_f, \varepsilon_\omega$ and ε_r.

4

Periodic solutions in problems of excitation of mechanical oscillations

4.1 Special form of notation for equations of motion and their solutions

To begin with, the systems under consideration are naturally split into two subsystems: a mechanical oscillating system and a vibration exciter (or exciters). These subsystems are prescribed to some extent arbitrarily, in particular, the same exciter can be related to several oscillatory systems. However, in each particular case the processes in the exciter and the motion of the oscillatory system influence each other. The forces created by the exciter affect the oscillatory system and there always exist some elements of the system on which these forces act directly. Alternatively, the reaction forces of the element comprise a part of the exciter and their motion affects the processes in the exciter. If the motion of the mentioned elements is known then the processes in the exciter are prescribed and the motion of the remaining elements of the oscillatory system is no longer needed.

Taking this into account we cast the equations for the values describing the processes in the exciter in such a way that they contain not the coordinates of the oscillatory system but auxiliary quantities (denoted by $\xi_1, \ldots, \xi_k$) which are the displacements or the rotation angles corresponding to the reaction forces in the elements. Let us also assume that the oscillatory system is linear, i.e. its motion is governed by linear differential equations with time-independent coefficients. If we denote the set of generalised coordinates describing the configuration of the oscillatory system by v, then v is either a vector of finite dimension or an element of a

Hilbert space G. Let us denote the "proper" coordinates of the exciter by $q = (q_1, \ldots, q_m)$ and write down the kinetic potential of the system in the form

$$L = L_1(q, \dot{q}, \xi, \dot{\xi}) + L_2(v, \dot{v}). \qquad (4.1.1)$$

Provided that dependence $\xi(t)$ is given, the equations of motion must provide us with the possibility to determine $q(t)$ regardless of the dimension of vector v and the time-history of its separate components. Thus, the form of $L_1(q, \dot{q}, \xi, \dot{\xi})$ must be independent of the way of introducing the coordinates v, their number etc. In other words, function L_1 must be invariant to the particular form of the oscillatory system. But if the oscillatory system is taken then $\xi_1, \ldots, \xi_k$ should be expressed in terms of the components of vector v. If the system is linear, then ξ_i must be linear functionals of v and thus be represented by the scalar products of the form

$$\xi_i = (v, v_i), \quad i = 1, \ldots, k, \qquad (4.1.2)$$

where v_i denote constant vectors or elements of the same space as v (however it may also be a larger space). The values of ξ_i describing the "backward" influence of the oscillations on the motion of the exciter do not necessarily belong to the generalised coordinates of the system. For this reason, they are referred to as the "functionals (or parameters) of the backward influence".

The Lagrange's equations are as follows

$$\frac{d}{dt} \frac{\partial L_1}{\partial \dot{q}} - \frac{\partial L_1}{\partial q} = R(q, \dot{q}) + E(t),$$

$$M\ddot{v} + B\dot{v} + Cv = \sum_{i=1}^{k} Q_i v_i. \qquad (4.1.3)$$

Here

$$Q_i = -\left(\frac{d}{dt} \frac{\partial L_1}{\partial \dot{\xi}_i} - \frac{\partial L_1}{\partial \xi_i} \right),$$

R and E describe non-potential forces in the exciter and the non-mechanical actions, respectively, M, B, C denote quadratic matrices or linear operators in space G. Guided by the physical meaning of the problem we take that M, B, C are symmetric and in addition to this M, C are positive-definite matrices whereas B is positive semi-definite.

One can see from eq. (4.1.3) that vectors v_i describe the distribution of the exciter's forces over the oscillatory system whereas coefficients Q_i in front of v_i determine the value of these forces. Additionally, the condition that v_i are constant values implies that the loading is referred to the undeformed oscillatory system. In contrast to quantities ξ_i whose mechanical meaning is solely determined by the exciter, vectors v_i will be determined only when the oscillatory system is prescribed.

The assumption that the oscillatory system is linear is related to the assumption of the smallness of its displacements compared with a certain typical dimension. However this does not mean that quantities Q_i and the first six equations in (4.1.3) can be linearised with respect to ξ, as the displacements can be compared with typical dimensions of the exciters. For instance, the linear elastic electromechanical systems possess this property. If the displacements are small in comparison with the latter dimensions, the problem, nonetheless, may remain non-linear (with a small parameter proportional to the ratio of the displacement to a typical size). In particular, this is the case for the oscillations excited by rotation due to an unbalanced rotor.

Eliminating ξ_i from eq. (4.1.3) by means of eq. (4.1.2) for any particular oscillatory system we can obtain the equations of motion in terms of the coordinates. However, there exists a number of cases in which one can study a class of system by using eq. (4.1.3). For example, let the equations in (4.1.3) be linear, then they are described by the Lagrange function

$$L = L_1 + L_2,$$

where

$$L_1 = \frac{1}{2}\sum_{r,s}^{m}(m_{rs}\dot{q}_r\dot{q}_s - c_{rs}q_rq_s) + \sum_{r}^{m}\sum_{i}^{k}(d_{ri}\dot{q}_r\dot{\xi}_i - h_{ri}q_r\xi_i),$$

$$L_2 = \frac{1}{2}(M\dot{v}, \dot{v}) - \frac{1}{2}(Cv, v).\tag{4.1.4}$$

Assuming for simplicity that the prescribed forces are harmonic and taking into account the damping we arrive at the following equations of motion

$$\sum_{s}^{m}(m_{rs}\ddot{q}_s + b_{rs}\dot{q}_s + c_{rs}q_s) = U_r \sin\omega t + V_r, (r = 1, \ldots, m),$$

$$M\ddot{v} + B\dot{v} + Cv = \sum_{i}^{k}Q_iv_i.\tag{4.1.5}$$

Here

$$V_r = -\sum_{i}^{k}(d_{ri}\ddot{\xi}_i + h_{ri}\xi_i), \quad Q_i = -\sum_{r}^{m}(d_{ri}\ddot{q}_r + h_{ri}q_r).\tag{4.1.6}$$

The quantities V_r can also be introduced in the general case by recasting the first m equations (4.1.3) in the following form

$$\frac{d}{dt}\frac{\partial L_{10}}{\partial \dot{q}} - \frac{\partial L_{10}}{\partial q} = R(q, \dot{q}) + E(t) + V,\tag{4.1.7}$$

where

$$L_{10} = L_1(q, \dot{q}, \xi \equiv 0, \dot{\xi} \equiv 0), \quad V = \frac{d}{dt} \frac{\partial(L_{10} - L_1)}{\partial \dot{q}} - \frac{\partial(L_{10} - L_1)}{\partial q}.$$

Putting $V = 0$ in eq. (4.1.7) we obtain equations of motion for the oscillatory system fixed in the equilibrium position. Thus, values of V characterise, to some extent, the effect of the backward influence of oscillations on the exciters. Quantities V can also be considered as the generalised forces of the same physical nature as the prescribed forces $E(t)$. With this in view, we refer to these forces as the "total generalised vibrational forces". Such quantities are encountered, for instance, in the problems of the dynamics of systems with mechanical exciters in which the values averaged over a period are called vibrational moments, [15].

Let us consider periodic solutions of system (4.1.5) having period $2\pi/\omega$. Looking for the solution in the form $q_r = a_r \sin(\omega t - \varepsilon_r)$ and inserting this into eq. (4.1.6), we can find Q_i as a function of $a = (a_1, \ldots, a_m)$ and $\varepsilon = (\varepsilon_1, \ldots, \varepsilon_m)$, i.e.

$$Q_i = Q_{i1}(a, \varepsilon) \sin(\omega t - \vartheta_i(a, \varepsilon)). \tag{4.1.8}$$

Let us introduce into consideration the components $k_1^{(ij)} \geq 0$ and $\psi_1^{(ij)}$ of the square $k \times k$ matrices K_1 and Ψ_1, determined as follows. Let a $2\pi/\omega$-periodic solution of equation

$$M\ddot{v}^{(i)} + B\dot{v}^{(i)} + Cv^{(i)} = v_i \sin \omega t \tag{4.1.9}$$

be given. Let us find the scalar products $(v^{(i)}, v_j)$. Their amplitude and phase angles are denoted in the following $k_1^{(ij)}, \psi_1^{(ij)}$

$$(v^{(i)}, v_j) = k_1^{(ij)} \sin(\omega t - \psi_1^{(ij)}). \tag{4.1.10}$$

Quantities $k_1^{(ij)}$ and $\psi_1^{(ij)}$ are determined from the solution of the problem of the forced oscillations of the oscillatory system subjected to prescribed harmonic forces and have the following physical meaning. The system is assumed to be subjected to a harmonic load of frequency ω with a unit amplitude ($Q_i = 1$) which is distributed over the oscillatory system as well as the part of the exciter's load corresponding to vector v_i. Let us determine the time-dependence of the $j - th$ functional of the backward influence ξ_j under forced harmonic oscillations caused by the mentioned load. The amplitude of ξ_j and the phase shift between ξ_j and the load are equal to $k_1^{(ij)}$ and $\psi_1^{(ij)}$ respectively.

If a system without damping is considered, then $k_1^{(ij)}$ are the influence coefficients. The reciprocity property $k_1^{(ij)} = k_1^{(ji)}$ as well as the property $\psi_1^{(ij)} = \psi_1^{(ji)}$ is also valid for the system with friction.

Having eq. (4.1.8) we can express the functional of the backward influence in terms of the amplitude a, angles ε and values $k_1^{(ij)}$, $\psi_1^{(ij)}$ as follows

$$\xi_j = \sum_i^k Q_{i1}(a, \varepsilon) k_1^{(ij)} \sin(\omega t - \vartheta_i(a, \varepsilon) - \psi_1^{(ij)}). \qquad (4.1.11)$$

The amplitudes and phases of the total vibrational forces can be expressed from eq. (4.1.6) in terms of the above values

$$V_r = V_{r1}(a, \varepsilon, K_1, \Psi_1) \sin(\omega t - \gamma_r(a, \varepsilon, K_1, \Psi_1)). \qquad (4.1.12)$$

Let us insert eq. (4.1.12) into the first m equations in (4.1.5). This allows us, in principle, to find the dependences $q_r = q_r(t, a, \varepsilon, K_1, \Psi_1)$. Determining the amplitudes and phases of q_r we arrive at the system for $a_1, \ldots, a_m$, $\varepsilon_1, \ldots, \varepsilon_m$ which allows us to determine these values as functions of the components of the matrices K_1 and Ψ_1. Making use of eq. (4.1.8), we finally obtain

$$a = a(K_1, \Psi_1), \ \varepsilon = \varepsilon(K_1, \Psi_1), \ Q_{i1} = Q_{i1}(K_1, \Psi_1), \ \vartheta_i = \vartheta_i(K_1, \Psi_1). \qquad (4.1.13)$$

The latter set of these relationships allows us to find the forces in terms of matrices K_1 and Ψ_1. Thus, if relationships (4.1.1) are constructed for an exciter, then the problem of the forced oscillation reduces to the determination of matrices K_1 and Ψ_1 as well as to using these relationships. In order to construct relationships (4.1.13) it is necessary to know only that part of Lagrange's function L_1, which is written in terms of the parameters of the backward influence.

Instead of amplitudes a and phases c one can determine the amplitudes of the sine and cosine components. The corresponding matrices of these amplitudes should also be introduced for the oscillatory system.

Let us consider now the case in which the equations of motion contain a small parameter μ and are reduced to the form

$$\dot{x} = X(x, t) + \mu Y(x, \xi, \dot{\xi}, t, \mu),$$

$$M\ddot{v} + B\dot{v} + Cv = \sum_i^k Q_i(x, \dot{x}) v_i + \mu \ldots . \qquad (4.1.14)$$

Here $x = (x_1, \ldots, x_p)$ are unknown variables which should be introduced instead of $q_1, \ldots, q_m$ in order to cast the equations in the form of (4.1.1).

The equations of this particular type are obtained in the case when the displacements in the oscillatory system can be understood as being small of order of μ in comparison with a typical dimension of the exciter. Particularly, this is the case for systems with mechanical vibration exciters [66] and synchronising systems [15]. In what follows we consider the case when the equations in (4.1.1) are Routh's equations, this case covering the problems of oscillations under action of electromagnets.

Let functions X and Y be $2\pi/\omega$−periodic with respect to t. Let us also assume that for any continuous $2\pi/\omega$−periodic function $\varphi(t)$ each of the equations

$$M\ddot{v} + B\dot{v} + Cv = \varphi(t)v_i \qquad (4.1.15)$$

admits a unique $2\pi/\omega$−periodic solution $v_\varphi^{(i)}(t)$ such that $\max\limits_t |(v_\varphi^{(i)}(t), v_i)|/$ $\max\limits_t |\varphi(t)| < h^{(i)}$, where a constant $h^{(i)}$ is coincident for all φ and $h^{(i)} = O(1)$. This assumption corresponds to the so-called "non-resonant" case, see the resonant case in which $h^{(i)} = O(1/\mu)$.

The generating solutions of eq. (4.1.1) are considered to be non-isolated. In the case of an isolated generating solution the influence of the oscillations results only in small corrections to variables x obtained without vibrational forces. Hence, the backward influence of the oscillations on the exciter's motion can not lead to qualitatively new effects and is of little interest from this perspective. For this reason, let us consider the case in which the system

$$\dot{x} = X(x, t) \qquad (4.1.16)$$

admits a family of $2\pi/\omega$−periodic solutions $x = x_0(t, \alpha), \alpha = (\alpha_1, \ldots, \alpha_n)$ depending upon n arbitrary parameters. Let functions $x_0(t, \alpha)$ be given, then functions $Q_i(x_0, \dot{x}_0)$ are also given functions of t, α. Therefore, the coefficients of the Fourier series

$$Q_i = \sum_\nu Q_{i\nu}(\alpha) \cos(\nu\omega t - \vartheta_{i\nu}(\alpha)) \qquad (4.1.17)$$

can be also treated as given functions of α. Let us introduce, as above, the matrices of the harmonic influence coefficients and phase shifts. These matrices should be introduced for all frequencies appearing in eq. (4.1.1). Let us denote these matrices and their components as K_ν, Ψ_ν and $k_\nu^{ij}, \psi_\nu^{ij}$, respectively. They are defined by the relationships

$$(v_\nu^i, v_j) = k_\nu^{(ij)} \cos(\nu\omega t - \psi_\nu^{(ij)}), \quad k_\nu^{(ij)} \geq 0, \qquad (4.1.18)$$

where $v_\nu^{(i)}$ denotes a $2\pi/\nu\omega-$ periodic solution of the equation

$$M\ddot{v}_\nu^{(i)} + B\dot{v}_\nu^{(i)} + Cv_\nu^{(i)} = v_i \cos\nu\omega t. \qquad (4.1.19)$$

The physical meaning of matrices K_ν, Ψ_ν for $\nu \neq 1$ is coincident with that of matrices K_1, Ψ_1, with the only difference being that the oscillatory system is subjected the unit load of frequency $\nu\omega$ rather than ω.

Now we can write down the expressions for the functionals of the backward influence in the generating approximation which contain the components of matrices K_ν, Ψ_ν and values $\alpha_1, \ldots, \alpha_n$ as parameters, i.e.

$$\xi_{j0} = \sum_i \sum_\nu^k Q_{i\nu}(\alpha) \cos(\nu\omega t - \vartheta_{i\nu}(\alpha) - \psi_\nu^{(ij)})k_\nu^{(ij)}. \qquad (4.1.20)$$

Let us also assume that we have $2\pi/\omega-$periodic solutions $z_1, \ldots, z_n$ of the system of linear equations with periodic coefficients which is referred to as the system conjugate to the variational system of equations for the generating solution

$$\dot{z} + Zz = 0, \qquad Z = \| (\partial X_s / \partial x_r)_0 \| . \qquad (4.1.21)$$

Here z denotes a vector with the same dimension as x and the zero subscript implies that the derivatives are taken at $x = x_0$. Then the equations for determination of $\alpha_1, \ldots, \alpha_n$ can be written such that they contain the components K_ν, Ψ_ν as parameters. They have the form

$$P_r(\alpha_1, \ldots, \alpha_n, K, \Psi) = \left\langle \sum_s^n z_{rs} Y_{s0}(t, \alpha, K, \Psi) \right\rangle = 0, \quad r = 1, \ldots, n .$$
$$(4.1.22)$$

Here $Y_{s0}(t, \alpha, K, \Psi) = Y_s(x_0, \xi_0, \dot{\xi}_0, t, 0)$ denotes the components of vector Y, whereas symbols K, Ψ denote the entire set of matrices K_ν, Ψ_ν.

Provided that solutions $\alpha = \alpha(K, \Psi)$ of system (4.1.22) are found as functions of K, Ψ and are substituted into the above relationships, then all of the required variables in the generating approximation are also found as functions of K, Ψ. As a result, determination of the oscillations in any linear oscillatory system excited by an exciter with given expression for Q_i reduces to the solution of the problem of the forced oscillation and using derived expressions. The elements of the determinant $|\partial P_r/\partial \alpha_s|$ and thus the stability conditions can be expressed in a similar form in terms of the components of matrices K, Ψ.

Relationships containing matrices K, Ψ are also useful in that they can be utilised when the equations of motion for the oscillating system are not yet derived, whereas these matrices are obtained, for example, experimentally.

The solution can be cast in this form when the system (4.1.1) is autonomous. In this case it is necessary to know the dependence $K(\nu\omega), \Psi(\nu\omega)$ in some intervals of the values of ω. Additionally, if system (4.1.1) is nearly conservative, then the stability conditions, see [42], contain not only $K(\nu, \omega), \Psi(\nu\omega)$ but also their derivatives with respect to ω.

Khodzhaev suggested solutions for a number of problems on oscillations under the action of electromagnets in [41], [43], [45] and solutions for the problems on synchronisation of mechanical vibrations in [42]. The same approach was suggested by Sperling [113] however the stability conditions for the second set (derived in [42]) are absent in this paper.

4.2   Integral stability criterion for periodic motions of electromechanical systems and systems with quasi-cyclic coordinates

Let us construct the equations governing the quasi-stationary oscillations of electromechanical systems with closed linear currents without assuming small displacements

$$\frac{d}{dt}\frac{\partial W}{\partial i_s} + \frac{\partial F}{\partial i_s} = E_s(t), \quad s = 1, \ldots, m,$$

$$\frac{d}{dt}\frac{\partial L_M}{\partial \dot{q}_{m+s}} - \frac{\partial L_M}{\partial q_{m+s}} - \frac{\partial W}{\partial q_{m+s}} = Q_{m+s}, \quad s = 1, \ldots, n - m. \qquad (4.2.1)$$

Here $i_1, \ldots, i_m$ denote currents, F is the "electric dissipation function", $E_1, \ldots, E_m$ are prescribed $2\pi/\omega-$periodic electromotive forces (emf), $q_{m+1}, \ldots, q_n$ are the generalised coordinates, L_M is the kinetic potential, and $Q_{m+1}, \ldots, Q_n$ are the non-potential generalised forces.

For inductively-connected non-branched loops we have

$$F = \frac{1}{2}\sum_{s}^{m} R_s i_s^2. \qquad (4.2.2)$$

Let us adopt that F has the same form in the general case which can be achieved by a linear transformation of the original currents.

Let us express the field energy W in terms of the inductances and currents

$$W = \frac{1}{2}\sum_{r,s}^{m} L_{rs}(q_{m+1}, \ldots, q_n)i_s i_r. \qquad (4.2.3)$$

Let E_*, L_* and R_* denote the characteristic values of the emf, inductance and resistance, respectively. Let us introduce characteristic current $i_* = E_*/\omega L_*$ and take $\tau = \omega t$. Then we can recast the first m equations in the following non-dimensional form

$$\frac{d}{d\tau}\sum_{r}^{s} l_{rs}\eta_r + \rho\beta_s\eta_s = e_s, \quad s = 1, \ldots, m. \qquad (4.2.4)$$

Here $l_{rs} = L_{rs}/L_*, \eta_r = i_r/i_*, \beta_s = R_s/R_*, e_s = E_s/(\omega L_* i_*), \rho = R_*/\omega L_*$. Assuming the electric dissipation to be small, that is the characteristic inductive (ωL_*) and active (R_*) resistances fulfill the relationship $\omega L_* \gg R_*$, we deem ρ to be a small parameter. Systems with spatial conductors are considered under analogous assumptions. Besides, in what follows we study a peculiar, but interesting from the technological perspective, case in which it is sufficient to assume that only few of the R_s are small. Thus, eq. (4.2.4) contains equations with small parameter ρ. Let η_r, q_{m+s} be obtained in the form of a power series in terms of ρ. If we return to the original dimensional variables, then the result must coincide with that obtained if the solution of eq. (4.2.1) is sought in the form of a power series in terms of R_*, provided that the value of R_* is conditionally taken as being small in comparison with the other values of any dimension. With this in view, we do not introduce the small parameter ρ explicitly and manipulate directly the equations in (4.2.1).

Averaging the first m equations in (4.2.1) over the period and denoting the mean values by $\langle\ \rangle$ we obtain

$$R_s\langle i_s\rangle = \langle E_s\rangle, \quad \langle i_s\rangle = \frac{\omega}{2\pi}\int\limits_0^{2\pi/\omega} i_s(t)dt. \tag{4.2.5}$$

It follows that the constant components $\langle E_s\rangle$ of the given emf should be small. Indeed, if $\langle E_s\rangle \sim E_*$, then $\langle i_s\rangle/i_* \sim 1/\rho$. In practice, these currents are approximately equal to the short circuit current under the nominal voltage. For the formal viewpoint, another procedure of Poincaré's method is needed. It is natural to take $\langle E_s\rangle$ as being small of the order of R_s, then $\langle i_s\rangle \sim i_*$. This corresponds to the nominal working regimes of the electromechanical facilities.

Let us express the ponderomotive force $\partial W/\partial q_{m+r}$ in terms of the magnetic fluxes $\Phi_r = \partial W/\partial i_r$. Then equations (4.2.1) can be integrated by means of the following scheme. Let us represent E_s in the form $E_s = U_s + \langle E_s\rangle$, $\langle U_s\rangle = 0$. Omitting the small terms in eq. (4.2.1) and denoting the generating approximation by a zero subscript we obtain

$$\Phi_{s0} = \alpha_s + V_s(t), \quad \dot V_s = U_s, \quad \langle V_s\rangle = 0. \tag{4.2.6}$$

Here α_s denote arbitrary constant components of yet undetermined magnetic fluxes. Let us substitute eq. (4.2.6) into the second set of equations in (4.2.1). Then we arrive at the equations for the oscillations of a mechanical system under the actions of forces which are prescribed functions of time, mechanical coordinates and parameters α_s. They allow us, in principle, to determine q_{m+r} as a function of t and α_s. This means that the family of generating solutions depending on m is obtained. Among these solutions, we should take those which are equal to the periodic solutions of the original system (with small terms) for $\rho = 0$. Only certain values of $\alpha_1, \ldots, \alpha_m$ correspond to these generating solutions. In this particular case these values are the solutions of the following system of equations

$$P_s(\alpha_1, \ldots, \alpha_m) \equiv \langle i_{s0}(\alpha_1, \ldots, \alpha_m)\rangle - i_{cs} = 0, \quad s = 1, \ldots, m. \tag{4.2.7}$$

Here $i_s = \partial W/\partial\Phi_s$ denote the currents which are considered as functions of the magnetic fluxes and the mechanical coordinates, i.e. $i_{cs} = \langle E_s\rangle/R_s$. For sufficiently small ρ, for each solution of system (4.2.7) there is a periodic solution of system (4.2.1). Taking one of them and inserting the values $\alpha_1, \ldots, \alpha_m$ into eq. (4.2.6) and into the expressions for $q_{m+s0}(t, \alpha_1, \ldots, \alpha_m)$, $i_{s0}(t, \alpha_1, \ldots, \alpha_m)$ we find the required variables with accuracy up to the order of ρ. The higher order terms can be found, if required. Obtaining the derivatives $\partial P_s / \partial \alpha_r$ one can analyse the stability of the regime under consideration, [78].

However, application of the described procedure is made difficult by the fact that the equations for determining $q_{m+s0}(t, \alpha_1, \ldots, \alpha_m)$ are in general non-linear. Provided that they are piecewise-linear, their solution can be obtained by the matching method. In general, one should use either approximate methods, or numerical integration. But this strategy, especially the numerical approaches, does not allow one to determine explicit dependences of q_{m+s0} on $\alpha_1, \ldots, \alpha_m$. Therefore, what remains is to prescribe various values of $\alpha_1, \ldots, \alpha_m$, determine solutions $q_{m+s0}(t)$ corresponding to each set of $\alpha_1, \ldots, \alpha_m$ and calculate the values of P_s aiming at approaching the solution of eq. (4.2.7) by changing $\alpha_1, \ldots, \alpha_m$.

The above calculations are simplified if it is known that the required values of $\alpha_1, \ldots, \alpha_m$ renders an extremum to a certain function $\Lambda(\alpha_1, \ldots, \alpha_m)$. Indeed, in this case, the sets of $\alpha_1, \ldots, \alpha_m$ can be taken by using the conventional search methods for an extremum. In the present chapter we indicate the cases in which such a function Λ exists and establish its physical meaning. Since equations (4.2.1) comprise a particular case of the equations of mechanics with quasi-cyclic coordinates we will consider the latter, more general equations.

The relations between the stable periodic solutions of the system with small parameter and the points of minima of a certain scalar function Λ was first established by Blekhman for synchronisation problems [15]. He referred to this principle as the integral stability criterion. This result was generalised in [89] for the class of systems called quasiconservative synchronising systems. Particular cases of the problem of synchronisation were studied by Lavrov in [70] by means of the integral criterion. These authors are of the opinion that the integral criterion is of interest itself (regardless of the usefulness for calculations) and link this with physical clarity and the fact that Λ can be constructed without detailed equations of motion.

It is interesting to clarify what systems, along with the above mentioned, obey the integral criterion. As follows from its derivation, see [89], [44], it is necessary that the equations with small parameter have the structure of mechanical equations and arbitrary constants appear in the generating approximation in a "natural" way. In mechanics, there exist two classes of systems whose periodic motions form families depending upon arbitrary constants. These are the conservative systems (for which the arbitrary parameters are initial phases and the energy constant or period) and the systems with cyclic coordinates (the parameters are cyclic momenta). The system with quasi-cyclic coordinates which is considered in what follows belongs to the second most natural class of the systems for which the integral criterion holds.

Let us consider a system with holonomic stationary constraints described by m quasi-cyclic $(q_1, q_2, \ldots, q_m)$ and $n - m$ positional $(q_{m+1}, \ldots, q_n)$ coordinates. It is assumed that the quasi-cyclic coordinates correspond to the generalised coordinates of two sorts, namely small forces of viscous damping and forces depending only on time, with the latter being assumed to be $2\pi/\omega$−periodic. Lagrange's equations for such a system has the form

$$\dot{p}_s + \mu b_s \dot{q}_s = U_s(t) + \mu U_{cs}, \quad s = 1, \ldots, m,$$

$$\frac{d}{dt}\frac{\partial L}{\partial \dot{q}_{m+s}} - \frac{\partial L}{\partial q_{m+s}} = N_{m+s}, \quad s = 1, \ldots, n - m, \qquad (4.2.8)$$

$$L = T(q_{m+1}, \ldots, q_n, \dot{q}_1, \ldots, \dot{q}_n) - \Pi(q_{m+1}, \ldots, q_n).$$

Here p_s denote the quasi-cyclic momenta, $\langle U_s \rangle = 0, U_{cs} = \text{const}, N_{m+s}$ are the non-potential generalised forces corresponding to the positional coordinates, L is the kinetic potential of the system, and μ is a small parameter. One can adopt that $\mu > 0$, then $b_s > 0$. A more general case in which the "quasi-cyclic" dissipative forces are given by the dissipation function of the form

$$F = \frac{1}{2}\sum_{r,s}^{m} b_{rs}\dot{q}_r\dot{q}_s$$

reduces to the previous case by a linear transformation of only quasi-cyclic coordinates.

Let us consider Routh's equations corresponding to eq. (4.2.8)

$$\dot{p}_s - \mu b_s \frac{\partial L_R}{\partial p_s} = U_s(t) + \mu U_{cs}, \quad s = 1, \ldots, m,$$

$$\frac{d}{dt}\frac{\partial L_R}{\partial \dot{q}_{m+s}} - \frac{\partial L_R}{\partial q_{m+s}} = N_{m+s}, \quad s = 1, \ldots, n - m. \qquad (4.2.9)$$

Here L_R denote Routh's kinetic potential related to Lagrange's function as follows

$$L_R = L - \sum_{s}^{m} p_s \dot{q}_s. \qquad (4.2.10)$$

L_R is inserted into eq. (4.2.9) as a function of $p_1, \ldots, p_m, \dot{q}_{m+1}, \ldots, \dot{q}_n, q_{m+1}, \ldots, q_n$. If the original Lagrange's function L is prescribed as a function of $\dot{q}_1, \ldots, \dot{q}_m$ and the positional coordinates and velocities, then it is necessary to express the quasi-cyclic velocities in terms of momenta and $\dot{q}_{m+1}, \ldots \dot{q}_n, q_{m+1}, \ldots, q_n$ from the linear equations for $\dot{q}_1, \ldots, \dot{q}_m$

$$\frac{\partial T}{\partial \dot{q}_s} = p_s, \quad s = 1, \ldots, m. \qquad (4.2.11)$$

For $\mu = 0$, the following system of m equations can be separated from eq. (4.2.9)

$$\dot{p}_{s0} = U_s(t), \quad s = 1, \ldots, m, \tag{4.2.12}$$

which yields, up to the constants $\alpha_1, \ldots, \alpha_m$, the quasi-cyclic momenta in the generating approximation

$$p_{s0} = \alpha_s + V_s(t), \quad s = 1, \ldots, m. \tag{4.2.13}$$

Here and in what follows, the antiderivatives $V_s, \dot{V}_s = U_s$ are taken such that $\langle V_s \rangle = 0$.

Let us insert eq. (4.2.13) into the last $n - m$ equations in (4.2.9). Then we obtain equations for q_{m+s0}, in which $\alpha_1, \ldots, \alpha_m$ appear as parameters. Let us assume that these equations admit a stable $2\pi/\omega$−periodic isolated solution (i.e. such a solution which has no new constants) q_{m+s0} for any $\alpha_1, \ldots, \alpha_m$ from a certain region A. The set of functions $p_{s0}(t, \alpha_s)$, $q_{m+s0}(t, \alpha_1, \ldots, \alpha_m)$ comprises a family of the generating solutions. Inserting p_{s0}, q_{m+s0} into the small terms of the first m equations in eq. (4.2.9) we obtain the equations for the constants $\alpha_1, \ldots, \alpha_m$ from the conditions of periodicity of the first approximation (p_{s1}) to the quasi-cyclic momenta. In this case, the periodicity conditions reduce to the requirement that the small terms do not contain constant components after substitution of $p_s = p_{s0}, q_{m+s} = q_{m+s0}$. Then we arrive at equations

$$P_s(\alpha_1, \ldots, \alpha_m) = -\left\langle \frac{\partial L_R}{\partial p_s} \right\rangle_0 - i_{cs} = 0, s = 1, \ldots, m. \tag{4.2.14}$$

Here $i_{cs} = U_{cs}/b_s$. Let system (4.2.1) admits the solution $\alpha_s = \alpha_{s*}, s = 1, \ldots, m$ belonging to A. Then the periodic regime corresponding to this solution is stable if roots $\lambda'_1, \ldots, \lambda'_m$ of the equation

$$\det \left\| \left(\frac{\partial P_r}{\partial \alpha_s} \right)_* + \lambda' \kappa_s \delta_{rs} \right\| = 0 \tag{4.2.15}$$

have negative real parts and is unstable if $\operatorname{Re} \lambda'_i > 0$. The case of zero and pure imaginary roots is beyond the scope of the present book. In eq. (4.2.1) $\kappa_s = 1/b_s$, δ_{rs} being the Kronecker delta.

Let us clarify when the conditions

$$P_s = -\frac{\partial}{\partial \alpha_s} \left(\langle L_R \rangle_0 + \sum_{s=1}^{m} i_{cs}\alpha_s \right), \quad s = 1, \ldots, m \tag{4.2.16}$$

are satisfied. Integrating by parts we obtain

$$\left\langle \frac{\partial L_R}{\partial p_s} \right\rangle_0 = \frac{\partial}{\partial \alpha_s} \langle L_R \rangle_0 - \sum_{r=1}^{n-m} \left\langle \frac{\partial L_{R0}}{\partial \dot{q}_{m+r0}} \frac{\partial \dot{q}_{m+r0}}{\partial \alpha_s} + \frac{\partial L_R}{\partial q_{m+r0}} \frac{\partial q_{m+r0}}{\partial \alpha_s} \right\rangle$$

$$= \frac{\partial}{\partial \alpha_s} \langle L_R \rangle_0 + \sum_{r=1}^{n-m} \left\langle \left[\frac{d}{dt} \frac{\partial L_R}{\partial \dot{q}_{m+r}} - \frac{\partial L_R}{\partial q_{m+r}} \right]_0 \frac{\partial q_{m+r0}}{\partial \alpha_s} \right\rangle.$$

$$(4.2.17)$$

It follows from Routh's equations that relationships (4.2.1) are valid provided that

$$\sum_{r=1}^{n-m} \left\langle N_{m+r0} \frac{\partial q_{m+r0}}{\partial \alpha_s} \right\rangle = 0, \quad s = 1, \dots, m. \qquad (4.2.18)$$

Point $(\alpha_{1*}, \dots, \alpha_{m*})$ is a stationary point of function

$$\Lambda(\alpha_1, \dots, \alpha_m) = -\langle L_R \rangle_0 - \sum_{s=1}^{m} i_{cs} \alpha_s. \qquad (4.2.19)$$

In addition to this, if conditions (4.2.1) are satisfied then matrix $\partial P_r / \alpha_s$ is symmetric. In this case the eigenvalues λ'_i are real-valued and their signs (under the condition that all $\kappa_s > 0$) do not depend on the values of κ_s. Thus, the stability can be judged by signs of the roots $\lambda_1, \dots, \lambda_m$ of the equation

$$\det \left\| \left(\frac{\partial P_r}{\partial \alpha_s} \right)_* + \lambda \delta_{rs} \right\| = 0. \qquad (4.2.20)$$

Finally, we conclude that values $\alpha_1 = \alpha_{1*}, \dots, \alpha_m = \alpha_{m*}$ corresponding to the stable solution delivers a minimum for function $\Lambda(\alpha_1, \dots, \alpha_m)$.

The latter assertion is a formulation of the integral criterion of stability for systems of the considered class. The criterion is obviously valid if all of the forces corresponding to the positional coordinates are potential. However, it can also be valid in the case of non-potential forces N_{m+s}.

For example, let N_{m+s} be represented as linear forms in the generalised velocities

$$N_{m+s} = \sum_{r}^{n-m} \beta_{rs} \dot{q}_{m+r}, \qquad (4.2.21)$$

whereas q_{m+s} can be cast in the form of a series

$$q_{m+s0} = \sum_{\nu} q_{m+s0}^{(\nu)}(\alpha_1, \dots, \alpha_m) \cos(\nu \omega t - \varphi_\nu),$$

where the phase shifts $\varphi_{m+s0}^{(\nu)} = \varphi_\nu, \nu = 1, \dots$ of harmonics (i.e. components) of q_{m+s0} do not depend on $\alpha_1, \dots, \alpha_m$ and are equal to each other for all q_{m+s0}, $s = 1, \ldots, n - m$. Then

$$
\sum_s \left\langle N_{m+s0} \frac{\partial q_{m+s0}}{\partial \alpha_r} \right\rangle = \sum_{s,k} \beta_{ks} \left\langle - \sum_\nu \nu \omega q_{m+k0}^{(\nu)} \sin(\nu\omega t - \varphi_\nu) \times \right.
$$

$$
\left. \sum_\nu \frac{\partial q_{m+s0}^{(\nu)}}{\partial \alpha_r} \cos(\nu\omega t - \varphi_\nu) \right\rangle = 0 .
$$

We refer to functions with equal phase shifts of the components in the Fourier expansion as component-synphase, whilst the conditions $\varphi_{m+10}^{(\nu)} = \ldots = \varphi_{n0}^{(\nu)} = \varphi_\nu$ are called the component-synphase conditions. The above means that for the integral criterion to exist in the case where N_{m+s} are linear forms of $\dot{q}_{m+1}, \ldots, \dot{q}_n$, it is sufficient that the phase shifts in the expansion of the positional coordinates determined in the generating approximation do not depend on $\alpha_1, \ldots, \alpha_m$, whereas these coordinates satisfy the component-synphase conditions. Since no condition is imposed on the properties of coefficients β_{rs}, this criterion is valid in this case, in particular, when N_{m+s} are viscous damping forces.

This case has no analogy in the problems of periodic motion of quasi-conservative systems since their constants $\alpha_1, \ldots, \alpha_m$ are the phase shifts of the object coordinates [15], [89] which appear in the solution in the combinations $\omega t + \alpha_s$, values $\varphi_{m+s0}^{(\nu)}$ depend necessarily on $\alpha_1, \ldots, \alpha_m$ and the integral criterion holds only for a carrying (oscillatory) system without dissipation, [89].

The integral criterion holds also in the case when the sums in eq. (4.2.1) are derivatives of some function with respect to α_s, and in particular when they do not depend on $\alpha_1, \ldots, \alpha_m$. Such a case is studied in Section 4.5.

Let us write down the expression for function Λ in detail. The kinetic energy has the following form: $T = T_1 + U + T_2$, where T_1 and T_2 are respectively the quadratic forms of the quasi-cyclic and positional generalised coordinates, and U is their bilinear form. Using this notation, the expression for L_R is cast in the form: $L_R = T_2 - \Pi - T_1 = L_2 - T_1$. Hence, $\Lambda = \langle T_1 \rangle_0 - \langle L_2 \rangle_0 - W_c$, where

$$
W_c = \sum_r^m i_{cr} \alpha_r . \tag{4.2.22}
$$

Function Λ also retains this form in the case in which the sums in eq. (4.2.22) are values i'_{cr} which are independent of $\alpha_1, \ldots, \alpha_m$. What is required is to replace i_{cr} by $i_{cr} - i'_{cr}$.

The generating approximation and the form of function Λ do not change under adding terms of the form $\mu()^\bullet$ into the first m equations in eq. (4.2.8) and any terms of order of μ into the last $n-m$ equations. Finally the dependence $\langle L_R \rangle_0$ on $\alpha_1, \ldots, \alpha_m$ does not change if the system gains such k additional degrees of freedom corresponding to the coordinates $q_{n+1}, \ldots, q_{n+k}$

that the expression for T is written down as follows

$$T = T_1 + U + T_2 + T_3, \qquad T_3 = T_3(q_{n+1}, \ldots, q_{n+k}, \dot{q}_{n+1}, \ldots, \dot{q}_{n+k})$$

$$T_1 = \frac{1}{2}\sum_{r,s}^{m} a_{rs}\dot{f}_r\dot{f}_s, \qquad U = \sum_r^{m}\sum_s^{n-m} a_{rm+s}\dot{f}_r\dot{q}_{m+s}. \qquad (4.2.23)$$

Here, as above, $a_{rs}, a_{r,m+s}$ depend only on $q_{m+1}, \ldots, q_n$. Moreover, T_2 has the previous form and

$$\dot{f}_r = \dot{q}_r + \sum_{j=1}^{k} n_{rj}\dot{q}_{n+j}, \qquad n_{rj} = \text{const}. \qquad (4.2.24)$$

Let us find the form of Routh's equations. Denoting the sum in eq. (4.2.2) as $\dot{g}_r$ we have $\partial T/\partial \dot{q}_r = \partial T/\partial \dot{f}_r$ and

$$\dot{q}_r = \dot{q}_r^{(0)}(p_1, \ldots, p_m, \dot{q}_{m+1}, \ldots, \dot{q}_n, q_{m+1}, \ldots, q_n) - \dot{g}_r, \qquad (4.2.25)$$

where $\dot{q}_r^{(0)}$ denotes the quasi-cyclic velocity in the system without additional arguments (i.e. $\dot{q}_r^{(0)}$ also depend on their arguments as $\dot{q}_r$ in the former system). The expression for Routh's kinetic potential is given by

$$L_R = -T_1(\dot{q}_r) + T_1(\dot{g}_r) + T_2(\dot{q}_{m+r}) + U(\dot{g}_r, \dot{q}_{m+s}) + T_3 - \Pi. \qquad (4.2.26)$$

Here only dependences on the generalised velocities are shown (in an abbreviated form). Removing $\dot{q}_r$ with the help of eq. (4.2.2) from (4.2.2) and applying the identity

$$\sum_{r,s}^{m} a_{rs}\dot{q}_r^{(0)}\dot{g}_s + \sum_s^{n-m}\sum_r^{m} a_{sm+r}\dot{q}_{m+r}\dot{g}_s = \sum_s^{m} p_s\dot{g}_s \qquad (4.2.27)$$

we obtain

$$L_R = L_R^{(0)}(p_1, \ldots, p_m, \dot{q}_{m+1}, \ldots, \dot{q}_n, q_{m+1}, \ldots, q_n) + \sum_s^{m} p_s\dot{g}_s + T_3. \qquad (4.2.28)$$

Here $L_R^{(0)}$ denotes Routh's kinetic potential for the system without additional coordinates. Thus, the generating equations for $p_1, \ldots, p_m, q_{m+1}, \ldots,$ q_n have the same form and the same solutions as those for the system without additional coordinates. The equations for the additional coordinates are as follows

$$\frac{d}{dt}\frac{\partial T_3}{\partial \dot{q}_{n+s}} - \frac{\partial T_3}{\partial q_{n+s}} = -\sum_r^{m} n_{rs}\dot{p}_r + N_{n+s}, \qquad s = 1, \ldots, k. \qquad (4.2.29)$$

System (4.2.2) in the generating approximation does not contain $\alpha_1, \ldots,$ α_m. Let us assume that, for $\dot{p}_r = U_r$, the system admits an isolated stable solution for which $\dot{q}_{n+10}, \ldots, \dot{q}_{n+k0}$ are $2\pi/\omega-$periodic functions of time. Then equations for determining $\alpha_1, \ldots, \alpha_m$, the stability conditions and function Λ differ from those of the case without additional coordinates only in that i_{cr} is replaced by

$$i_{cr} + \sum_{j=1}^{k} n_{rj}\langle \dot{q}_{n+j0} \rangle .$$

The above results are also generalised to the rotational motions, that is, to the motions with some coordinates obeying the law $q_{m+s} = \omega t + \psi(t)$, where $\psi(t)$ is a $2\pi/\omega-$periodic function of time. Functions T, Π and N_{m+s} should be $2\pi-$periodic with respect to the corresponding coordinates q_{m+s} or contain only their differences $q_{m+s} - q_{m+r}$.

In the case of a linear positional coordinate the integral criterion can be modified. Routh's equations corresponding to the positional coordinates are given by

$$M\ddot{v} + Cv = Q + N, \qquad Q = \frac{d}{dt}\frac{\partial T_1}{\partial \dot{v}} - \frac{\partial T_1}{\partial v} . \qquad (4.2.30)$$

Here $v = (q_{m+1}, \ldots, q_n)$, $N = (N_{m+1}, \ldots, N_n)$, M, C are symmetric $(n - m) \times (n - m)$ matrices with constant entries. Denoting the scalar products by parentheses we obtain

$$L_2 = \frac{1}{2}(M\dot{v}, \dot{v}) - \frac{1}{2}(Cv, v) . \qquad (4.2.31)$$

Let us take equation (4.2.3) in the generating approximation, multiply both its parts by $\partial v_0/\partial \alpha_r$ and average over the period. Under the condition (4.2.1) we obtain

$$\left\langle \left(M\ddot{v}_0, \frac{\partial v_0}{\partial \alpha_r} \right) \right\rangle + \left\langle \left(Cv_0, \frac{\partial v_0}{\partial \alpha_r} \right) \right\rangle = \left\langle \left(Q_0, \frac{\partial v_0}{\partial \alpha_r} \right) \right\rangle . \qquad (4.2.32)$$

Differentiating eq. (4.2.3) with respect to α_r and calculating the scalar product of the result and v_0 yields

$$\left\langle \left(M\frac{\partial \ddot{v}_0}{\partial \alpha_r}, v_0 \right) \right\rangle + \left\langle \left(C\frac{\partial v_0}{\partial \alpha_r}, v_0 \right) \right\rangle = \left\langle \left(\frac{\partial N_0}{\partial \alpha_r}, v_0 \right) \right\rangle + \left\langle \left(\frac{\partial Q_0}{\partial \alpha_r}, v_0 \right) \right\rangle .$$
$$(4.2.33)$$

We sum up relationships (4.2.32) and (4.2.33) term by term and integrate the terms with $\ddot{v}_0$ by parts. The result is

$$\frac{\partial}{\partial \alpha_r}\langle L_2 \rangle_0 = \frac{\partial}{\partial \alpha_r}(V_N + V_Q)_0 . \qquad (4.2.34)$$

Let us refer to quantities $V_Q = -1/2\langle\langle Q, v \rangle\rangle$ and $V_N = -1/2\langle\langle N, v \rangle\rangle$ as the virial of the forces of action of the quasi-cyclic subsystem on the positional subsystem and the virial of the non-potential generalised forces respectively. Using eq. (4.2.3) we can remove $\langle L_2\rangle_0$ from the expression for and express Λ in terms of the following virials

$$\Lambda = \langle T_1\rangle_0 - V_{Q0} - V_{N0} - W_c. \tag{4.2.35}$$

If $N_m+1,\ldots,N_n$ are linear forms of $\dot{q}_{m+1},\ldots,\dot{q}_{m+n}$, whilst $q_{m+10},\ldots,q_{n0}$ are component synphase with phases independent of $\alpha_1,\ldots,\alpha_m$, then $V_{N0} = 0$, eq. (4.2.1) is fulfilled and

$$\Lambda = \langle T_1\rangle_0 - V_{Q0} - W_c.$$

This representation is useful due to the following fact. In a number of problems of vibration excitations (see the next section) it is convenient to express T_1 such that it contains some linear functionals $\xi_i = (v, v_i), v_i = $ const rather than the coordinates v themselves. The dependence of $T_1(p,\xi)$ will be "invariant" to the form of the positional subsystem. If additionally $V_{N0} = 0$, then the form of the averaged functions in eq. (4.2.3) as functions of p, ξ is immaterial to the details of the positional subsystem. However, the form of $\xi(t, \alpha_1, \ldots, \alpha_m)$ depends essentially on this positional subsystem.

4.3 Energy relationships for oscillations of current conductors

Given a system of bodies including m linear conductors subjected to external $2\pi/\omega-$periodic electromotive forces let us consider the case where the magnetic field can be taken to be quasi-stationary for frequencies $\omega,\ldots,\nu_*\omega$, where ν_* is sufficiently high. In general, the forthcoming analysis is valid with accuracy up to the high-frequency "tails" of the required functions effective from an $(\nu_* + 1) - th$ harmonic. This is due to the fact that the dynamic effects in the material are neglected. The relations between $\mathbf{B}$ and $\mathbf{H}$ in the material is considered to be linear and the active resistances of the conductors are assumed to be small in comparison with the inductive resistances at frequency ω. The Lagrange-Maxwell equations for the system under consideration are given by

$$\dot{\Phi}_r + \mu R_r i_r = U_r(t) + \mu U_{cr}, \quad r = 1,\ldots,m$$
$$\frac{d}{dt}\frac{\partial L_2}{\partial \dot{q}_{m+r}} - \frac{\partial L_2}{\partial q_{m+r}} = N_{m+r} + Q_{m+r}, \quad r = 1,\ldots,n-m, \tag{4.3.1}$$

where $i_r = \dot{q}_r, r = 1,\ldots,m$ denote currents in the conductors, $q_{m+1},\ldots,q_n$ are the mechanical generalised coordinates, with the coordinates (charges) q_r being quasi-cyclic. We introduce the notation for the ponderomotive forces

$$Q_{m+r} = \frac{\partial}{\partial q_{m+r}}W(i_1,\ldots,i_m,q_{m+1},\ldots,q_n) \tag{4.3.2}$$

and the magnetic fluxes through the conductor loops

$$\Phi_r = \frac{\partial}{\partial i_r} W(i_1, \ldots, i_m, q_{m+1}, \ldots, q_n), \qquad (4.3.3)$$

with W denoting the energy of the magnetic field.

In eq. (4.3.1) μR_r denotes active resistance of the conductors, whereas U_r and μU_{rc} denote variable and constant parts of the external emf, respectively. The latter are assumed to be small, so that no currents $i = O(1/\mu)$ are observed in a stationary regime.

The parameters of the generating solution $\alpha_1, \ldots, \alpha_m$ are the constant components of the magnetic fluxes calculated with accuracy up to small terms

$$\Phi_{r0} = \alpha_r + V_r(t), \qquad r = 1, \ldots, m. \qquad (4.3.4)$$

The value

$$W_c = \sum_{r=1}^{m} i_{cr} \alpha_r, \qquad i_{cr} = \frac{U_{cr}}{R_r} \qquad (4.3.5)$$

is referred to as the bias energy and have the following physical meaning. Let all $U_r = 0$, and the $r-th$ loop be subjected to a constant flux α_r. Then the current in the $r-th$ loop is $i_r = i_{cr}$, and the energy of the system of direct currents (bias currents) is equal to W_c (the field is considered as being external for the currents).

Relationship (4.2.1) allows us to formulate the following statement: if a system has no non-potential forces or these forces fulfill condition (4.2.1), then, under a stable periodic motion, the constant components of the magnetic fluxes (up to small values) render a minimum to the function of these components which is equal to the energy of the magnetic field minus the mechanical kinetic potential and bias energy, all energies (the mechanical kinetic potential included) being averaged over the period.

In the case where L_2 corresponds to a linear oscillating system, the mechanical kinetic potential can be replaced by a sum of the virials of the non-potential mechanical and ponderomotive forces due to eq. (4.2.3).

Let us assume that other linear conductors are located near the mentioned conductors, so that if a magnetic induction line encloses the "original" conductor then it also encloses the whole adjacent set of the additional conductors. The resistances of the additional conductors are assumed not to be small, otherwise the currents are $i = O(1/\mu)$). Then the charges transmitted along the additional conductors are additional coordinates, see Sec. 3.2. The quantities f_r are as follows

$$f_r = q_r + \sum_{j=1}^{l_r} w_{rj} q_j^{(r)}. \qquad (4.3.6)$$

Here l_r denotes the number of additional conductors located near the $r-th$ original conductor, $l_1 + \ldots + l_m = l$, $q_j^{(r)}$ is the charge in $j-th$ conductor of the $r-th$ set. The rational numbers w_{rj} are defined as follows. Let the loop of the $r-th$ original conductor follow a certain line $n_1^{(r)}$ times, whereas the loop of $j^{(r)}-th$ conductor follows a line $n_j^{(r)}$ times (in fact $n_j^{(r)}$ is the number of turns). Then $w_{jr} = n_j^{(r)}/n_1^{(r)}$.

The influence of additional conductors on the motion of the system, regardless of the types of emf, connections and connected elements (coils, capacitors, rectifiers etc.) is taken into account by replacing i_{cr} in the expression for W_c by

$$i'_{cr} = i_{cr} + \sum_{j=1}^{l_r} w_{rj} \langle i_{j0}^{(r)} \rangle. \tag{4.3.7}$$

As there always exist induction lines enclosing the original loop and not enclosing the additional loops, then the present derivation is valid only under the condition that this "difference" can be described only by terms of the order of μ in the expression for W. The first m equations in eq. (4.3.1) gain terms of the type $\mu()^\bullet$, and the generating solution does not change with any addition into the equations for the mechanical coordinates.

4.4 On the relationship between the resonant and non-resonant solutions

Let us consider a finite-dimensional system

$$\dot{x} = X(x,t) + \mu Y(x,\xi,\dot{\xi},t,\mu),$$

$$M\ddot{v} + B\dot{v} + Cv = f\left[\sum_i^k Q_i(x)v_i + \mu \ldots\right], \tag{4.4.1}$$

which differs from eq. (4.1.1) by the scalar multiplier f on the right hand side of the equations of motion of the oscillatory system. In the non-resonant case $f = O(1)$ and, as above, this multiplier can be included in Q_i.

While studying systems (4.4.1), in addition to the assumptions of Section 4.3.1, one uses the assumptions corresponding to the resonant case. Their essence is as follows. Let us take $M = M_0 + \delta M_1$, $C = C_0 + \sigma C_1$, $B = B_0 + \gamma B_1$ and consider the equation

$$M_0\ddot{v} + C_0 v = 0. \tag{4.4.2}$$

Its eigenvalues, i.e. the roots of the polynomial $|C_0 - \lambda^2 M_0|$ are denoted as λ_ρ^2. The resonant case assumes that one of these values is $\lambda_\rho = \omega$, whilst the other eigenvalues differ from $\nu\omega$ (ν is an integer) in non-small values and $\delta, \sigma, \gamma, f$ are small values of the order μ, with δ, σ, γ not being equal to zero simultaneously.

We describe now the procedure of determining the periodic solutions in the resonant case. Let us take $\delta = \mu\delta_1, \sigma = \mu\sigma_1$ and so on. Equation (4.4.1) takes the form

$$\dot{x} = X(x,t) + \mu Y(x,\xi,\dot{\xi},t,\mu), \qquad (4.4.3)$$

$$M_0\ddot{v} + C_0 v = \mu\left[-\delta_1 M_1\ddot{v} - \gamma_1 B_1\dot{v} - \sigma C_1 v + f_1\sum_i^k Q_i(x)v_i + \mu\ldots\right].$$

In the generating approximation x, v are determined independently. Let us assume for simplicity that only one eigenvector corresponds to the eigenvalue $\lambda = \omega$. We denote this eigenvector as $v^{(1)}$ however ω is not necessarily the first eigenvalue. The generating solution is given by

$$x = x_0(t,\alpha_1,\ldots,\alpha_n),$$
$$v = v_0(t,A_1,A_2) = (A_1\cos\omega t + A_2\sin\omega t)\,v^{(1)}. \qquad (4.4.4)$$

It contains not n, but $n+2$ constants $\alpha_1,\ldots,\alpha_n, A_1, A_2$. The equations for their determination are

$$P_r(\alpha_1,\ldots,\alpha_n,A_1,A_2) = \left\langle\sum_s^p z_{rs}Y_{s0}\right\rangle = 0, \quad r = 1,\ldots,n,$$

$$P_{n+1}(\alpha_1,\ldots,\alpha_n,A_1,A_2) = A_1\left((\sigma_1 C_1 - \omega^2\delta_1 M_1)v^{(1)}, v^{(1)}\right) +$$

$$+\gamma_1 A_2(B_1 v^{(1)}, v^{(1)}) - f_1\sum_i^k Q_{i1}\cos\vartheta_{i1}(v_i, v^{(1)}) = 0, \qquad (4.4.5)$$

$$P_{n+2}(\alpha_1,\ldots,\alpha_n,A_1,A_2) = -\gamma_1 A_1(B_1 v^{(1)}, v^{(1)}) +$$

$$+A_2((\sigma_1 C_1 - \omega^2\delta_1 M_1)v^{(1)}, v^{(1)}) - f_1\sum Q_{i1}\sin\vartheta_{i1}(v_i, v^{(1)}) = 0,$$

Here z_{rs} are the same functions as in eq. (4.1.21), $Y_{s0} = Y_s(x_0,\xi_0.\dot{\xi}_0,t,0)$, $\xi_{i0} = (v_0, v_i), Q_{i1}, \vartheta_{i1}$ denote the amplitude and phase of the first harmonic $Q_i(t)$ (see eq. (4.1.17)). The latter two equations in eq. (4.4.5) have the following meaning. Inserting the generating approximation into the terms of the order of μ in the second equation in (4.4.3) and expanding the result as a Fourier series yields vectors v_c and v_s which are the coefficients of $\cos\omega t$ and $\sin\omega t$. The equations under consideration express the condition that v_c and v_s are orthogonal to $v^{(1)}$.

Provided that the equations in (4.4.5) are resolvable, one can find, in principle, that $\alpha_1,\ldots,\alpha_n, A_1, A_2$, which allows one to determine the generating solution, analyse its stability etc. Under the analogous assumptions one can also seek the periodic solutions for autonomous systems.

Generally speaking, both the non-resonant and resonant cases should be studied for each exciter. Both cases are considered in particular in the problems of the oscillations caused by mechanical exciters [15], [66]. It can be shown that the non-resonant solution is more general and the resonant solution can be obtained from the non-resonant one.

Let us refer to the region in the parameter space as resonant (non-resonant) if the assumptions of the resonant (non-resonant) case are valid there. In the non-resonant region we have the following equations

$$P_r(\alpha_1,\ldots,\alpha_n, fK_0, fK_1, \Psi_1,\ldots) = 0, \quad r = 1,\ldots,n \qquad (4.4.6)$$

for determining $\alpha_1,\ldots,\alpha_n$. It is easy to see that matrices K_ν and parameter f appear in these equations only in the form of a product fK_ν. Matrices K_ν exist also at points of the resonant region however $K_1 = O(1/\mu), fK_\nu = O(\mu), \nu \neq 1$ here. In both regions $fK_1 = O(1)$. Therefore, equations (4.4.6) can also be written for the resonant region. Let us take a point in this region and construct equations (4.4.6). Alternatively, let us construct equations (4.4.5) for this point, express A_1, A_2 in terms of $\alpha_1,\ldots,\alpha_n$ from the last two equations and insert the results into the first n equations. Let us show that the obtained equations for $\alpha_1,\ldots,\alpha_n$ coincide with the equations in (4.4.6) up to the values of order μ.

To this end, we consider the equation for v_0 in the non-resonant case

$$M\ddot{v}_0 + B\dot{v}_0 + Cv_0 = f\sum_{i}^{k} Q_i(x_0(t,\alpha))v_i\,. \qquad (4.4.7)$$

Solving these equations for v_0, inserting it into $Y(x_0, \xi_0, \dot{\xi}_0, t, 0)$ and averaging the result yields eq. (4.4.6) corresponding to the given values of M, B and C.

In general, matrices C_0, M_0, as well as matrices C, M should be taken as being positive definite. For this reason, the eigenvectors $v^{(\rho)}$ of equation $(C - \lambda^2 M)v = 0$ form a basis in the space of configurations of the oscillatory system. This allows one to seek the solution of eq. (4.4.7) in the form

$$v_0 = \sum_{\nu}\sum_{\rho}(C_{\nu\rho}\cos\nu\omega t + D_{\nu\rho}\sin\nu\omega t)v^{(\rho)}\,. \qquad (4.4.8)$$

The equations for $C_{\nu,\rho}, D_{\nu,\rho}$ are as follows

$$\sum_{\rho}((C - \nu^2\omega^2 M)v^{(\rho)}, v^\kappa)C_{\nu,\rho} + \nu\omega(Bv^{(\rho)}, v^\kappa)D_{\nu,\rho}$$

$$= f\sum_{i}^{k} Q_{i\nu}\cos\vartheta_{i\nu}(v_i, v^{(\kappa)}),$$

$$\sum_{\rho}-\nu\omega(Bv^{(\rho)}, v^{(\kappa)})C_{\nu,\rho} + ((C - \nu^2\omega^2 M)v^{(\rho)}, v^{(\kappa)})D_{\nu,\rho}$$

$$= f\sum_{i}^{k} Q_{i\nu}\sin\vartheta_{i\nu}(v_i, v^{(\kappa)}), \quad \kappa = 1, 2,\ldots\,. \qquad (4.4.9)$$

Because $\delta_1, \sigma_1, \gamma_1$ do not vanish simultaneously, equations (4.4.9) are resolvable in the resonant region. Let us elucidate the form of the solutions.

Let $\nu \neq 1$. Assuming $v^{(\rho)}$ to be orthonormalised, i.e. $\left(M_0 v^{(\rho)}, v^{(\kappa)}\right) = \delta_{\rho,\kappa}$, we obtain

$$(\lambda_\kappa^2 - \nu^2 \omega^2) C_{\nu\kappa} + \mu \sum_\rho ((\sigma_1 C_1 - \nu^2 \omega^2 \delta_1 M_1) v^{(\rho)}, v^{(\kappa)}) C_{\nu,\rho} +$$

$$\nu \omega \gamma_1 (B v^{(\rho)}, v^{(\kappa)}) D_{\nu\rho} = \mu f_1 \sum_i^k Q_{i\nu} \cos \vartheta_{i\nu} (v_i, v^{(\kappa)}), \qquad (4.4.10)$$

$$(\lambda_\kappa^2 - \nu^2 \omega^2) D_{\nu\kappa} + \mu \ldots = \mu f_1 \sum_i^k Q_{i\nu} \sin \vartheta_{i\nu} (v_i, v^{(\kappa)}).$$

One can see that all $C_{\nu\kappa}, D_{\nu\kappa} = O(\mu), \nu \neq 1$. For $\nu = 1$ we have

$$(\lambda_\kappa^2 - \omega^2) C_{1\kappa} + \mu \sum_\rho ((\sigma C_1 - \omega^2 \delta_1 M_1) v^{(\rho)}, v^{(\kappa)}) C_{1\rho} +$$

$$\mu \omega \gamma_1 (B v^{(\rho)}, v^{(\kappa)}) D_{1\rho} = \mu f_1 \sum_i^k Q_{i1} \cos \vartheta_{i1} (v_i, v^{(\kappa)}),$$

$$(\lambda_\kappa^2 - \omega^2) D_{1\kappa} + \mu \sum_\rho -\omega \gamma_1 (B v^{(\rho)}, v^{(\kappa)}) C_{1\rho} + \qquad (4.4.11)$$

$$((\sigma_1 C_1 - \omega^2 \delta_1 M_1) v^{(\rho)}, v^{(\kappa)}) D_{1\rho} = \mu f_1 \sum_i^k Q_{i1} \sin \vartheta_{i1} (v_i, v^{(\kappa)}).$$

Let us write down the equation corresponding to $\kappa = 1$

$$\sum_\rho ((\sigma_1 C_1 - \omega^2 \delta_1 M_1) v^{(1)}, v^{(1)}) C_{11} + \omega \gamma_1 (B_1 v^{(1)}, v^{(1)}) D_{11} -$$

$$f_1 \sum_i^k Q_{i1} \cos \vartheta_{i1} (v_i, v^{(1)}) + \mu \ldots = 0. \qquad (4.4.12)$$

If we denote here $C_{11} = A_1, D_{11} = A_2$, then eq. (4.4.12) coincides with the $(n+1) - th$ equation in (4.4.5) with accuracy up to the terms of order μ. Similarly, the second equation in (4.4.1) for $\kappa = 1$ coincides with the $(n+2) - th$ equation in eq. (4.4.5). Thus, dependences $C_{11}(\alpha_1, \ldots, \alpha_n)$, $D_{11}(\alpha_1, \ldots, \alpha_n)$, obtained from eq. (4.4.9) are coincident with dependences $A_1(\alpha_1, \ldots, \alpha_n)$, $D_1(\alpha_1, \ldots, \alpha_n)$ obtained from the last two equations in (4.4.5) with accuracy up to the values of the order $O(\mu)$. The functions $P_r(\alpha_1, \ldots, \alpha_n)$ obtained by substituting the solution of eq. (4.4.8) for the taken point of the region into Y_r and further averaging (i.e. functions P_r obtained for the resonant region by the non-resonant procedure) coincide with the above accuracy with functions $P_r(\alpha_1, \ldots, \alpha_n)$ obtained for the same point after removing A_1, A_2 from (4.4.5). If $|\partial P_r / \partial \alpha_s| \neq 0$, then $\alpha_1, \ldots, \alpha_n$, determined from eqs. (4.4.6) and (4.4.5) differ by values of the order μ. However, the latter means that the resonant generating solution can be obtained from the non-resonant one and the non-resonant solution can be utilised in the resonant region.

This conclusion is also valid in the cases of multiple roots as well as in the cases of the roots $\lambda_\rho = \nu \omega, \nu \neq 1$ etc.

Thus, if only the generating solution is of interest, the special consideration of the resonant case is unnecessary.

This can not be extended to the stability conditions. However, in the problem of oscillations excited by a rotating unbalanced body when the vibrator is considered as a nearly conservative object, the stability conditions in the non-resonant case [89] coincide with those in the resonant case, the latter being obtained via an asymptotic method by Kononenko in [66].

In contrast to this, if the resonant solution is used in the non-resonant region, the result is coincident with that which is obtained by omitting all of the harmonics in eq. (4.4.8) except the first, and all the modes except $v^{(1)}$. This strategy is often admissible. In principle, the resonant assumptions allows us to solve the problem in the cases in which eq. (4.4.7) is non-linear, i.e. when the oscillatory system is non-linear or Q_i depend on ξ. In this case, utilising the resonant solution in the non-resonant region is equivalent to applying the method of harmonic balance for solving eq. (4.4.7).

4.5 Routh's equations which are linear in the positional coordinates

Even if the kinetic potential L_2 corresponds to a linear system, the equations for the positional coordinates are, generally speaking, non-linear by virtue of the dependence of actions Q on the positional coordinates. There exist however two cases which result in linear equations. Under stationary constraints, the structure of Routh's function $R = L + \Pi$ is as follows, see [76],

$$
R = \frac{1}{2} \sum_{r,s=1}^{n-m} (A_{m+r\,m+s} + D_{m+r\,m+s}) \dot{q}_{m+r} \dot{q}_{m+s} + \sum_{r=1}^{n-m} \sum_{s=1}^{m} D_{m+rs} p_s \dot{q}_{m+r} -
$$

$$
\frac{1}{2} \sum_{r,s=1}^{m} A^{(rs)} p_r p_s, \quad (\| A^{(rs)} \| = \| A_{rs} \|^{-1}). \tag{4.5.1}
$$

Expressions for the forces of action of a quasi-cyclic subsystem on a positional one are given by

$$
Q_{m+r} = -\frac{d}{dt} \sum_{s=1}^{n-m} D_{m+r\,m+s} \dot{q}_{m+s} + \frac{1}{2} \sum_{i,s=1}^{n-m} \frac{\partial D_{m+i\,m+s}}{\partial q_{m+r}} \dot{q}_{m+i} \dot{q}_{m+s} -
$$

$$
\sum_{s=1}^{m} p_s \sum_{i+1}^{n-m} \dot{q}_{m+1} \left(\frac{\partial D_{m+rs}}{\partial q_{m+i}} - \frac{\partial D_{m+is}}{\partial q_{m+r}} \right) - \sum_{s=1}^{m} D_{m+rs} \dot{p}_s -
$$

$$
\frac{1}{2} \sum_{i,s=1}^{m} \frac{\partial A^{(is)}}{\partial q_{m+r}} p_i p_s, \quad (r = 1, \dots, n-m). \tag{4.5.2}
$$

To obtain linear equations in the generating approximation for the positional coordinates, we take that the non-potential forces corresponding to the positional coordinates are linear forms in $\dot{q}_{m+1}, \ldots, \dot{q}_n$ with constant coefficients. As follows from eq. (4.5.2), two cases are possible.

1. *Quasi-harmonic generating system.* Quasi-harmonic equations, i.e. the linear equations with periodic coefficients, are obtained if $D_{m+rm+s} = $ const $(r, s = 1, \ldots, n - m)$, $D_{m+rs}(r = 1, \ldots, n - m, s = 1, \ldots, m)$ are sums of the constant values and linear forms of the positional coordinates, whereas $A^{(rs)}(r, s = 1, \ldots, m)$ are sums of constant values, linear and quadratic forms of the positional coordinates. The constant terms in $A^{(rs)}$ do not affect the form of Q_r, whilst the second term on the right hand side of eq. (4.5.2) vanishes. The system of generating equations is inhomogeneous if at least one of two conditions is satisfied: 1) $D_{m+r,s}$ contains constant terms and 2) $A^{(rs)}$ contains linear terms. This system is homogeneous if all $D_{m+r,s}$ are linear forms and $A^{(rs)}$ is a sum of a constant value and a quadratic form.

Let us notice one specific case. Let the products of the quasi-cyclic and positional coordinates be absent in the expression for the kinetic energy, i.e. $U = 0$. Then all of $D_{m+rs} = 0, D_{m+rm+s} = 0$. Let also $A^{(rs)}$ contain no linear terms. Splitting T_1 into the energy of quasi-cyclic subset for the fixed positional subsystem, T_1^*, and an "additional" energy, ΔT_1,

$$T_1 = T_1^* + \Delta T_1, \ T_1^* = \frac{1}{2} \sum_{r,s=1}^{m} A_*^{(rs)} p_r p_s, \ \Delta T_1 = \frac{1}{2} \sum_{r,s=1}^{m} \Delta A^{(rs)} p_r p_s,$$

$$(4.5.3)$$

where $\Delta A^{(rs)}$ denote the quadratic forms of q_{m+r}, we obtain

$$V_Q = \Delta T_1 . \tag{4.5.4}$$

Function Λ has the following form

$$\Lambda = \langle T_1^* \rangle - V_Q - W_c . \tag{4.5.5}$$

If additionally $V_{Q0} = 0$, then function Λ has the form corresponding to the case of $q_{m+1}, \ldots, q_n = 0$, i.e. under the fixed positional subsystem. Hence, in this case the positional subsystem does not affect the motion of the cyclic one with accuracy up to small terms (however motion of the positional subsystem essentially depends upon the motion of the quasi-cyclic one). In the problems of the vibration excitation this implies that the backward influence of oscillations on the exciter is not essential despite the presence of the family of generating solutions and the importance of small terms depending on the positional coordinates.

If $T_1^* = 0$ and at least one $U_{cs} \neq 0$, then the solutions of the considered type do not exist at all. If $U_{cs} = 0$, then we arrive at the special case of the method of small parameters ($P_r \equiv 0$) which requires consideration of the terms of the order μ^2 in the solutions sought.

2. *Generating system with constant coefficients.* Equations with constant coefficients are obtained if $D_{m+rs} = $ const, $D_{m+rm+s} = $ const, whilst $A^{(rs)}$

are linear form of the positional coordinates. The positional coordinates in the generating approximation are determined from the solution of the problem of the forced oscillations of a linear system subjected to forces which are prescribed functions of time and parameters $\alpha_1, \ldots, \alpha_m$.

Let the expression for the kinetic energy of the system with quasi-cyclic coordinates contain no products of the quasi-cyclic and positional velocities, i.e. their bilinear form $U = 0$. If the expression for T_1 can be written such that it contains the parameters of the backward influence (i.e. in the form invariant to the particular form of the oscillatory system), then equations for the parameters of the generating solution and the stability conditions can be cast in the form containing the harmonic influence coefficients for the oscillatory system as parameters.

Let us represent the expression for Routh's kinetic potential L_R in terms of the functional of the backward influence

$$
L_R = L_2(v, \dot{v}) - \frac{1}{2} \sum_{r,s=1}^{m} \left(A_{rs} + \sum_{j=1}^{k} \Delta A_{rs}^{(i)} \xi_i \right) p_r p_s = L_2 - T_1^* - \Delta T_1 \,.
$$

(4.5.6)

Keeping the assumptions of Sec. 4.3.1, we write down Routh's equations

$$
\dot{p}_s - \mu \beta_s \frac{\partial L_r}{\partial p_s} = U_s(t) + \mu U_{cs}, \quad s = 1, \ldots, n,
$$

$$
M\ddot{v} + B\dot{v} + Cv = \sum_{i}^{k} Q_i v_i \,.
$$

(4.5.7)

The generalised forces Q_i are given by the relationships

$$
Q_i = -\frac{1}{2} \sum_{r,s+1}^{m} \Delta A_{rs}^{(i)} p_r p_s \,.
$$

(4.5.8)

In the generating approximation we have

$$
p_{s0} = \alpha_s + V_s(t), \quad \dot{V}_s = U_s, \quad \langle V_s \rangle = 0,
$$

$$
Q_{i0} = -\frac{1}{2} \sum_{r,s=1}^{m} \Delta A_{rs}^{(i)} (\alpha_r \alpha_s + 2\alpha_r V_s + V_r V_s) \,.
$$

(4.5.9)

Given Q_{i0}, we can write the following expressions for ξ_{j0}

$$
\xi_{j0} = -\frac{1}{2} \sum_{i}^{k} \sum_{r,s}^{m} \Delta A_{rs}^{(i)} \left[k_0^{(ij)} (\alpha_r \alpha_s + V_{rs}^{(0)}) + \right.
$$

$$
\left. +2 \sum_{\nu, \nu \neq 0} k_\nu^{(ij)} \alpha_r V_{s\nu} \cos(\nu \omega t - \vartheta_{s\nu} - \psi_\nu^{(ij)}) \right] + \xi_{*j}, \quad j = 1, \ldots, k \,.
$$

(4.5.10)

The adopted notation corresponds to the equalities

$$V_s = \sum_{\nu,\nu\neq 0} V_{s\nu}\cos(\nu\omega t - \vartheta_{s\nu}),$$

$$V_{rs} = V_{rs}^{(0)} + \sum_{\nu,\nu\neq 0} V_{rs}^{(\nu)}\cos(\nu\omega t - \vartheta_{rs}^{(\nu)}),$$

$$\xi_{*j} = -\frac{1}{2}\sum_{i}^{k}\sum_{\nu,\nu\neq 0} k_\nu^{(ij)}\sum_{r,s}^{m}\Delta A_{rs}^{(i)}V_{rs}^{(\nu)}\cos(\nu\omega t - \vartheta_{rs}^{(\nu)} - \psi_\nu^{(ij)}). \tag{4.5.11}$$

Thus ξ_{*j} are the parts of ξ_j which are independent of $\alpha_1,\ldots,\alpha_m$ and $\langle\xi_{*j}\rangle = 0$.

Inserting ξ_{j0} from eq. (4.5.10) into the relationships

$$\dot{q}_r = -\frac{\partial L_R}{\partial p_r} = \sum_{s=1}^{m}\left[A_{rs} + \sum_{i}^{k}\Delta A_{rs}^{(i)}\xi_i\right]p_s \tag{4.5.12}$$

and averaging the result we obtain equations for $\alpha_1,\ldots,\alpha_m$

$$P_r(\alpha_1,\ldots,\alpha_m) \equiv \sum_{s,u,z=1}^{m} a_{rsuz}\alpha_s\alpha_u\alpha_z + \sum_{s=1}^{m} a_{rs}\alpha_s - e_r = 0, \quad r = 1,\ldots,m. \tag{4.5.13}$$

Here

$$a_{rsuz} = -\frac{1}{2}\sum_{i,j}^{k}\Delta A_{rz}^{(j)}\Delta A_{su}^{(i)}k_{(0)}^{(ij)},$$

$$a_{rs} = A_{rs} + \sum_{u,z=1}^{m} a_{rsuz}V_{uz}^{(0)} -$$

$$\frac{1}{2}\sum_{i,j}^{k}\sum_{u,z}^{m}\sum_{\nu,\nu\neq 0}\Delta A_{ru}^{(j)}\Delta A_{sz}^{(i)}V_{u\nu}V_{z\nu}K_\nu^{(ij)}\cos(\vartheta_{u\nu} - \vartheta_{z\nu} - \psi_{(\nu)}^{(ij)}), \tag{4.5.14}$$

$$e_r = i_{cr} - \sum_{s}^{m}\sum_{j}^{k}\Delta A_{rs}^{(j)}\langle\xi_{*j}V_s\rangle, \quad i_{cr} = U_{cr}/\beta_r.$$

The dependences $\alpha_r(K,\Psi)$ can be found and analysed sufficiently simply only in particular cases, for example, for $k,m = 1,2$. Generally speaking, $\alpha_1,\ldots,\alpha_m$ should be determined from equations in eq. (4.5.13), the particular values of K,Ψ having been substituted into these equations.

The further calculation of $\alpha_1,\ldots,\alpha_m$ is simplified if the integral criterion holds. Let us find the sufficient conditions of its existence. Let us construct the derivatives

$$\frac{\partial P_r}{\partial \alpha_s} = \sum_{u,z}^{m}(a_{rsuz} + a_{rusz} + a_{ruzs})\alpha_u\alpha_z + a_{rs}. \tag{4.5.15}$$

It follows from the relationships $k_0^{ij} = k_0^{ji}$ and evident equalities $\Delta A_{rs}^{(i)} = \Delta A_{sr}^{(i)}$ that the coefficients a_{rsuz} do not change if we interchange the extreme (left and right) subscripts, as well as the middle subscripts, and simultaneously interchange the subscripts in the first and second pairs. Indeed,

$$a_{rusz} = -\frac{1}{2} \sum_{i,j}^{k} \Delta A_{rz}^{(i)} \Delta A_{su}^{(j)} k_0^{(ij)} = -\frac{1}{2} \sum_{i,j}^{k} \Delta A_{su}^{(i)} \Delta A_{rz}^{(j)} k_0^{(ij)} = a_{srzu} \,.$$

(4.5.16)

Hence, $a_{rsuz} = a_{rusz}$, that is, the first two coefficients in the sum in eq. (4.5.15) are equal. Let us interchange the subscripts r and s in eq. (4.5.15). The third coefficient in the sum does not change. In addition to this,

$$\sum_{u,z} a_{sruz} \alpha_u \alpha_z = \sum_{u,z} a_{aszu} \alpha_u \alpha_z = \sum_{u,z} a_{rsuz} \alpha_u \alpha_z \,.$$

(4.5.17)

Hence, the sum in eq. (4.5.15) itself does not change. The first two terms in the expression for a_{rs} possess the same property. Using the property of the reciprocity of the harmonic influence coefficients and phases $\psi_\nu^{(ij)}$ and interchanging the subscripts i, j, u, z yields

$$a_{rs} - a_{sr} = -\frac{1}{2} \sum_{i,j}^{k} \sum_{u,z}^{m} \sum_{\nu,\nu \neq 0} \Delta A_{ru}^{(j)} \Delta A_{sz}^{(i)} V_{u\nu} V_{z\nu} \times$$

$$k_\nu^{(ij)} \left[\cos(\vartheta_{u\nu} - \vartheta_{z\nu} - \psi_\nu^{(ij)}) - \cos(\vartheta_{z\nu} - \vartheta_{u,\nu} - \psi_\nu^{(ij)}) \right] \,.$$

(4.5.18)

Hence, $a_{rs} = a_{sr}$, if $\vartheta_{u\nu} = \vartheta_{z\nu}$ where $u, z = 1, \ldots, m$. Thus the equalities $\partial P_r / \partial \alpha_s = \partial P_s / \partial \alpha_r$ hold in the case when the generalised forces corresponding to the quasi-cyclic coordinates are component-synphase. Then, $P_r = \partial \Lambda / \partial \alpha_r$, that is, the condition of the component-synphase coordinates is the sufficient condition for the existence of the integral criterion. Another sufficient condition is $\psi_\nu^{(ij)} = 0$ which is satisfied when all of the generalised forces corresponding to the positional coordinates are potential forces. The forces $U_s(t)$ are not necessarily component-synphase.

Under component-synphase $U_s(t)$ the integral criterion is valid because the non-potential forces in the generating approximation satisfy the relationships

$$\left\langle \sum_{s}^{n-m} N_{m+s0} \frac{\partial q_{m+s0}}{\partial \alpha_r} \right\rangle \equiv -\left\langle \left(B\dot{v}, \frac{\partial v_0}{\partial \alpha_r} \right) \right\rangle = i'_{cr} \,,$$

(4.5.19)

where i'_{cr} are not zero but do not depend on $\alpha_1, \ldots, \alpha_m$. Indeed, the external forces

$$\sum_{i}^{k} Q_{i0} v_i = -\frac{1}{2} \sum_{i}^{k} \sum_{r,s}^{m} \Delta A_{rs}^{(i)} \left[\alpha_r \alpha_s + 2\alpha_r \sum_{\nu,\nu \neq 0} V_{s\nu} \cos(\nu\omega t - \vartheta_\nu) + V_r V_s \right] v_i$$

(4.5.20)

cause oscillations of the form $v_0 = v_c + v_g + v_*$, where v_c is time-independent, v_g is represented by the following expansion

$$v_g = \sum_{\nu, \nu \neq 0} v_\nu^{(1)} \cos(\nu\omega t - \vartheta_\nu) + v_\nu^{(2)} \sin(\nu\omega t - \vartheta_\nu) , \qquad (4.5.21)$$

with $v_\nu^{(1)}, v_\nu^{(2)}$ being linear forms of $\alpha_1, \ldots, \alpha_m$ of the type $v_\nu^{(1)} = L_\nu(\alpha) v_{\nu*}^{(1)}$, $v_\nu^{(2)} = L_\nu(\alpha) v_{\nu*}^{(2)}$, whilst v_* does not depend on $\alpha_1, \ldots, \alpha_m$. Therefore,

$$\left\langle \left(B\dot{v}_0, \frac{\partial v_0}{\partial \alpha_r} \right) \right\rangle = \left\langle \left(B(v_g + v_*)^\bullet, \frac{\partial v_c}{\partial \alpha_r} + \frac{\partial v_g}{\partial \alpha_r} \right) \right\rangle . \qquad (4.5.22)$$

Since the derivatives do not contain a constant term, and v_ν is independent of $\alpha_1, \ldots, \alpha_m$, then

$$\left\langle \left(B(v_g + v_*)^\bullet, \frac{\partial v_c}{\partial \alpha_c} \right) \right\rangle , \qquad \left\langle \left(B\dot{v}_g, \frac{\partial v_g}{\partial \alpha_r} \right) \right\rangle = 0 .$$

Finally we have

$$\left\langle \left(B\dot{v}_0, \frac{\partial v_0}{\partial \alpha_r} \right) \right\rangle = \left\langle \left(B\dot{v}_*, \frac{\partial v_g}{\partial \alpha_r} \right) \right\rangle . \qquad (4.5.23)$$

Clearly, the value on the right hand side of eq. (4.5.23) does not depend on $\alpha_1, \ldots, \alpha_m$. Let us indicate the particular form of the integral criterion corresponding to the case under consideration. Let us present the kinetic energy of the exciter in the form $T_1 = T_e + \Delta T$. Here T_e denotes the energy for $\xi_i \equiv 0$, i.e. for an oscillatory system which is immovable in the undeformed state. ΔT denotes an "additional" energy

$$T_e = \frac{1}{2} \sum_{r,s}^m A_{rs} p_r p_s, \quad \Delta T = \frac{1}{2} \sum_{r,s}^m \sum_i^k \Delta A_{rs}^{(i)} \xi_i p_r p_s . \qquad (4.5.24)$$

Since ΔT is linear in ξ_i, the virial of the external forces is related to the additional energy of the exciter by the relationship

$$V_Q = \frac{1}{2} \langle \Delta T \rangle . \qquad (4.5.25)$$

Integrating by parts we obtain

$$\langle (B\dot{v}_0, v_0) \rangle = -\langle (Bv_0, \dot{v}_0) \rangle . \qquad (4.5.26)$$

Matrix (or operator) B is symmetric, thus

$$-\langle (Bv_0, \dot{v}_0) \rangle = -\langle (B\dot{v}_0, v_0) \rangle, \qquad \langle (B\dot{v}_0, v_0) \rangle = 0 , \qquad (4.5.27)$$

that is, the virial V_N of the non-potential forces in the oscillatory system is zero if these forces are those of viscous damping. By using eq. (4.2.2) we arrive at the following expression for Λ

$$\Lambda = \langle T_e \rangle_0 + \frac{1}{2} \langle \Delta T \rangle_0 - W_c, \quad W_c = \sum_r^m (i_{cr} - i'_{cr}) \alpha_r. \tag{4.5.28}$$

Entering the total kinetic energy T_1, we have

$$\Lambda = \langle T_1 \rangle_0 - \frac{1}{2} \langle \Delta T \rangle_0 - W_c. \tag{4.5.29}$$

Along with the more general representation of function Λ in terms of the averaged Routh's kinetic potential or the kinetic potential of the oscillatory system

$$\Lambda = \langle T_1 \rangle_0 - \langle L_2 \rangle_0 - W_c \tag{4.5.30}$$

expressions (4.5.28) and (4.5.29) provide us with three forms of the integral criterion, each utilising two of four functions $T_1, T_e, \Delta T, L_2$.

It follows from eqs. (4.5.29) and (4.5.30) that $\langle L_2 \rangle_0 = 1/2 \langle \Delta T \rangle_0$. Calculating the scalar product of the equation of motion for the oscillatory system and v, and averaging over the period we can obtain a more general relationship $\langle L_2 \rangle = 1/2 \langle \Delta T \rangle$. Let us derive an explicit expression for function Λ. It is presented by a sum of a form of fourth degree, a quadratic and a linear form in $\alpha_1, \ldots, \alpha_m$, and it can also contain an arbitrary term Λ_c which does not depend on $\alpha_1, \ldots, \alpha_m$. If Λ is defined according to eq. (4.5.28)-(4.5.30), then Λ_c is not zero. Then we have

$$\Lambda = \frac{1}{4} \sum_{r,s,u,z} a_{rsuz} \alpha_r \alpha_s \alpha_u \alpha_z + \frac{1}{2} \sum_{r,s}^m a_{rs} \alpha_r \alpha_s - \sum_r^m (i_{cr} - i'_{cr}) \alpha_r + \Lambda_c. \tag{4.5.31}$$

In order to calculate coefficients a_{rsuz}, a_{rs} there is no need to use their representation in terms of the Fourier coefficients. We can use, for example, the following notation. Let us introduce the matrix impulse-frequency characteristic of the oscillatory system $K(t) = \| K_{ij}(t) \|, i, j = 1, \ldots, k$, which is defined as follows. Let a single $2\pi/\omega-$periodic load $f(t) v_j$ act on the oscillatory system. Then for pure forced oscillations, the dependence of the functional ξ_i on time can be written in the form

$$\xi_i(t) = \frac{\omega}{2\pi} \int_0^{2\pi/\omega} K_{ij}(t - \tau) f(\tau) d\tau, \tag{4.5.32}$$

where K_{ij} does not depend upon the particular form of the $2\pi/\omega-$periodic function $f(t)$. This enables us to rewrite eq. (4.5.10) in the following form

$$\xi_{j0} = \frac{\omega}{2\pi} \sum_i^k \int_0^{2\pi/\omega} K_{ij}(t - \tau) Q_{i0}(\tau) d\tau \tag{4.5.33}$$

and the expression for $\langle \Delta T \rangle_0$ as follows

$$\langle \Delta T \rangle_0 = -\frac{\omega^2}{4\pi^2} \sum_{i,j}^{k} \int_0^{2\pi/\omega} dt \int_0^{2\pi/\omega} d\tau Q_{j0}(t) K_{ij}(t-\tau) Q_{i0}(\tau). \qquad (4.5.34)$$

The above expressions are obtained from this equation with the help of the following relationship

$$K_{ij}(t) = k_0^{(ij)} + 2 \sum_{\nu, \nu \neq 0} k_\nu^{(ij)} \cos(\nu\omega t - \psi_\nu^{(ij)}). \qquad (4.5.35)$$

5

Oscillations caused by electromagnets

5.1 Equations for determining the constant components of magnetic fluxes

The case in which the Routh function is linear in the position coordinates corresponds to the problems of vibrations of the linear mechanical systems with the electromagnets attracting a ferromagnetic body (an anchor) under translation or rotation. Let us first consider the systems with a translating anchor. They consist of a mechanical oscillator, i.e. an aggregate of elastic bodies or bodies elastically connected, electromagnets (ferromagnetic cores with windings) and anchors near the cores. In practice, the magnets with the U-type or E-type cores are used whereas the anchor has either a form similar to the form of the core or a flat surface facing the core. The anchors and cores of the electromagnets are considered as rigid bodies. The element of the oscillating system attached to the anchor should move relative to the element fixed at the core. The ponderomotive forces attracting the anchor to the core and vice versa cause vibrations of these elements and the rest of the oscillating system. The anchor is assumed to translate relative to core only in the direction perpendicular to the plane of the end surface while the other relative motions are not feasible by virtue of the corresponding constraints.

This schematisation is used for modelling a number of technical devices for transportation of granular materials (feeders and conveyers), orientation of details, separation of materials (shaking sieves) concrete consolidation (platform vibrator) etc. A majority of such devices utilises a single

a   b

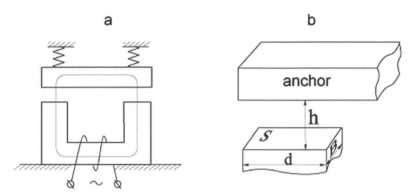

FIGURE 5.1.

electromagnet. A simple system with one electric and one mechanical degree of freedom is presented in Fig. 5.1. There are devices having two electromagnets, attracting a common anchor (Fig. 5.2) and the devices with many electromagnets. An example of a multivibrator is the electrovibrating conveyers. They consist of a plant, electromagnets, the so-called reactive masses attached to the anchors and the elastic elements c and c_0 (Fig. 5.3). Under the action of electromagnets the plant and the reactive masses vibrate in the direction indicated by the arrow in Fig. 5.3. The plant is usually schematized [14, 16] as a beam, therefore in this case the vibrating system is a system with distributed parameters.

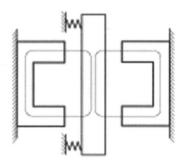

FIGURE 5.2.

Thus, the problems of vibrations excited by electromagnets are of the keen interest for engineering. For this reason, the present chapter contains indication as to how to choose the parameters for design of the electromagnets. In addition, the investigation of the systems assuming a simple schematization (for example, presented in Fig. 5.1) presents an interest for electromechanics irrespective of the technical applications.

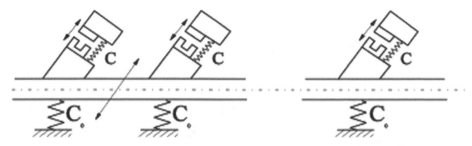

FIGURE 5.3.

In what follows we also consider the electromagnets with the turning anchor (Fig. 5.4). In this case the constraints of the oscillating system must admit angular displacements of the anchor relative to the core and prohibit any other relative displacements.

Let us write down the expression for energy of the magnetic field. We assume that the sizes of the core cross-section b, d (Fig. 5.1b) or the characteristic values of another shape of the cross-section are large in comparison with the distance between the anchor and the core but small in comparison with the length of the field lines (one of the lines is shown in Fig. 5.1). This allows one to assume that the scalar potential is constant over the cross-section and the field in the gap between the anchor and the core is homogeneous. The field in the ferromagnet can be taken into account by introducing the magnetic resistance $R_M^{(s)}$ of the core and the anchor. Provided that the ferromagnet can be considered as being ideal the relations between by the sizes of the core are not important, and in this case the results are obtained by assuming $R_M^{(s)} = 0$. We have

$$W = \frac{1}{2} \sum_{s=1}^{m} R_M^{(s)} \Phi_s^2 + \frac{2}{\mu_0 S_s}(h_s + \xi_s)\Phi_s^2. \qquad (5.1.1)$$

Here s is the number of the electromagnet, m is the total number of electromagnets in the system, Φ_s is the magnetic flux through the cross-section of the core or the anchor, S_s is the cross-sectional area of the end surface of the core (Fig. 5.1b), μ_0 is the magnetic permeability of the air, h_s is the distance between the core and the anchor in the undeformed oscillating system and ξ_s is the change of this distance measured toward the increase in the gap.

The anchors and the cores of the electromagnet are the elements of the system reacting the forces created by the exciter (by the magnetic field). As the mechanical elements (having a mass etc.) they are considered as a part of the oscillating system however they also belong to the exciter. Hence, the expression for W is invariant with respect to the type of the oscillating system, and values ξ_s play the role of functionals of the backward influence.

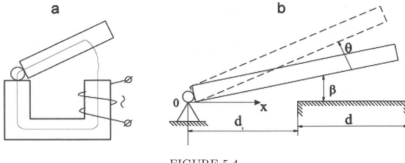

FIGURE 5.4.

The energy of field in eq. (5.1.1) is written for the U-shaped magnet. The same expression for W is valid for electrically and geometrically symmetric E-shaped magnet, that is, the magnet with a symmetry plane and consisting of two identical U-shaped magnets. In this case one needs to write down eq. (5.1.1) for a half of E-shaped magnet and double the result.

The expression (5.1.1) is constructed under the assumption of the same magnetic flux through any section of the magnetic conductor and both gaps between the core and the anchor. In the present chapter, unless otherwise stated, the value $R_M^{(s)}$ is considered to be constant, that is, neither hysteresis nor saturation in ferromagnet is taken into account. Finally, our consideration is limited only to solving the generating solution. For sufficiently powerful electromagnets these factors result in the errors of the order of the single-phase transformers. For practical purposes, the errors do not exceed the limits of uncertainties of the mechanical parameters, for instance in the mining industry the mass of the plant includes the added mass of vibrating granular media which is rather indefinite.

We designate the number of windings of $s-th$ electromagnet by z_s, and the currents in the windings by $i_s, i_{s2}, \ldots, i_{s,z_s}$, the current in the winding in the circuit with small active resistance being designated by i_s. This current corresponds to the quasi-cyclic generalized velocity while the other currents correspond to the additional generalized velocities, see Section 4.2. Under an appropriate choice of the positive directions for the currents the relation between the flux and the currents can be written in the form

$$n_s i_s + \sum_{j=2}^{z_s} n_{sj} i_{sj} = \left[R_M^{(s)} + \frac{2}{\mu_0 S_s}(h_s + \xi_s) \right] \Phi_s . \qquad (5.1.2)$$

Here n_{sj} denotes the number of turns of the $j-th$ winding of the $s-th$ electromagnet. The relation (5.1.2) being a special case of relation (4.2.25) expresses the Ohm law for the magnetic circuit of the $s-th$ electromagnet, the value in the left hand side of eq. (5.1.2) being equal to the magnetomotive force.

From eq. (5.1.1) we obtain the expression for the ponderomotive forces

$$Q_s = -\frac{1}{\mu_0 S_s} \Phi_s^2 . \tag{5.1.3}$$

The generalized momentum is Φ_s rather than the flux linkage $n_s \Phi_s$. Therefore one set of equations is given by

$$n_s \dot{\Phi}_s + R_s i_s = U_s(t) + U_{cs}, \quad s = 1, \ldots, m, \tag{5.1.4}$$

where $\langle U_s \rangle = 0$ and $U_{cs} = \text{const}$. There is no need to write down the other equations since it is possible to introduce matrices K_ν and Ψ_ν. Nevertheless we consider the equations of motion in order to demonstrate an example of determining v_s and calculating $k_\nu^{(rs)}$ and $\psi_\nu^{(rs)}$.

The motion of the system does not change if the surface ponderomotive forces acting on the anchor are replaced by their resultant force designated by $\check{Q}_s$. Let us also introduce the resultant of the ponderomotive forces $\check{Q}_{*s}$ acting on the core. Clearly, $\check{Q}_s = -\check{Q}_{*s}$. Let us consider a constant unit force which is applied to the anchor in opposition to $\check{Q}_s$. This unit force gives rise to a number of generalized forces in the equations of motion of the oscillating systems. Let f_s denote an abstract vector which is constructed of the above generalized forces and has the order equal to the number of degrees of freedom of the oscillating system. By analogy we define the vector f_{*0} corresponding to a unit force applied to the core and directed oppositely to $\check{Q}_s$. Then $v_s = f_s + f_{*s}, \xi_s = (v, v_s)$ and the equations of motions are

$$M\ddot{v} + B\dot{v} + Cv = -\sum_s^m \frac{1}{\mu_0 S_s} \Phi_s^2 v_s . \tag{5.1.5}$$

Let us proceed to determination of the periodic solutions. The subscript "zero" corresponding to the generating approximation is omitted in what follows. As shown in the previous chapter the resonance solution is obtained from the non-resonance one and presents less interest as a system with the high quality factor ($k_\nu = O(1/\rho)$) is difficult to realize in practice. In the generating approximation we have

$$\Phi_s = \alpha_s + V_s(t), \quad s = 1, \ldots, m, \tag{5.1.6}$$

where V_s is the antiderivative of U_s/n_s such that $\langle V_s \rangle = 0$.

The forces acting on the anchor and the core of $s - th$ electromagnet are equal to

$$Q_s = -\frac{1}{\mu_0 S_s} (\alpha_s^2 + 2\alpha_s V_s + V_s^2). \tag{5.1.7}$$

Let us introduce matrices $K_\nu = \| k_\nu^{(rs)} \|$ and $\Psi_\nu = \| \psi_\nu^{(ij)} \|$ defined in the following manner. Let the electromagnets be disconnected from the sources of emf and all circuits are open (then all currents and all ponderomotive forces are absent). Let us assume that the core and the anchor of $r - th$ electromagnet are subjected to two harmonic forces which are equal in the magnitude (equal to unity) and opposite in direction, these forces being directed along the line of action of forces $\check{Q}_r$ and $\check{Q}_{*r}$. Let the frequency of the forces be equal to $\nu\omega$. Let us determine the stationary (forced) vibration of the oscillating system under action of these forces, the amplitude of change in the distance between the core and the anchor of $s - th$ electromagnet as well as the phase shift between the above relative motion and the forces. The amplitude and the phase shift are equal to k_ν^{rs} and $\psi_\nu^{(rs)}$ respectively.

In other words, $k_\nu^{(rs)}, \psi_\nu^{(rs)}$ are determined by the relations

$$(v_\nu^{(r)}, v_s) = k_\nu^{(rs)} \cos\left(\nu\omega t - \psi_\nu^{(rs)}\right), \quad k_1^{(rs)}, k_2^{(rs)}, \ldots \geq 0, \qquad (5.1.8)$$

where $v_\nu^{(r)}$ denotes the $2\pi/\nu\omega-$periodic solution of the equation

$$M\ddot{v}_\nu^{(r)} + B\dot{v}_\nu^{(r)} + Cv_\nu^{(r)} = v_r \cos\nu\omega t. \qquad (5.1.9)$$

Utilising eq. (5.1.7) one can write down the expressions for ξ_s in the generating approximation

$$\xi_s = -\sum_r^m \frac{1}{\mu_0 S_r}[k_0^{(rs)}(\alpha_r^2 + \sigma_r^2)+ \qquad (5.1.10)$$

$$+2\alpha_r \sum_r V_{r\nu} k_\nu^{(rs)} \cos(\nu\omega t - \varphi_{r\nu} - \psi_\nu^{(rs)})] + \xi_{*s}, \quad s = 1, \ldots, m\,.$$

This expression contains $k_\nu^{(rs)}$ and $\psi_\nu^{(rs)}$ as parameters. The denotations in the latter expression are as follows

$$V_r = \sum_\nu V_{r\nu} \cos(\nu\omega t - \varphi_{r\nu}),$$

$$V_r^2 = \sigma_r^2 + \sum_{\nu, \nu \neq 0} V_{r\nu}^{(2)} \cos(\nu\omega t - \varphi_{r\nu}^{(2)}),$$

$$\xi_{*s} = -\sum_r^m \frac{1}{\mu_0 S_r} \sum_{\nu, \nu \neq 0} V_{r\nu}^{(2)} k_\nu^{(rs)} \cos(\nu\omega t - \varphi_{r\nu}^{(2)} - \psi_\nu^{(rs)})\,. \qquad (5.1.11)$$

Each electromagnet should have only one winding of this sort. In the generating approximation the other currents $i_{sj}, \ldots, i_{sz_s}, s = 1, \ldots, m$ do not depend on $\alpha_1, \ldots, \alpha_m$ and are not affected by the mechanical vibrations. While determining these currents the electromagnet should be considered as an ideal transformer, the winding with current i_s being considered as primary one and the other windings being viewed as secondary ones. This allows one to replace the secondary windings by the sources of emf equal to $-(n_{sj}/n_s)U_s$ in series with the resistances equal to the active resistances of the windings. As a result, for determining $i_{s2}, \ldots, i_{sz_s}$ one obtains a standard problem of electrical engineering of the stationary regime in the circuits with given parameters and emf. Its solution and consequently $i_{s2}, \ldots, i_{sz_s}$ are considered to be known in the following.

Inserting $i_{s2}, \ldots, i_{sz_s}$ in eq. (5.1.2) we find the currents $i_1, \ldots, i_m$ in the generating approximation as functions of time, $\alpha_1, \ldots, \alpha_m$ and $k_\nu^{(rs)}, \psi_\nu^{(rs)}$. Averaging the small terms in eq. (5.1.4) we arrive at the equations for $\alpha_1, \ldots, \alpha_m$

$$P_r(\alpha_1, \ldots, \alpha_m) = \alpha_r \left(a_r - \sum_s^m a_{rs}\alpha_s^2 \right) - \sum_s^m b_{rs}\alpha_s - e_r = 0, r = 1, \ldots, m.$$

(5.1.12)

Here

$$a_r = R_M^{(r)} + \frac{2h_r}{\mu_0 S_r} - \frac{2}{\mu_0 S_r} \sum_s^m \frac{1}{\mu_0 S_s} \sigma_s^2 k_0^{(rs)} a_{rs} = \frac{2k_0^{(rs)}}{\mu_0^2 S_r S_s},$$

$$b_{rs} = \frac{2}{\mu_0^2 S_r S_s} \sum_{\nu, \nu \neq 0} V_{r\nu} V_{s\nu} k_\nu^{(rs)} \cos(\varphi_\nu^{(r)} - \varphi_\nu^{(s)} - \psi_\nu^{(rs)}),$$

$$e_r - \sum_{j=2}^{z_r} n_{rj}\langle i_{rj} \rangle + \frac{U_{cr}}{R_r}n_r - \frac{2}{\mu_0 S_r} \langle \xi_{*r} V_r \rangle.$$

(5.1.13)

The periodic regime corresponding to a solution of equations (5.1.12) is stable if all roots $\lambda_1', \ldots, \lambda_m'$ of the equation

$$det \parallel \partial P_r / \partial \alpha_s + \lambda \kappa_r \delta_{rs} \parallel = 0$$

(5.1.14)

have negative real parts and unstable if the real part of at least one root is positive. The cases of multiple, zero or pure imaginary roots are not considered here. In eq. (5.1.14) we denote $\kappa_r = n_r^2/R_r$.

Usually the electromagnets in the system are connected to a single emf and $U_1(t) = \ldots = U_m(t)\varphi_\nu^{(1)} = \ldots = \varphi_\nu^{(m)}$, that is, each component is synphase. Then matrix $\partial P_r/\partial \alpha_s$ is symmetric and one can put $\kappa_r = 1$ in eq. (5.1.14). In addition to this, in this case (and also in the case in which $\Psi_\nu = 0$ for all ν) the above formulations of the integral criterion are valid.

Let us also consider the electromagnets with the turning anchor (Fig. 5.4). Let β_s designate the angle between the core and the anchor in the undeformed oscillating system (Fig. 5.4), and let ϑ_s be the change of this angle under the vibrations. The difference in the scalar potentials φ_s at all points of the surface of the core and the anchor is assumed to be constant, and the field in ferromagnet is taken into account by means of the magnetic resistance. Assuming that $x = 0$ on the rotation axis we obtain (Fig. 5.4)

$$H_s = \frac{\varphi_s}{h(\beta_s + \vartheta_s)}, \quad \Phi_s = b_s \int_{d_{1s}}^{d_{1s}+d_s} \frac{\mu_0 \varphi_s}{h(\beta_s + \vartheta_s)} dx = \frac{\mu_0 b_s \varphi_s}{\beta_s + \vartheta_s} \ln\left(1 + \frac{d_s}{d_{1s}}\right),$$

(5.1.15)

where b in the width of the pole, see Fig. 5.1b. Integrating $B_s H_s$ over the region between the core and the anchor and adding the energy of field in the ferromagnet we have

$$W = \frac{1}{2} \sum_s^m R_M^{(s)} \Phi_s^2 + \frac{\mu_0 b_s \varphi_s^2}{\beta_s + \vartheta_s} \ln\left(1 + \frac{d_s}{d_{1s}}\right).$$

(5.1.16)

Expressing here φ_s in terms of Φ_s with the help of eq. (5.1.15) we obtain

$$W = \frac{1}{2} \sum_s^m R_M^{(s)} \Phi_s^2 + \frac{\beta_s + \vartheta_s}{\mu_0 b_s \ln(1 + d_s/d_{1s})} \Phi_s^2.$$

(5.1.17)

The derivative $\partial W / \partial \Phi_s$ is equal to the magnetomotive force. This results relates the flux to the currents

$$n_s i_s + \sum_{j=2}^{z_s} n_{sj} i_{sj} = \left[R_M^{(s)} + \frac{\beta_s + \vartheta_s}{\mu_0 b_s \ln(1 + d_s/d_{1s})}\right] \Phi_s.$$

(5.1.18)

The ponderomotive forces acting on the anchor yields the moment about the axis of rotation

$$M_s = -\frac{\partial W}{\partial \vartheta_s} = -\frac{1}{2\mu_0 b_s \ln(1 + d_s/d_{1s})} \Phi^2.$$

(5.1.19)

It follows from eqs. (5.1.17-5.1.19) that the solution of the problem of vibration caused by electromagnets can be obtained from the solution for the electromagnets with attracting anchor provided that ξ_s, h_s and $2/\mu_0 S_s$ are replaced by ϑ_s, β_s and $1/[\mu_0 b_s \ln(1 + d_s/d_{1s})]$ respectively. In addition to this, matrices K_ν and Ψ_ν must be determined from the problem of the forced vibrations caused by two oppositely directed harmonic moments of the unit magnitude, one being applied to the anchor and the other acting on the core.

5.2   Systems with a single electromagnet

Let us consider the systems with a single electromagnet. First, we demonstrate two simple examples of determining values of k_ν and ψ_ν (the index

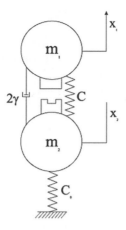

FIGURE 5.5.

indicating the number of the electromagnet is omitted). For the system shown in Fig. 5.1 v is a scalar value, $f_1 = 1, f_{*1} = 0, v_1 = f_1 = 1, \xi = v$ and

$$k_\nu = \frac{1}{\sqrt{(s - \nu^2\omega^2 m)^2 + 4\nu^2\omega^2\gamma^2}},$$

$$\tan\psi_\nu = \frac{2\gamma\nu\omega}{s - \nu^2\omega^2 m}, \quad k_0 = \frac{1}{c}. \tag{5.2.1}$$

The denotations here correspond to Fig. 5.1 and the following equation of motion

$$m\ddot{v} + 2\gamma\dot{v} + cv = -Q. \tag{5.2.2}$$

Let us now study the oscillating system shown in Fig. 5.5. We designate an abstract (in this case two-dimensional) vector by its components in the square brackets. Then we have $v = [x_1, x_2]$, where x_1, x_2 are the displacements (Fig. 5.5), $f_1 = [1, 0], f_{*1} = [0, -1], v = [1, -1], \xi = (v, v_1) = x_1 - x_2$. Equations of motion have a kind

$$m_1\ddot{x}_1 + 2\gamma(\dot{x}_1 - \dot{x}_2) + c(x_1 - x_2) = -Q,$$
$$m_2\ddot{x}_2 + 2\gamma(x_2 - \dot{x}_1) + c(x_2 - x_1) + c_0 x_2 = Q. \tag{5.2.3}$$

These two equations can be written in the form of a single vector equation, that is, we have eq. (5.1.5). The system in Fig. 5.5 is an idealization of the "double mass" vibration devices which are actively engaged in practice. The requirement to them is that the force acting on the immovable base is minimum. Therefore the rigidity c_0 is small and can be neglected for calculation of k_ν and ψ_ν. Then k_ν, ψ_ν are to be determined by relations (5.2.1) provided that $m = m_1 m_2/(m_1 + m_2)$.

Let us consider the case in which the magnet has one winding connected to the alternating current power (such a magnet is termed reactive). The

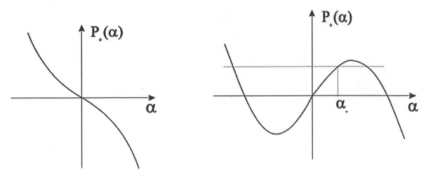

FIGURE 5.6.

voltage of the power is denoted as $U \sin \omega t$. According to Section 5.1 we have

$$\Phi = \alpha - \frac{U}{n\omega} \cos \omega t,$$

$$\xi = -\frac{k_0}{\mu_0 S} \left(\alpha^2 + \frac{U^2}{2n^2\omega^2} \right) + \frac{2U\alpha k_1}{n\omega\mu_0 S} \cos(\omega t - \psi_1) -$$

$$- \frac{U^2 k_2}{2n^2\omega^2\mu_0 S} \cos(2\omega t - \psi_2). \tag{5.2.4}$$

In order to obtain α one needs to solve the equation of the third degree

$$P(\alpha) \equiv \alpha(a - b - a_1\alpha^2) = 0, \tag{5.2.5}$$

where

$$a = R_M + \frac{2h}{\mu_0 S} - \frac{k_0 U^2}{n^2\omega^2\mu_0^2 S^2},$$

$$a_1 = \frac{2}{\mu_0^2 S^2}, \quad b = \frac{2U^2 k_1 \cos \psi_1}{n^2\omega^2\mu_0^2 S^2}. \tag{5.2.6}$$

The parameters denoted by a_1, a_{11}, b_{11} in eq. (5.1.13) are designated in eq. (5.2.5) as a, a_1, b respectively.

Since $k_0^{(rr)} > 0$ by virtue of the property of the influence coefficients, then $a_1 > 0$. It is clear that the constant component of ξ must be greater than $-h$. Comparing the expression for this component in eq. (5.2.4) with the expression for a we conclude that $a > 0$.

The periodic regime corresponding to some solution $\alpha = \alpha_*$ of eq. (5.2.5) is stable if $dP/d\alpha > 0$ and unstable if $dP/d\alpha < 0$ at $\alpha = \alpha_*$.

If $a < b$ then equation (5.2.5) has one root $\alpha = 0$ (Fig. 5.6a) and the only periodic regime is unstable (since $dP/d\alpha < 0$ at $\alpha = 0$). If $a > b$ then polynomial $P(\alpha)$ has three roots (Fig. 5.6b), one being equal to zero

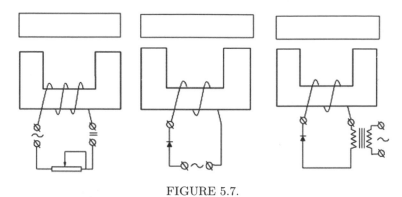

FIGURE 5.7.

and the others differing only in sign. The regime corresponding to the root $\alpha = 0$ is stable whereas the other two regimes are unstable. However at $\alpha = 0$

$$\xi = -\frac{U^2}{2n^2\omega^2\mu_0 S}[k_0 + k_2 \cos(2\omega t - \psi_2)], \qquad (5.2.7)$$

that is, the displacements contain only a constant component and a second harmonic. Therefore the reactive electromagnet can be only used for excitation of vibrations with the double frequency of the power.

This conclusion and the condition $\alpha = 0$ are usually considered to be evident, see [62, 32]. However the question of existence and stability of other regimes modes is not discussed. The above analysis confirms this conclusion and simultaneously indicates that other modes exist though they are not realized because of instability. The question arises as to whether these regimes can be stabilized by means of the factors not yet considered here. It can happen indeed and it is shown in Section 5.6.

In practice, the vibrations of frequency ω are rather required than those of 2ω. To obtain these, the electromagnets with more difficult circuits are used. For example, the circuits shown in Fig. 5.7a,b are utilised. The first one possesses the shortcoming that the constant component of the current passes through the alternating emf whereas the variable components passes through the source of constant emf, that is, the circuits alternating and constant currents are coupled. Analogously, in the case shown in Fig. 5.7b the network is supplied by a constant component of the current. This shortcoming can be removed by means of a decoupling transformer, Fig. 5.7c. The circuit of Fig. 5.7b is also common in practice. The relevant problems possess the following specific feature. At the small active resistance in the generating approximation, one obtains the problem of electric oscillation in the circuit with inductance and rectifier (Fig. 5.8). The resistance of opened rectifier is considered to be equal to zero while that of the closed rectifier is infinite. Let the valve be locked. Then the voltage applied to the valve is equal to the power voltage $U\sin\omega t$. When the sign of this voltages

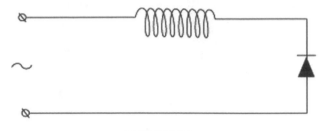

FIGURE 5.8.

changes, the rectifier opens and causes a nonzero current in the circuit. This happens, in particular, at $t = 0$. Then we obtain

$$n\dot{\Phi} = U \sin \omega t, \quad \Phi(0) = 0. \tag{5.2.8}$$

From here we have

$$\Phi = \frac{U}{n\omega}(1 - \cos \omega t). \tag{5.2.9}$$

The valve is closed when the current in the circuit vanishes. Because Φ and i have the same sign, cf. (5.1.2), this happens at $t = 2\pi/\omega$. At this time instant the voltage changes the sign and valve opens again. Thus eq. (5.2.9) holds true at any t. Up to the small values the flux, and in turn force Q, do not depend on the mechanical vibrations.

The stable regime in the systems with the reactive electromagnet possesses the same feature. In this case the backward influence of vibration on the exciter is considerable and, in particular, it can lead to instability.

In comparison, for example, with the mechanical vibration exciters the main advantage of electromagnets is that the amplitude can be adjusted by changing the electric parameters. In this sense the circuit of Fig. 5.9 with the compensating transformer is more convenient than that in Fig. 5.7. The number of turns of the compensating transformer and their directions are chosen such that the voltage of the second winding of the transformer is equal in value and opposite in sign to the voltage induced in the additional winding of the electromagnet (referred to as the bias winding). The current in this winding is $i_2 = U_c/R_2$. For determining α we have the equation

$$P(\alpha) \equiv P_0(\alpha) - e \equiv \alpha(a - b - a_1\alpha^2) - e = 0, \tag{5.2.10}$$

where n_2 is the number of turns of the additional winding and $e = n_2 i_2$ denotes the bias mmf. A stable regime exists under the condition that equation (5.2.10) has three roots. This regime corresponds to the root with the minimum absolute value $\alpha = \alpha_*$, see Fig. 5.6b. The polynomial $P(\alpha)$ has three roots if $a > b$ and

$$(a - b)^{3/2} > \frac{3}{2}|e|\sqrt{3a_1}. \tag{5.2.11}$$

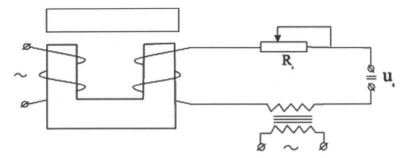

FIGURE 5.9.

The flux and vibrations are described by relations (5.2.4) at $\alpha = \alpha_*$. It follows from eq. (5.2.10) that $e \to 0$ as $\alpha_* \to 0$, then α_* is an odd function of e, the sign of α_* coincides with the sign of e and $|\alpha_*|$ increases as $|e|$ increases. It is typical for technical facilities that frequency ω is close to one of the eigenfrequencies and the first harmonic of the displacement is much "greater" than the constant component and the second harmonic. Therefore the maximum (within the period) displacement of some point of the oscillating system, which is approximately equal to the amplitude of the first harmonic, depends on e similar to the dependence $\alpha_*(e)$. The dependence on b is similar provided that $b > 0$. Since b is proportional to $k_1 \cos \psi_1$, the form $\alpha_*(b)$ suggests the dependence on parameters of the oscillating system. Similar to the problem of equilibrium, it can happen that not each value of e or b on these curves can be realized since the value $\xi = -h$ corresponding to impact of the anchor on the core can be achieved before the limiting point.

It is useful to have the expression for current in the winding with small resistance. Introducing

$$I_1 = \frac{4\alpha^2 U k_1 \cos \psi_1}{\gamma} - \frac{U}{n\omega}\left[R_M + \frac{2(h - \xi_s)}{\mu_0 S}\right] + \frac{U^3 k_2 \cos \psi_2}{2n^2\omega^2\gamma},$$

$$I_1^* = \frac{4\alpha^2 U k_1 \sin \psi_1}{\gamma} + \frac{u^3 k_2 \sin \psi_2}{2n^2\omega^2\gamma}, \quad \xi_c = \frac{k_0}{\mu_0 S}\left(\alpha^2 + \frac{U^2}{2n^2\omega^2}\right),$$

$$I_2 = \frac{2U^2 k_1 \cos \psi_1}{n\omega\gamma} + \frac{\alpha U^2 k_2 \cos \psi_2}{n\omega\gamma},$$

$$I_2^* = \frac{2U^2 k_1 \sin \psi_1}{n\omega\gamma} + \frac{\alpha U^2 k_2 \sin \psi_2}{n\omega\gamma},$$

$$I_3 = \frac{U^3 k_2 \cos \psi_2}{n^2\omega^2\gamma}, \quad I_3^* = \frac{U^3 k_2 \sin \psi_2}{n^2\omega^2\gamma}, \tag{5.2.12}$$

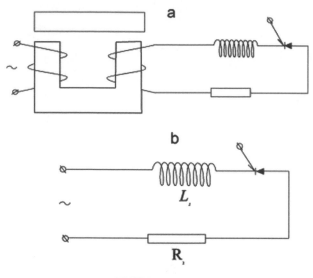

FIGURE 5.10.

where $\gamma = nw\mu_0^2 S^2$ we obtain from eq. (5.1.12)

$$ni(t) = I_1 \cos \omega t + I_1^* \sin \omega t - I_2 \cos 2\omega t - I_2^* \sin 2\omega t +$$
$$+ I_3 \cos 3\omega t + I_3^* \sin 3\omega t. \qquad (5.2.13)$$

Given the current, it is possible to find the consumed power of the alternating current

$$N = \langle U(t)i(t) \rangle = \frac{2a^2 U^2 k_1 \sin \psi_1}{n\gamma} + \frac{U^4 k_2 \sin \psi_2}{2n^3 \omega^2 \gamma}. \qquad (5.2.14)$$

Usually, the aim of devices with the bias electromagnets is to obtain maximum amplitudes of the first harmonic of displacement. To this end, these devices use the resonance effect, that is, the frequency of free vibrations is taken close to ω. Then the terms with k_2 are small in comparison with the terms k_1 and can be neglected. This is not in conflict with the non-resonant assumptions of Chapter 4 forasmuch as it is possible to assume $k_1 = O(1)$ rather than $O(1/\rho)$. In practice, typical values are $k_1/k_0 = 5-6, k_0/k_2 \sim 3, 1/\rho = 25 \div 100)$. Dealing with the reactive electromagnet one should put $\alpha = 0$ in eq. (5.2.13). In this case the terms with k_2 are essential because the corresponding devices are "tuned" to frequency 2ω.

A weak side of the system in Fig. 5.9 is that it is necessary to have a constant emf. A circuit with rectifier or thyristor in the bias circuit (Fig. 5.10) is got rid of this shortcoming and presumably is the best one for devices with a single electromagnet. Let us find the bias current. For this purpose we consider the circuit of Fig. 5.10b, where $U_2 = -(n_2/n_1)U \sin \omega t$. Let the thyristor open when the voltage phase is equal to φ_0, that is, at $t = t_0 = \varphi_0/\omega$. For $t > t_0$ the circuit is governed by the equation

$$L_2 \dot{i}_2 + R_2 i_2 = -\frac{n_2}{n_1} U \sin \omega t, \quad i_2(t_0) = 0. \tag{5.2.15}$$

Let us assume for definiteness that the thyristor is opened if $i_2 > 0$. Then one can accept that $-\pi/\omega < t_0 < 0$. For $t > t_0$ we have

$$i_2 = -\frac{n_2}{n} \frac{U}{z_2} [\exp\left(-\frac{t - t_0}{T_2}\right) \sin(\varphi_2 - \varphi_0) + \sin(\omega t - \varphi_2)]. \tag{5.2.16}$$

Here

$$T_2 = \frac{L_2}{R_2}, \quad \varphi_2 = \arctan \omega T_2, \quad z_2 = \sqrt{\omega^2 L_2^2 + R_2^2}. \tag{5.2.17}$$

The time instant of cutoff t_1 is determined by the equation

$$\exp\left(-\frac{t_1 - t_0}{T_2}\right) \sin(\varphi_2 - \varphi_0) + \sin(\omega t - \varphi_2) = 0. \tag{5.2.18}$$

Further $i_2 \equiv 0$ in the interval $t_1 < t < 2\pi + t_0$ etc. The constant component of the current is equal to

$$\langle i_2 \rangle = \frac{n_2}{n} \frac{U}{2\pi R_2} (\cos \varphi_1 - \cos \varphi_0), \quad \varphi_1 = \omega t_1. \tag{5.2.19}$$

Equation for determining α is obtained if one takes $\langle i_2 \rangle$ from eq. (5.2.19) and puts $e = n_2 \langle i_2 \rangle$ in eq. (5.2.10). The expressions for Φ and ξ remain unchanged. The dependence of amplitude of vibration on e does not change, too. But one can change e by varying φ. This allows one to influence the vibrations by increasing or decreasing the weak currents in the thyristor circuits without changing parameters of the power circuits.

Current $i(t)$ is determined differently than in the previous problem. We have

$$i(t) = \frac{1}{n} \sum_{\nu}^{3} I_\nu \cos \nu \omega t + I_\nu^* \sin \nu \omega t - \frac{n_2}{n} [i_2(t) - \langle i_2 \rangle], \tag{5.2.20}$$

where I_ν, I_ν^* are determined from eq. (5.2.13). Using eq. (5.2.20) one can find $i(t)$ either by means of the curve (5.2.20) versus time, or, vice versa, by expanding $i_2(t)$ in a Fourier series.

The solution for the circuit with a single diode rectifier in the bias circuit is obtained from the solution for the circuit with thyristor at $\varphi_0 = -\pi$.

5.3   System with electromagnets in the differentiating circuit

Technical difficulties caused by the necessity to "decouple" the circuits of alternating and direct currents can be surmounted if one uses two electromagnets attracting a common anchor, see Fig. 5.11. The cores of electromagnets are rigidly fixed to each other. The electromagnets with two rigidly fixed anchors are also used (Fig. 5.12). The bias windings of both electromagnets are placed such that the alternating emf variables induced are mutually compensated. These electromagnets are referred to as electromagnets in differential circuit. They can be considered as a single electromagnet termed the "two-gap" electromagnet.

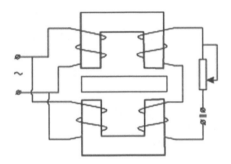

FIGURE 5.11.

One can see that the circuit of Fig. 5.11 is obtained from the circuit of Fig. 5.9 if the transformer is replaced by another electromagnet. In other words, the electric materials in the transformer are now used for creating ponderomotive forces. In comparison with the circuit of Fig. 5.10 the circuit of Fig. 5.11 has the advantage that neither inductance L_2 nor resistance R_2 is needed (the resistance R_2 in Fig. 5.11 can be much smaller than that in Fig. 5.10). A shortcoming of the differential circuit is a large number of the air gaps. A source of direct current is not obligatory, for example, it is possible to use a circuit of Fig. 5.13 where the bias circuit contains a rectifier and an additional winding with a small number of turns (and accordingly a low additional resistance).

Let us consider the circuit in Fig. 5.11 or Fig. 5.12. We assume that the magnetic resistance of the anchor (in the case of Fig. 5.11) or the common part of two U-cores (Fig. 5.12) is small in comparison with the sum of magnetic resistances of the air gaps and the other areas of the U-core. This allows one to assume that the field lines are closed, as shown in Fig. 5.11, and neglect the part of the flux shown by the dashed-line in Fig. 5.11. We have $\xi_1 = -\xi_2$ and $k_0^{11} = k_0^{22} = -k_0^{(12)}$, $k_\nu^{(11)} = k_\nu^{(22)} = k_\nu^{(12)}$, $\psi_\nu^{(11)} =$

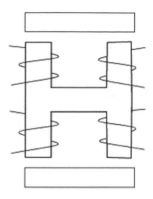

FIGURE 5.12

$\psi_\nu^{(22)} = \psi_\nu^{(12)} + \pi$. The electromagnets are assumed to be identical, then we obtain

$$\Phi_1 = \alpha_1 - \frac{U}{n\omega}\cos\omega t, \quad \Phi_2 = \alpha_2 - \frac{U}{n\omega}\cos\omega t,$$

$$\xi_1 = -\frac{1}{\mu_0 S}\left[k_0(\alpha_1^2 - \alpha_2^2) - 2(\alpha_1 - \alpha_2)\frac{U}{n\omega}k_1\cos(\omega t - \psi_1)\right]. \qquad (5.3.1)$$

The relations between the fluxes and currents are given by

$$ni_1 + n_2 i_{12} = \left(R_M + \frac{2}{\mu_0 S}(h + \xi_1)\right)\Phi_1,$$

$$ni_2 - n_2 i_{12} = \left(R_M + \frac{2}{\mu_0 S}(h - \xi_1)\right)\Phi_2. \qquad (5.3.2)$$

Signs of the terms ni_{12} are different[1] because the alternating current windings and the bias windings are placed differently in two electromagnets (in order to compensate the induction emf in the bias windings). Designating $e = n_2\langle i_{12}\rangle$ we obtain the equations for determining α_1, α_2

$$P_1(\alpha_1, \alpha_2) \equiv \alpha_1[a - a_1(\alpha_1^2 - \alpha_2^2)] - b(\alpha_1 - \alpha_2) - e = 0,$$

$$P_2(\alpha_1, \alpha_2) \equiv \alpha_2[a - a_1(\alpha_2^2 - \alpha_1^2)] - b(\alpha_2 - \alpha_1) - e = 0. \qquad (5.3.3)$$

Here a, a_1, b are determined by relations (5.2.6), only in expression for a in (5.2.6) the term with k_0 is cast aside.

System (5.3.3) assumes the solution of the sort $\alpha_2 = -\alpha_1$ referred to as the symmetric solution. This solution is

$$\alpha_1 = \frac{e}{a - 2b}. \qquad (5.3.4)$$

---

[1]In Section 5.1 the positive directions of the currents are chosen such that all currents appear with a plus sign in the expression for mmf. The mmf of the second electromagnet should be written in the form $ni_2 + n_2 i_{22}$ and for the bias circuit one should take into account that $i_{22} = -i_{12}$.

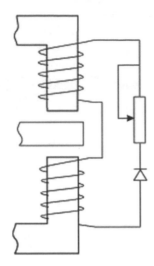

FIGURE 5.13.

Let us obtain the derivatives

$$\frac{\partial P_1}{\partial \alpha_1} = a - b - a_1(\alpha_1^2 - \alpha_2^2) - 2a_1\alpha_1^2,$$

$$\frac{\partial P_2}{\partial \alpha_2} = a - b - a_1(\alpha_2^2 - \alpha_1^2) - 2a_1\alpha_2^2,$$

$$\frac{\partial P_1}{\partial \alpha_2} = \frac{\partial P_2}{\partial \alpha_1} = 2a_1\alpha_1\alpha_2 + b. \tag{5.3.5}$$

The symmetric regime (that is, the regime corresponding to the symmetric solution) is stable if the roots of the polynomial

$$\Delta(\lambda) = \begin{vmatrix} a - b - 2a_1\alpha_1^2 + \lambda & b - 2a_1\alpha_1^2 \\ b - 2a_1\alpha_1^2 & a - b - 2a_1\alpha_1^2 + \lambda \end{vmatrix} \tag{5.3.6}$$

are negative. Transforming eq. (5.3.6)

$$\Delta(\lambda) = (a - 2b + \lambda)(a - 4a_1\alpha_1^2 + \lambda) \tag{5.3.7}$$

we obtain that this regime is stable provided that the inequality

$$a - 2b > 2|e|\sqrt{a_1/a} \tag{5.3.8}$$

is valid, otherwise it is unstable.

Let us consider stability of the other regimes. To this end we construct the sum $P_1 + P_2$. Cancelling out $\alpha_1 + \alpha_2$ (it is allowed for the non-symmetrical solutions) we obtain

$$a - a_1(\alpha_1 - \alpha_2)^2 = 0. \tag{5.3.9}$$

Using eq. (5.3.9) one can find the non-symmetrical solutions. There can be either four solutions or no solutions at all. For stability it is necessary (and sufficient) that the coefficients of polynomial $\Delta(\lambda)$ are positive. Let us transform the free term of $\Delta(\lambda)$ which is equal to $|\partial P_r/\partial\alpha_s|$. Let us sum the first and the second columns of the determinant $|\partial P_r/\partial\alpha_s|$ and then sum the first and the second rows of the obtained determinant. Using eq. (5.3.9) we obtain

$$|\partial P_r/\partial\alpha_s| = \Delta(0) = -[2a_1(\alpha_1^2 - \alpha_2^2)]^2 < 0. \tag{5.3.10}$$

Hence the non-symmetrical regimes are unstable.

For the symmetric regime the first harmonics of the electromagnets' forces acting on the anchor are added whereas the other harmonics vanish. Therefore

$$\xi_1 = \frac{4eUk_1}{n\omega\mu_0 S(a - 2b)} \cos(\omega t - \psi_1), \tag{5.3.11}$$

that is, the vibrations are harmonic. For $e = 0$ we have $\xi_1 \equiv 0$. In other words, the vibrations of the doubled frequency degenerate into the mechanical equilibrium. The stability condition for this equilibrium is $a - 2b > 0$.

Let us find the currents in the windings in the circuit with a small resistance. It follows from eq. (5.3.2) that

$$ni_1 = I_1 \cos\omega t + I_1^* \sin\omega t - I_2 \cos 2\omega t - I_2^* \sin 2\omega t,$$
$$ni_2 = I_1 \cos\omega t + I_1^* \sin\omega t + I_2 \cos 2\omega t + I_2^* \sin 2\omega t. \tag{5.3.12}$$

Here

$$I_1 = \frac{8\alpha_1^2 U k_1}{\gamma} \cos\psi_1 - \frac{U}{n\omega}\left(R_M + \frac{2h}{\mu_0 S}\right),$$

$$I_1^* = \frac{8\alpha_1^2 U k_1}{\gamma} \sin\psi_1,$$

$$I_2 = \frac{4\alpha_1 U^2 k_1}{n\omega\gamma} \cos\psi_1, \quad I_2^* = \frac{4\alpha_1 U^2 k_1}{n\omega\gamma} \sin\psi_1. \tag{5.3.13}$$

The current consumed by the magnet from the power supply

$$i = i_1 + i_2 = \frac{2}{n}(I_1 \cos\omega t + I_1^* \sin\omega t) \tag{5.3.14}$$

contains no higher harmonics and this is a merit of these devices.

The solution of the above problems is remarkable because of the closed form expressions for dependences of all sought quantities on the problem parameters.

If an arbitrary periodic voltage $U(t)$ (rather than sinusoidal) without constant component is applied to the electromagnet windings, then only symmetric mode is stable. The total force acting on the anchor (or the body composed of rigidly fixed cores) is as follows

$$Q = \frac{4e}{\mu_0 S(a - 2b)} V(t), \quad \dot{V} = \frac{1}{n} U(t), \quad \langle V \rangle = 0. \tag{5.3.15}$$

Here a is as above and

$$b = \frac{2}{\mu_0^2 S^2} \sum_\nu V_\nu^2 k_\nu \cos \psi_\nu . \tag{5.3.16}$$

Consequently, the electromagnetic forces acting on the oscillating system change in time as an integral over the voltage on the windings with a small resistance. Let the displacement of the oscillating system follows, to some extent, the external forces. Then the system is a certain integrator (with "electric input" and "mechanical output") for periodic signals in regard of reproducing the signal shape. This property is not typical for nonlinear systems. However the gain factor depends on the input amplitude in terms of b. Once again, this fact stresses that the system is essentially nonlinear.

In practice, there are devices with several pairs of electromagnets in the differential circuit. Let us assume that we have m pairs. Let one magnet have the number $r, r = 1, \ldots, m$, whereas the other have number $m - r$.

We suppose for simplicity, that all electromagnets are identical and connected to one source of sinusoidal voltages $U \sin \omega t$. We obtain

$$\Phi_r = \alpha_r - \frac{U}{n\omega} \cos \omega t,$$

$$\xi_r = -\frac{1}{\mu_0 S} \sum_s^m k_0^{sr} (\alpha_s^2 - \alpha_{m+s}^2) - 2(\alpha_s - \alpha_{m-s}) \frac{U}{n\omega} k_1 \cos \left(\omega t - \psi_1^{(sr)} \right),$$

$$\xi_{m+r} = -\xi_r, \quad r = 1, \ldots, m . \tag{5.3.17}$$

Equations for determining $\alpha_1, \ldots, \alpha_{2m}$ are as follows

$$P_r(\alpha_1, \ldots, \alpha_{2m}) \equiv$$

$$\equiv \alpha_r (a_r - \sum_s^m a_{rs}(\alpha_s^2 - \alpha_{m+s}^2)) - \sum_s^m b_{rs}(\alpha_s - \alpha_{m+s}) - e_r = 0,$$

$$P_{m+r}(\alpha_1, \ldots, \alpha_{2m}) \equiv \alpha_{m+r} (a_r - \sum_s^m a_{rs}(\alpha_{m+s}^2 - \alpha_s^2)) -$$

$$- \sum_s^m b_{rs}(\alpha_{m+s} - \alpha_s) + e_r = 0, \quad (r = 1, \ldots, m). \tag{5.3.18}$$

Here

$$a_r = R_M + \frac{2h}{\mu_0 S}, \qquad a_{rs} = \frac{2}{\mu_0^2 S^2} k_0^{(rs)},$$

$$b_{rs} = \frac{2U k_1^{(rs)}}{n\omega \mu_0^2 S^2} \cos \psi_1^{(rs)}. \tag{5.3.19}$$

In practice, the bias currents in different electromagnets are usually adjusted separately, therefore e_r are not assumed to be identical for all r.

Equations (5.3.18) assume a symmetric solution of the kind $\alpha_{m+s} = -\alpha_s$. The corresponding values $\alpha_r, r = 1, \ldots, m$ are determined from the linear system

$$P'_r(\alpha_1, \ldots, \alpha_m) = a_r \alpha_r - 2 \sum_s^m b_{rs} \alpha_s - e_r = 0. \tag{5.3.20}$$

In comparison to the general case the analysis of stability of the symmetric mode is simplified. Let us construct the derivative of P_r, P_{m+r} with respect to α_s, α_{m+s} and put $\alpha_{m+r} = -\alpha_r$. Then we obtain

$$\frac{\partial P_r}{\partial \alpha_s} = \frac{\partial P_{m+r}}{\partial \alpha_{m+s}} = a_r \delta_{rs} - 2a_{rs} \alpha_r \alpha_s - b_{rs},$$

$$\frac{\partial P_{m+r}}{\partial \alpha_s} = \frac{\partial P_r}{\partial \alpha_{m+s}} = -2a_{rs}\alpha_r\alpha_s + b_{rs}. \tag{5.3.21}$$

Here the polynomial $\Delta(\lambda)$ is the determinant of the order $2m \times 2m$

$$\Delta(\lambda) = \begin{vmatrix} \partial P_r/\partial \alpha_s + \lambda \delta_{rs} & \partial P_r/\partial \alpha_{m+s} \\ \partial P_{m+r}/\partial \alpha_s & \partial P_{m+r}/\partial \alpha_{m+s} + \lambda \delta_{rs} \end{vmatrix}. \tag{5.3.22}$$

The elements are conditionally replaced by the square blocks of the order $m \times m$.

We subtract $r - th$ line from each $m + p - th$ line of determinant (5.3.22) and in the obtained determinant we add each $m + r - th$ and $r - th$ column. The result is the determinant whose lower left minor $m \times m$ consists solely of zeros. This allows us to present $\Delta(\lambda)$ in the form

$$\Delta(\lambda) = \Delta_1(\lambda)\Delta_2(\lambda), \tag{5.3.23}$$

where Δ_1 and Δ_2 are polynomials of λ of order m

$$\Delta_1 = |(a_r + \lambda)\delta_{rs} - 2b_{rs}|,$$
$$\Delta_2 = |(a_r + \lambda)\delta_{rs} + 4a_{rs}\alpha_r\alpha_s|. \tag{5.3.24}$$

Thus, the analysis of stability of the symmetric mode is reduced to determining the sign of the roots of two equations of degree m, rather than those of a single equations of degrees $2m$, as in the general case.

The roots of polynomial $\Delta_1(\lambda)$ do not depend on $\alpha_1, \ldots, \alpha_m$. This means that if $\Delta_1(\lambda)$ has positive roots, then the symmetric mode is unstable regardless of values of the vibration amplitude. In particular, the position of mechanical equilibrium is unstable at $e_1 = \ldots = e_m = 0$. If all roots $\Delta_1(\lambda)$ are negative, then stability of the symmetric mode can always be achieved by reducing $|e_1|, \ldots, |e_m|$.

Comparing eq. (5.3.24) to (5.3.20), we obtain

$$\Delta_1(\lambda) = |\partial P_r'/\alpha_s + \lambda \delta_{rs}|. \qquad (5.3.25)$$

Hence, the requirement of negativeness of the roots of $\Delta_1(\lambda)$ expresses the stability condition for the system, for which the equalities $\alpha_{m+s} = -\alpha_s$ are artificially satisfied.

5.4   Interaction effects. Synthesis of electromagnets for vibration excitation

Let us list some qualitative effects in the systems with electromagnets, caused by interaction of the electromagnets with the oscillating system. In other words, we consider the case in which the ponderomotive forces depend on properties of the oscillating system. Under such effects we understand the features of vibrations of the considered systems in comparison to vibrations for which the mechanical motions are absent and $\xi_1 \equiv \ldots \equiv \xi_m \equiv 0$. These effects are listed in what follows.

1. *Amplification or attenuation of vibrations.* In order to estimate the influence of vibration on the electromagnets we introduce the coefficient k_a (referred to as the interaction coefficient) as a relation of some values (for example, maximum displacement, amplitude of the first harmonics of the force or $\xi(t)$) for real motion to this value under action of the force found under the assumption that the mechanical vibration is absent. The inequality $k_a > 1$ means that the interaction increases vibrations while $k_a < 1$ implies the attenuation effects. As an example we consider the system with two electromagnets in the differential circuit and connected to a source of sinusoidal emf. It is convenient to determine k_a in terms of the amplitude of vibration or force. From eq. (5.3.11) we obtain

$$k_a = \frac{a}{a - 2b}.$$

Using expression for vector b in (5.2.5) we find, that $k_a > 1$ if $\cos \psi_1 > 0$ and $k_a < 1$ if $\cos \psi_1 < 0$. For the oscillating system with a single degree of freedom the vibrations increase at the under-resonant frequencies (i.e. $\omega < \lambda$, where λ is the natural frequency of the undamped system) and the vibrations decrease at the frequencies higher than the resonance one. This effect becomes more noticeable while approaching the resonant frequency.

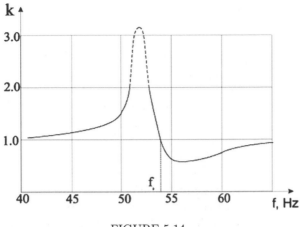

FIGURE 5.14.

Interestingly enough, this effect disappears ($k_a = 1$) exactly at the resonance, that is at $\omega = \lambda$. A characteristic dependence $k_a(\omega)$ is presented in Fig. 5.14. The dash line denotes the part of curve in the instability region. The case $a = 2b$ and $k_a = \infty$ is also possible, however the system is unstable a fortiori.

It is also noteworthy that this effect can be considerable not only if amplitude $\xi(t)$ is close to h. Indeed, for the differential circuit k_a does not depend on e, therefore by reducing the bias and taking the appropriate parameters we can obtain any amplification of vibration for arbitrary small amplitudes.

The interaction coefficient can be introduced for other problems of vibration excitation. To some extent it is similar to the dynamic amplification factor k_d. For example, in the problems of vibration excitation, the formula $A = k_a k_d \delta_{st}$ is a generalization of the formula for the amplitude of forced vibrations $A = k_d \delta_{st}$, where δ_{st} denotes the static displacement.

2. *Shift of the maximum amplitude from resonance.* Let us consider the dependence of the amplitude of $\xi(t)$ on frequency, i.e. $A(\omega)$ and $A_0(\omega)$, in the real motion and in the case when the force is calculated without consideration of mechanical vibrations. For simplicity, let us limit our consideration by the differential circuit. Up to small values of order ρ the curve $A_0^*(\omega) = \omega A_0(\omega)/\omega_*$, where ω_* is a fixed frequency, reproduces the amplitude-frequency characteristics $k_1(\omega)$ (see eq. (5.3.11)). Its maximum is not coincident with the resonance $\omega = \lambda$ due to friction. The maximum of the curve $A^*(\omega) = \omega A(\omega)/\omega_*$ is additionally moved due to interaction. It is shifted to low frequencies for the single degree of freedom systems or in the frequency interval near the first natural frequency. This effect is especially important for determination of dissipation by measuring of deviation of maximum $k_1(\omega)$ from the resonance.

3. *Instability of vibrations.* Instability can be explained only by the interaction forasmuch as vibration is always stable if the force is calculated without interaction. The instability appears as sudden impacts of the anchor against the core. This instability can be observed experimentally by increasing voltage and simultaneously decreasing bias such that no collision occurs before the loss of stability. It is essential that stability is possible at the amplitudes which are small in comparison with h. For example, in the case of differential circuit the mechanical equilibrium, i.e. the vibration with zero amplitude, becomes unstable as U increases.

4. *Birth of new modes.* One of the effects of interaction is that there are solutions $\alpha_1, \ldots, \alpha_m$ which, as $k_\nu^{(rs)}$ decreases, do not turn into the solutions for $k_{nu}^{(rs)} = 0$. These stable modes exist only in the case of some additional factors and are considered in Section 5.6.

The relations of Sections 5.1-5.3 allow us to determine the vibrations if the system parameters are given. Another problem (the problem of synthesis) is of greater interest for practice. This problem is to determine parameters of the electromagnet which is able to excite the required vibrations in a given oscillating system. Let us study this problem and set out the method of synthesis, based upon the above solution to nonlinear problems.

Let us consider the systems with one electromagnet connected to the source of sinusoidal emf. Let the vibrations of frequency ω be required. Then the electromagnet must have a bias by means of one of the circuits of Section 5.2. Presumably, the best circuit is shown in Fig. 5.10. The problem of synthesis prescribes k_0, k_1, ψ_1 and the required amplitude Q_1 of the first harmonic of force Q or the required amplitude ξ_1 of the first harmonics of vibrations (it is easy to find Q_1 in terms of ξ_1). In addition, the admitted value of the maximum induction $[B]$, the voltage amplitude U and the network frequency ω are prescribed.

The solution of this problem is not unique since the requirement

$$\frac{2\alpha U}{n\omega\mu_0 S} = -Q \qquad (5.4.1)$$

can be satisfied if one takes two of three parameters α, n, S arbitrarily within the limits given by the inequality

$$\alpha + \frac{U}{n\omega} \leq [B]S. \qquad (5.4.2)$$

Let us designate $\Phi_1 = U/n\omega, p = \alpha/\Phi_1, \sigma = (\alpha + \Phi_1)/[B]S$. According to eq. (5.4.2) $\sigma \leq 1$. If we set p and σ then the other parameters of the electromagnet are determined as follows. We have from eq. (5.4.1)

$$\Phi_1 = \frac{\mu_0 Q_1}{2[B]}\frac{1+p}{\sigma p}, \quad \alpha = p\Phi_1, \quad S = \frac{\Phi_1(1+p)}{\sigma[B]}, \quad n = \frac{U}{n\Phi_1}. \qquad (5.4.3)$$

Inserting the above values of these parameters in eq. (5.2.4) we determine the constant component ξ_0 and the amplitude of the second harmonics ξ_2 of displacement ξ. This allows us to determine the gap $h = k_h(\xi_0 + \xi_1 + \xi_2)$, where the coefficient k_h allows one to achieve the amplitudes greater than nominal ξ_1 by means of increase in bias. It is rational to take $k_h = 1.2 - 1.3$.

Further the value R_M should be prescribed. Usually the value R_M is about $0.2 - 0.3$ of the gap resistance, the inaccuracy playing no role within these limits.

Now we proceed to equation (5.2.10). All values are known, that is, the magnetomotive force of bias e can be obtained from this equation. It is remarkable that there is no need to solve equation of the third degree, that is, the problem of synthesis is simpler than the problem of analysis.

When the value e is found, the parameters of the bias circuit should be determined. This problem is not specific for electromagnets because the bias current does not depend on the vibration. It is reasonable to come from the condition that inductance L_2 or resistance R_2 (in the case of Fig. 5.10) is minimal. Calculating the bias circuit from eq. (5.2.2) it is possible to find the currents, to choose the winding wire etc. In addition to this, the stability is to be proved by means of eq. (5.2.11).

The arbitrariness in the choice of p, σ will be removed if the values of parameters obey some additional optimum condition $C = \min$ where C is a function of parameters of the electromagnet. In order to determine the optimum values of p, σ it is possible to set p, σ, find the parameters by means of the above method, calculate C and reach the minimum with the help of the standard methods. A simple result is obtained if the optimality criterion is the cross-section area S. We have

$$S = \frac{\mu_0 Q_1}{2[B]^2} \frac{(1+p)^2}{\sigma^2 p}. \tag{5.4.4}$$

Clearly, it is necessary to take $\sigma = \sigma_{\max} = 1$. Function $(1+p)^2/p$ achieves a minimum at $p = 1$. Therefore the condition $S = \min$ yields $p = \sigma = 1$.

The problem of synthesis for the differential circuit etc. are solved by analogy. If the optimum values of the parameters do not satisfy the stability condition then this condition should be considered as an additional restriction in the form of an inequality. The optimum problem should be solved under this restriction.

5.5 Generating solution for the induction - field strength nonlinearity

Let us now study the vibration caused by electromagnets in the case when B and H are so considerable that the curvature of curve $B(H)$ becomes significant. Here and in what follows B, H denote respectively the projections of vectors $\mathbf{B}, \mathbf{H}$ on the parallel axis, therefore B, H can have any sign. Function $B(H)$ is supposed to be single-valued (hysteresis is not taken into account) and odd, that is $B(-H) = -B(H)$. Additionally, the curvature of curve $B(H)$ is negative for $H > 0$. We suppose also that the ferromagnetic is not saturated and

$$dB/dH \sim (dB/dH)_{H=0} \gg \mu_0, \qquad (5.5.1)$$

otherwise the currents are $O(1/\rho)$. The sizes of the core section are assumed to be small in comparison to the length of field line. This allows one to assume that induction B is constant over the cross-section and to ignore the difference in the length of the field lines. Next, we have $\Phi_s = S_s B_s$ where s is the number of electromagnet in the system. We introduce the following value

$$W = \int\limits_V dV \int\limits_0^B H dB = \sum_s l_s \int\limits_0^{\Phi_s} H_s(\Phi_s) d\Phi_s + \frac{1}{2}\frac{2}{\mu_0 S_s}(h_s + \xi_s)\Phi_s^2. \quad (5.5.2)$$

Here V is the region where the field is taken into account, l_s is the length of the field line in the ferromagnetic, the denotations of Section 5.1 being taken for the remaining parameters. Dependences $H_s(\Psi_s)$ obtained from dependences $H_s(B)$ can be different for different s. They are assumed to be prescribed in the following. The expression for the ponderomotive forces are

$$Q_s = -\frac{\partial W}{\partial \xi_s} = -\frac{1}{\mu_0 S_s}\Phi_s^2, \qquad (5.5.3)$$

that is, they coincide with those in the "magnetic-linear" case. Only the relation between the magnetic fluxes and currents takes a new form

$$n_s i_s + \sum_{j=2}^{z_s} n_{sj} i_{sj} = \frac{\partial W}{\partial \Phi_s} = l_s H_s + \frac{2}{\mu_0 S_s}(h_s + \xi_s)\Phi_s. \qquad (5.5.4)$$

Relationship (5.5.4) expresses the law of total current.

Since the equations for the currents in windings with small resistance and the expressions for the ponderomotive forces are given, as before, by eqs. (5.1.4) and (5.1.5), the sought variables in the generating approximation are determined by relations (5.1.6), (5.1.7) and (5.1.10). The currents i_{sj} should be determined as it is explained above. The equations for determining $\alpha_1, \ldots, \alpha_m$ are now obtained by averaging relations (5.1.4) and is written down in the form

$$P_r(\alpha_1, \ldots, \alpha_m) = p_r(\alpha_r) + \alpha_r(a_r - \sum_s^m a_{rs}\alpha_s^2) - \sum_s^m b_{rs}\alpha_s - e_s = 0,$$

$$r = 1, \ldots, m, \qquad (5.5.5)$$

where

$$p_r = l_r \langle H_r(\alpha_r + V_r) \rangle, \quad a_r = \frac{2h_r}{\mu_0 S_r} - \frac{2}{\mu_0 S_r} \sum_s^m \frac{k_0^{(sr)}}{\mu_0 S_s} \sigma_s^2. \tag{5.5.6}$$

The other values in eq. (5.5.6) coincide with those in eq. (5.1.14). Parameter a_r differs from a_r in eq. (5.1.14) in absence of the term $R_M^{(r)}$.

When α_s have been found it is possible to determine $H_s(t)$, and then currents i_s from eq. (5.5.4). The conditions of stability are written down in the form given in Section 5.3. The conditions of the integral criterion of stability are kept with the only difference that instead of the field energy it is necessary to take W from eq. (5.5.2).

5.6 Oscillation of the feed frequency in system with reactive electromagnet

The influence of nonlinearity in the ferromagnetic is twofold. Firstly, in the magnetic-nonlinear system there can exist modes which are qualitatively coincident with the modes in the magnetic-linear system and converting to the latter when the nonlinearity parameter in ferromagnetic tends to zero. In these cases one speaks about determining the quantitative corrections caused by "the magnetic nonlinearity". They can be calculated, for example, by the method of successive approximation, in which "the magnetic-linear" solution is taken as the first approximation. Secondly, one can expect that the magnetic nonlinearity leads to origination of qualitatively new modes which do not convert to the modes existing in the magnetic-linear case. For example this is the case when the roots of eq. (5.5.6) tend to infinity as the parameter characterizing ferromagnetic nonlinearity decreases. In what follows we consider these new solutions.

The existing vibration facilities seem to be satisfactorily described under the assumption of the linear relationship between B and H. However the requirement to decrease the magnet size may lead to corrections of magnetic nonlinearity. Qualitative new modes can be expected only under principally different choice of the system parameters.

Let us consider the system with a single magnet having a winding with an external emf $U \sin \omega t$. In this case the flux and the vibration are described by relation (5.2.4).

For determining α we have the equation

$$P(\alpha) = p(\alpha) - a_1 \alpha^3 + (a - b)\alpha = 0. \tag{5.6.1}$$

Here

$$p(\alpha) = l \left\langle H \left(\alpha - \frac{U}{n\omega} \cos \omega t \right) \right\rangle \tag{5.6.2}$$

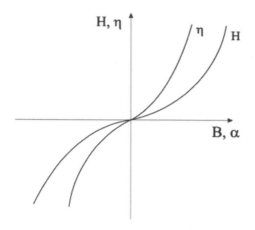

FIGURE 5.15.

and the other denotations are given in Sections 5.1 and 5.2, except for the
omitted indices of a_1, a_{11}, b_{11} (as a result a_1, a_{11}, b_{11} are denoted now by
a, a_1, b). Curve $p(\alpha)$ is "similar" to curve $H(B)$; in particular $p(\alpha)$ at $\alpha > 0$
is convex downward and $p(0) = 0, p(-\alpha) = -p(\alpha)$, see Fig. 5.15.

The mode corresponding to some solution $\alpha = \alpha_*$ of eq. (5.6.1) is stable,
if the inequality $dP/d\alpha > 0$ holds for $\alpha = \alpha_*$.

If the magnet and anchor are immobile, then eq. (5.5.1) takes the form

$$P(\alpha) = p(\alpha) + \frac{2h\alpha}{\mu_0 S} = 0 \qquad (5.6.3)$$

and assumes the only (evidently stable) solution $\alpha = 0$. Thus, the frequency
of the electromagnetic force is 2ω.

When the vibration is considered eq. (5.6.3) is equivalent to neglecting interaction between the magnet and the oscillating system. Thus, in
the systems with the sinusoidal external emf the mechanical vibrations of
frequency ω are impossible if the interaction or the nonlinearity in the
dependence $B(H)$ is neglected.

Let us consider equation (5.6.1). Let

$$p'_0 + a - b < 0, \quad p'_0 = (dp/d\alpha)_{\alpha=0}. \qquad (5.6.4)$$

Then solution $\alpha = 0$ of eq. (5.5.1) is unstable. By definition, $p(\alpha)$ increases
rapidly beginning at some values of α (saturation region), such that eq.
(5.6.1) has another solutions $\alpha \neq 0$. The corresponding modes, however,
can be unrealizable since the condition $dB/dH \gg \mu_0$ or the condition
of absence of collisions $\min \xi(t) > -h$ are violated. For this reason, on
example we show how to choose the parameters for which these conditions
are satisfied for nontrivial roots of eq. (5.5.1).

Let a part of curve $B(H)$ can be considered as a straight line. Let another part can be viewed as curvilinear and $dB/dH \gg \mu_0$. We denote the

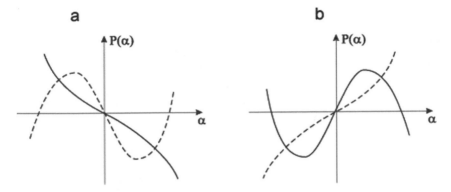

FIGURE 5.16.

upper bounds of these parts as B_p and B_m respectively. We also designate $\Phi_p = B_p S$ and $\Phi_m = B_m S$. Curve $p(\alpha)$ can also have rectilinear and
curvilinear parts; the "shorter" the first, the larger $U/n\omega$. For $U/n\omega = \Psi_p$
the rectilinear parts of $p(\alpha)$ disappears. Let us take $u/n\omega = \Phi_p$ and such
mechanical parameters that inequality (5.6.4) holds and $p - \alpha p_0' > a_1 \alpha^3$
for any $\alpha > 0$. The graph $P(\alpha)$ is presented in Fig. 5.16a by a dash line.
Changing the parameters such that $P_0' = p_0' + a - b < 0$ tends to zero we
obtain nontrivial roots of equation $P(\alpha) = 0$ with arbitrary small absolute
values. Therefore the condition $|\alpha| + U/n\omega < \Phi_m$ can be satisfied for these
roots.

Condition (5.6.4) requires that value $k_1 \cos \psi_1$ is sufficiently large. The
latter is mostly possible when ω is close to one of natural frequencies of
the oscillating system. In this case $k_1 \gg k_2$ and amplitude $\xi(t)$ is "nearly"
proportional to $|\alpha|$. This means that the condition $\min \xi(t) > -h$ is also
satisfied as $|\alpha|$ decreases.

According to Fig. 5.16a, the nontrivial roots correspond to the stable
modes. Under the above conditions $p_0' + a - b < 0, p - p_0' \alpha > a_1 \alpha^3$ these
modes can be described by means of the simplest approximation of $B(H)$.
More often than not, we take

$$H(B) = B/\mu_c + \gamma B^3. \qquad (5.6.5)$$

Then

$$P(\alpha) = \left(\frac{l\gamma}{S^3} - \frac{k_0}{\mu_0^2 S^2} \right) \alpha^3 + \left[\frac{2h}{\mu_0 S} + \frac{l}{\mu_0 S} - \left(\frac{2k_1 \cos \psi_1}{\mu_0^2 S^2} + \right. \right.$$
$$\left. \left. + \frac{k_0}{\mu_0^2 S^2} - \frac{3}{2} \frac{l\gamma}{S^3} \right) \frac{U^2}{n^2 \omega^2} \right] \alpha = (p_3 - a_1)\alpha^3 + (a - b + p_0')\alpha. \qquad (5.6.6)$$

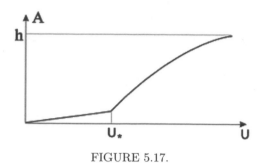

FIGURE 5.17.

The vibration of frequency ω corresponds to

$$\alpha = \pm\sqrt{\frac{b - a - p_0'}{p_3 - a_1}}. \tag{5.6.7}$$

Consequently, the vibration exists and is stable when $p_3 - a_1 > 0, a - b + p_0' < 0$, two these modes having different sign of α.

The dependence of the coefficient at α on U is shown in eq. (5.6.6) and it allows us to study the development of vibrations under the change of voltage. Let us consider the case when the coefficient at U is positive and $p_3 - a_1 > 0$. The vibrations of double frequency corresponding to the root $\alpha = 0$ are stable for small U. At some $U = U_*$ the equality $a - b + p_0' = 0$ holds, the solution $\alpha = 0$ becomes unstable but the nontrivial solutions appeared at $U = U_*$ become stable. A further increase in U excites vibrations of frequency ω (Fig. 5.17, where $A = \max|\xi|$). Its amplitude increases continuously with the growth of voltage unless collisions occur. This corresponds to transition from the dash curve of Fig. 5.16b to the dash curve of Fig. 5.16a.

The solid line in Fig. 5.16a denotes the graph $P(\alpha)$ for the case in which the magnetic nonlinearity is small and $p_3 < a_1$. As voltage increases, the system "converts" from the solid curve in Fig. 5.16b the solid curve in Fig. 5.16a. Thus the vibrations of the double frequency become unstable whereupon stable vibrations without collisions (as in the magnetic-linear case) are no more possible .

Because $H(B)$ differs from a polynomial of third degree the stable nontrivial solutions can be obtained in other regions of the space of parameters. For example, if we start from the representation

$$H(B) = B/\mu_c + \gamma B^3 + \varepsilon B^5 \tag{5.6.8}$$

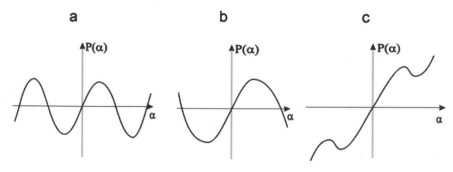

FIGURE 5.18.

we have

$$P(\alpha) = p_5\alpha^5 + (p_3 - a_1)\alpha^3 + (a - b + p_0')\alpha,$$

$$p_5 = \frac{l\varepsilon}{S^5} \, p_3 = \frac{l\gamma}{S^3} + \frac{l\varepsilon}{S^5}\frac{U^2}{n^2\omega^2},$$

$$p_0' = \frac{l}{\mu_c S} + \frac{3}{2}\frac{l\gamma}{S^3}\frac{U^2}{n^2\omega^2} + \frac{15}{8}\frac{l\varepsilon}{S^5}\frac{U^4}{n^4\omega^4}. \tag{5.6.9}$$

For $p_3 - a_1 > 0$ we obtain the same process of the vibration progress. However if $p_3 - a_1 < 0$ in some interval of U, and the coefficient at α in eq. (5.6.9) decreases with growth U, then the following behaviour of the system is possible. When U increases from $U = 0$ to $U - U_*$ such that $a - b + p_0' = 0$, the vibrations of double frequency occur. At $U = U_*$ they become unstable and the system jumps to vibrations of frequency ω corresponding to the values

$$\alpha = \pm\sqrt{\frac{a_1 - p_3}{p_5}}. \tag{5.6.10}$$

If we begin to decrease U then the vibrations remain stable at $U = U_*$ and "transform" into vibrations of double frequency only at $U = U_{**} < U_*$. The quenching of vibrations at $U = U_*$ and $U = U_{**}$ corresponds to transition from the curves shown in Fig. 5.18a to the curves in Fig. 5.18b and Fig. 5.18c respectively. Figure 5.19 displays this development of vibrations with a "voltage pulling". In this case at $U_{**} < U < U_*$ there are three stable modes: one trivial and two nontrivial with opposite signs of α.

Under the above approximation $H(B)$ and $p_3 - a_1 > 0$ the vibration develops similarly in any system for which $p(\alpha) - p_0'(\alpha) > a_1\alpha^3$ at $\alpha > 0$. Clearly, the nontrivial mode is simpler to realize experimentally if we ensure that the latter inequality holds and then increase voltage up to the values $U > U_*$, where U_* is determined from the equation $a - b + p_0' = 0$. The modes corresponding to the approximation $H(B)$ by a polynomial of fifth degree in the case $p_3 - a_1 < 0$, appear more difficult for observation. In

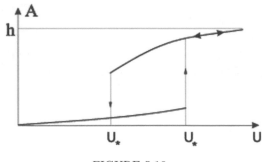

FIGURE 5.19.

any case more detailed information about dependences $H(B)$ and $p(\alpha)$ is
needed.

Existence of stable vibrations of frequency ω in the systems with the reactive electromagnets presents an effect caused by the combination of "interaction plus ferromagnetic nonlinearity". Indeed, this effect disappears,
if we neglect at least one of these nonlinear factors.

For excitation of vibrations with the frequency of power (more required in
practice than the vibration of double frequency) only the electromagnets
with bias were used earlier. They require additional windings for extinguishing resistances etc. The above effect gives a possibility to use simple
electromagnets for this purpose.

5.7   Oscillations in systems with two electromagnets. Asymmetrical modes

Let us consider the systems with two identical electromagnets in the differential circuit and connected only to the source of sinusoidal emf (Fig. 5.20).
As above, the fluxes and mechanical vibrations are described by relations
(5.3.1). For determining α_1, α_2 we obtain the following system

$$P_1(\alpha_1, \alpha_2) = p(\alpha_1) + a\alpha_1 - a_1\alpha_1(\alpha_1^2 - \alpha_2^2) - b(\alpha_1 - \alpha_2) = 0,$$
$$P_2(\alpha_1, \alpha_2) = p(\alpha_2) + a\alpha_2 - a_1\alpha_2(\alpha_2^2 - \alpha_1^2) - b(\alpha_2 - \alpha_1) = 0. \qquad (5.7.1)$$

Here

$$a = \frac{2k_0}{\mu_0^2 S^2}, \quad b = \frac{2U^2 k_1 \cos \psi_1}{n^2 \omega^2 \mu_0^2 S^2}, \quad s = \frac{2h}{\mu_0 S}. \qquad (5.7.2)$$

The vibration is stable if both roots λ_1, λ_2 of the equation

$$\Delta(\lambda) \equiv |\partial P_r/\partial \alpha_s + \lambda \delta_{rs}| = 0 \qquad (5.7.3)$$

are negative, and unstable, if there is a positive root. The derivative in eq.
(5.7.3) is calculated at α_1, α_2 from system (5.7.1).

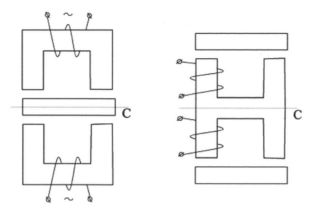

FIGURE 5.20.

Equations (5.7.1) have the solution $\alpha_1 = \alpha_2 = 0$. In this case $\xi_1 \equiv 0$, that is, the vibrations of the double frequency degenerate into the mechanical equilibrium. In the magnetic-linear case, when $p(\alpha) = \alpha l / \mu_r S$, only this solution can be stable. However it can be viewed as a "degenerated" symmetric solution $\alpha_1 = -\alpha_2$ at $e = 0$ (see Section 5.1). Therefore one can expect that system (5.7.1) has also other solutions. Let us consider them. For α_1 we obtain the following equation

$$P(\alpha_1) \equiv P_1(\alpha_1, -\alpha_1) \equiv -P_2(\alpha_1, -\alpha_1) \equiv p(\alpha_1) + (a - 2b)\alpha_1 = 0. \quad (5.7.4)$$

For $p_0' + a - 2b > 0$ this equation assumes only the trivial solution $\alpha_1 = 0$ (let us remind that at $\alpha > 0$ function $p(\alpha)$ can not be convex upwards). For $p_0' + a - 2b < 0$, along with $\alpha_1 = 0$ there are two additional nontrivial solutions of the opposite signs. In contrast to eq. (5.2.1) having an additional term is $a_1 \alpha^3$, equation (5.7.4) has no extra solution for any $p(\alpha)$ of the above sort. The nontrivial symmetric modes are stable if the inequalities

$$dp/d\alpha_1 + a - 2b > 0, \quad dp/d\alpha_1 + a - 4a_1\alpha_1^2 > 0 \qquad (5.7.5)$$

hold true. The first condition in eq. (5.7.5) can be written in the form $dP/d\alpha_1 > 0$ and, similar to eq. (5.7.4), does not contain a_1. As follows from the second condition in (5.7.5) the value of a_1 is important for realisation of this mode. Since the nontrivial solutions exist only for $p_0' + a - 2b < 0$, they are feasible only when the trivial solution is unstable.

The asymmetrical modes will be considered by means of approximation of $H(B)$ by a polynomial of the third degree according to Section 5.6. This allows one to find the regions in the spaces of parameters in which they can be realised with relative ease. Starting from eq. (5.6.5), we obtain

$$P_1(\alpha_1, \alpha_2) \equiv p_3\alpha_1^3 + (p_0' + a)\alpha_1 - a_1\alpha_1(\alpha_1^2 - \alpha_2^2) - b(\alpha_1 - \alpha_2) = 0,$$
$$P_2(\alpha_1, \alpha_2) \equiv p_3\alpha_2^3 + (p_0' + a)\alpha_2 - a_1\alpha_2(\alpha_2^2 - \alpha_1^2) - b(\alpha_2 - \alpha_1) = 0.$$
$$(5.7.6)$$

Here p_3 is determined according to eq. (5.6.6).

The solutions of eq. (5.7.6) are obtained in closed form. Let us construct the following combinations $\alpha_1 P_1 - \alpha_2 P_2$ and $\alpha_2 P_1 - \alpha_1 P_2$. Cancelling the multiplier $\alpha_1^2 - \alpha_2^2$, we have

$$\alpha_1^2 + \alpha_2^2 = \frac{b - a - p_0'}{p_3 - a_1}, \quad \alpha_1 \alpha_2 = -\frac{b}{2a_1 - p_3}, \tag{5.7.7}$$

$$\alpha_1 = \pm\frac{1}{2}\left[\frac{b - a - p_0'}{p_3 - a_1} - \frac{2b}{2a_1 - p_3}\right]^{1/2} \pm \frac{1}{2}\left[\frac{b - a - p_0'}{p_3 - a_1} + \frac{2b}{2a_1 - p_3}\right]^{1/2},$$

$$\alpha_2 = \pm\frac{1}{2}\left[\frac{b - a - p_0'}{p_3 - a_1} - \frac{2b}{2a_1 - p_3}\right]^{1/2} \mp \frac{1}{2}\left[\frac{b - a - p_0'}{p_3 - a_1} + \frac{2b}{2a_1 - p_3}\right]^{1/2}.$$
$$\tag{5.7.8}$$

The four possible combinations of signs in eq. (5.7.8) yield four asymmetrical solutions of system (5.7.6). Either no solution exists at all or all four solutions exist. The condition of existence is written down in the form

$$\frac{b - a - p_0'}{p_3 - a_1} \geq \left|\frac{2b}{2a_1 - p_3}\right|. \tag{5.7.9}$$

Given one solution, the other three can be obtained by permutation of α_1 and α_2 or changing the signs. Therefore the mechanical vibrations in various modes differ only in the sign of the constant term or in the shift π of the phases of the first harmonic.

Next we consider eq. (5.7.3). Using eq. (5.7.7) we obtain

$$\Delta(\lambda) = \lambda^2 + p_3\frac{b - a - p_0'}{p_3 - a_1}\lambda + 2(p_3 - a_1)(2a_1 - p_3)(\alpha_1^2 - \alpha_2^2)^2 = 0. \tag{5.7.10}$$

The considered modes are asymptotically stable if the coefficients in eq. (5.7.10) are positive and they are unstable if at least one coefficient is negative.

It follows from eq. (5.7.9) that if such a mode exists then the coefficient at λ is positive. Considering the free term we find the following condition of stability $a_1 < p_3 < 2a_1$. However in this case it should be $b - a - p_0' > 0$. Hence $b > 0$ and the condition of existence (5.7.9) takes the form

$$b(4a_1 - 3p_3) > (2a_1 - p_3)(a + p_0'). \tag{5.7.11}$$

Consequently, the stable asymmetrical modes are possible provided that $3/4p_3 < a_1 < p_3$.

The condition $b - a - p_0' > 0$ causes the condition $p_0' + a - 2b < 0$, that is, both stable asymmetrical modes and nontrivial symmetric modes are possible only if the mechanical equilibrium is stable.

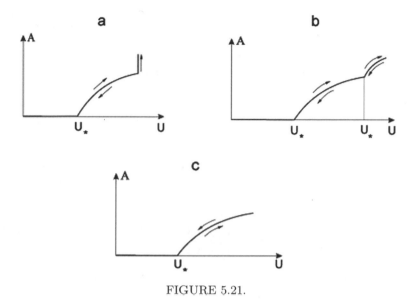

FIGURE 5.21.

Let us study the development of vibrations as the power voltage U increases (this is equivalent to increase of $|b|$). For the above approximation we write down the nontrivial symmetric solutions

$$\alpha_1 = \pm\sqrt{\frac{2b - a - p_0'}{p_3}}\tag{5.7.12}$$

and the conditions of their existence and stability, obtained from eq. (5.7.5)

$$2b - a - p_0' > 0, \quad b(4a_1 - 3p_3) < (2a_1 - p_3)(a + p_0').\tag{5.7.13}$$

For $b < 0$, that is for $\cos\psi_1 < 0$, only the trivial solution is stable. Let $b > 0$. There are three subcases: $a_1 > p_3, a_1 < p_3 < 4/3a_1$ and $4/3a_1 < p_3$. In the first one $U < U_*$ (where U_* is such that $b = b_* = 1/2(a + p_0')$) the trivial mode is stable. For $U = U_*$ the point of bifurcation is achieved and this gives birth to two nontrivial symmetric modes with the opposite signs of α. These two modes are stable unless $U = U_{**}$, where U_{**} is obtained for the equality

$$b = b_{**} = \frac{2a_1 - p_3}{4a_1 - 3p_3}(a + p_0').\tag{5.7.14}$$

As U increases from U_* to U_{**} the amplitude of vibration increases from zero. For $U > U_{**}$ the stable vibrations are impossible, see Fig. 5.21a.

The subcase $a_1 < p_3 < 4/3a_1$ is more interesting. For $U < U_{**}$ the vibrations develop as in the first case however at $U = U_{**}$ another point of bifurcation is achieved which gives rise to four stable asymmetrical modes. The symmetric mode becomes unstable and system passes to one of the asymmetrical modes. At the point of bifurcation the values of α_1 and α_2 in eq. (5.7.8) differ only in sign and coincide with their values in eq. (5.7.12). The system "selects" the asymmetrical solution which has the set of signs α_1, α_2 of the antecedent symmetric solution. This ensures the concordance of the phases of vibrations before and after the bifurcation point. As a result we observe no jump in the constant component (equal to zero in the symmetric mode) and the first harmonic of vibrations. Under further increase of U the asymmetrical mode remains stable and the amplitude grows unless the collisions, cf. Fig. 5.21b.

In the third subcase, for $U < U_*$ again we have only a stable trivial mode. At $U = U_*$ and $U > U_*$ the nontrivial symmetric modes become stable, see Fig. 5.21c.

Thus, in the considered systems the combination of "interaction plus ferromagnetic nonlinearity" generates the new symmetric modes and for $a_1 < p_3 < 4/3a_1$ results in establishing the asymmetrical modes. However the latter are characterised by the asymmetric distribution of the electric and mechanical variables in space relative to the plane of symmetry C, cf. Fig. 5.20. It is worth noting that it is possible in the systems which are electrically and mechanically symmetric.

All is needed to determine the fluxes, forces and vibrations is the dependence $p(\alpha)$. For this reason, the above conclusion is also valid for ferromagnetics with essential hysteresis provided that the character of the dependence $p(\alpha)$ remains the same. However in this case the currents in windings with the small resistance change considerably. The same is valid for the influence of Foucault's currents in ferromagnetics. If they are taken into account then we have a system with several volumetric (massive) conductors in addition to the above mentioned linear conductors. Following [90] we expand their current density in series in terms of the corresponding functions of spatial coordinates and introduce a countable set of currents. As a result we obtain that the above equations are added by an infinite (however countable) group of equations of the form

$$n_{Fs}\dot{\Phi} + R_{Fs}i_{Fs} = 0, \quad s = 1, 2, \ldots, \tag{5.7.15}$$

where currents i_{Fs} are related to the flux by the relations similar to those for the linear conductors. Clearly, values R_{Fs} in eq. (5.7.15) are not small. Then in the generating approximation the Foucault's currents i_{Fs} can be found in terms of the known value of $\dot{\Phi}$. They do not contain the constant component and thus have no influence on α, forces and vibrations.

For practical purpose, the Foucault's currents seem to be taken into account rather accurate by assuming that the electromagnet has an additional winding described by the equation

$$n_F\dot{\Phi} + R_F i_F = 0. \tag{5.7.16}$$

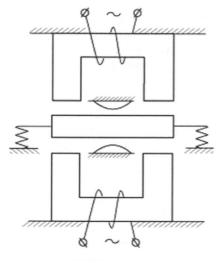

FIGURE 5.22.

Here n_F and R_F denote the equivalent number of coils and the resistance, which, generally speaking, must be determine experimentally. For determining the current in the winding with small resistance it is necessary to account for i_F and the other currents.

5.8   Vibrations in system with collisions. Influence of mechanical nonlinearity

If the oscillating system is nonlinear, then functions $P_r(\alpha_1, \ldots, \alpha_m)$ can not be set in the explicit form as it was done before for linear systems with the help of influence coefficients. Therefore the systems with various nonlinearities and diverse numbers of degrees of freedom should be considered individually. In result there appears a great number of problems of vibrations excited by electromagnets in nonlinear mechanical systems. The solution (obtained in cooperation with M.M.Vetyukov [124]) to one of problems of such sort is shown in what follows.

Let us consider a rigid body in the elastic suspension, whose displacement in both directions being limited by two symmetric stops. The vibrating body collides against these stops. The body is subjected to forces of attraction of two identical and symmetrically located electromagnets, Fig. 5.22. Let us assume that a sinusoidal emf $U \sin \omega t$ is applied to the magnets and the magnetic nonlinearity is negligible. We show that the stable mechanical vibration of frequency ω is possible.

Our analysis will be limited by the case in which the elastically suspended body can be considered as a system with a single degree of freedom, then ξ

is the only mechanical coordinate. Friction in the oscillating system is not taken into account. We assume that it is enough to introduce a coefficient of restitution R in order to describe the collision. Thus the velocity $\dot{\xi}_-$ before the collision and the velocity $\dot{\xi}_+$ after it are related as follows $\dot{\xi}_+ = -R\dot{\xi}_-, 0 \leq R \leq 1$. We restrict our consideration by the search of $2\pi/\omega$-periodic modes of a certain type. Namely, we search for the modes for which $\alpha_1 = -\alpha_2$, within the period the body collides each stop only once and the body velocity before collisions against the first and the second stops is the same. The sought modes, consequently, are symmetric in the explained sense; in addition, they are symmetric in time (about the time instant in the middle between the collisions).

The solution of the formulated problem is rather cumbersome, that is why we will show only the basic relationships. For brevity, we introduce the dimensionless variables $\tau = \omega t, \varphi_1 = \Phi_1/\Phi_*, \varphi_2 = \Phi_2/\Phi_*, \zeta = \xi/h_1; \Phi_* = U/n\omega$ where h_1 is the displacement of the body from the position in the undeformed elastic constraints up to the contact with one of the supports. We arrive at the equations

$$\varphi_1' + \rho(\sigma + \zeta)\varphi_1 = \sin\tau,$$
$$\varphi_2' + \rho(\sigma + \zeta)\varphi_2 = \sin\tau, \qquad (5.8.1)$$
$$\zeta'' + \nu^2\zeta = -f(\varphi_1^2 - \varphi_2^2),$$

describing vibration in the intervals between the collisions. A prime in eq. (5.8.1) means the differentiation with respect to τ. In addition we introduce the following quantities

$$\rho = \frac{R_1}{\omega L_0}, \quad L_0 = \frac{n^2\mu_0 S}{2h_1}, \quad \sigma = \frac{1}{h_1}\left(\frac{1}{2}R_M\mu_0 S + h\right),$$
$$\nu^2 = \frac{c}{m\omega^2}, \quad f = \frac{\Phi_*^2}{\mu_0 Smh_1\omega^2}, \qquad (5.8.2)$$

where c denotes the rigidity of the elastic suspensions and m is the body mass of body. The remaining denotations are coincident with those in Section 5.3.

The collisions take place at $\zeta = \pm 1$ and the relation between the velocities before and after the collision still has the form $\zeta_+' = -R\zeta_-'$.

The generating solution will be determined by the scheme of the continuous case, that is, we find φ_1 and φ_2 omitting the small terms, then we determine forces in terms of the obtained φ_1, φ_2, then we determine the displacements in terms of the forces. After this we insert obtained values of φ_1, φ_2 and ζ into the small terms, average them and construct the equations for determining constant components of the fluxes. According to [65, 88, 91] for every real-valued solution of these equations[2] there exists a periodic mode in the original piece-continuous system with the same number of collisions for period that in the generating approximation. This mode is close to the generating one in the sense that the difference in both solutions for $\varphi_1, \varphi_2, \zeta$ and the time instants of collision is small. However the velocities ζ' in the small intervals near the moments of collision differ in considerable values since the difference between the "exact" and "generating" time instants of collision is small and nevertheless not equal to zero.

Let us take $\varphi_1 = \alpha - \cos\tau, \varphi_2 = -\alpha - \cos\tau$ in the generating approximation. If such solutions exist, then the equation for α obtained by averaging small terms in the first equation in eq. (5.8.1) coincides with the equation obtained by averaging small terms in the second equation. It will be proved in what follows. In addition, by virtue of symmetry of the system there should exist a solution $\alpha = -\alpha_*$ along with the solution $\alpha = \alpha_*$. As it is sufficient to consider one of these solutions we can take $\alpha > 0$.

For determining ζ we obtain the equation

$$\zeta'' + \nu^2\zeta = 4f\alpha\cos\tau \tag{5.8.3}$$

with the above conditions of collision.

Unlike the continuous case, the oscillating system with collisions can have periodic modes of different types (with a different number of collisions, with different instants in the period etc.) for the same type force. The symmetric in time mode with two impacts is the simplest. Let τ_0 designate the collision impact at $\zeta = -1$ within the interval $0 \leq \tau \leq 2\pi$, and let ζ_0' denote the velocity after the collision. The conditions of periodicity for the motions of this type can be taken in the form $\zeta(\tau_0 + \pi) = 1, \zeta'(\tau_0 + \pi - 0) = \zeta_0'/R$. We write down the general solution of eq. (5.8.3)

$$\zeta = C_1\sin\nu(\tau - \tau_0) + C_2\cos\nu(\tau - \tau_0) + A\cos\tau, \tag{5.8.4}$$

where $A = 4f\alpha/(\nu^2 - 1)$. Eliminating C_1, C_2 and ζ_0' from the conditions $\zeta(\tau_0) = -1, \zeta'(\tau_0) = \zeta_0'$ and conditions of periodicity we obtain

$$A\sin\tau_0\frac{1-R}{1+R}\frac{1}{\nu}\tan\frac{\nu\pi}{2} - A\cos\tau_0 = 1, \quad 0 \leq \tau_0 \leq 2\pi. \tag{5.8.5}$$

The displacement $\zeta(\tau)$ is determined by the relations

$$\zeta(\tau) = \frac{1}{\nu}\frac{1-R}{1+R}\frac{1}{\cos(\nu\pi/2)}A\sin\tau_0\sin\left[\nu(\tau-\tau_0) - \frac{\nu\pi}{2}\right] + A\cos\tau,$$
$$\tau_0 \leq \tau \leq \tau_0 + \pi, \tag{5.8.6}$$

$$\zeta(\tau) = -\zeta(\tau - \pi), \quad \tau_0 + \pi \leq \tau \leq \tau_0 + 2\pi.$$

---

[2]Strictly speaking, under the conditions that the generating solution is not on the boundary of the region of existence and the Jacobian $(\partial P/\partial\alpha)$ does not vanish.

Given the force amplitude, relations (5.8.5) and (5.8.6) allow one to find τ_0 and the displacement. In this case these relations determine the dependences τ_0 and ζ on α.

Let us limit our consideration by the case $1 < \nu < 2$. The value $\nu < 1$ corresponds to $k_1 \cos \psi_1 < 0$. However for $k_1 \cos \psi_1 < 0$ the position of mechanical equilibrium is stable. Therefore for $\nu < 1$ the vibrations, if exist, are not self-excited.

Let us consider the conditions of existence of shock vibrations. It should be $\zeta_0' > 0$. However

$$\zeta_0' = \frac{2R}{1-R} A \sin \tau_0. \tag{5.8.7}$$

Because $A > 0$ for $\nu > 1$ then $\sin \tau_0 > 0$. It can be proved that in this case eq. (5.8.5) is satisfied only if $A \geq 1, \pi/2 < \tau_0 \leq \pi$, each $A \geq 1$ corresponding a unique value τ_0 in the interval $\pi/2 < \tau_0 \leq \pi$.

In addition, it should be $|\zeta| < 1$ for $\tau_0 < \tau < \tau_0 + \pi$. This requirement leads to some limitations imposed on A and the system parameters. They determine some region in the space of parameters such that $\max |\zeta(\tau)| = 1$ on its boundary. In the theory of piece-continuous systems this boundary is called C−boundary. The relations describing C−boundary are very difficult in this case. Keeping in mind that such modes exist and the region of their existence is not "too small" we can limit our search to narrower regions lying in the C−boundary. In particular, ζ' does not change sign in such regions if $\tau_0 < \tau < \tau_0 + \pi$. To find their boundary we consider a "border" mode $\zeta_g(\tau)$ for which ζ_g' does not change sign however $\zeta_g' = 0$ at $\tau = \tau_g$. In this case $\zeta''(\tau_g) = 0$. But the equalities $\zeta'(\tau_g) = 0, \zeta''(\tau_g) = 0$ can be simultaneously executed only if the values of parameters are related to each other, this relationship describing the sought boundary. In order to obtain this relationship we exclude sine and cosine function of the argument $\nu(\tau_g - \tau_0) - \nu\pi/2$ from the above equalities. As a result we obtain that the "internal" points of the considered region are characterised by the inequality

$$1 < A < \left[\sqrt{1 - \frac{\cos^2(\nu\pi/2)}{\nu^2} \left(\frac{1-R}{1+R} \right)^{-1}} - \frac{\sin \nu\pi/2}{\nu^2} \right]^{-1}. \tag{5.8.8}$$

Thus, if condition (5.8.8) holds then $\zeta'(\tau)$ does not change sign during the half-period and the inequality $\max |\zeta(\tau)| < 1$ is automatically executed.

Let us consider $P_r(\alpha)$. We have

$$P_1(\alpha, -\alpha) = \langle (\sigma + \zeta)(\alpha - \cos \tau) \rangle,$$
$$P_2(\alpha, -\alpha) = \langle (\sigma - \zeta)(-\alpha - \cos \tau) \rangle. \tag{5.8.9}$$

Since $\zeta(\tau)$ is symmetric with respect to time we have $P_1(\alpha) = -P_2(\alpha)$, that is, both equations $P_1 = 0, P_2 = 0$ can be satisfied by the same value of α. This means that the magnetic-symmetric solutions are possible.

Inserting eq. (5.8.6) into eq. (5.8.9) and using the relation between A and α, we obtain

$$\sin 2\tau_0 = \frac{1-R}{1+R} \frac{\nu^2-1}{2} \pi \left(\frac{\nu^2-1}{2f} \sigma - 1 \right). \qquad (5.8.10)$$

This equation yields τ_0 and one can find A and α from eq. (5.8.5). All sought values can be found without solving transcendent equations. In this sense the considered problem of vibrations subjected to the unknown force by magnets is simpler than that in the case of the prescribed force.

Let us study stability of this mode. In the problems of such type one can construct some linear relations, for example, difference equations for the time instant of the collision and the values of $\varphi_1, \varphi_2, \zeta'$ at the collision which can be used as variational equations in the case of continuous systems. These relations allow us to find four values (corresponding to the order of the system) which are analogues of the characteristic indices. The signs of their real parts define stability of the synchronous mode. These values will be the functions of parameter ρ, thus two of them (critical values) turn to zero at $\rho = 0$ while two others (noncritical values) have real parts which do not vanish at $\rho - 0$. The stability conditions corresponding to the noncritical indices can be found if we consider only the generating solution. They are the conditions of stability of the vibrations of the system with stops obtained under the assumption that force $4f\alpha \cos \tau$ is prescribed[3]. Let us find these conditions.

Let $\zeta_*(\tau)$ designate the displacement in the perturbed motion and let $\delta\tau_k$ and $\delta\zeta_k'$ denote the increases in time of $k - th$ collision at $\zeta_* = -1$ and the velocities after this collision. Let $k - th$ collision of the unperturbed motion be a collision at the moment $\tau_0 + 2k\pi$. By analogy with eq. (5.8.4) we obtain

$$\zeta_* = (C_1 + \delta C_{1k}) \sin \nu(\tau - \tau_0 - 2k\pi - \delta\tau_k) +$$
$$+ (C_2 + \delta C_{2k}) \cos \nu(\tau - \tau_0 - 2k\pi - \delta\tau_k) + A \cos \tau. \qquad (5.8.11)$$

Keeping the terms which are not higher than the first order of the increment we find from the conditions at $\tau = \tau_0 + 2k\pi + \delta\tau_k$ that

$$\delta C_{2k} = A \sin \tau_0 \delta\tau_k, \qquad \delta C_{1k} = \frac{1}{\nu}(\delta\zeta_k' + A \cos \tau_0 \delta\tau_k). \qquad (5.8.12)$$

Under small initial perturbations the collisions against the first and second stops in the perturbed motion will alternate as in the unperturbed motion,

---

[3]In the case of a damped linear systems these vibrations are always stable. If damping is not taken into account or small (of the order of ρ) stability is determined by additional terms of order ρ to pure imaginary characteristic indices obtained for the system in variations for the generating solution. These stability conditions are referred to as the "additional" ones, cf. [15]. In the problems under consideration (see Chapter 7) they are always satisfied in the non-resonant region, [15].

at least within some time interval. We designate the time instant of the following collision (at $\zeta_* = 1$) as $\tau_0 + (2k+1)\pi + \delta\tau_{k+1}$ and the velocity before collision as $(\zeta_0' + \delta\zeta_{k+1}')/R$. After transformations we obtain with the above accuracy that

$$\frac{2A\sin\tau_0}{1-R}\delta\tau_{k+1} = \left[A\sin\tau_0\left(\frac{1+R}{1-R} - \cos\nu\pi\right) - \frac{1}{\nu}\sin\nu\pi A\cos\tau_0\right]\delta\tau_k -$$
$$- \frac{1}{\nu}\sin\nu\pi\delta\zeta_k',$$

$$\frac{1}{R}\delta\zeta_{k+1}' = -\left(A\sin\tau_0\nu\tan\frac{\nu\pi}{2} + A\cos\tau_0\right)\delta\tau_k + \delta\zeta_k' +$$
$$+ \left(A\sin\tau_0\nu\tan\frac{\nu\pi}{2} + A\cos\tau_0\right)\delta\tau_{k+1}. \quad (5.8.13)$$

The relations (5.8.13) determine a linear operator Z mapping $\delta\tau_k, \delta\zeta_k'$ into $\delta\tau_{k+1}, \delta\zeta_{k+1}'$. We designate the moment of the next collision at $\zeta_* = -1$ as $\tau_0 + 2(k+1)\pi + \delta\tau_{k+2}$ and the velocity after collision as $\zeta_0' + \delta\zeta_{k+2}'$. Based on the symmetry of the initial mode it is possible to show that the relation between $\delta\tau_{k+2}, \delta\zeta_{k+2}'$ and $\delta\tau_{k+1}, \delta\zeta_{k+1}'$ is described by the same operator Z. Consequently the stability is determined by the eigenvalues of this operator. Provided that their absolute values is less than one, then the vibration is asymptotically stable under the prescribed force. If the absolute value of at least one of them is larger than one, then the vibration is unstable.

For determining the eigenvalues of operator Z we put $\delta\tau_k = z^k\delta\tau, \delta\zeta_k' = z^k\delta\zeta'$. Then we arrive at the characteristic equation of the system of finite-difference equations (5.8.13)

$$z^2 + \frac{1}{2}\left[\frac{1-R^2}{\nu}\sin\nu\pi\cot\tau_0 + (1-R^2)\cos\nu\pi - (1-R)^2\right]z + R^2 = 0,$$
$$(5.8.14)$$

whose roots z_1, z_2 are the sought-for eigenvalues. If $R < 1$ and

$$\frac{1}{2(1+R^2)}\left|\frac{1-R^2}{\nu}\sin\nu\pi\cot\tau_0 + (1-R^2)\cos\nu\pi - (1-R)^2\right| < 1,$$
$$(5.8.15)$$

then $|z_1|, |z_2|$ and the vibration is stable. If inequality (5.8.15) holds with an opposite sign of inequality then the vibration is unstable. The case in which one of the inequalities transforms into equality corresponds to the boundary of the instability region and will be not considered in what follows.

Getting rid of the absolute value we can set inequality (5.8.15) in the form of two equivalent inequalities. Using eq. (5.8.5) we can write these two inequalities in the form of a single inequality

$$
A > \left\{ 1 + \left(\frac{1+R}{1-R} \right)^2 \frac{\nu^2}{\sin^2 \nu \pi} \left[2 + (1 - \cos \nu \pi) \left(\frac{1-R}{1+R} \right)^2 \right]^2 \right\} \times
$$

$$
\times \left(\frac{1+R}{1-R} \right)^{-1} \left\{ \frac{1}{\nu} \tan \frac{\nu \pi}{2} - \frac{\nu}{\sin \nu \pi} \left[2 + (1 - \cos \nu \pi) \left(\frac{1-R}{1+R} \right)^2 \right] \right\}^{-1}.
$$

$$(5.8.16)$$

Let us now find the first approximation to the critical characteristic indices and construct the corresponding conditions of stability. These approximation are determined by analogy to the continuous case [88]. Consequently, if the functions $P_1(\alpha_1, \alpha_2), P_2(\alpha_1, \alpha_2)$ were known, they would be the roots of the equation $|\partial P_r / \partial \alpha_s + \lambda \delta_{rs}| = 0$. However as only the symmetric mode is considered it is necessary to proceed to the equations for perturbed motion.

Let $\delta \varphi_1, \delta \varphi_2$ designate the difference in φ_1, φ_2 between the perturbed and unperturbed motions, whereas ζ_* denotes the displacement in the perturbed motion. In the generating approximation $\delta \varphi_1, \delta \varphi_2$ are constant. In the continuous case the first approximation to the characteristic indices can be found as follows. We construct the variational equations and introduce the new variables $\delta \varphi_1 = \eta \exp(\rho \lambda_1 \tau), \delta \varphi_2 = \eta_2 \exp(\rho \lambda_1 \tau), \delta \zeta = \vartheta \exp(\rho \lambda_1 \tau)$. Then λ_1 is determined from the condition that $\eta_1, \eta_2, \vartheta$ are 2π−periodic functions of τ. To calculate λ_1 up to the terms of order ρ it is sufficient to find ϑ in the generating approximation. Under this approximation for ϑ we obtain a linear equation with 2π−periodic right part which is a linear form of the constant values η_{10}, η_{20}. Inserting the result into small terms of two other equations we obtain a linear homogeneous system for η_{10}, η_{20} from the conditions for periodicity of η_{11}, η_{21}. Then we can determine λ_1 from the condition $\eta_{10}, \eta_{20} \neq 0$.

Now we will act in the same manner, with the only difference that we determine ζ_* instead of quantity $\delta \zeta = \zeta_* - \zeta$ having in the generating approximation four discontinuities of the derivative in the period. Because function $\delta \zeta$ must be 2π−periodic at $\rho = 0$ it is sufficient to retain the values not higher than the first order in η_{10}, η_{20}. Finally, in the intervals of continuity of the derivative we obtain the following equation for ζ_*

$$
\zeta_*'' + \nu^2 \zeta_* = [4f\alpha + 2f(\eta_{10} - \eta_{20})] \cos \tau - 2f\alpha(\eta_{10} + \eta_{20}). \qquad (5.8.17)
$$

A periodic solution ζ_* can be found by analogy with the original vibrations $\zeta(\tau)$ with the difference that the time-symmetry is broken because of the second term in the right hand side of eq. (5.8.17). For this reason, it is necessary to consider both time intervals between the collisions and impose the periodicity conditions for the whole period. Generally speaking, the constant term in the right hand side of eq. (5.8.17) considerably complicates the solution. The latter is very bulky even despite the fact that the higher degrees of η_{10}, η_{20} are omitted. Let us show the final result. The time instant of the collision at $\zeta_* = -1$ is determined by the relations

$$\tau_{*0} = \tau_0 + \delta\tau_0,$$

$$\delta\tau_0 = 2f(\eta_{10} - \eta_{20}) \left[A^2(1 - \nu^2) \left(\sin\tau_0 + \frac{1-R}{1+R}\frac{1}{\nu}\tan\frac{\nu\pi}{2}\cos\tau_0 \right) \right]^{-1} -$$

$$- 2f\alpha \tan\frac{\nu\pi}{2}(\eta_{10} + \eta_{20}) \left[A\sin\tau_0\frac{2\nu^2}{\sin\nu\pi} \left[\sin^2\frac{\nu\pi}{2} + \left(\frac{1-R}{1+R} \right)^2 \right] -$$

$$- \nu A\cos\tau_0\frac{1-R}{1+R} \right]^{-1} \quad (5.8.18)$$

and the velocity ζ'_{*0} just after the collision is equal to

$$\zeta'_{*0} = \zeta'_0 + \delta\zeta'_0,$$

$$\delta\zeta'_0 = \frac{4fR}{1-R}(\eta_{10} - \eta_{20}) \left[(1 - \nu^2) \left(\sin\tau_0 + \frac{1+R}{1-R}\frac{1}{\nu}\tan\frac{\nu\pi}{2}\cos\tau_0 \right) \right]^{-1}$$

$$- \frac{4fR\alpha}{\nu(1-R)}\tan\frac{\nu\pi}{2}(\eta_{10} + \eta_{20}) \left(-\sin\tau_0\frac{2\nu}{\sin\nu\pi}\frac{1+R}{1-R} + \cos\tau_0 \right) \times$$

$$\times \left\{ \sin\tau_0\frac{2\nu}{\sin\nu\pi} \left[\sin^2\frac{\nu\pi}{2} + \left(\frac{1+R}{1-R} \right)^2 \right] - \cos\tau_0\frac{1+R}{1-R} \right\}^{-1}. \quad (5.8.19)$$

The moment of the next collision at $\zeta_* = 1$ and the velocity after the collision are related to $\delta\tau_0, \delta\zeta'_0$ by the formulae

$$\tau_{*1} = \tau_0 + \pi + \delta\tau_1, \quad (5.8.20)$$

$$\delta\tau_1 = \frac{1-R}{2} \left[\frac{1+R}{1-R} - \cos\nu\pi - \frac{1}{\nu}\sin\nu\pi\cot\tau_0 \right]\delta\tau_0 -$$

$$- \frac{1-R}{2}\frac{\sin\nu\pi}{\sin\tau_0}\frac{1}{\nu A}\delta\zeta'_0 - \frac{1-R}{2} \left(\frac{\sin\nu\pi}{\nu} - 2\cos^2\frac{\nu\pi}{2}\cot\tau_0 \right) \times$$

$$\times \frac{2f}{A(\nu^2 - 1)}(\eta_{10} - \eta_{20}) + \frac{\sin^2(\nu\pi/2)}{\nu^2\sin\tau_0}\frac{2f\alpha}{A}(\eta_{10} + \eta_{20}),$$

$$\zeta'_{*1} = -(\zeta'_{*0} + \delta\zeta'_{*1}),$$

$$\delta\zeta'_{*1} = \frac{A}{R}\left(\cos\tau_0\cos\nu\pi - \nu\sin\tau_0\sin\nu\pi + \nu\frac{1+R}{1-R}\sin\tau_0\tan\frac{\nu\pi}{2}\right)\delta\tau_0 +$$

$$+ \frac{1}{R}\cos\nu\pi\delta\zeta'_0 + \frac{A}{R}\left(\cos\tau_0 - \nu\frac{1+R}{1-R}\sin\tau_0\tan\frac{\nu\pi}{2}\right)\delta\tau_1 +$$

$$+ \frac{2f}{R(\nu^2-1)}\left(2\cos^2\frac{\nu\pi}{2}\sin\tau_0 + \nu\sin\tau_0\cos\tau_0\right)(\eta_{10} - \eta_{20}) +$$

$$+ 2f\alpha\frac{\sin\nu\pi}{\nu R}(\eta_{10} + \eta_{20}).$$

The perturbed motion in the intervals between the collisions is described by the equalities

$$\zeta_*(\tau) = \frac{1}{\nu}\frac{1+R}{1-R}\frac{A\sin\tau_0}{\cos(\nu\pi/2)}\sin\left[\nu(\tau-\tau_0) - \frac{\nu\pi}{2}\right] + A\cos\tau -$$

$$- \frac{1+R}{1-R}\frac{A\sin\tau_0}{\cos(\nu\pi/2)}\delta\tau_0\cos\left[\nu(\tau-\tau_0) - \frac{\nu\pi}{2}\right] +$$

$$+ \left[\frac{2f}{\nu^2-1}(\eta_{10} - \eta_{20})\sin\tau_0 + A\cos\tau_0\delta\tau_0 + \delta\zeta'_0\right]\sin\nu(\tau-\tau_0) -$$

$$- \left[\frac{2f}{\nu^2-1}(\eta_{10} - \eta_{20})\cos\tau_0 - A\sin\tau_0\delta\tau_0 - \frac{2f\alpha}{\nu}(\eta_{10} + \eta_{20})\right] \times$$

$$\times\cos\nu(\tau-\tau_0) + \frac{2f}{\nu^2-1}(\eta_{10} - \eta_{20})\cos\tau - \frac{2f\alpha}{\nu^2}(\eta_{10} + \eta_{20}),$$

$$\tau_{10} < \tau < \tau_{*1}.$$

$$\zeta_*(\tau) = -\frac{1}{\nu}\frac{1+R}{1-R}\frac{A\sin\tau_0}{\cos(\nu\pi/2)}\sin\left[\nu(\tau-\tau_0) - \frac{2\pi\nu}{2}\right] + A\cos\tau +$$

$$+ \frac{1+R}{1-R}\frac{A\sin\tau_0}{\cos(\nu\pi/2)}\delta\tau_0\cos\left[\nu(\tau-\tau_0) - \frac{3\nu\pi}{2}\right] -$$

$$- \frac{1}{\nu}\left[\frac{2f}{\nu^2-1}(\eta_{10} - \eta_{20})\sin\tau_0 + A\cos\tau_0\delta\tau_1 + \delta\zeta'_1\right]\sin\nu(\tau-\tau_0-\pi) +$$

$$+ \left[\frac{2f}{\nu^2-1}(\eta_{10} - \eta_{20})\cos\tau_0 - A\sin\tau_0\delta\tau_1 + \frac{2f\alpha}{\nu^2}(\eta_{10} + \eta_{20})\right] \times$$

$$\times\cos\nu(\tau-\tau_0-\pi) + \frac{2f}{\nu^2-1}(\eta_{10} - \eta_{20})\cos\tau - \frac{2f\alpha}{\nu^2}(\eta_{10} + \eta_{20}),$$

$$\tau_{*1} < \tau < \tau_{*0} + 2\pi. \qquad (5.8.21)$$

The obtained solution $\zeta_*(\tau)$ should be inserted into the equations derived from the variational equations. With the required accuracy they can be written down in the form

$$\eta'_1 + \rho\lambda_1\eta_1 + \rho\zeta_*\varphi_1 - \rho\zeta\varphi_1 + \rho(\sigma+\zeta)\eta_1 = 0,$$

$$\eta'_2 + \rho\lambda_1\eta_2 - \rho\zeta_*\varphi_2 + \rho\zeta\varphi_2 + \rho(\sigma-\zeta)\eta_2 = 0. \qquad (5.8.22)$$

The first approximation η_{11}, η_{21} are $2\pi-$periodic if the small terms in eq. (5.8.22) do not contain any constant component. This condition yields a system of linear homogeneous algebraic equations for η_{10}, η_{20}. Because $\eta_{10}, \eta_{20} \neq 0$ is required, the system determinant must be equal to the zero. This leads to a square equation which determines λ_1. Negative real parts are required for stability. In this case the roots are real-valued and they are determined by the expressions without radicals. After lengthy rearrangements the condition of stability corresponding to the requirement that one of the roots is negative reduces to the form

$$8\frac{\alpha^2}{\nu^2}\frac{f}{\pi}\tan\frac{\nu\pi}{2}\left(\frac{1}{\nu}\cos\tau_0 - \frac{2\sin\tau_0}{\sin\nu\pi}\frac{1+R}{1-R}\right)\times \tag{5.8.23}$$

$$\times\left[\frac{2\nu\sin\tau_0}{\sin\nu\pi}\left(\frac{1-R}{1+R}\sin^2\frac{\nu\pi}{2} + \frac{1+R}{1-R}\right) - \cos\tau_0\right]^{-1} + 4f\frac{\alpha_2}{\nu^2} < \sigma.$$

If one does not take into account that τ_0 satisfies eq. (5.8.10) then the second condition obtained by calculation of the second root is written down in the form similar to (5.8.23) as it also contains a bulky fraction. However if we take into account eq. (5.8.10) and replace this bulky fraction by the following one

$$\frac{1-R}{1+R}\frac{1}{\nu}\tan\frac{\nu\pi}{2}$$

according to eq. (5.8.5), then we obtain the inequality

$$\cos 2\tau_0 > 0. \tag{5.8.24}$$

Taking into account the limitations indicated above , we obtain from this inequality that $3/4\pi < \tau_0 < \pi$.

It is clear that condition (5.8.23) can be satisfied by taking sufficiently large values of σ, for example, by means of reducing the distance between the supports. If, in addition to this, the inequalities are satisfied

$$2\frac{1+R}{1-R} - \frac{2\sin^2\frac{\nu\pi}{2}}{\frac{2}{\nu\pi}\tan\frac{\nu\pi}{2} - 1}\frac{1-R}{1+R} < \frac{1}{\nu}\sin\nu\pi\cot\tau_0 <$$

$$< 2\frac{1+R}{1-R} + 2\sin^2\frac{\nu\pi}{2}\frac{1-R}{1+R}, \tag{5.8.25}$$

then inequality (5.8.23) is satisfied for any σ, because its left hand side is negative. Thus the condition of stability (5.8.16) is also satisfied since it can be rewritten in the form

$$\frac{1}{\nu}\sin\nu\pi\cot\tau_0 < 2\frac{1+R}{1-R} + 2\sin^2\frac{\nu\pi}{2}\frac{1-R}{1+R}. \tag{5.8.26}$$

Hence, the conditions of stability (5.8.16) and (5.8.23) are not contradictory in any case. It is also evident that they do not contradict eq. (5.8.24).

Finally, rewriting sufficient condition for existence of the mode (5.8.8) in the form

$$\frac{1+R}{1-R}\frac{\sin \tau_0}{\cos \dfrac{\nu\pi}{2}} > -\frac{1}{\nu} \tag{5.8.27}$$

we arrive at the inequality

$$\frac{1}{\nu}\sin \nu\pi \cot \tau_0 > -2\frac{1+R}{1-R}\cos \tau_0 \sin \frac{\nu\pi}{2}. \tag{5.8.28}$$

As $\tan(\nu\pi/2) < 0$, the right hand side of this inequality is smaller than the left hand side of the first inequality in eq. (5.8.25). Thus there are no contradictions between all these limitations which yield the sufficient conditions for existence and stability of the obtained modes. They determine a region in the space of parameters which ensures feasibility of the considered vibrations. A simpler way to achieve it is to take a value of ν sufficient close to one and simultaneously decrease the value of f such that $f(\nu^2 - 1)^{-1}$ is not very large.

As follows from eq. (5.8.10) the inequality $3/4\pi < \tau_0 < \pi$ can be satisfied only if

$$\frac{2f}{\sigma(\nu^2 - 1)} > 1. \tag{5.8.29}$$

In the denotations of this chapter the latter inequality presents the condition of instability of the mechanical equilibrium. This condition is necessary for existence of the obtained vibration. It reveals similarity of the systems with mechanical and magnetic nonlinearity, cf. Section 5.3.

The conclusion about the existence of stable vibrations of frequency ω in the system with collisions suggests another opportunity (in addition to using a magnetic nonlinearity) for exciting vibrations of this frequency with the help of simple electromagnets connected only to the sources of alternating current.

Although the general theory of Poincare's method for piece-continuous systems has been developed in detail [65, 88, 91], the above problem is probably a unique example in which the calculations of Poincare's method for the system discontinuous in the generating approximation, are carried out fully analogous to the continuous case. Therefore previous study may be of interest for the theory of piece-continuous systems, too.

Stable vibrations of frequency ω when the electromagnets are fed only from a power of alternating current seem to be excited not only in the systems with impacts, but also in any nonlinear oscillating system with "hard" elastic characteristics. However if the system with a elastic characteristic is nonlinear due to nonlinear friction such vibrations are hardly possible. In any case, they can not be realised in technical systems intended for displacement of granular materials, although the influencing of material is somehow equivalent to action of the nonlinear friction described, for example, by term $\beta\dot{\xi}^3$.

The electromagnets considered in the present chapter are described by Routh's equations which are linear with respect to mechanical (positional) coordinates in the generating approximation. When the system is supplied only by alternating current, stable vibrations of frequency ω are possible only because of combination "interaction plus nonlinearity in the ferromagnetics or the oscillating system" and impossible, if magnetic and mechanical nonlinearities are negligible. It is interesting to find out whether this particular condition is also valid for the systems with other field geometry when the ponderomotive forces, as functions of magnetic fluxes, depend on the mechanical coordinates and the equations of motion of the oscillating system are nonlinear. In the case of a single electric degree of freedom this question leads to the following problem: to establish whether the system

$$\dot{\Phi} + RL^{-1}\Phi = U\sin\omega t,$$

$$M\ddot{v} + B\dot{v} + Cv = -\frac{1}{2}\Phi^2\frac{\delta}{\delta v}L^{-1} \qquad (5.8.30)$$

has stable $2\pi/\omega-$periodic solutions such that expansion of v in Fourier series contains a non-small first harmonic. In eq. (5.8.30) the inductance $L > 0$ is a functional of v or a function of the components of vector v in the finite-dimensional case whereas the value of R is assumed to be proportional to a small parameter.

6

Charged particle in electromagnetic field

6.1 Motion due to fast oscillating forces

Although the solution of this problem is known and set out in "Mechanics" by L.D. Landau and E.M. Lifshits [73] as Kapitsa's method, in what follows this problem is analysed anew [97]. It pursues two objectives.

Firstly, [73] does not presents the conditions under which the generalization of the conclusion of potentiality of the mean forces in the first approximation for systems with many degrees of freedom. The conclusion that the amplitudes of rapidly oscillating forces have a potential is the most simple and natural. Let us notice that under this condition the potentiality appears to be considerably less unexpected than without it inasmuch as the initial system is Hamiltonian.

Secondly, in [73] the small parameter is not explicitly introduced and the procedure different from the well-known averaging methods is used. This allowed Yu.D. Daletskiy and M.G.Kreyn [27] to notice that the reasoning of [73] remained incomprehensible. Therefore it is useful to write down the equations with a small parameter and obtain the solution by the averaging method. This solution can be viewed as an answer to the above remark. This also gives a possibility to calculate the higher terms of approximation.

A standard way of averaging is that the averaged system remains Hamiltonian. However it makes sense to try another method of averaging such that for any approximation the expression for the kinetic energy of the averaged system is coincident with that of the original system. For example, after averaging the equation of motion of a material point describes the point motion however under non-potential forces. In what follows we consider both methods of averaging under the assumption of potentiality of the amplitudes of rapidly oscillating forces.

Let us consider the system with a single degree of freedom and let q, p designate the generalized coordinate and the momentum respectively. The inertia coefficients in the expression for kinetic energy are assumed to be independent of q. Let the system be subjected to the force of the type $F_1(q) \cos \nu t + F_2(q) \sin \nu t$ and the potential forces with potential $\Pi(q)$. The equations of motion in the form of Hamilton's equations with the generalized forces are given by

$$\dot{q} = \frac{1}{m} r, \quad \dot{p} = F_1(q) \cos \nu t + F_2(q) \sin \nu t - \frac{d\Pi}{dq}, \qquad (6.1.1)$$

where m is the inertia coefficient. We designate the characteristic values of coordinate q, the amplitudes of forces F_1 and F_2 and time by $[q], [F]$ and $[t] = 1/\nu$ respectively. Thus the characteristic values of the momentum and the inertia forces are $[p] = m[q]\nu$ and $[Q] = m[q]\nu^2$ respectively. If frequency ν is assumed to be high, as in [73], the ratio $[F]/[Q]$ can be considered as a small parameter ε.

On the other side, the value $[F]/m\nu^2$ is a characteristic amplitude of fast vibrations of frequency ν in the system subjected to the force $F_1 \cos \nu t + F_2 \sin \nu t$. Therefore the assumption about smallness of ε is equivalent to the assumption of smallness of the amplitude of fast vibrations in comparison with $[q]$. In [73] the smallness of these amplitudes is considered as the basic assumption without an explicit introducing a small parameter.

In the following we are interested only in those motions of the system which essentially depend on rapidly oscillating forces in spite of the fact that these forces is small in comparison with the characteristic value of inertia force. For this purpose we assume that the initial value of momentum equals ε, i.e. $p_0/[p] \sim \varepsilon$, otherwise the motion reduces mainly to the motion by inertia and the rapidly oscillating forces are hardly noticeable. The forces $d\Pi/dq$ must be also small in order to ensure that their influence on motion is the same as that of the rapidly oscillating forces. As it is shown below, these forces are of the order of ε^2.

Now we introduce the following dimensionless variables: time νt, coordinate $q/[q]$ and momentum $p/\varepsilon[p]$. Under the above assumptions, the dimensional momentum is of the order of ε in comparison with the characteristic value of the momentum whereas the dimensionless momentum has the order of unity at any time instant of the motion. Assuming that the dimensional and dimensionless variables have the same denotations we cancel ε out of the second equations and arrive at the following system

$$\dot{q} = \varepsilon r, \quad \dot{p} = F_1(q) \cos t + F_2(q) \sin t - \varepsilon \frac{d\Pi}{dq}. \qquad (6.1.2)$$

System (6.1.2) differs from the equations in [73] in the explicit small parameter ε.

By changing the variables

$$p = g + F_1(q)\sin t - F_2(q)\cos t \qquad (6.1.3)$$

we lead this system to the standard form

$$\dot{q} = \varepsilon(g + F_1(q)\sin t - F_2(q)\cos t),$$
$$\dot{g} = -\varepsilon\left(\frac{dF_1}{dq}\sin t + \frac{dF_2}{dq}\cos t\right)(g + F_1\sin t - F_2\cos t) - \varepsilon\frac{d\Pi}{dq}. \quad (6.1.4)$$

With the help of the standard change of variables

$$q = \xi + \varepsilon U_1(\xi, \eta, t) + \varepsilon^2 U_2(\xi, \eta, t) + \dots,$$
$$g = \eta + \varepsilon V_1(\xi, \eta, t) + \varepsilon^2 V_2(\xi, \eta, t) + \dots \qquad (6.1.5)$$

system (6.1.4) is led to the autonomous system

$$\dot{\xi} = \varepsilon\Theta_1(\xi, \eta) + \varepsilon^2\Theta_2(\xi, \eta) + \dots,$$
$$\dot{\eta} = \varepsilon H_1(\xi, \eta) + \varepsilon^2 H_2(\xi, \eta) + \dots. \qquad (6.1.6)$$

It is assumed that the right hand sides of equations (6.1.2) have a sufficient number of bounded derivative and $U_1, V_1, U_2, V_2, \dots$ in eq. (6.1.5) are bounded functions of t.

In the first approximation we can consider $q = \xi$, $g = \eta$, and the equations for ξ, η can be obtained by averaging the right hand sides of equations (6.1.4) over t. Then results in the equations

$$\dot{\xi} = \varepsilon\eta, \qquad \dot{\eta} = -\varepsilon\frac{dU_{ef}}{d\xi}, \qquad (6.1.7)$$

where

$$U_{ef} = \Pi + \frac{1}{4}(F_1^2 + F_2^2). \qquad (6.1.8)$$

The equations (6.1.7) describe the motion of the system in the potential field with the potential U_{ef}, the result coinciding with that in [73]. However in contrast to [73] the present result is obtained by a standard procedure of the averaging method.

This conclusion about potentiality of the forces averaged in the first approximation can be generalised onto a mechanical system with n degrees of freedom. The motion of such system is described by the equations

$$\dot{q} = A^{-1}(q)r,$$
$$\dot{p} = -\frac{1}{2}r^T\frac{\partial A^{-1}}{\partial q}p + F_1(q)\cos\nu t + F_2(q)\sin\nu t - \frac{\partial\Pi}{\partial q}. \qquad (6.1.9)$$

Here q, p are $[1 \times n]$ vectors-columns of the generalized coordinates and momenta, A is the matrix of the inertia coefficients and F_1, F_2 are $[1 \times n]$ vectorial functions.

Adopting the above assumptions of small parameters and introducing the dimensionless variables of the one-dimensional problem yields the following equations

$$\dot{q} = \varepsilon A^{-1}(q)r,$$

$$\dot{p} = F_1(q)\cos t + F_2(q)\sin t - \varepsilon \left[\frac{1}{2}p^T \frac{\partial A^{-1}}{\partial q}p + \frac{\partial \Pi}{\partial q} \right]. \qquad (6.1.10)$$

By replacement of variables $p = g + F_1 \sin t - F_2 \cos t$ system (6.1.10) is recast in the standard form

$$\dot{q} = \varepsilon A^{-1}(q)(g + F_1 \sin t - F_2 \cos t),$$

$$\dot{g} = -\frac{1}{2}\varepsilon g^T \frac{\partial A^{-1}}{\partial q} g - \varepsilon(F_1 \sin t - F_2 \cos t)^T \frac{\partial A^{-1}}{\partial q} g -$$

$$- \frac{1}{2}\varepsilon(F_1 \sin t - F_2 \cos t) \frac{\partial A^{-1}}{\partial q}(F_1 \sin t - F_2 \cos t) - \qquad (6.1.11)$$

$$- \varepsilon \left(\frac{\partial F_1}{\partial q} \sin t - \frac{\partial F_2}{\partial q} \cos t \right) A^{-1}(g + F_1 \sin t - F_2 \cos t) - \varepsilon \frac{\partial \Pi}{\partial q}.$$

By the new replacement (6.1.5), where ξ, η are vectors, we obtain the equations of the first approximation

$$\dot{\xi} = \varepsilon A^{-1}(\xi)\eta,$$

$$\dot{\eta} = -\frac{1}{2}\eta^T \frac{\partial A^{-1}}{\partial \xi}\eta - \varepsilon \frac{\partial \Pi}{\partial \xi} -$$

$$- \frac{\varepsilon}{2}\left(\frac{1}{2}F_1^T \frac{\partial A^{-1}}{\partial \xi}F_1 + \frac{1}{2}F_2^T \frac{\partial A^{-1}}{\partial \xi}F_2 + \frac{\partial F_1}{\partial \xi}A^{-1}F_1 + \frac{\partial F_2}{\partial \xi}A^{-1}F_2 \right) =$$

$$= -\frac{1}{2}\varepsilon\eta^T \frac{\partial A^{-1}}{\partial \xi}\eta - \varepsilon G(\xi). \qquad (6.1.12)$$

Provided that functions $F_1(q)$ and $F_2(q)$ are arbitrary, the vector-function $G(\xi)$ in eq. (6.1.12) can not be set in the form $-\partial U_{ef}/\partial \xi$ and the conclusion about potentiality of the averaged forces (made in [73]) is not valid. The most simple and natural conclusion ensuring potentiality of the averaged forces in the first approximation is the existence of potentials for the vectorial amplitudes F_1 and F_2. We suppose in the following that there are scalar functions $U_1(q)$ and $U_2(q)$ such that

$$F_1(q) = -\frac{\partial U_1}{\partial q}, \qquad F_2(q) = -\frac{\partial U_2}{\partial q}. \qquad (6.1.13)$$

Then matrices $\partial F_1/\partial q$ and $\partial F_2/\partial q$ are symmetric and

$$G(\xi) = \frac{\partial U_{ef}}{\partial \xi}, \qquad U_{ef} = \Pi + \frac{1}{4}(F_1^T A^{-1} F_1 + F_2^T A^{-1} F_2). \qquad (6.1.14)$$

In this case we obtain the equations

$$\dot{\xi} = \varepsilon A^{-1}(\xi)\eta, \quad \dot{\eta} = -\varepsilon \frac{\partial}{\partial \xi}\left(\frac{1}{2}\eta^T A^{-1}\eta + U_{ef}\right), \qquad (6.1.15)$$

which are similar to equations (6.1.7) and have the potential derived in [73]. However the conclusion about potentiality of the averaged forces in the first approximation is made under the condition (6.1.13) and seems to be not unexpected since the amplitudes F_1, F_2 are potential.

The problem of motion of a conducting rigid body in a fast-oscillating magnetic field considered in Section 2.1 demonstrates that the mean electromagnetic forces acting on the body are also potential although the original system is not Hamiltonian.

Introduction of small parameter in the explicit form in equations (6.1.10) allows one to obtain the higher approximations, too. These approximations are necessary not only for accurate calculation of the mean forces. There are more interesting cases in which the higher approximations determine the motion qualitatively. This is the case, for example, when the effective potential U_{ef} does not depend on one or more generalized coordinates. Thus the "drift" of the systems toward the coordinates which are absent in the potential can be exposed only by means of higher approximations.

Let us consider the case, when the condition (6.1.13) holds and assume for the sake of simplicity that A — const, for instance, a rigid body in a plane motion. Thus the system (6.1.11) is non-autonomous, Hamiltonian and has the Hamilton function

$$H - \frac{\varepsilon}{2}(g + F_1 \sin t - F_2 \cos t)^T A^{-1}(g + F_1 \sin t - F_2 \cos t). \qquad (6.1.16)$$

Averaging of these systems is usually conducted such that the system remains Hamiltonian under any approximation. We looks for the change of variables

$$\xi = \frac{\partial S(q, \eta, t)}{\partial \eta}, \quad g = \frac{\partial S(q, \eta, t)}{\partial q}, \qquad (6.1.17)$$

where

$$S = q^T \eta + \varepsilon S_1(q, \eta, t) + \dots . \qquad (6.1.18)$$

For arbitrary function $S(q, \eta, t)$ we have the equation

$$H + \frac{\partial S}{\partial t} = K,$$

where K is a new Hamilton function. In the second approximation we obtain

$$S_1 = (F_1 \cos t + F_2 \sin t)^T A^{-1} \eta - \frac{1}{4} F_1^T A^{-1} F_2 \cos 2t +$$

$$+ \frac{1}{8}(F_1^T A^{-1} F_1 - F_2^T A^{-1} F_2) \sin 2t, \tag{6.1.19}$$

$$K = \varepsilon \left(\frac{1}{2} \eta^T A^{-1} \eta + U_{ef} \right) + \frac{1}{2} \varepsilon^2 \left(F_1^T A^{-1} \frac{\partial F_2}{\partial \xi} - F_2^T A^{-1} \frac{\partial F_1}{\partial \xi} A^{-1} \eta \right).$$

This yields the equations

$$\dot{\xi} = \varepsilon A^{-1} \eta + \frac{1}{2} \varepsilon^2 A^{-1} (\frac{\partial F_2}{\partial \xi} A^{-1} F_1 - \frac{\partial F_1}{\partial \xi} A^{-1} F_2), \tag{6.1.20}$$

$$\dot{\eta} = -\varepsilon \frac{\partial U_{ef}}{\partial \xi} - \frac{1}{2} \varepsilon^2 \frac{\partial}{\partial \xi} \left[\left(F_1^T A^{-1} \frac{\partial F_2}{\partial \xi} - F_2^T A^{-1} \frac{\partial F_1}{\partial \xi} \right) A^{-1} \eta \right].$$

The merit of this system is that it is Hamiltonian. However it is difficult to find a particular mechanical system governed by equations (6.1.20). For example, if the initial system is a material point, then equations (6.1.20) describe another system with three degrees of freedom, whose structure depends on the sort of forces with amplitudes F_1 and F_2 acting on the point.

Therefore it is reasonable to apply another method of averaging, for which the expression for kinetic energy of the averaged system in any approximation coincides with that of the original system. In particular, after the averaging the equations of motion of the material point remain the equations of motion of the point. Naturally, that the Hamiltonian structure is lost and the mean forces acting on the point in higher approximations are no more potential.

The second method of deriving the averaged equations reduces to an appropriate choice of arbitrary functions in the expressions for U_1, V_1, U_2, V_2 etc. in eq. (6.1.5). For any approximation we need to keep the following form of the first equation relating ξ and η

$$\dot{\xi} = \varepsilon A^{-1} \eta. \tag{6.1.21}$$

In the second approximation it is necessary to choose function V_1 such that the relation

$$\Theta_2 = \frac{1}{2} A^{-1} \left(\frac{\partial F_2}{\partial \xi} A^{-1} F_1 - \frac{\partial F_1}{\partial \xi} A^{-1} F_2 \right) + A^{-1} \langle V_1 \rangle = 0 \tag{6.1.22}$$

holds true. In other words, $\langle V_1 \rangle$ should be equal to

$$\langle V_1 \rangle = \frac{1}{2} \left(\frac{\partial F_1}{\partial \xi} A^{-1} F_2 - \frac{\partial F_2}{\partial \xi} A^{-1} F_1 \right). \tag{6.1.23}$$

In eqs. (6.1.22) and (6.1.23) the angular brackets imply the mean value over the period. No special requirement is imposed on function U_1, and eq. (6.1.22) is written under the assumption that U_1 has a zero mean value. The second group of equations of motion in the second approximation has the form

$$\dot{\eta} = -\varepsilon \frac{\partial U_{ef}}{\partial \xi} + \varepsilon^2 M A^{-1} \eta ,$$

where the matrix

$$M = \frac{\partial F_2}{\partial \xi} A^{-1} \frac{\partial F_1}{\partial \xi} - \frac{\partial F_1}{\partial \xi} A^{-1} \frac{\partial F_2}{\partial \xi}$$

is skew-symmetric. Taking into account eq. (6.1.21) we have

$$\dot{\eta} = -\varepsilon \frac{\partial U_{ef}}{\partial \xi} + \varepsilon M \dot{\xi} , \qquad (6.1.24)$$

thus the mean forces appearing in the second approximation are gyroscopic forces.

By an appropriate choice of function V_2 it is possible to keep equation (6.1.21) in the third approximation etc.

Equations (6.1.21) and (6.1.24) describe motion of the initial system with the kinetic energy given by matrix A under the action of forces $-\varepsilon \partial U_{ef}/\partial \xi$ and $\varepsilon M \dot{\xi}$.

The second (non-potential) force is conditioned by interaction of sine and cosine harmonics of fast-oscillating forces and disappears (as well as terms of the second order in eq. (6.1.20)) if these forces consist only of a single harmonic. Then for analysis of the drifts toward the coordinates which are explicitly absent in the expression for U_{ef} we need to calculate another approximation. These drifts are observable only within the time intervals of the order of T/ε^2 and greater. For a qualitative study of such motions one can apply the approach expounded in Appendix D.

6.2   Motion of the charged particle in high-frequency electromagnetic field

The first approximation to the averaged equations of motion of the charged particle in high-frequency electromagnetic field are obtained in [81], and [74] is concerned with the higher approximations. This problem is considered here again for the following reasons.

Firstly, the equations of motion are reduced to the standard Krylov-Bogolyubov form by means of the procedure different from that in [74]. This allows one to obtain more simple equations and consider a more general case than in [74]. Secondly, it is shown that for averaging there is no need to expand the expression for the law of induction in terms of the parameter. In a special case considered in [74] this leads to an erroneous result. In addition to this, in [74] the evolutional components of the initial variables are not introduced and both averaged and original equations in the standard form are written in terms of the same variables, however it is correct only in the first approximation. Thirdly, in [74] it is not indicated how the arbitrary functions of slow variables can be chosen for construction of higher approximations, however the choice of these functions essentially affect the structure of the averaged equations.

The arbitrary functions are chosen such that the averaged equations have the form of the equations of motion for the material point or the form of Hamilton equations. The second and third (for the special case considered in [74]) approximations are obtained. In the first approximation the averaged Hamilton function is obtained in [23], however this paper does not contain the form of the original relativistic Hamilton function allowing derivation of equations in the standard form.

The higher approximations considered are necessary not only for refinements of calculations, but they also can provide one with a qualitative description of the motion. For example, they are required when the mean force in the first approximation does not depend on one or two coordinates, for the analysis of drift toward these "absent" coordinates.

New problems are formulated here, in particular, the problem of motion of the charged particle in a high-frequency field with slowly varying amplitude and phase.

Equations of motion of the charged particle in the field of monochromatic electromagnetic wave have the form

$$\ddot{\mathbf{r}} = \frac{e}{m}\sqrt{1 - \frac{1}{c^2}\dot{\mathbf{r}}^2}\left[\mathbf{E} + \frac{1}{c}\dot{\mathbf{r}} \times \mathbf{H} - \frac{1}{c^2}\dot{\mathbf{r}}(\dot{\mathbf{r}}\mathbf{E})\right]. \qquad (6.2.1)$$

Here

$$\mathbf{E} = \mathbf{E_1}(\mathbf{r})\cos\omega t + \mathbf{E_2}(\mathbf{r})\sin\omega t,$$
$$\mathbf{H} = \mathbf{H_1}(\mathbf{r})\cos\omega t + \mathbf{H_2}(\mathbf{r})\sin\omega t. \qquad (6.2.2)$$

Let $[E], [H], L$ designate respectively the characteristic values of intensity of the electric and magnetic fields and the characteristic distance at which $\mathbf{E_1}(\mathbf{r})/[E], \mathbf{H_1}(\mathbf{r})/[H]$ etc. are essentially changing. Let us enter the dimensionless time ωt and the dimensionless radius-vector $\mathbf{r}/L$, however the dimensionless variables and dimensionless relations $\mathbf{E_1}(\mathbf{r}), \mathbf{H_1}(\mathbf{r})$ etc. have the denotations of the original dimensional variables. Then we arrive at the equation

$$\ddot{\mathbf{r}} = \frac{e[E]}{m\omega^2 L}\sqrt{1 - \left(\frac{\omega L}{c}\right)^2 \dot{\mathbf{r}}^2}\left[\mathbf{E_1}(\mathbf{r})\cos t + \mathbf{E_2}(\mathbf{r})\sin t +\right.$$

$$+ \frac{[\mathbf{H}]}{[\mathbf{E}]}\frac{\omega L}{c}\dot{\mathbf{r}}(\mathbf{H_1}(\mathbf{r})\cos t + \mathbf{H_2}(\mathbf{r})\sin t) -$$

$$\left. - \left(\frac{\omega L}{c}\right)^2\dot{\mathbf{r}}\left(\dot{\mathbf{r}}\mathbf{E_1}(\mathbf{r})\cos t + \dot{\mathbf{r}}\mathbf{E_2}(\mathbf{r})\sin t\right)\right]. \tag{6.2.3}$$

In what follows we also need the equations for $\mathbf{E_1}, \mathbf{H_1}, \mathbf{E_2}, \mathbf{H_2}$ which are obtained from the induction law written in the non-dimensional form

$$\nabla \times \mathbf{E_1} = -\frac{[H]}{[E]}\frac{\omega L}{c}\mathbf{H_2}, \qquad \nabla \times \mathbf{E_2} = \frac{[H]}{[E]}\frac{\omega L}{c}\mathbf{H_1}. \tag{6.2.4}$$

In various electromagnetic fields the relations $[H]/[E]$ and $\omega L/c$ can have values of the different orders [74]. In the following we consider the "normal" case [74] in which these relations are of the order of unity.

The ratio of the characteristic Coulomb forces to the characteristic force of inertia

$$\varepsilon = e[E]/m\omega^2 L \tag{6.2.5}$$

is considered to be small. But the value of $e[E]/m\omega^2$ is the characteristic amplitude of rapid oscillation of the particle with frequency ω subjected to a characteristic Coulomb force. Therefore the assumption about smallness of ε is equivalent to the assumption about smallness of the amplitudes of rapid oscillations in comparison with L. In [81] and [74] the smallness of these amplitudes is the basic assumption without introducing parameter ε by means of relation (6.2.5).

Motion of particle is further considered as non-relativistic, i.e. ratio $|\dot{\mathbf{r}}|/c$ in eq. (6.2.1) is assumed to be small, however the relativistic corrections in the equations of motion are taken into account. Presumably, the most interesting case is that in which $|\dot{\mathbf{r}}|/c = O(\varepsilon)$. This case is considered in what follows.

In the problem under consideration we are interested in the case in which the electromagnetic forces essentially affect motion of the particle despite the fact that they are small in the above sense. This is the case only under the condition that the dimensionless speed of the particle $|\dot{\mathbf{r}}|$ is small within the whole time interval needed for the particle to pass the distance of order L, otherwise the particle passes this interval by inertia. As it is shown below the sought motion takes place if $|\dot{\mathbf{r}}| = O(\varepsilon)$. For this reason we introduce a new independent variable $\mathbf{v} = \dot{\mathbf{r}}/\varepsilon$. We can choose the characteristic values $[E]$ and $[H]$ such that $[E] = [H], \omega L = c$. Then we arrive at the system

$$\dot{\mathbf{r}} = \varepsilon\mathbf{v}, \quad \dot{\mathbf{v}} = \sqrt{1 - \varepsilon^2 v^2}[\mathbf{E_1}\cos t + \mathbf{E_2}\sin t + \varepsilon\mathbf{v}\times(\mathbf{H_1}\cos t + \mathbf{H_2}\sin t) -$$

$$- \varepsilon^2\mathbf{v}(\mathbf{v}\mathbf{E_1})\cos t - \varepsilon^2\mathbf{v}(\mathbf{v}\mathbf{E_2})\sin t]. \tag{6.2.6}$$

Equations (6.2.4) take the form

$$\nabla \times \mathbf{E_1}(\mathbf{r}) = -\mathbf{H_2}(\mathbf{r}), \qquad \nabla \times \mathbf{E_2}(\mathbf{r}) = \mathbf{H_1}(\mathbf{r}) . \qquad (6.2.7)$$

Let us transform equations (6.2.6) to the standard Krylov-Bogolyubov form. Keeping in mind the desire to construct the second approximation we decompose the square root in eq. (6.2.6) and retain only two first terms. In the new system we introduce a new variable $\mathbf{u}$ by the relationship

$$\mathbf{v} = \mathbf{u} + \mathbf{E_1} \sin t - \mathbf{E_2} \cos t . \qquad (6.2.8)$$

We arrive at the system in standard form

$$\dot{\mathbf{r}} = \varepsilon(\mathbf{u} + \mathbf{E_1} \sin t - \mathbf{E_2} \cos t), \qquad (6.2.9)$$
$$\dot{\mathbf{u}} = \varepsilon[(\mathbf{u} + \mathbf{E_1} \sin t - \mathbf{E_2} \cos t) \times (\mathbf{H_1} \cos t + \mathbf{H_2} \sin t) +$$
$$+ ((\mathbf{u} + \mathbf{E_1} \sin t - \mathbf{E_2} \cos t)\nabla)(\mathbf{E_2} \cos t - \mathbf{E_1} \sin t)] - \varepsilon^2 (\mathbf{u} + \mathbf{E_1} \sin t -$$
$$- \mathbf{E_2} \cos t) [(\mathbf{u} + \mathbf{E_1} \sin t - \mathbf{E_2} \cos t)(\mathbf{E_1} \cos t + \mathbf{E_2} \sin t)] -$$
$$- \frac{1}{2}\varepsilon^2(\mathbf{u} + \mathbf{E_1} \sin t - \mathbf{E_2} \cos t)^2(\mathbf{E_1} \cos t + \mathbf{E_2} \sin t) .$$

According to the averaging method we introduce the evolutional components $\boldsymbol{\xi}, \boldsymbol{\eta}$ of the variables $\mathbf{r}, \mathbf{u}$ by the formulae

$$\mathbf{r} = \boldsymbol{\xi} + \varepsilon\mathbf{v_1}(\boldsymbol{\xi}, \boldsymbol{\eta}, t) + \varepsilon^2\mathbf{v_2}(\boldsymbol{\xi}, \boldsymbol{\eta}, \mathbf{t}) + \dots ,$$
$$\mathbf{u} = \boldsymbol{\eta} + \varepsilon\mathbf{u_1}(\boldsymbol{\xi}, \boldsymbol{\eta}, t) + \varepsilon^2\mathbf{u_2}(\boldsymbol{\xi}, \boldsymbol{\eta}, \mathbf{t}) + \dots . \qquad (6.2.10)$$

The oscillating terms $\mathbf{u_1}, \mathbf{v_1}$ etc. in the decompositions (6.2.10) determined up to arbitrary functions of variables $\boldsymbol{\xi}, \boldsymbol{\eta}$ are chosen such that the evolutional equations have the appearance of the equations of motion of the material point subjected to certain effective forces. In other words, we construct the equations or $\boldsymbol{\xi}, \boldsymbol{\eta}$ in the form

$$\dot{\boldsymbol{\xi}} = \varepsilon\boldsymbol{\eta}, \quad \dot{\boldsymbol{\eta}} = \varepsilon\mathbf{F_1}(\boldsymbol{\xi}, \boldsymbol{\eta}) + \varepsilon^2\mathbf{F_2}(\boldsymbol{\xi}, \boldsymbol{\eta}) . \qquad (6.2.11)$$

Substituting eq. (6.2.10) into eq. (6.2.9) we obtain the following equations for functions $\mathbf{u_1}$ and $\mathbf{v_1}$

$$\frac{\partial \mathbf{v_1}}{\partial t} = \mathbf{E_1} \sin t - \mathbf{E_2} \cos t,$$
$$\mathbf{F_1} + \frac{\partial \mathbf{u_1}}{\partial t} = (\boldsymbol{\eta} + \mathbf{E_1} \sin t - \mathbf{E_2} \cos t) \times (\mathbf{H_1} \cos t + \mathbf{H_2} \sin t) -$$
$$- ((\boldsymbol{\eta} + \mathbf{E_1} \sin t - \mathbf{E_2} \cos t)\nabla)(\mathbf{E_1} \sin t - \mathbf{E_2} \cos t). \quad (6.2.12)$$

We determine $\mathbf{F_1}$ from the condition of boundedness of $\mathbf{u_1}$ as a function of t

$$\mathbf{F_1} = \frac{1}{2}[\mathbf{E_1} \times \mathbf{H_2} - \mathbf{E_2} \times \mathbf{H_1} - (\mathbf{E_1}\nabla)\mathbf{E_1} - (\mathbf{E_2}\nabla)\mathbf{E_2}] . \qquad (6.2.13)$$

By the induction law (6.2.7) the expression (6.2.13) is transformed to the form

$$\mathbf{F}_1 = -\nabla\Phi, \qquad \Phi = \frac{1}{4}(E_1^2 + E_2^2) \tag{6.2.14}$$

coinciding with the well-known expression for the mean force in the first approximation [74].

In eq. (6.2.13) and in what follows the character ∇ stands for the differentiation with respect to ξ and the law of induction is used in the following form: $\nabla \times \mathbf{E}_1(\xi) = -\mathbf{H}_2(\xi)$. Thus in contrast to [74] there is no need to expand expression (6.2.7) in terms of degrees of ε.

Next, from eq. (6.2.12) we find $\mathbf{u}_1$ and $\mathbf{v}_1$

$$\mathbf{v}_1 = -(\mathbf{E}_1 \cos t + \mathbf{E}_2 \sin t), \tag{6.2.15}$$

$$\mathbf{u}_1 = \eta \times (\mathbf{H}_1 \sin t - \mathbf{H}_2 \cos t) + (\eta\nabla)(\mathbf{E}_1 \cos t + \mathbf{E}_2 \sin t) -$$

$$- \frac{1}{4}[\mathbf{E}_1 \times \mathbf{H}_1 - \mathbf{E}_2 \times \mathbf{H}_2 - (\mathbf{E}_2\nabla)\mathbf{E}_1 + (\mathbf{E}_1\nabla)\mathbf{E}_2] \cos 2t -$$

$$- \frac{1}{4}[\mathbf{E}_1 \times \mathbf{H}_2 + \mathbf{E}_2 \times \mathbf{H}_1 - (\mathbf{E}_1\nabla)\mathbf{E}_1 + (\mathbf{E}_2\nabla)\mathbf{E}_2] \sin 2t + \varphi_1(\xi,\eta) .$$

Here φ_1 is still an arbitrary function of ξ, η A similar terms can be introduced in $\mathbf{v}_1$ however it does not lead to a simplification and for this reason function $\mathbf{v}_1$ is taken such that its mean value vanishes.

Let us consider the terms of order ε^2 in the first equation (6.2.9)

$$\frac{\partial\mathbf{v}_2}{\partial t} = (\eta\nabla)(\mathbf{E}_1 \cos t + \mathbf{E}_2 \sin t) -$$

$$- ((\mathbf{E}_1 \cos t + \mathbf{E}_2 \sin t)\nabla) \times (\mathbf{E}_1 \sin t - \mathbf{E}_2 \cos t) + \mathbf{u}_1 . \tag{6.2.16}$$

The standard procedure of the averaging method requires that the conditions of boundedness of $\mathbf{v}_2$ as a function of t yields, after averaging eq. (6.2.16), the second term in the first equation (6.2.11). However, in the given problem this term vanishes if the drift correction φ_1 to the velocity is given by the following expression

$$\varphi_1 = \frac{1}{2}[(\mathbf{E}_2\nabla)\mathbf{E}_1 - (\mathbf{E}_1\nabla)\mathbf{E}_2] . \tag{6.2.17}$$

The terms of order ε^2 in the second equation (6.2.9) determines the coefficient $\mathbf{F}_2(\xi,\eta)$ in eq. (6.2.11). Finally the equations of the second approximation are as follows

$$\dot{\xi} = \varepsilon\eta,$$

$$\dot{\eta} = -\frac{\varepsilon}{4}\nabla\Phi - \varepsilon^2\frac{1}{2}[\eta \times (\mathbf{H}_1 \times \mathbf{H}_2 - (\mathbf{E}_1\nabla)\mathbf{H}_1 - (\mathbf{E}_2\nabla)\mathbf{H}_2) - \tag{6.2.18}$$

$$\mathbf{H}_1 \times ((\eta\nabla)\mathbf{E}_1) - \mathbf{H}_2 \times ((\eta\nabla)\mathbf{E}_2) - ((\eta \times \mathbf{H}_1)\nabla)\mathbf{E}_1 - ((\eta \times \mathbf{H}_2\nabla)\mathbf{E}_2 +$$

$$(\eta\nabla)((\mathbf{E}_1\nabla)\mathbf{E}_2 - (\mathbf{E}_2\nabla)\mathbf{E}_1) + (\mathbf{E}_2\nabla)((\eta\nabla)\mathbf{E}_1) - (\mathbf{E}_1\nabla)((\eta\nabla)\mathbf{E}_2)] .$$

In the case considered in [74] $\mathbf{E}_2 = 0$, $\mathbf{H}_1 = 0$ we have $\mathbf{F}_2 = 0$ (this result is distinguished from that obtained in [74]). In this case it is necessary to calculate the next approximation.

While deriving the evolutional equations (6.2.18) we choose the arbitrary function φ_1 of the slow variables such that these equations have the form of the equations of motion of the material point. The same can be done for any approximation. This allows one to interpret the right hand side of the second equation (6.2.18) as an effective force that facilitates the analysis of possible motions. However the analytical investigation of the obtained equations is complicated by that the forces acting on the point are functions of the generalized coordinates and velocities. Therefore it is useful to consider another method of choosing the arbitrary functions that results in the canonical equations. It is more comfortable to proceed from the canonical equations equivalent to equations (6.2.1) rather than from these equations. It is convenient to take the Hamilton function in the form

$$\mathcal{H} = \sqrt{m^2 c^4 + c^2 \left(\mathbf{P} - \frac{e}{c} \mathbf{A} \right)^2} - mc^2 , \qquad (6.2.19)$$

which differs from the usual Hamilton function (see, for example [73]) in the constant term outside the radical. In eq. (6.2.19) $\mathbf{P}$ denotes the generalized momentum, $\mathbf{A}$ is the vectorial potential whereas scalar potential is taken to be equal to zero. Moreover

$$c\omega \mathbf{A}(\mathbf{r}, t) = \mathbf{E}_2(\mathbf{r}) \cos \omega t - \mathbf{E}_1(\mathbf{r}) \sin \omega t . \qquad (6.2.20)$$

Transforming the Hamilton function to the dimensionless variables yields

$$\mathcal{H} = 1/\varepsilon \cdot \sqrt{1 + \varepsilon^2 (\boldsymbol{\pi} - \mathbf{A})^2} - 1/\varepsilon , \qquad (6.2.21)$$

where $\boldsymbol{\pi}$ is the dimensionless generalized momentum. The denotation for the dimensionless vectorial potential is not changed.

We seek the canonical transformation of variables $\mathbf{r}, \boldsymbol{\pi}$ into variables $\boldsymbol{\xi}, \mathbf{r}$

$$\boldsymbol{\xi} = \frac{\partial S(\mathbf{r}, \mathbf{p}, t)}{\partial \mathbf{p}}, \qquad \boldsymbol{\pi} = \frac{\partial S(\mathbf{r}, \mathbf{p}, t)}{\partial \mathbf{r}} , \qquad (6.2.22)$$

where the generating function has the form of the asymptotic expansion

$$S = \mathbf{r}\mathbf{p} + \varepsilon S_1(\mathbf{r}, \mathbf{p}, t) + \dots . \qquad (6.2.23)$$

The new Hamilton function $\mathcal{K}$ is related to the generating function S by the relation

$$\mathcal{K} = \mathcal{H} + \frac{\partial S}{\partial t} . \qquad (6.2.24)$$

The correction $S_1, S_2, \dots$ to the generating function are chosen such that the new Hamilton function does not explicitly depend on time. The new Hamilton function $\mathcal{K}$ is also sought for as the asymptotic expansion

$$\mathcal{K} = \varepsilon \mathcal{K}_1(\boldsymbol{\xi}, \mathbf{p}) + \varepsilon^2 \mathcal{K}_2(\boldsymbol{\xi}, \mathbf{p}) + \dots . \qquad (6.2.25)$$

Inserting eqs. (6.2.21),(6.2.23),(6.2.25) into eq. (6.2.24) and considering eq. (6.2.22) we obtain in the first approximation

$$\frac{\partial S_1}{\partial t} + \frac{1}{2}(\mathbf{r} - \mathbf{A})^2 = \mathcal{K}_1 \qquad (6.2.26)$$

and from the condition of boundedness of S_1 as a function of t we have

$$\mathcal{K}_1 = \frac{1}{2}\mathbf{p}^2 + \frac{1}{4}(\mathbf{E}_1^2 + \mathbf{E}_2^2) . \qquad (6.2.27)$$

The expression for S_1 is obtained by quadrature

$$S_1 = \mathbf{p}(\mathbf{E}_2 \sin t + \mathbf{E}_1 \cos t) - \frac{1}{2}\mathbf{E}_1\mathbf{E}_2 \cos 2t - \frac{1}{8}(\mathbf{E}_2^2 - \mathbf{E}_1^2)\sin 2t . \quad (6.2.28)$$

Arbitrary function of $\boldsymbol{\xi},p$ appearing at the integration is chosen such that the mean value of S_1 vanishes.

In the second approximation

$$\mathcal{K}_2 = \frac{1}{2}\mathbf{p}[(\mathbf{E}_1\nabla)\mathbf{E}_2 - (\mathbf{E}_2\nabla)\mathbf{E}_1] . \qquad (6.2.29)$$

Comparing the Hamilton equations with the Hamilton function $\varepsilon\mathcal{K}_1 + \varepsilon^2\mathcal{K}_2$ we obtain

$$\dot{\boldsymbol{\xi}} = \varepsilon\mathbf{r} + \varepsilon^2\boldsymbol{\varphi}_1, \quad \dot{\mathbf{r}} = -\varepsilon\nabla\Phi - \frac{\varepsilon^2}{2}\{\mathbf{r}\nabla)[(\mathbf{E}_1\nabla)\mathbf{E}_2 - (\mathbf{E}_2\nabla)\mathbf{E}_1]+$$
$$+ \mathbf{p} \times [\nabla \times ((\mathbf{E}_1\nabla)\mathbf{E}_2 - (\mathbf{E}_2\nabla)\mathbf{E}_1)]\} . \qquad (6.2.30)$$

Attention should be paid to coincidence of the term of order ε^2 in the first equation in (6.2.30) with the drift correction to the velocity obtained from eq. (6.2.18). Equations (6.2.30) have an obvious integral $\varepsilon\mathcal{K}_1 + \varepsilon^2\mathcal{K}_2 =$ const. However a physical interpretation of the system described by these equations presents a problem.

In the particular case considered in [74] in which $\mathbf{E}_2 = 0$ we obtain $\mathcal{K}_2 = 0$ and it is necessary to calculate the further approximation

$$\mathcal{K}_3 = -\frac{1}{2}\left[\frac{1}{4}\mathbf{p}^4 + \frac{3}{4}\mathbf{r}^2\mathbf{E}_1^2 + \frac{3}{32}\mathbf{E}_1^2 - \frac{1}{2}(\nabla(\mathbf{p}\mathbf{E}_1))^2 - \right.$$
$$- \frac{1}{16}(\nabla\mathbf{A}^2)^2 + (\mathbf{E}_1\nabla)(\mathbf{p}\nabla(\mathbf{p}\mathbf{E}_1)) - \frac{3}{16}\mathbf{E}_1\nabla(\mathbf{E}_1^2)+$$
$$\left. + \frac{1}{4}\left((\mathbf{E}_1\nabla)^2\mathbf{E}_1^2 - ((\mathbf{E}_1\nabla)\mathbf{E}_1)\nabla\mathbf{E}_1^2\right)\right] . \qquad (6.2.31)$$

Generally speaking, for the problems under consideration there are two most rational procedures of averaging resulting either in the equations of motion of the material point or in the Hamilton equations for any approximation.

Consideration of the higher approximations is required not only for refinement of calculations by the averaging method but it can be necessary
for the qualitative analysis of motion. This is the case in which potential Φ
in eq. (6.2.14) is independent of one or two Cartesian or curvilinear coordinates whereas the force acting in the direction of these coordinates vanishes
in the first approximation. Then the motion in the direction of these coordinates can be determined only by means of the higher approximations.
It is especially important to calculate the drift correction to velocity φ_1.
To this end, it is necessary to consider the motion within the time interval
of order of $1/\varepsilon^2$ or greater, the general theorems of the averaging method
being insufficient to this aim, see Appendix D.

The considered problem possesses one characteristic spatial scale of the
order of the wave length L. But of interest are the problems with another,
essentially greater scale of the order of L/ε caused, for example, by slow
changes in amplitude and phase of high-frequency vibrations. For such
problems $\mathbf{E}_1$ and $\mathbf{E}_2$ are functions of $\mathbf{r}$ and $\varepsilon\mathbf{r}$, and for analysis of the
influence of slow change in the field it is necessary to consider motion in
the time interval of the order of $1/\varepsilon^2$.

6.3   Particle in electromagnetic field with high-frequency and constant components

Let us consider motion of the non-relativistic charged particle in the electromagnetic field, the latter consisting of high-frequency electromagnetic field
and a homogeneous constant magnetic field $\mathbf{B}$, [9]. It is common knowledge
[72] that the equations of motion can be set as the canonical equations of
mechanics with the Hamilton function

$$\mathcal{H}(\mathbf{r}, \mathbf{P}, t) = \frac{1}{2m} \left(\mathbf{P} - \frac{e}{c} \mathbf{A}(\mathbf{r}, t) \right)^2 . \qquad (6.3.1)$$

In this case the vectorial potential $\mathbf{A}$ can be taken in the form

$$\mathbf{A}(\mathbf{r}, t) = \frac{1}{2} (\mathbf{B} \times \mathbf{r}) + \frac{c}{\omega} (\mathbf{E}_2(\mathbf{r}) \cos \omega t - \mathbf{E}_1(\mathbf{r}) \sin \omega t) . \qquad (6.3.2)$$

Let us introduce in eqs. (6.3.1), (6.3.2) the dimensionless variables which
are equal to the ratio of dimensional variables to their characteristic values.
The choice of characteristic values of time $[t]$, distance $[L]$, magnetic $[B]$
and electric $[E]$ fields is discussed below.

Let the dimensionless variables have the denotation of the corresponding
dimensional variables, then the dimensionless Hamilton function is given
by

$$H = \frac{1}{2} \left\{ \mathbf{P} - \beta \left[\frac{1}{2} \mathbf{B} \times \mathbf{r} + \varepsilon (\mathbf{E}_2(\mathbf{r}) \cos \nu t - \mathbf{E}_1(\mathbf{r}) \sin \nu t) \right] \right\}^2 , \qquad (6.3.3)$$

where $\beta = e[B][t]/mc, \varepsilon = c[E]/\omega[L][B]$ and $\nu = \omega[t]$.

Using the Hamilton function (6.3.3), one can write down the canonical equations for the dimensionless coordinates and momenta. However in what follows we use another canonically conjugated variables introduced as the action-angle variables for describing motions in the homogeneous magnetic field. The reason is that the high-frequency field is considered as a perturbation of the basic constant field. These variables are recommended to be used in such cases (and in general for study of perturbed Hamiltonian systems), see for example [68]. Then the phases of the multifrequency system are immediately selected, the slow variables are easily entered in the resonance case and so on.

Let the Cartesian coordinates of the particle be orthogonal generalized coordinates. We introduce new variables I, φ, x_L, y_L instead of the dimensionless variables x, y, P_x, P_y (z and P_z are not affected)

$$x = \frac{1}{\sqrt{\beta B}}(\sqrt{2I} \sin \varphi + x_L), \quad P_x = \frac{\sqrt{\beta B}}{2}(\sqrt{2I} \cos \varphi - y_L),$$

$$y = \frac{1}{\sqrt{\beta B}}(\sqrt{2I} \cos \varphi + y_L), \quad P_y = \frac{\sqrt{\beta B}}{2}(-\sqrt{2I} \sin \varphi + x_L). \quad (6.3.4)$$

This change is canonical. When the particle moves in the constant magnetic field B (axis z is directed along the field $\mathbf{B}$), variables I, φ are the action-angle variables and the new generalized momentum x_L and new generalized coordinate y_L are the coordinates of the point where the trajectory (helix) intersects plane xy. Motions with $I = 0$ are not considered here (for the case when I is small or large, see below).

Let us write down the new Hamilton function of the original problem by using transformations (6.3.4)

$$\mathcal{K} = \beta B I + \frac{1}{2}P_z^2 + \frac{1}{2}\varepsilon^2(E_2^2 \cos^2 \nu t - \mathbf{E}_1 \cdot \mathbf{E}_2 \sin 2\nu t + E_1^2 \sin^2 \nu t) -$$

$$- \varepsilon P_z[(\mathbf{E}_2 \mathbf{e}_3) \cos \nu t - (\mathbf{E}_1 \mathbf{e}_3) \sin \nu t] -$$

$$- \varepsilon \sqrt{2I}[(\mathbf{E}_2 \mathbf{e}_1) \cos \varphi \cos \nu t - (\mathbf{E}_1 \mathbf{e}_1) \cos \varphi \sin \nu t] +$$

$$+ \varepsilon \sqrt{2I}[(\mathbf{E}_2 \mathbf{e}_2) \sin \varphi \cos \nu t - (\mathbf{E}_1 \mathbf{e}_2) \sin \varphi \sin \nu t]. \quad (6.3.5)$$

Here $\mathbf{e}_1, \mathbf{e}_2, \mathbf{e}_3$ are the unit vectors of axes of the Cartesian systems of coordinates and the coordinates x, y in functions $\mathbf{E}_1(\mathbf{r}), \mathbf{E}_2(\mathbf{r})$ must be expressed in terms of the new variables.

The Hamilton function $\mathcal{K}$ is a periodic function of φ and $\psi = \nu t$, hence these arguments are understood as phases.

Let us assume that the Coulomb force due to the high-frequency field be small in comparison with Lorentz force acting on the particle in the constant magnetic field, and the Larmor frequency at the motion in the constant field is comparable with the frequency of the variable field. Thus one should take $[B] = B[t] = mc/e$, then $\beta = 1, \mathbf{B} = \mathbf{e}_3$ and the parameter

$\varepsilon \ll 1$. Instead of the characteristic distances $[L]$ it is more convenient to take the characteristic value of the transverse speeds v_1, then $[L] = v_1 me/cB$.

In the important cases which are considered further the amplitude of the high-frequency field is a functions of the work $\varepsilon_1 \mathbf{r}$, where $\varepsilon_1 = \omega[L]/c$. Forasmuch as the motion is assumed to be non-relativistic, then the parameter $\varepsilon_1 \ll 1$. The form of the expression for ε suggests that the characteristic value $[E]$ of the amplitude of the variable electric field must be of the second order of smallness in comparison with B. The most interesting situation from a physical perspective is that when parameters $\varepsilon < \varepsilon_1$ are of the same order of smallness.

Let us consider the resonance, i.e. the case when parameter $\varepsilon_2 = 1 - \nu$ is small and let the latter be of the order of the other small parameters. We introduce a new coordinate which is the difference of phases $\delta = \varphi - \nu t$ instead of φ . This replacement is canonical. The new Hamilton function is related to the old one by the formula

$$\mathcal{K}' = \mathcal{K}(I, \delta, t) - \nu I . \qquad (6.3.6)$$

The variables δ, I, P_z, h_L, u_L are slow, with $\dot{\delta}, \dot{I} \sim \varepsilon$ and $\dot{P}_z, \dot{h}_L, \dot{u}_L \sim \varepsilon^2$. A rapid variable is coordinate z. It can take on large values $z \sim 1/\varepsilon$ at $t \sim 1/\varepsilon$, however it is not crucial since z appears only in functions $\mathbf{E}_1(\varepsilon_1 \mathbf{r})$ and $\mathbf{E}_2(\varepsilon_1 \mathbf{r})$. In the first approximation of the averaging method one can consider that $z = z(t_0) + P_z(t - t_0), P_z > h_L, u_L = \text{const}$, and $\varepsilon_1 \mathbf{r} = P_z r \mathbf{e}_3, r = \varepsilon(t - t_0)$. The averaged equations of motion in the first approximation are equations only for I, δ. These equations are Hamilton with the Hamilton function obtained by averaging eq. (6.3.6) over the explicitly appearing time

$$\mathcal{K} = \varepsilon_2 I + \frac{1}{2}P_z^2 - \frac{\varepsilon}{2}\sqrt{2I}[(\mathbf{E}_2\mathbf{e}_1)\cos\delta + (\mathbf{E}_1\mathbf{e}_1)\sin\delta] +$$
$$+ \frac{\varepsilon}{2}\sqrt{2I}[(\mathbf{E}_2\mathbf{e}_2)\sin\delta + (\mathbf{E}_1\mathbf{e}_2)\cos\delta] . \qquad (6.3.7)$$

The equations corresponding to the Hamilton functions $\mathcal{K}$ will be autonomous by virtue of dependences $\mathbf{E}_1, \mathbf{E}_2$ from the slow time τ.

Let us consider motion of the particle when the homogeneous field is perturbed by the field of a monochromatic, elliptically polarized wave propagating along axis with the unit vector $\mathbf{n}_3$. In this case the electric field is given by

$$\mathbf{E} = \mathbf{n}_1 E_1 \cos(\varepsilon_1 \mathbf{n}_3 \cdot \mathbf{r} - \nu t) + \mathbf{n}_2 E_2 \sin(\varepsilon_1 \mathbf{n}_3 \cdot \mathbf{r} - \nu t) . \qquad (6.3.8)$$

Formula (6.3.8) implies that the unit vectors $\mathbf{n}_1, \mathbf{n}_2, \mathbf{n}_3$ form a right Cartesian trihedron, and the direction of axes x, y are chosen such that the unit vector $\mathbf{n}_3$ lies in plane xz and the unit vector $\mathbf{n}_2$ is directed along axis y, then the unit vector $\mathbf{n}_1$ also lies in plane xz. From eq. (6.3.8) we obtain

$$\mathbf{E}_1(\varepsilon_1\mathbf{r}) = \mathbf{n}_1 E_1 \cos(\varepsilon_1\mathbf{n}_3 \cdot \mathbf{r}) + \mathbf{n}_2 E_2 \sin(\varepsilon_1\mathbf{n}_3 \cdot \mathbf{r}),$$
$$\mathbf{E}_2(\varepsilon_1\mathbf{r}) = \mathbf{n}_1 E_1 \sin(\varepsilon_1\mathbf{n}_3 \cdot \mathbf{r}) + \mathbf{n}_2 E_2 \cos(\varepsilon_1\mathbf{n}_3 \cdot \mathbf{r}). \tag{6.3.9}$$

Inserting eq. (6.3.9) into eq. (6.3.7) we obtain

$$\mathcal{K} = \varepsilon_2 I + \frac{1}{2}P_z^2 - \frac{\varepsilon}{2}\sqrt{2I}(E_1\cos\theta + E_2)\sin(\delta + P_z r\cos\theta),$$
$$\cos\theta = \mathbf{n}_3 \cdot \mathbf{e}_3. \tag{6.3.10}$$

Taking into account that in the first approximation $P_z = $ const and introducing a new slow variable $\alpha = \delta + P_z r \cos\theta$ we arrive at the new Hamiltonian system (which is now autonomous) with the new Hamilton function

$$G(I,\alpha) = (\varepsilon_2 + \varepsilon_1 P_z \cos\theta)I - \frac{\varepsilon}{2}\sqrt{2I}(E_1\cos\theta + E_2)\sin\alpha, \tag{6.3.11}$$

where the constant term $1/2P_z^2$ is omitted. For the averaged equations $G(I,\alpha)$ is an integral of motion. If one introduces the original variables $\mathbf{r},\mathbf{P}$ in eq. (6.3.11), then G is the adiabatic invariant (rather than I, as is sometimes assumed).

Equations of motion of the system with Hamilton function G have the form

$$\dot{I} = \frac{\varepsilon}{2}\sqrt{2I}(E_1\cos\theta + E_2)\cos\alpha,$$
$$\dot{\alpha} = (\varepsilon_2 + \varepsilon_1 P_z\cos\theta) - \frac{\varepsilon}{2\sqrt{2I}}(E_1\cos\theta + E_2)\sin\alpha. \tag{6.3.12}$$

Axis $I = 0$ is the locus of discontinuities of system (6.3.12). As mentioned above, it is not possible to explore the motion with $I \sim \varepsilon$ by the above method. But another strategy is applicable. As one can see from the previous analysis, the Lagrange equations in the first approximation are linear equations with slowly varying coefficients $E_1(r), E_2(r)$. Therefore one can use for analysis not only the averaging method but also the WKB-method, where the case of small I is not singular. However for small values of I the averaging method appears to be essentially simpler.

The solutions of equations (6.3.12) on the cylinder are periodic. In such cases the exact solutions of the equations of motion can qualitatively differ from the first approximation due to the terms of higher order of smallness of (in particular, the terms describing the variable magnetic field are of the order of ε^2). The terms of higher order are noticeable, firstly, in the time intervals $\sim 1/\varepsilon^2$ and greater and, secondly, they can be taken into account only in the higher approximations of the averaging method. This allows one to pose a problem of calculation of the higher approximations and study of motion in time intervals greater than $1/\varepsilon$.

The problem of the resonance motion of a particle in the field of a plane wave and a constant magnetic field has already been considered in [60, 61]. In these papers the velocity of the particle was not assumed to be small in comparison with the phase velocity of the wave. Therefore the formal results of [60, 61] are more general than the results obtained above. To obtain the results of [60] it is necessary to introduce a small parameter ε in the corresponding equations in [60, 61], introduce variable I, replace Bessel functions by the first terms of the expansion in terms of the degrees of argument and so on. The result is that it is simpler to study the case of small velocity of the particle (that is not directly considered in [60, 61]) by itself. In addition to this, the use of results [60, 61] is made difficult by that no small parameters are introduced explicitly and no fast and slow variables are indicated in these papers.

Let the Larmor frequency in the constant field and the wave frequency be of the same order but essentially different. Let us take the above ratio of amplitudes of the variable and constant fields. In this case the system with the Hamilton function $\mathcal{K}$ (6.3.5) is a system with two fast phases. It is essential that the fast phases appear in eq. (6.3.5) only in the form of pairwise products like $\cos \varphi \cos \nu t$ etc. In this case no small denominators appear in the first approximation. Averaging the equations of motion corresponding to the Hamilton function (6.3.5) over on both fast phases we obtain that $I = 0$ in the first approximation etc. In other words, in the case of no resonance the high-frequency field does not qualitatively influence the motion.

Let us also consider the case in which the ratio of the amplitudes of the variable fields to the value of constant field is of the order of unity. The asymptotic method is applicable only to the subcase $\omega \gg \omega_L \omega_L = eB/mc$. The characteristic values of the sought variable and prescribed fields should be introduced according to Section 6.2.

As a result we arrive at the system with the following Hamilton function

$$\mathcal{H} = \frac{\varepsilon}{2} \left(\mathbf{q} - \frac{1}{2}\mathbf{e}_3 \times \mathbf{r} - \mathbf{E}_2 \cos t + \mathbf{E}_1 \sin t \right)^2 . \tag{6.3.13}$$

Here a new dimensionless non-small momentum $\mathbf{q} = \mathbf{P}/\varepsilon$ is introduced, however, in contrast to the above expression, the relativistic corrections are omitted. Unlike formulae (6.3.8), functions $\mathbf{E}_1$ and $\mathbf{E}_2$ in eq. (6.3.13) are now dependent on $\mathbf{r}$, rather than on $\varepsilon_1 \mathbf{r}$.

Conducting averaging similar to that in Section 6.2 we obtain, in the first approximation, the system with the following Hamilton function

$$\mathcal{K}(\mathbf{r},\mathbf{q}) = \frac{\varepsilon}{2} \left(\mathbf{q} - \frac{1}{2}\mathbf{e}_3 \times \mathbf{r} \right)^2 + \frac{1}{4}(E_1^2 + E_2^2) , \tag{6.3.14}$$

which describes motion of the particle in the original constant magnetic field and the electric field, the latter being variable in space and given by the potential $U_{ef} = 1/4(E_1^2 + E_2^2)$.

Let us also point out the case in which amplitudes of the variable fields has one less order of smallness than the value of the constant field (in contrast to the resonance case considered above in which the amplitudes are two orders less). In this case parameter ε in the Hamilton function (6.3.3) is not small, but parameter ε_1 is small as above. The non-small terms of the Hamilton function correspond to motion of the particle in the field obtained by superposition of the constant magnetic field and the high-frequency electric field with the amplitudes slowly changing in space. The small terms describe the influence of the variable magnetic field. The methods for solving this problem have not yet been developed.

In conclusion we notice that no one of the considered cases yields the results that obtained in [81] and indicated in [34].

6.4 Relativistic charged particle in essentially heterogeneous crossed electric and magnetic fields

As a rule, exact integration of the equations of motion of the charged particles in heterogeneous electric and magnetic fields is not feasible. At the same time, numerical computation is extremely laboured if the particle rotates and completes a great number of revolutions around the line of the magnetic field.

A scheme of averaging the equations of motion of the charged particle in arbitrary weakly heterogeneous electric and magnetic fields over the fast phase of rotation was constructed and rigorously substantiated in [17], the total Coulomb force being considered small in comparison to the Lorentz force. The present section is concerned with motion of the relativistic charged particle in electric $\mathbf{E}$ $(E_x(x), E_y = \mathrm{const}, 0)$ and magnetic $\mathbf{H}$ $(H_x = \mathrm{const}, 0, H_z(x))$ crossed fields. Only y-component of the Coulomb force is considered to be small in comparison with the Lorentz force and no limitation is laid on the rate of change of the fields, so that the obtained results are applicable also to the fields with jumps. The latter situation is typical for the collisionless mhd-shock-waves.

Motion of the particle is considered in weakly heterogeneous field of the same configuration. This problem formally belongs to the class studied by the ordinary theory but it is degenerated and requires calculations of the second approximations.

Let us consider motion of a charged particles in external electric $\mathbf{E}$ and magnetic $\mathbf{H}$ fields. Let us denote the radius-vector of the particle by $\mathbf{r}$, its mass by m and its charge by $e > 0$. The equation of motion is given by eq. (6.2.1). The electromagnetic field is considered to be given, thus the components of the electric and magnetic fields are

$$\mathbf{E} = E_x\mathbf{i} + E_y\mathbf{j}, \quad E_x = E(x), \quad E_y = \text{const} > 0,$$
$$\mathbf{H} = H_x\mathbf{i} + H_z\mathbf{k}, \quad H_x = \text{const}, \quad H_z(x) > 0, \tag{6.4.1}$$

where $\mathbf{i}, \mathbf{j}, \mathbf{k}$ are the unit vectors of the Cartesian axes. The case $e < 0$, $H_z(x) > 0$ etc. are considered by analogy. Let us introduce in the equation of motion the dimensionless variables equal to ratio of the dimensional values to their characteristic values. The characteristic value are as follows: the velocity of light c for velocity, a value $[H]$ (determined for any particular problem) for intensity of the magnetic field, $mc^2/e[H]$ for coordinate and $mc/e[H]$ for time. Keeping the former denotations for the variables and projecting the equation of motion in the coordinate axes we obtain

$$\dot{x} = u,$$
$$\dot{u} = \sqrt{1 - V^2}[e(x) + vh(x) - u^2e(x) - \varepsilon uv],$$
$$\dot{v} = \sqrt{1 - V^2}[\varepsilon wh_x - uh(x) + (1 - v^2)\varepsilon - vue(x)],$$
$$\dot{w} = -\sqrt{1 - V^2}[\varepsilon vh_x + \varepsilon wv + wue(x)]. \tag{6.4.2}$$

Here

$$v = \dot{y}, \quad w = \dot{z}, \quad \varepsilon = E_y/[H], \quad \varepsilon h_x = H_x/[H],$$
$$h(x) = H_z(x)/[H], \quad e(x) = E_x(x)/[H], \quad V^2 = u^2 + v^2 + w^2.$$

The small parameter is ε, that is, two assumptions are made. Firstly, the y-component of the Coulomb forces is considered to be small in comparison to the Lorentz force. Secondly, the angle of slope of the lines of magnetic field to axis $0Z$ is considered to be small, too. The existing adiabatic theory of motion of the charged particles (see, for example [20, 92]) suggests that the total Coulomb force is small in comparison with the Lorentz force and the magnetic field intensity is assumed to be slightly changing within the distances of order of the particle gyroradius. These basic limitations of the existing theory no longer exist in the present analysis. Only one component of vector $\mathbf{E}$ is assumed to be small and $H_z(x)$ can essentially vary within the distances of order of the particle gyroradius.

We introduce the new variables in eq. (6.4.2)

$$p_x = \gamma u, \qquad p_z = \gamma w,$$
$$p_y = \gamma v + A(x), \quad A(x) = \int_0^x h(s)ds, \quad \gamma = (1 - V^2)^{1/2}. \tag{6.4.3}$$

Then system (6.4.2) takes on the form

$$
\begin{aligned}
\dot{x} &= p_x/\gamma, \\
\dot{p}_x &= e(x) + h(x)(p_y - A(x))/\gamma, \\
\dot{p}_y &= \varepsilon + \varepsilon h_x p_z/\gamma, \\
\dot{p}_z &= -\varepsilon h_x (p_y - A(x))/\gamma .
\end{aligned}
\tag{6.4.4}
$$

The system of the first two equations can be considered as the equation of motion of a certain conditional material point with the following Hamilton function

$$
\mathcal{H} = \{1 + p_x^2 + (p_y - A(x))^2 + p_z^2\}^{1/2} + \varphi(x) ,
\tag{6.4.5}
$$

where $d\varphi/dx = -e(x)$ and p_y, p_z are given by the second subsystem in eq. (6.4.4).

As is generally known, if functions p_y and p_z in system (6.4.3) depend only on the slow time $\tau = \varepsilon t$ then this system has an adiabatic invariant (see, for example [5]). In the case $h_x \neq 0$ the system under consideration does not belong to this type. Nonetheless we show that this system also possesses an adiabatic invariant which is the action variable in the corresponding problem with $p_y, p_z = $ const.

The action variable is proportional to area of the phase trajectory limited by the phase curve $\mathcal{H} = E = $ const, i.e.

$$
I = \frac{1}{\pi} \int_{x_1}^{x_2} [(E - \varphi(s))^2 - 1 - p_z^2 - (p_y - A(s))^2]^{1/2} ds .
\tag{6.4.6}
$$

Here $x_1(E, p_y, p_z)$ and $x_2(E, p_y, p_z); x_1 < x_2$ are the roots of the equation

$$
(E - \varphi(s))^2 - 1 - p_z^2 - (p_y - A(s))^2 = 0.
$$

The integral in eq. (6.4.6) and in the similar cases below is calculated at $E, p_y, p_z = $ const. It is also assumed that the phase trajectories of system (6.4.3) at $E, p_y, p_z = $ const are closed. In particular this is the case if $\varphi(s) \equiv 0$.

In the case $h_x \neq 0$ we introduce the action-angle variables in the same manner as they are introduced in system (6.4.3) with the Hamilton function (6.4.5) for $E, p_z, p_y = $ const. We obtain the system for I, p_y, p_z, ψ with the single fast phase ψ. Replacing differentiation with respect to t by the differentiation with respect to ψ we obtain the system in the standard

Krylov-Bogolyubov form

$$
\frac{dI}{d\psi} = \frac{\varepsilon}{\omega} \left\{ \left[\frac{\partial I}{\partial E}(p_y - \tilde{A}) - \frac{\partial I}{\partial p_z} h_x(p_y - \tilde{A}) + \frac{\partial I}{\partial p_y} h_x p_z \right] \tilde{\gamma}^{-1} + \right.
$$
$$
\left. + \frac{\partial I}{\partial p_y} \right\} + \varepsilon^2 \ldots,
$$
$$
\frac{dp_y}{d\psi} = \frac{\varepsilon}{\omega} \{ 1 + h_x p_z / \tilde{\gamma} \} + \varepsilon^2 \ldots,
$$
$$
\frac{dp_z}{d\psi} = -\frac{\varepsilon}{\omega} h_x(p_y - \tilde{A}) / \tilde{\gamma} + \varepsilon^2 \ldots. \tag{6.4.7}
$$

Here $\omega(I, p_y, p_z) = (\partial I/\partial E)^{-1} \neq 0$ and $\tilde{A}, \tilde{\gamma}$ are functions of $A(x), \gamma(h, p_x)$ expressed in terms of the new variables I, ψ. In accordance with the averaging method we introduce the new slow (evolutional) variables $J(\tau), \kappa(\tau)$ and $\mu(\tau)$ with the help of the relationships

$$
I = J + \varepsilon u_1(J, \kappa, \mu, \psi) + \varepsilon^2 \ldots,
$$
$$
p_y = \kappa + \varepsilon v_1(J, \kappa, \mu, \psi) + \varepsilon^2 \ldots,
$$
$$
p_z = \mu + \varepsilon w_1(J, \kappa, \mu, \psi) + \varepsilon^2 \ldots, \tag{6.4.8}
$$

where J, κ, μ satisfy the equations

$$
\frac{dJ}{d\psi} = \varepsilon \Xi_1(J, \kappa, \mu) + \varepsilon^2 \ldots,
$$
$$
\frac{d\kappa}{d\psi} = \varepsilon H_1(J, \kappa, \mu) + \varepsilon^2 \ldots,
$$
$$
\frac{d\mu}{d\psi} = \varepsilon Z_1(J, \kappa, \mu) + \varepsilon^2 \ldots. \tag{6.4.9}
$$

Here functions u_1, v_1, w_1 are chosen such their mean values averaged over the fast phase ψ are equal to zero, and the coefficients Ξ_1, H_1, Z_1 of the averaged (evolutional) equations are equal to the mean values of right parts of the corresponding equations of system (6.4.7).

Performing the replacements (6.4.8) and (6.4.9) and calculating the mean values of the coefficients in the right hand sides of equations (6.4.7) we obtain

$$
\langle \tilde{\gamma} \rangle = \frac{1}{\mu} \frac{\partial \mathcal{E}}{\partial \mu},
$$
$$
\langle (\kappa - \tilde{A}) / \tilde{\gamma} \rangle = \frac{\partial \mathcal{E}}{\partial \kappa}. \tag{6.4.10}
$$

Here $\mathcal{E}$ is the evolutional component of E introduced by the relationship $E = \mathcal{E} + \varepsilon \zeta_1(J, \kappa, \mu, \psi)$ and the angular brackets denote averaging over the fast phase ψ.

It follows from eq. (6.4.10) that $\Xi \equiv 0, |J(t) - J(0)| < C\varepsilon$ in the time interval $t \sim 1/\varepsilon$. According to the definition [5] this proves the adiabatic invariance of the evolutional component of the action variable J

$$J = \frac{1}{\pi} \int_{x_1}^{x_2} \left[(\mathcal{E} - \varphi(s))^2 - 1 - \mu^2 - (\kappa - A(s))^2 \right]^{1/2} ds . \qquad (6.4.11)$$

In the case $h_x = 0$ the system (6.4.3) and (6.4.4) belongs to the systems considered in [5] and has the adiabatic invariant given by eq. (6.4.11).

Relationship (6.4.9) allows one to write down a slow subsystem of the evolutional equations corresponding to eq. (6.4.4) in the following form

$$\dot\kappa = \varepsilon \left[1 + \frac{h_x \mu}{\mathcal{L}(J, \kappa, \mu)} \right] + \varepsilon^2 \ldots ,$$

$$\dot\mu = -\varepsilon h_x \frac{\partial \mathcal{E}(J, \kappa, \mu)}{\partial \kappa} + \varepsilon^2 \ldots . \qquad (6.4.12)$$

Here $\mathcal{E}(J, \kappa, \mu)$ is the function obtained by inverting dependences (6.4.11), and system (6.4.12) is integrated at $J = \text{const}$.

For the sake of simplicity we take $e(x) \equiv 0$. Let us consider motion of the particle in the electric and magnetic fields studied above, but now we assume that z-component of the magnetic field $h(x)$ is slightly changing in distances of the order of $mc^2/e[H]$.

Let $[r]$ denote the characteristic scale of change in the magnetic field intensity. It was assumed earlier that $[r] \sim mc^2/c[H]$ and, if the field intensity is discontinuous, then $[r] \ll mc^2/e[H]$. Provided that $h(x)$ is a slowly changing function then the problem belongs to the case $[r] \sim mc^2/e[H]\varepsilon$ studied by the existing adiabatic theory (see, e.g. [20, 92]). However the considered problem is degenerated and requires a special consideration.

Introducing in eqs. (6.4.3) and (6.4.4) variables $\rho = \varepsilon h p^2 = \gamma^2(u^2 + v^2)$, $\lambda = \gamma(1 + wh_x)$ and the angle ψ such that $\cos\psi = \gamma u/p$, $\sin\psi = -\gamma v/p$ we obtain the system with a single fast phase ψ

$$\dot\rho = \varepsilon p \cos\psi/\gamma,$$
$$\dot p = -\varepsilon\lambda \sin\psi/\gamma,$$
$$\dot\lambda = -\varepsilon p(1 - hx^2)\sin\psi/\gamma,$$
$$\dot\psi = [h(\rho) - \varepsilon\lambda \cos\psi/p]/\gamma . \qquad (6.4.13)$$

Replacing differentiation with respect to t by differentiation with respect to ψ we arrive at the system in the standard form

$$\frac{d\rho}{d\psi} = \varepsilon \frac{p\cos\psi}{h(\rho)}\left[1 + \varepsilon\frac{\lambda\cos\psi}{ph(\rho)}\right] + \varepsilon^3\ldots,$$

$$\frac{dp}{d\psi} = -\varepsilon\frac{\lambda\sin\psi}{h(\rho)}\left[1 + \varepsilon\frac{\lambda\cos\psi}{ph(\rho)}\right] + \varepsilon^3\ldots,$$

$$\frac{d\lambda}{d\psi} = -\varepsilon(1 - hx^2)\frac{p\sin\psi}{h(\rho)}\left[1 + \varepsilon\frac{\lambda\cos\psi}{ph(\rho)}\right] + \varepsilon^3\ldots.$$

Averaging over ψ in the first order leads to a trivial result that ρ and p are singly conserved in the time intervals of order of $t \sim 1/\varepsilon$. Therefore the fundamental conclusion on conserving the adiabatic invariant $p_\perp/h \simeq p^2/h(p = p_\perp + \varepsilon\ldots)$ is trivial for the given class of problems.

Thus it is necessary to find the solution of system (6.4.13) in the second approximation. The evolutional (drift) components $\bar\rho, \bar p, \bar\lambda$ of functions ρ, p, λ are introduced by the relations

$$\rho = \bar\rho + \varepsilon u_1(\bar\rho, \bar p, \bar\lambda, \psi) + \varepsilon^2 u_2(\bar\rho, \bar p, \bar\lambda, \psi) + \varepsilon^3\ldots,$$

$$p = \bar p + \varepsilon v_1(\bar\rho, \bar p, \bar\lambda, \psi) + \varepsilon^2 v_2(\bar\rho, \bar p, \bar\lambda, \psi) + \varepsilon^3\ldots,$$

$$\lambda = \bar\lambda + \varepsilon w_1(\bar\rho, \bar p, \bar\lambda, \psi) + \varepsilon^2 w_2(\bar\rho, \bar p, \bar\lambda, \psi) + \varepsilon^3\ldots, \qquad (6.4.14)$$

where $\bar\rho, \bar p, \bar\lambda$ satisfy the following system of equations

$$\dot{\bar\rho} = \varepsilon\Xi_1(\bar\rho, \bar p, \bar\lambda) + \varepsilon^2\Xi_2(\bar\rho, \bar p, \bar\lambda) + \varepsilon^3\ldots,$$

$$\dot{\bar p} = \varepsilon H_1(\bar\rho, \bar p, \bar\lambda) + \varepsilon^2 H_2(\bar\rho, \bar p, \bar\lambda) + \varepsilon^3\ldots,$$

$$\dot{\bar\lambda} = \varepsilon W_1(\bar\rho, \bar p, \bar\lambda) + \varepsilon^2 W_2(\bar\rho, \bar p, \bar\lambda) + \varepsilon^3\ldots. \qquad (6.4.15)$$

Replacing the variables with the help of eqs. (6.4.14), (6.4.15) and equating the coefficients in front of the first degree of ε yields

$$\Xi_1 + \frac{\partial u_1}{\partial\psi} = \frac{p}{h(\bar\rho)}\cos\psi,$$

$$H_1 + \frac{\partial v_1}{\partial\psi} = -\frac{\lambda}{h(\bar\rho)}\sin\psi,$$

$$W_1 + \frac{\partial w_1}{\partial\psi} = -(1 - hx^2)\frac{p}{h(\bar\rho)}\sin\psi. \qquad (6.4.16)$$

Using eq. (6.4.16) and averaging we obtain

$$\Xi_1 = 0, \quad u_1 = \frac{p}{h(\bar\rho)}\sin\psi,$$

$$H_1 = 0, \quad v_1 = \frac{\lambda}{h(\bar\rho)}\cos\psi,$$

$$W_1 = 0, \quad w_1 = (1 - hx^2)\frac{\bar p}{h(\bar\rho)}\cos\psi.$$

The problem is degenerated and it is necessary to consider higher approximations. In the second approximation we have

$$
\Xi_2 + \frac{\partial u_2}{\partial \psi} = \frac{\bar{\lambda}}{h(\bar{\rho})} \cos^2 \psi - \frac{h'(\bar{\rho})}{2h^3(\bar{\rho})} \bar{p}^2 \sin 2\psi + \frac{\bar{\lambda}}{h^2(\bar{\rho})} \cos^2 \psi,
$$

$$
H_2 + \frac{\partial v_2}{\partial \psi} = -\frac{\bar{p}}{2h(\bar{\rho})} \sin 2\psi + \frac{h'(\bar{\rho})}{h^3(\bar{\rho})} \bar{\lambda} \bar{p} \sin^2 \psi - \frac{\bar{\lambda}^2}{2h^2(\bar{\rho})\bar{p}} \sin 2\psi,
$$

$$
W_2 + \frac{\partial w_2}{\partial \psi} = -(1 - hx^2)\frac{\bar{\lambda}}{2h^2(\bar{\rho})} \sin 2\psi + (1 - hx^2)\frac{h'(\bar{\rho})}{h^3(\bar{\rho})} \bar{p}^2 \sin^2 \psi -
$$

$$
- (1 - hx^2)\frac{\bar{\lambda}}{2h^2(\bar{\rho})} \sin 2\psi . \tag{6.4.17}
$$

Averaging eq. (6.4.17) we find the coefficients Ξ_2, H_2, W_2 of the drift equations. Returning to differentiation with respect to t we obtain the final system of the drift equations

$$
\dot{\rho} = \varepsilon^2 \frac{\bar{\lambda}}{h(\mu)} + \varepsilon^3 \dots ,
$$

$$
\dot{p} = \varepsilon^2 \bar{\lambda} \bar{p} \frac{h'(\bar{\rho})}{2h^3(\bar{\rho})} + \varepsilon^3 \dots ,
$$

$$
\dot{\lambda} = \varepsilon^2 (1 - hx^2)\bar{p}^2 \frac{h'(\rho)}{2h^3(\rho)} + \varepsilon^3 \dots . \tag{6.4.18}
$$

The system (6.4.18) has an integral

$$
J = \frac{\bar{p}^2}{h(\bar{\rho})} = \text{const} , \tag{6.4.19}
$$

which coincides with the well-known transverse adiabatic invariant. However in this case the result has two specific features. First, the adiabatic invariant in the second approximation is exactly $\bar{p}^2/h(\rho)$ which was not evident beforehand. In the general case of motion of the particle the value $\bar{p}^2/h(\bar{\rho})$ is a nontrivial adiabatic invariant in the first approximation which is related to change in the magnetic field intensity because of the motion along the field line. In the given case this drift is strongly suppressed and the adiabatic invariant appears in the second approximation. The fact that this invariant has a well-known expression is just a accidental coincidence caused by the above choice of the field geometry. Secondly, conserving the adiabatic invariant in the time intervals $t \sim 1/\varepsilon^2$ is essential here.

One of the possible mechanicsms of acceleration of particles in the outer space is interaction of the energetic particles with the fronts of collisionless mhd-shock-waves, the whose existence being proved experimentally. Interaction of solitary energetic non-relativistic particles with the front of the oblique mhd-shock-wave of infinitesimally small thickness was considered in [2] [121]. Here we consider interaction of the relativistic particles with front of the mhd-shock-wave of finite thickness.

Let the mhd-shock-wave propagates in the negative direction of axis x with velocity u_f. As is generally known [71], the magnetic field in the mhd-shock-wave is plane. Let $H_x = \text{const}, H_y = 0, H_z = H(x) > 0$ in the frame of resting front. In the case of piston shock waves, $H(x)$ is sharply changing from the constant value H_1 before the front to the constant value H_2 after the front. For the explosive shock waves, the jump of magnetic field at the front is accompanied by a suction wave, in which the magnetic field is slowly changing and return to the unperturbed value H_1. The angle of slope of the magnetic field lines to axis z is assumed to be small, i.e. $H_x/(H_x^2 + H_1^2)^{1/2} = \varepsilon h_x$.

In the frame of resting front there appears the constant electric field $E_x = 0, E_y = u_f H_1/c = \text{const}, E_z = 0$.

The characteristic velocities of propagating shock fronts in the interplanetary space are $u_f \sim 10^5 \div 10^6$ m/s, hence $u_f/c \sim 10^{-2} \div 10^{-3}$ and the equations of motion of the relativistic charged particles in the field of the collisionless mhd-shock-wave take the form of eq. (6.4.2), where $\varepsilon = (u_f/s)\cos\alpha_1, \cos\alpha_1 = H_1/(h_x^2 + H_1^2)^{1/2}$ and $[H] = (H_x^2 + H_1^2)^{1/2}$.

Let us show that $\text{sgn}(\partial\mathcal{E}/\partial h) = \text{sgn}(dh/dx)$ under the above assumptions. For the sake of definiteness we assume $dh/dx > 0$. Introducing the abbreviation $2\eta = \mathcal{E}^2 - 1 - \mu^2$ we obtain from eq. (6.4.11)

$$\frac{\partial J}{\partial \mathcal{E}} = \frac{\mathcal{E}}{\pi} \int_{-\pi/2}^{\pi/2} \frac{d\theta}{h(A^{-1}(\kappa - \sqrt{2\eta}\sin\theta))} > 0, \quad \frac{\partial J}{\partial \kappa} = \frac{\sqrt{2\eta}}{\pi} \times \qquad (6.4.20)$$

$$\times \int_0^{\pi/2} \left\{ \frac{1}{h(A^{-1}(\kappa - \sqrt{2\eta}\sin\theta))} - \frac{1}{h(A^{-1}(\kappa + \sqrt{2\eta}\sin\theta))} \right\} \sin\theta d\theta < 0.$$

In the second inequality A^{-1} denotes the function inverse to $A(s)$ and we took into account that functions $h(x), A(s)$ are monotonically increasing. Using inequality (6.4.20) and taking into account relation $\partial\mathcal{E}/\partial\kappa = (-\partial J/\partial\kappa)(\partial\mathcal{E}/\partial J)$ we obtain the above assertion.

Using eqs. (6.4.3),(6.4.4),(6.4.10),(6.4.11) it is easy to obtain the equation for the evolutional component of the kinetic energy $\mathcal{K}$, introduced by relation $K = \mathcal{K} + \varepsilon\varphi_1(J, \kappa, \mu, t)$ and related to variable $\mathcal{E}$ by the relation $\mathcal{K} = \mathcal{E} - 1$

$$\dot{\mathcal{K}} = \varepsilon\frac{\partial\mathcal{K}}{\partial\kappa} = -\frac{1}{h_x}\dot{\mu} . \qquad (6.4.21)$$

Thus, in motion in the area of increasing field $dh_x/dx > 0$ (region of the shock wave front) the averaged kinetic energy monotonically increases with time, μ monotonically decreasing. Analogously, the averaged kinetic energy decreases in the region of suction wave where $dh/dx < 0$,

The abscissa of the velocity of the leading center $\dot{x}_c$ satisfies the equation

$$\dot{x}_c = \frac{\dot{\kappa}}{h(x_c)} = \frac{\varepsilon}{h(x_c)}\left(1 + h_x\frac{\mu}{\varepsilon}\right) \tag{6.4.22}$$

and is the velocity of maximum of the integrand (center of vibrations) in eq. (6.4.11).

It is clear that in the case of pure transverse wave $h_x = 0$ and $\dot{x}_c$ does not change sign, that is the particle moves without reflections. In the case of $h_x \neq 0$ the reflection is possible. As one can see from eq.(6.4.22) the condition of reflection is the equality $\dot{x}_c = 0$, i.e. the ratio $\mu/\mathcal{E}$ reaches the value $-1/h_x$.

Let us consider the case when the particle moved initially to the front, i.e. $\dot{x}_{c0} > 0$ and consequently $\mu_0/\mathcal{E}_I > -1/h_x$. Further, when moving in the front region μ can only decrease. It is not difficult to show that the ratio $\mu/\mathcal{E}$ also decreases. The following cases are possible: $-1/h_x < \mu_0/\mathcal{E}_I \leq 0$ and $-1/h_x < 0 < \mu_0/\mathcal{E}_I$. In the first case κ first grows and then decreases after reaching a certain critical value $\kappa = \kappa(\mathcal{E}_I, \mu_0, J_n)$. Thus x_c behaves like κ (let us remind that $h(x) > 0$), that is, it increases to some value and then decreases, i.e. the particle is reflected. Two variants are possible in the second case: (i) $\mu/\mathcal{E}$ decreases monotonically, passes through zero and achieves the value $-1/h_x$, that is, the particle is reflected and (ii) $\mu/\mathcal{E}$ decreases monotonically and remains positive. At the same time x_c permanently increases and the particle crosses the front and leaves it. If the particle moves from region after the front to the front then $\dot{x}_{c0} < 0$ and $\mu_0/\mathcal{E}_0 < -1/h_x < 0$. Further $\mu/\mathcal{E}$ is only decreasing and the particle crosses the front and moves into unperturbed region. The motion in the region of the suction wave is considered by analogy.

Here one can see a fundamental difference of properties of the front of transverse collisionless mhd-shock-waves from other waves of magnetic field that was first reported in [93]. A sharp change in the properties takes place already for angles ~ 0.001 rad.

Let us now calculate the change in the mean kinetic energy of the particle in all considered cases. As follows from eq. (6.4.21)

$$\mathcal{K}_1 - \mathcal{K}_0 = -\frac{1}{h_x}\int_{t_0}^{t_1}\dot{\mu}dt = \frac{\mu_0 - \mu_1}{h_x}, \tag{6.4.23}$$

and because μ only decreases in the front region, the kinetic energy increases in all considered cases of motion of the particle through the front.

Equating the values of invariant J (6.4.11) in the regions of weakly heterogeneous field before and after the front we obtain

$$(\mathcal{E}_0^2 - 1 - \mu_0^2)/h(x_{c0}) = (\mathcal{E}_1^2 - 1 - \mu_1^2)/h(x_{c1}) . \tag{6.4.24}$$

This result formally corresponds to conserving the transverse adiabatic invariant however it can be strictly proved only in the framework of the approach developed here.

Let us now determine the change of energy in the case of reflection of the particle from the front in detail. It is convenient to consider the expression for the kinetic energy of the particle after the reflection in plane $x_{c1} = x_{c0}$. Then $h(x_{c0}) = h(x_{c1})$ and by using eqs. (6.4.23) and (6.4.24) we have

$$\frac{\mathcal{K}_1}{\mathcal{K}_0} = \frac{h_x + w_0}{h_x + \bar{w}_1} + \frac{1}{\mathcal{K}_0} \frac{w_0 - \bar{w}_1}{h_x + \bar{w}_1} . \tag{6.4.25}$$

Here w_0 and $\bar{w}_1$ are the initial value and the evolutional part of z-component of the velocity of the particle at the time instants t_0 and t_1 respectively. The relationship between w_0 and $\bar{w}_1$ is given by the formula

$$\bar{w}_1 = -(2h_x + w_0 + h_x^2 w_0)/(1 + h_x^2 + 2h_x w_0) . \tag{6.4.26}$$

In the non-relativistic case, eqs. (6.4.25) and (6.4.26) yield the following equation for the particles reflected in the region of the homogeneous field after the front

$$\frac{\mathcal{K}_1}{\mathcal{K}_0} = 1 + \frac{4m\gamma}{\gamma^2 + 1}(m\gamma - 1) , \tag{6.4.27}$$

where $m = h_2/h_1 - 1, \gamma = \tan^2\theta$ and θ denotes the pitch angle of the particle. Calculating maximum of this expression with respect to γ for a typical ratio $h_2/h_1 = 3$ we obtain $\mathcal{K}_1/\mathcal{K}_0 = 4m^2 + 1$, which is in agreement with [121].

In the other cases the calculation is carried out by analogy, i.e. one obtains $\mathcal{K}_1(\mathcal{K}_0, w_0)$ and $w_1(\mathcal{K}_0, w_0)$ from the system (6.4.23) and (6.4.24).

In the case of explosive shock wave, the field experiences a sharp jump at the front, then changes slowly in the suction wave $(dh/dx < 0)$ and return to the unperturbed value h_1. At first μ decreases, and then increases to the initial value. It follows from eq. (6.4.23) that the particle energy does not change at all in this approximation.

In conclusion we notice that the developed theory allows one to perform a detailed study of change in the power spectrum of particles at passing and reflection of particles from the collisionless mhd-shock-wave, dispersion due to fluctuations of the magnetic field etc.

6.5   Motion of the non-relativistic particle in heterogeneous magnetic field

Solution of the problem considered in the previous section and analysis of the results is considerably simplified for the non-relativistic particle. Undoubtedly, all formulas of the present section can be obtained by a limiting process from the corresponding relations for the relativistic particle. However the solution of the problem for the non-relativistic case is simpler and more transparent than that for the relativistic one. Besides, the results related to the non-relativistic case, are essentially used in the physically important case of motion of the particle in the field of mhd-shock-wave. Because of this, the non-relativistic problem is solved again although a part of this solution was known earlier.

Motion of the non-relativistic charged particle in given electric $\mathbf{E}$ and magnetic $\mathbf{H}$ fields are described by eq. (6.2.1) in which the terms of order $\mathbf{v}^2/c^2$ are omitted

$$\dot{\mathbf{r}} = \mathbf{v}, \qquad \dot{\mathbf{v}} = \frac{e}{m}\left[\mathbf{E} + \frac{1}{c}\mathbf{v} \times \mathbf{H}\right] . \qquad (6.5.1)$$

The components of electric and magnetic fields are given by formulae (6.4.1).

Let us introduce the dimensionless variables in equations (6.5.1) but unlike the relativistic case, the characteristic velocity is the initial velocity of particle $v_0 \ll c$ and the characteristic coordinate is $mv_{\perp 0}c/e[H]$. Let the dimensionless variables have the denotations of the dimensional variables. Projecting equations (6.5.1) onto the coordinate axes we obtain

$$\dot{h} = u,$$
$$\dot{u} = h(x)v + e(x),$$
$$\dot{v} = \varepsilon - h(x)u + \varepsilon h_x w,$$
$$\dot{w} = -\varepsilon h_r v . \qquad (6.5.2)$$

Here $\dot{y} = v, \dot{z} = w, \varepsilon = E_y c/v_{\perp 0}[H], \varepsilon h_x = H_x/[H], h(x) = H_z(x)/[H],$ $e(x) = cE_x/v_{\perp 0}[H]$. The value ε is assumed to be the small parameter, i.e. as above two assumptions are made. Firstly, the y−component of the Coulomb forces is considered small in comparison to the Lorentz force. Secondly, the angle of slope of the magnetic field lines to axis z is considered to be small, too.

Instead of v, w we introduce new variables $\lambda = h_x w + 1$ and $\sigma = v + A(x)$ where $A(x) = \int_0^x h(s)ds$. System (6.5.2) takes the form

$$\dot{h} = u,$$
$$\dot{u} = h(x)\left(\sigma - A(x)\right) + e(x), \qquad (6.5.3)$$

$$\dot{\sigma} = \varepsilon \lambda,$$
$$\dot{\lambda} = -\varepsilon h_x^2(\sigma - A(x)). \qquad (6.5.4)$$

The subsystem (6.5.3) can be considered as equation for one-dimensional motions of a conditional material point in the potential field that is changing in time (with the change σ) and described by the Hamilton function

$$\mathcal{H} = \frac{1}{2}u^2 + \frac{1}{2}(\sigma - A(x))^2 + \varphi(x) , \qquad (6.5.5)$$

where $d\varphi(x)/dx = -e(x)$. System (6.5.3),(6.5.4) has an integral coinciding with the total energy of the particle

$$U = \frac{1}{2}u^2 + \frac{1}{2}(\sigma - A(x))^2 + \frac{\lambda^2}{2h_x^2} + \varphi(x) = \text{const} \qquad (6.5.6)$$

up to inessential constant. It is generally known that if function σ in system (6.5.3) depends only upon the slow time $\tau = \varepsilon t$ then this system has an adiabatic invariant which is an action variable in the corresponding problem with $\sigma = \text{const}$ [5].

In the case of $h_x \neq 0$ the considered system does not belong to this type. Nevertheless we show that this system has also an adiabatic invariant of the same type as in the known cases. We assume that the phase trajectories of system (6.5.3) are closed at $\sigma = \text{const}$. Instead of x, u we introduce the action-angle variables I, ψ similar to those in system (6.5.3) with Hamilton function (6.5.5) at $\sigma = \text{const}$.

The action variable is proportional to the area of the phase trajectory bounded by the phase curve $\mathcal{H} = E$, i.e.

$$I = I(E, \sigma) = \frac{1}{\pi} \int\limits_{x_1}^{x_2} [2E - (\sigma - A(s))^2 - 2\varphi(s)]^{1/2}ds . \qquad (6.5.7)$$

Here $x_1(E, \sigma), x_2(E, \sigma), x_1 < x_2$ are the roots of the equation

$$(\sigma - A(x))^2 + 2\varphi(x) = 2E.$$

The phase trajectories of system (6.5.3) at $\sigma = \text{const}$ are closed by assumption, i.e. this system describes vibrations. The position of the center, amplitude and frequency of these vibrations are slowly changing. The double-amplitude of vibrations is equal to $x_2 - x_1$, the frequency is $\omega = \dot{\psi} = \partial E/\partial I \neq 0$ and the center of vibrations is the root of the equation

$$h(x_c)(\sigma - A(x_c)) + e(x_c) = 0 \qquad (6.5.8)$$

up to values of the higher order.

In the unperturbed system at $\varepsilon = 0$ we have $\dot{I} = 0, \dot{\psi} = \omega(I, \sigma) = \text{const}$. In the perturbed system we have

$$\dot{I} = \frac{\partial I}{\partial E}\dot{E} + \frac{\partial I}{\partial \sigma}\dot{\sigma} = \varepsilon\lambda\left[\frac{\partial I}{\partial E}(\sigma - A(x)) + \frac{\partial I}{\partial \sigma}\right],$$

$$\dot{\psi} = \omega(I, \sigma) + \varepsilon \dots . \qquad (6.5.9)$$

Taking into account relations (6.5.9) the system (6.5.3),(6.5.4) is written down as a system with a single fast phase. Changing the independent variable t by phase ψ this system can be presented in the following form

$$\frac{dI}{d\psi} = \varepsilon\frac{\lambda}{\omega}\left[\frac{\partial I}{\partial E}(\sigma - \tilde{A}) + \frac{\partial I}{\partial \sigma}\right] + \varepsilon^2\ldots,$$

$$\frac{d\sigma}{d\psi} = \varepsilon\frac{\lambda}{\omega} + \varepsilon^2\ldots,$$

$$\frac{d\lambda}{d\psi} = -\varepsilon\frac{h_x^2}{\omega}. \tag{6.5.10}$$

Following the averaging method we introduce new (evolutional) variables $J(\tau), \kappa(\tau), \mu(\tau)$ by means of the relations

$$I = J(\tau) + \varepsilon u_1(J, \kappa, \mu, \psi) + \varepsilon^2\ldots,$$

$$\sigma = \kappa(\tau) + \varepsilon v_1(J, \kappa, \mu, \psi) + \varepsilon^2\ldots,$$

$$\lambda = \mu(\tau) + \varepsilon w_1(J, \kappa, \mu, \psi) + \varepsilon^2\ldots, \tag{6.5.11}$$

where $\tau = \varepsilon t$ and variables J, κ, μ satisfy the system of averaged equations which is similar to (6.4.9). The asymptotic procedure of the averaging method yields the system of evolutional equations for the particle motion

$$\dot{J} = 0 + \varepsilon^2\ldots,$$

$$\dot{\kappa} = \varepsilon\mu + \varepsilon^2\ldots,$$

$$\dot{\mu} = -\varepsilon h_x^2\frac{\partial\mathcal{E}(J, \kappa)}{\partial\kappa} + \varepsilon^2\ldots, \tag{6.5.12}$$

where $\mathcal{E}$ is the evolutional component of variable E. System (6.5.12) has two integrals

$$J = J(\mathcal{E}, \kappa) = \frac{1}{\pi}\int[2(\mathcal{E} - \varphi(s)) - (\kappa - A(s))]^{1/2}ds,$$

$$\mathcal{E} + \frac{\mu^2}{2h_x^2} = U = \text{const}, \tag{6.5.13}$$

that proves the adiabatic invariance of the action variable.

In order to obtain the expression for $\mathcal{E}(J, \kappa)$ it is necessary to invert the first equation in (6.5.13) which is possible, in general, only for a few functions $h(x)$ and $e(x)$. However some conclusions about the particle motion can be drawn without knowledge of the explicit form of function $\mathcal{E}(J, \kappa)$. For simplicity we limit our consideration by the case $e(x) \equiv 0$. First, we show that $\text{sgn}(\partial\mathcal{E}/\partial\kappa) = \text{sgn}(dh/dx)$. We have from eq. (6.5.13) that $\partial J/\partial\mathcal{E} > 0$ if $dh/dx > 0$ or $\partial J/\partial\mathcal{E} < 0$ if $dh/dx < 0$ and $\partial J/\partial\kappa \leq 0$ for any sign dh/dx. This yields the above statement provided that one takes into account the relation $J(\mathcal{E}, \kappa) = \text{const}$.

Let the particle move in the region of growing field $dh/dx > 0$. It follows from eq. (6.5.8) that $\dot{\kappa} = h(\bar{x}_c)\dot{\bar{x}}_c$, i.e. the evolutional component of the "transverse" part of the kinetic energy increases or decreases simultaneously with the evolutional component of $\bar{h}_c$ which is the abscissa of the leading center.

The equation for the total kinetic energy T is obtained from eqs. (6.5.2)-(6.5.4)

$$\dot{T} = \varepsilon v = \varepsilon(\sigma - A(x)) \ . \tag{6.5.14}$$

After averaging we obtain

$$\dot{\mathcal{T}} = \varepsilon \frac{\partial \mathcal{E}}{\partial \kappa} \ , \tag{6.5.15}$$

where $\mathcal{T}$ denotes the slow (evolutional) component of the kinetic energy.

Thus, when the particle moves in the region of growing function $h(x)$ the average total kinetic energy increases in time regardless of the direction of motion.

Let us consider reflection of the particle. We limit ourselves by the case when initially $\mu > 0$ and after some time $\mu = 0$ and then $\mu < 0$. The variables κ, x_c which first increase according to eq. (6.5.12) then begin to decrease (the particle is reflected). Let us proceed to the integral of total energy in eq. (6.5.13).

It is clear that the value of $\mathcal{E}$ achieves the maximum $\mathcal{E} = U$ for the particle if $\mu = 0$. Therefore determining the condition of reflection reduces to search of the maximum of function $\mathcal{E}(J, \kappa)$ with respect to κ at $J = \text{const}$ and subsequent comparison with U.

A simple condition for reflection is obtained for the following fields. Let $h(x)$ increase to the value h^* whereupon the magnetic field becomes weakly heterogeneous. In such a field function $\mathcal{E}(J, \kappa)$ as a function of κ has a form similar to the form of $h(x)$, i.e. it first increases to the value $\mathcal{E} = Jh^*$ and then it becomes nearly constant. Therefore for any particle with $U > Jh^*$ the equalities $U = \mathcal{E}$ and $\mu = 0$ are impossible, i.e. such particles are not reflected, whereas the particles with $U < Jh^*$ reflect. Consequently, the condition of reflection is $U/J < h^*$.

6.6   Motion of the non-relativistic particle in perturbed axisymmetric field

In this section we consider a number of cases admitting separation of motion. The unperturbed electromagnetic field is assumed to be axisymmetric. Therefore the equation of motions of particle (6.5.1) can be projected into the axes of the cylindrical coordinates (r, φ, z). The system of equations of motion written in terms of the nondimensional variables of the previous section is as follows

$$\ddot{r} - r\dot{\varphi}^2 = \varepsilon_1 e_r + r\dot{\varphi}h_z - \dot{z}h_\varphi,$$

$$\frac{1}{r}(r^2\dot{\varphi})^{\cdot} = \varepsilon_2 e_\varphi + \dot{z}h_r - \dot{r}h_z,$$

$$\ddot{z} = \varepsilon_3 e_z + \dot{r}h_\varphi - r\dot{\varphi}h_r. \tag{6.6.1}$$

Here $\varepsilon_i = [E_i]c/v_{\perp 0}[B]$ and $[E_i]$ are the characteristic value of $i-th$ projection of the vector of electric field intensity.

Let us consider the case in which $\varepsilon_1 = 1, \varepsilon_2 e_\varphi = \varepsilon/r, e_z = 0, e_r = e(r)$, $h_\varphi = h_r = 0, h_z = h(r)$. In this case the generating system of equations for (6.6.1) is given by

$$\ddot{r} - r\dot{\varphi}^2 = e(r) + r\dot{\varphi}h(r),$$

$$\frac{1}{r}(r^2\dot{\varphi})^{\cdot} = -\dot{r}h(r),$$

$$\ddot{z} = 0. \tag{6.6.2}$$

The longitudinal motion (along the magnetic field) is separated and is insignificant for the further treatment. Introducing the flux functions

$$G(r) = \int_0^r sh(s)ds \tag{6.6.3}$$

one can transform the equations for transverse motion in eq. (6.6.2) to the following form

$$\dot{r} = u,$$

$$\dot{u} = \frac{1}{r^3}(\sigma - G(r))^2 + \frac{1}{r}(\sigma - G(r))h(r) + e(r),$$

$$\dot{\sigma} = (r^2\dot{\varphi} + G(r))^{\cdot} = 0, \tag{6.6.4}$$

describing the one-dimensional motion of the conditional material point in the field with the Hamilton function

$$\mathcal{H}(u, r, \sigma) = \frac{1}{2}u^2 + \frac{1}{2r^2}(\sigma - G(r))^2 + \psi(r), \tag{6.6.5}$$

where $e(r) = d\psi/dr$.

Let us assume that the material point corresponding to system (6.6.4) vibrates, i.e. the phase trajectories of system (6.6.4) are closed if $\sigma = \text{const}$. The perturbed system of equations of the transverse remains Hamiltonian

$$\dot{r} = \partial\mathcal{H}/\partial u,$$

$$\dot{u} = -\partial\mathcal{H}/\partial r,$$

$$\dot{\sigma} = \varepsilon, \tag{6.6.6}$$

though it is no longer autonomous and possesses the adiabatic invariant which is the action variable of the unperturbed problem

$$J(E, \sigma) = \frac{1}{\pi} \int\limits_{r_1}^{r_2} \left[2E \frac{1}{s^2}(\sigma - G(s))^2 - \psi(S) \right]^{1/2} ds. \tag{6.6.7}$$

Here E is a new variable defined by the relationship

$$E = \frac{1}{2}u^2 + \frac{1}{2r^2}(\sigma - G(r))^2 + \psi(r) \tag{6.6.8}$$

and $r_1, r_2 (r_1 < r_2)$ are the root of equations obtained from eq. (6.6.8) at $u = 0$ and $\sigma, E = $ const.

The coordinate of the "leading" centre r_c with an accuracy of small values is the root of the equation

$$\frac{1}{r_c^3}(\sigma - G(r_c))^2 + \frac{1}{r_c}(\sigma - G(r_c))h(r_c) + e(r_c) = 0. \tag{6.6.9}$$

If $e(r) = 0$ then r_c is the root of the equation

$$\sigma - G(r_c) = 0. \tag{6.6.10}$$

In addition to this, provided that the magnetic field is weakly heterogeneous we obtain

$$G(s) = \int\limits_0^s \tau h(\tau)D\tau = h(\xi)\frac{s^2}{2}, \tag{6.6.11}$$

where ξ is the evolutional component of the coordinate of the "leading" centre r_c introduced by the relationship $r_c = \xi + \varepsilon u_1 \ldots$. Estimating the integral in eq. (6.6.7) yields

$$J(E, \xi) = \frac{1}{\pi} \int\limits_{r_1}^{r_2} \left[2E\frac{h^2(\xi)}{4s^2}(\xi^2 - s^2)^2 \right]^{1/2} ds = \frac{E}{h(\xi)}, \tag{6.6.12}$$

which is coincident with the well-known transverse invariant of the existing adiabatic theory of motion of charged particles in a weakly heterogeneous field.

Now let the magnetic field be given by

$$\mathbf{H}(r) = h(r)\mathbf{e}_z + \frac{\varepsilon}{r}h_r\mathbf{e_r}, \quad h_r = \text{const}, \tag{6.6.13}$$

whereas the electric field is the same as before. In this case the components of the vectorial potential $\mathbf{A}(r)$ are equal to

$$A_r = 0, \quad A_\varphi = \frac{1}{r}\int rh(r)dr, \quad A_z = h_r\varphi. \tag{6.6.14}$$

The equation of motions of the particle (6.6.1) take the form

$$\dot{r} = u,$$
$$\dot{u} = r\omega^2 + e(r) + r\omega h(r),$$
$$\dot{\sigma} = \varepsilon(1 + wh_r),$$
$$\dot{w} = -\varepsilon\omega h_r, \qquad\qquad (6.6.15)$$

where $\omega = \dot{\varphi}$ and $w = \dot{z}$. Inserting a new variable $\lambda = 1 + h_r w$ in system (6.6.15) we arrive at the system

$$\dot{r} = u,$$
$$\dot{u} = \frac{1}{r^3}(\sigma - G(r))^2 \frac{1}{r}(\sigma - G(r)) + e(r),$$
$$\dot{\sigma} = \varepsilon\lambda,$$
$$\dot{\lambda} = -\varepsilon h_r^2 \frac{\sigma - G(r)}{r^2}. \qquad\qquad (6.6.16)$$

This system possesses the following energy integral

$$E = \frac{1}{2}\left[u^2 + (r\omega)^2 + \frac{\lambda^2}{h_r^2}\right] + \psi(r) = \text{const}. \qquad (6.6.17)$$

Here one can observe a full analogy with the problem considered in the previous section. System (6.6.16) possesses the adiabatic invariant

$$J(\mathcal{E}, \kappa) = \frac{1}{\pi}\int_{r_1}^{r_2}\left[2\mathcal{E} - \frac{1}{s^2}(\kappa - G(s))^2 - 2\psi(s)\right]^{1/2} ds, \qquad (6.6.18)$$

where $J, \mathcal{E}, \kappa$ are the evolutional components of variables I, E, σ respectively. The system of evolutionary equations is presented in the form

$$\dot{\kappa} = \varepsilon\mu,$$
$$\dot{\mu} = -\varepsilon h_r^2 \frac{\partial \mathcal{E}(J, \kappa)}{\partial \kappa}. \qquad\qquad (6.6.19)$$

Here μ is the evolutional component of variable λ, and function $\mathcal{E}(J, \kappa)$ is obtained by inversion of the function in (6.6.18). Equations (6.6.19) describe a new one-dimensional motion with the potential $\varepsilon h_r^2 \mathcal{E}(J, \kappa)$.

For the evolutional component ξ of the radial displacement of the "leading" centre r_c one can obtain the following equations

$$\dot{\xi} = \varepsilon\frac{\mu}{K(\xi, \mu)},$$
$$\dot{\mu} = \varepsilon\frac{h_r^2}{K(\xi, \mu)}\frac{\partial E^*(\xi, \mu)}{\partial \xi}, \qquad\qquad (6.6.20)$$

where function $K(\xi, \mu)$ is obtained by differentiation of the equation

$$\frac{1}{\xi^3}(\kappa - G(\xi))^2 + \frac{1}{\xi}(\kappa - G(\xi))h(\xi) + e(\xi) = 0 \qquad (6.6.21)$$

with respect to time. In the case $e(r) = 0$

$$K(\xi) = -\xi[h(\xi) + \xi h'(\xi)] \qquad (6.6.22)$$

and function $K(\xi)$ is independent of μ in contrast to function $E^* h(x_c)$ of Section 6.4. By introducing a "local" time ϑ: $d\vartheta = dt/K(\xi)$ system (6.6.20) reduces to the Hamilton function

$$\frac{d\xi}{d\vartheta} = \varepsilon\mu,$$

$$\frac{d\mu}{d\vartheta} = -\varepsilon h_r^2 \frac{\partial E^*(J, \xi)}{\partial \xi}. \qquad (6.6.23)$$

Unlike the previous treatment (in Cartesian frame of axis) this transformation is possible only for $e(r) = 0$. Besides, one observes an essential distinction in the form of function $K(\xi)$ in these two cases.

7

Some problems of nonlinear magnetoelasticity

7.1   Statement of "elastic-linear" problems of nonlinear magnetoelasticity

The present chapter deals with several variants of the following problem. An undeformed state of the system of elastic or elastically connected rigid bodies (there are conductors and ferromagnetic bodies among them) is given. The field intensity of the external electromotive forces is also prescribed. We look for the current density $\mathbf{j}$ (or the linear currents i), the magnetic field (namely the vector of magnetic induction $\mathbf{B}$ and the field intensity $\mathbf{H}$) and the elastic displacement $\mathbf{u}$ in statics and dynamics.

Let us limit our consideration by the case when various particular physical effects (magnetostriction, Hall effect etc.) can be ignored. Specifically, in the expression for density of the ponderomotive forces $\mathbf{f}$ in the magnetic-linear media we keep two terms

$$\mathbf{f} = \mathbf{j} \times \mathbf{B} - \frac{1}{2} H^2 \nabla \mu, \qquad (7.1.1)$$

where μ denotes the permeability of the medium. Proceeding to limit in eq. (7.1.1) determines the density of surface forces on the surface of discontinuity of μ. In what follows we consider only the surface force caused by the term $1/2 H^2 \nabla \mu$.

Let us first consider the mechanical equilibrium. The problem is to solve the equations for magnetic field and the equations of the theory of elasticity. They form a coupled system since the shape of the field region and the distribution of the density of currents in the space are dependent on displacements, and the displacements, in turn, are defined by the forces depending on the currents and the field.

Let us suppose that the displacements are small. In other words, we assume that, if the ponderomotive forces (generally speaking they depend on the displacements) are prescribed, the displacement can be determined from the equations of the linear theory of elasticity with prescribed volumetric and surface forces. Let us describe the cases in which it is necessary to consider the dependence of the field and ponderomotive forces on the displacement. We shall consider two cases: (i) attraction of the ferromagnetic bodies and (ii) the interaction of conductors with currents. In the first case the force is equal to the term $1/2H^2\nabla\mu$ in eq. (7.1.1) while in the second case the force is equal to $\mathbf{j} \times \mathbf{B}$.

Let us consider elastic bodies in space such that the distance between the certain parts of their surfaces is small. We first assume that all three dimensions of the bodies are of the same order and the sizes of the closely adjacent parts of surfaces are comparable with the characteristic dimension of the bodies. The closely adjacent parts should have "appropriate" shapes (such that the surfaces are close to each other in a significant area), for instance, these surfaces are flat. We also assume that at least one of the closely adjacent parts of surfaces is not fixed. Let the bodies be magnetic-homogeneous (μ =const) and their permeability differs from the permeability μ_0 of the surrounding inelastic medium. Then the surface ponderomotive force act on the bodies in magnetic field. Let us consider the equilibrium of the system subjected to these forces when interaction of the currents is negligible.

Let us introduce the following denotations for the characteristic parameters: l is the dimension of the bodies, h is the distance between the closely adjacent surfaces in the undeformed state (the distance is counted along the normal to one of the surfaces), a is the elastic displacement, μ is the magnetic permeability of the bodies, B_0 and B denote respectively the induction in the space between the closely adjacent surfaces and in the bodies, B_2 is the induction in the medium in the distance of the order of l from the surfaces.

As announced above, the ratio a/l is assumed to be small and serves as a criterion for smallness of the displacement.

Let us suppose that the characteristic values satisfy the following requirements: $a/l = O(h/l)$, that is, the displacements (also the relative displacement of closely adjacent surfaces along the normals to them) are comparable with the initial distance between the surfaces; $\mu_0/\mu = O(h/l)$, that is, the ratio of the medium permeability to the permeability of the bodies is small (the smaller values of this ratio are admitted, for example, $\mu_0/\mu \sim (h/l)^2$ etc.), $B_2/B_0 = O(h/l)$, that is, the field in the surrounding medium on distances of the bodies' dimension is small compared with the field in the gap between the bodies.

The requirement $\mu_0/\mu = O(a/l)$ is feasible only if the elastic body is ferromagnetic. The latter requirement imposes limitations onto the shape of bodies and the geometry of currents. An example of such a system is two halves of the ferromagnetic torus cut in the plane normal to the axial circle. The gap between the faces of the fixed halves should be small. The torus is coated by a winding whose axis is coincident with the axial circle of the torus. The system can consist of a single elastic body, a torus with an indent being an example.

In this case, small displacements of the points of the closely adjacent surfaces essentially change both the field and the ponderomotive force. Formally this can be shown, for example, in such a way. Let a contour C have approximately the shape of the characteristic field line. We consider the circulation $\mathbf{H}$ in the existing field and in the conditional field generated by the same current in the undeformed system. We have

$$\int_C \mathbf{H} \cdot d\tilde{\mathbf{n}} = \int_C \mathbf{H}_* \cdot d\tilde{\mathbf{n}} = I, \qquad (7.1.2)$$

where $\mathbf{H}_*$ belongs to the undeformed state and I denotes the total current embraced by contour C (change in the geometry of currents under deformation is supposed not to affect I).

From eq. (7.1.2) we obtain the equality

$$B_0 \frac{h-a}{\mu_0} + B \frac{l+a}{\mu} = B_{0*} \frac{h}{\mu_0} + B_* \frac{l}{h}. \qquad (7.1.3)$$

Here $h, h-a, l$ denote the lengths of parts of the contour in the nonmagnetic medium and in the ferromagnet respectively and B_0, B_{0*} etc. are the corresponding mean values of induction. We replace $l+a$ by l and transform eq. (7.1.3) to the form

$$\Delta B_0 + \frac{l}{h-a} \frac{\mu}{\mu_0} \Delta B = \frac{a}{h-a} B_{0*},$$

$$\Delta B_0 = B_0 - B_{0*}, \quad \Delta B = B - B_*. \qquad (7.1.4)$$

The factors in front of ΔB and B_{0*} in eq. (7.1.4) are of the order of one. Therefore, at least one of the increments $\Delta B, \Delta B_0$ should be comparable with B_{0*}. In general case $\Delta B \sim B_{0*}$ and $\Delta_0 \sim B_{0*}$, that is, the change in the field (and the ponderomotive forces) caused by the displacements of the close surfaces will be of the same order as the field itself (or the force).

The same conclusion is valid in the case in which the nonlinearity of dependence $B(H)$ in ferromagnet should be accounted for, however the characteristic value of dB/dH in the medium should be understood as μ. In addition to this, the zones where $dB/dH \sim \mu_0$ (provided that they exist) should not affect the described location of zones with large and small values of dB/dH.

An interesting case is that when conditions $Bl/\mu, B_*l/\mu$ in eq. (7.1.3) are small in comparison with two remaining terms. Then the ferromagnet can be viewed as being ideal ($\mu = \infty$). It leads to the significant simplifications, as the problem of estimation of the field in the bodies is no more relevant.

The basic situation described above admits some modifications. Specifically, the sizes of ferromagnetic bodies in three directions can be very different. Hence the characteristic value of induction in the ferromagnet and the gap between closely adjacent surfaces can vary. Let l, b, h_0 denote respectively the characteristic value of length, width and thicknesses of the body. Then for a plate or a shell $l \sim b \gg h_0$ and for a bar $l \gg b \sim h_0$. We will treat the situations when h and a are small compared to the minimum dimension of the closely adjacent surfaces, and $B_0 h/\mu_0 \gg Bl_c/\mu$ where l_c denotes the characteristic length of the field line. In any particular case, from the latter relationship we can receive the estimates relating the permeability to the dimensions. For example, for rods attracted by the lateral surfaces we should have $l_c \sim l, B \sim B_0 l/h_0, \mu/\mu_0 \sim l^2/hh_0$ or $\mu/\mu_0 \gg l^2/hh_0$.

It was earlier required that $B_0 \sim B$ for the three-dimensional bodies. However it is possible that $B \gg B_0$ provided that almost all line of the additional field are closed in one body. A similar case is possible in the system of ferromagnetic rods etc.

Let us describe another situation when the ponderomotive force are essentially dependent on the small displacements. We consider two conductors whose sizes b, h_0 of the cross-section are small as compared with length l. Let the conductors be located such that the distance between their lateral surfaces on the entire length (or on a considerable part of the length) be comparable with the dimensions of the cross-section. We assume that the conductors are either attracted or repelled. Let the lateral surface of the conductors either approach or depart on the distances comparable to the initial one. Then the forces acting on the deformed conductors differ from the forces calculated for the undeformed state in the values of the order of forces themselves.

The substantiation is as follows. We select an elementary thread of current in the conductor and consider the field generated in the second conductor by this thread. The induction in this field is approximately coincident with that of the field of infinite linear current tangential to the thread. However the induction in the latter field is inverse proportional to the distance to the current and, thus, essentially varies if the distance is comparable with the initial one. Therefore the overall field (as a sum of the contributions of all threads) at a certain point of the second conductor is essentially dependent on the distance to the first conductor. The same conclusion is valid for the forces acting on the second conductor. For this reason, if we consider the forces acting on the first conductor we obtain that the force of interaction of the conductors is essentially dependent on the distance between their lateral surfaces.

The described condition naturally leads to the problem of bending of two attracted or repelled thin rods with nearly parallel axes and to the problem of bending of a rod in the neighborhood of a ferromagnet. In the latter case we can assume that the current in the rod interacts with the "mirror current", that is, with its reflection in the ferromagnet. Similar to the previous problems these problem are "linear-elastic" since the linear equation of the statics of thin rods for small values $b/l, h_0/l$ can be utilised when the deflections are comparable with the cross-sectional size. The problem of equilibrium of strings or massive threads are stated by analogy however for these problems it is typical that the cross-section sizes are small in comparison with the displacement.

The rods can be not necessarily parallel. Then the interaction force can be replaced by a concentrated force in the first approximation.

For another shape of conductors the dependence of ponderomotive forces on small displacements is insignificant (if the bodies with a great number of crests and peaks are not considered). For example, the field in the neighbourhood of a thin layer with nonsmall curvature radii slowly varies under change in distance comparable with the original small value. Therefore the forces acting on another thin layer in the neighbourhood of the first one slowly vary under small displacements. The same conclusion is also valid for the interaction of closely adjacent volumetric conductors provided that the displacements are small in comparison with the radius of curvatures of closely adjacent surfaces and the body thickness.

Thus, there exist two basic classes of "linear-elastic" problems of the nonlinear magnetoelasticity for the equilibrium of closely adjacent ferromagnetic bodies and closely adjacent thin conductors[1].

Such situations are natural for practice. In particular, they appear in the cases in which the attraction of ferromagnetic bodies or the interaction of conductors with currents is used for displacement of the elastically attached or deforming flexible structural elements (electromagnetic actuating mechanisms, commutators etc.). In such devices the ferromagnetic bodies (e.g. cores and anchors of electromagnets) are usually fixed on distances which are approximately equal to the necessary elastic displacements. The currents which are necessary for creating the required displacements will be essentially smaller than those in the case when bodies are very distant from each other. Small gaps between the surfaces of ferromagnetic bodies or conductors are rather typical for electrical engineering. Therefore the "linear-elastic" problems present interest, for example, in connection with the strength design and in the cases in which any deformation caused by the field is undesirable.

---

[1]Field can be dependent on small displacements because the displacements change the values of currents. It is realised with the help of special systems, such as interrupters, sliding contacts etc. Such systems are not considered in what follows.

Generally speaking, determination of stresses in the ferromagnetic body presents a more intricate problem because of the magnetostriction. It is assumed in what follows that the striction stresses do not affect the considered "macrodisplacement". Correspondingly, the permeability μ is viewed as being independent of the deformations and only the surface forces on the surfaces of discontinuities μ, that is, on the boundaries of ferromagnetic bodies are considered.

7.2   Equilibrium of ferromagnetic bodies. Nonlinear boundary-value problems

The system of equations of linear-elastic problems of the nonlinear magnetoelasticity consists of the equations of problem of the stationary distribution of currents, the equations of magnetostatics and the equations of linear theory of elasticity. The equations of theory of elasticity contain the ponderomotive force in the configuration of the undeformed body. Specifically, for a three-dimensional body the boundary conditions are stated on the undeformed boundary.

We will find the field and the forces with the accuracy sufficient for estimation of the elastic displacements. Since in the linear theory of elasticity only lowest terms of a/l are kept then only the lowest terms should be retained for the estimation of currents, fields and ponderomotive forces. Besides one can ignore the forces acting in small zones provided that these forces are not too large. Hence the currents can be determined for the undeformed state by ignoring the small deformations of conductors. However the field and the values of forces should be determined for the deformed system even in the considered approximation.

Now the following question arises: what simplifications admit the problem of magnetostatics if we are interested only in the lowest terms of a/l. We consider first the system of closely adjacent ferromagnetic bodies. The standard conditions of continuity of the normal $\mathbf{B}$ and tangential $\mathbf{H}$ components should be set on these surfaces. On the parts of the surfaces near the other ferromagnetic bodies these conditions must be written for the deformed state and this is the way of modelling the dependence of the field on the displacements. The rest of the boundary between the ferromagnet and the non-magnetic medium can be viewed as being coincident with the surface of the undeformed bodies. Let us show now that, with a required accuracy, this problem is reduced to the problem in which all boundary condition are set on the undeformed surfaces however some conditions incorporate displacements. To this end, we express the field between the bodies in terms of the quantities characterising the field in the bodies.

Let us consider a layer between the closely adjacent parts of two surfaces. Let S and S_* denote these parts. We denote the normal vector to S at a point M inward the layer by $\mathbf{n}(M)$ while the normal vector to S_* at a point M_* (the point of intersection of $\mathbf{n}$ and S_*) outward the layer is denoted as $\mathbf{n}_*(M)$. The length of $\mathbf{n}$ between M and M_* in the undeformed and deformed state is denoted as $h(M)$ and $h(M) - u(M)$, respectively. In the considered approximation the geometric characteristics of the layer can be prescribed up to the highest terms in h/l. Therefore describing part S of the boundary can have an error of the order of h^2. In other words, accounting for deformation of the surfaces we are allowed to disregard their tangential displacements. Let us assume that radii of curvature of S and S_* are everywhere of the order of l^2/h. Then the angle between $\mathbf{n}$ and $\mathbf{n}_*$ is of the order of h/l (otherwise S and S_* could not be adjacent within distances of the order of l).

Let us denote the scalar potential on S and S_* in the deformed state as $\varphi(M)$ and $\varphi_*(M)$ respectively (there is a point-to-point correspondence between M and M_* therefore we can write $\varphi_*(M)$ instead of $\varphi_*(M_*)$). The many-valuedness of the potential is of no importance. From the above assumptions about the shape and the geometry of bodies we conclude that for bodies with sufficiently smooth surfaces the change in the potential on the surface within the distances of the order of h is small compared with its change within the layer. This leads to the condition

$$\left|\frac{\partial\varphi}{\partial s}\right| : \left|\frac{\varphi - \varphi_*}{h}\right| \sim \frac{h}{l}, \tag{7.2.1}$$

where $\partial\varphi/\partial s$ is the derivative with respect to either of the tangents to the surface. The second and third derivatives of φ with respect to s can not be large, at least in zones of the distance of the order of l from the zones with given currents. Therefore on S and S_* the following conditions

$$\left|\frac{\partial^2\varphi}{\partial s^2}\right| l \sim \left|\frac{\partial\varphi}{\partial s}\right|, \quad \left|\frac{\partial^3\varphi}{\partial s^3}\right| l \sim \left|\frac{\partial^2\varphi}{\partial s^2}\right| \tag{7.2.2}$$

hold. In addition to this, the potential near the layer edges can not be greater than $\varphi(M)$ by one order.

In the homogeneous medium, i.e. in the layer, potential φ satisfies Laplace equation

$$\triangle\varphi = 0. \tag{7.2.3}$$

Let us introduce coordinate ξ which is the distance from M along $\mathbf{n}$, $0 \leq \xi \leq h(M) - u(M)$. Under the above conditions the solution of equation (7.2.3) can be presented in the form

$$\varphi = \varphi(M) + \frac{\varphi_*(M) - \varphi(M)}{h(M) - u(M)}\xi + O(h^2/l^2) \tag{7.2.4}$$

everywhere in the layer except for the zones in the neighborhood of the edges. The symbol $O(h^2/l^2)$ means that the ratio of the next higher term to the second term in the right-hand side is of the order of h^2/l^2. Then we can be found the two first terms in the expansion of $\nabla\varphi$ and the field intensity in the layer. Taking into account eq. (7.2.2) we obtain

$$\mathbf{H}_0 = -\nabla\varphi = \frac{\varphi(M) - \varphi_*(M)}{h(M) - u(M)}\mathbf{n} + \xi\frac{\partial}{\partial M}\frac{\varphi(M) - \varphi_*(M)}{h(M) - u(M)} -$$
$$- \frac{\partial\varphi(M)}{\partial M} + O(h^2/l^2). \tag{7.2.5}$$

Here and in what follows $\partial/\partial M$ denotes the gradient of a function given on S. This gradient is a vector in the plane tangent to S at point M.

It follows from eq. (7.2.5) that the derivative $\partial\varphi/\partial n$ on the side S toward the layer is equal to

$$\frac{\partial\varphi_0}{\partial n} = \frac{\varphi_*(M) - \varphi(M)}{h(M) - u(M)} + O(h^2/l^2). \tag{7.2.6}$$

Omitting the correction terms $O(h^2/l^2)$ we write down the formula relating φ and φ_* to $\partial\varphi_0/\partial n$

$$\varphi_*(M) = \varphi(M) + [h(M) - u(M)]\frac{\partial\varphi_0(M)}{\partial n}. \tag{7.2.7}$$

If we introduce the normal derivative on sides S and S_* toward the "body"

$$\frac{\partial\varphi}{\partial n} = \frac{\mu_0}{\mu}\frac{\partial\varphi_0}{\partial n}, \quad \frac{\partial\varphi_*}{\partial n} = \frac{\mu}{\mu_*}\frac{\partial\varphi_0}{\partial n} \tag{7.2.8}$$

then we obtain

$$\varphi_*(M) = \varphi(M) + [h(M)u(M)]\frac{\mu}{\mu_0}\frac{\partial\varphi(M)}{\partial n},$$
$$\frac{\partial\varphi_*(M)}{\partial n} = \frac{\mu}{\mu_*}\frac{\partial\varphi(M)}{\partial n}. \tag{7.2.9}$$

Equations (7.2.9) relate the values of the scalar potential and its normal derivative on the closely adjacent surfaces. The equivalent equations relating the normal components of the induction and the tangential component of the intensity of the magnetic field are given by

$$B_n = B_{*n},$$
$$\mathbf{H}_{*s} = \mathbf{H}_s + \frac{\partial}{\partial M}\frac{(h - u)B_n}{\mu_0}. \tag{7.2.10}$$

The second relationship in eq. (7.2.10) is obtained from the terms of the order of h/l in eq. (7.2.5). While deriving this equation we take into account that $\mathbf{n}$ and $\mathbf{n}_*$ have different directions. In what follows, this difference plays no role.

At the edges of the layer, i.e. in the narrow zones in the neighborhood of the lines bounding S and S_* these relationships are not satisfied. However near the edges the induction is much greater than the induction far from the edges. Therefore these zones similar to all small zones with "not too large" values of induction can be neglected in the accepted approximation.

Relationships (7.2.9) or (7.2.10) should be considered as the boundary conditions of the problem of magnetostatics on S and S_*. With the required accuracy they can be set for the points of the undeformed surfaces, $u(M)$ being kept in the right hand side. As a result, the problem of magnetostatics reduces to determining the field in the bodies and in the environment, while the change in this zone under deformation is not considered. The space between the bodies are excluded from the treatment. For the sake of clarity we can consider these spaces as being filled by ferromagnet such that a single body appears and there are fixed barriers in this body coinciding with S or S_*, and conditions (7.2.9) or (7.2.10) are satisfied on these barriers.

Conditions (7.2.9), (7.2.10) depend significantly on displacement $u(M)$. On the other hand, this fact explains the dependence of the field on the displacement. The remaining boundary conditions have the appearance which is typical for the problems of magnetostatics.

In some cases the "thickness" of conductors (or coils) can be neglected and one can assume that the surface currents are prescribed on a part of the boundary of ferromagnet. Then the problem of magnetostatics is reduced to determining the solution of Laplace equation for φ only "in" the bodies under conditions (7.2.9), given values of the tangential component $\nabla\varphi$ on the subsurfaces with the surface currents and condition $\partial\varphi/\partial n = 0$ on the whole surface of the ferromagnet but S and S_*. The latter condition implies that "almost" all field lines do not leave the body possessing a greater permeability and the geometry described in Section 7.1 (e.g. a torus shape).

The problem for φ is also obtained in the cases in which on can put $\mu = \infty$ in some part of the body near the zone with the currents. Then some "piece" of the ferromagnet can be conditionally cut out and φ is considered as having a prescribed value on these new surfaces. In other cases the problem of magnetostatics can not be reduced only to determining φ because the currents rather than the potential are prescribed. In such cases one should utilise condition (7.2.10) or the conditions for vector potential obtained from eq. (7.2.10).

Provided that surfaces S and S_* are not smooth, in small zones near the discontinuity in $h(M)$ (for example, in the neighborhood of a step on S or S_*) the induction calculated under the assumption of a magneto-linear medium can be very large. Nevertheless we have every reason to discard these zones similar to zones in the neighborhood of edge of the layer. Indeed, because of the saturation of ferromagnet the actual value of induction is limited and the surface forces which are approximately proportional to $1 - \mu_0/\mu$, where μ is the characteristic value of dB/dH, are essentially smaller than that in the unsaturated ferromagnet since μ decreases. As a result, the load in the regions under consideration has a little influence on the system deformation. However it is necessary that the number of discontinuities $h(M)$ (that is, crests, steps etc.) is small in comparison with h/l. Subject to this remark we will make no difference between smooth and piecewise smooth surfaces.

Let us now proceed to the problem of theory of elasticity. The surface forces act on the closely adjacent surfaces of ferromagnetic bodies and the force density q can be found from the expression $1/2H^2\nabla\mu$ by proceeding to the limit in the case of discontinuity μ. With accuracy up to the higher terms in μ_0/μ we arrive at the following expression

$$\mathbf{q}(M) = \frac{1}{2}\mu_0 H_0^2 \mathbf{n} = \frac{1}{2}\left[\frac{\varphi(M) - \varphi_*(M)}{h(M)u(M)}\right]^2 \mathbf{n}. \qquad (7.2.11)$$

It allows one to set the boundary condition on the closely adjacent surfaces

$$\mathbf{p}_n = q(M)\mathbf{n}, \quad \mathbf{p}_{*n} = -q(M)\mathbf{n}. \qquad (7.2.12)$$

Here $\mathbf{p}_n$ and $\mathbf{p}_{*n}$ are the stress vectors on the elementary surfaces with the normals $\mathbf{n}$ and $\mathbf{n}_*$ respectively. As mentioned above, these conditions are set on the surfaces of the undeformed body.

Thus the problem of magneto-elasticity for the three-dimensional ferromagnetic bodies is reduced to determining the solutions of the linear equations of magnetostatics and theory of elasticity in the domain occupied by the undeformed body under nonlinear boundary conditions (7.2.9) or (7.2.10) and eq. (7.2.12). These conditions incorporate both displacements and the scalar potential or induction and field intensity, i.e. the mechanical and magnetic unknown variables need to be determined together.

Earlier we study more simple cases in which only one unknown function φ appears along with the problem of theory of elasticity. The maximum simplifications are achieved for ideal ferromagnetic bodies ($\mu = \infty$ everywhere in the ferromagnet). Then φ is the same at all points on surfaces S and S_*. For the systems of the above configuration with a single gap between the ideal ferromagnetic bodies the difference $\varphi - \varphi_*$ can be found at once with the required accuracy. This difference is equal to the value of the total current I embracing the field line taken with the corresponding sign. It results in the nonlinear boundary-value problem for the equations of theory of elasticity with boundary conditions (7.2.12), where

$$q = \frac{\lambda}{[h(M) - u(M)]^2}, \quad \lambda = \frac{1}{2}\mu_0 I^2. \qquad (7.2.13)$$

Provided that the field lines intersect several gaps between the bodies, φ and φ_* are functionals of displacements $u(M)$. They are determined in the following way. Let us consider a system with two interspaces (Fig. 7.1).

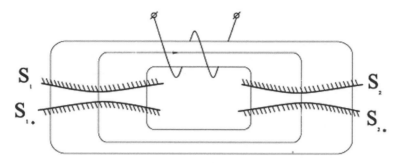

FIGURE 7.1.

On S_1 we put $\varphi = 0$, then $\varphi = I$ on S_2. Let the value of φ on S_{1*} and S_{2*} be denoted as φ_2. We make use of the fact that the magnetic fluxes in both gaps can be considered as being equal. With the help of eq. (7.2.4) we obtain

$$\varphi_2 = \frac{\sigma_2}{\sigma_1 + \sigma_2} I,$$

$$\sigma_1 = \int_{S_1} \frac{dM}{h_1(M) - u_1(M)}, \quad \sigma_2 = \int_{S_2} \frac{dM}{h_2(M) - u_2(M)}. \tag{7.2.14}$$

Similar relationships can be constructed for a greater number of gaps.

Let us consider the problems for one- and two-dimensional bodies (rods, plates etc.). We suppose that a corresponding technical theory is sufficient for description of the equilibrium. Then the nonlinearity appears in the equations (rather than in the boundary conditions as above). We demonstrate a few examples of equilibrium of the ideally ferromagnetic bodies. First we formulate the problem of bending of a ferromagnetic membrane by an electromagnet. Let the membrane be located near a planar surface of the ferromagnetic body (the poles of electromagnet, Fig. 7.2); the membrane itself can be considered as a second pole. It is assumed that all induction lines are embraced by the same total current I and are closed, as shown in Fig. 7.2. The lines pass the gap between the membrane and the pole, then along the membrane to its boundary and then along a ferromagnetic body (magnetic circuit) back to the pole. All the above-mentioned bodies are considered as being ideally ferromagnetic. Then the load per unit area of the membrane is defined by expression (7.2.13). The equilibrium equation and the boundary condition have the form

$$\triangle u + \frac{1}{2} \frac{\mu_0 I^2}{T} \frac{1}{(h - u)^2} = 0, \quad u|_\Gamma = 0. \tag{7.2.15}$$

Here u is the displacement, T is the membrane tension per unit length, h is the distance between the pole and the undeformed membrane and Γ is the membrane contour.

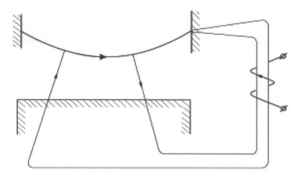

FIGURE 7.2.

Let us put $v = u/h$, $\kappa^2 = \mu_0 I^2/2Th^3$ and introduce the nondimensional coordinates obtained by multiplication of the corresponding dimensional coordinate by κ. We arrive at the following boundary-value problem

$$\Delta v + \frac{1}{(1-v)^2} = 0, \quad v|_\gamma = 0, \tag{7.2.16}$$

where the Laplace operator and contour γ correspond to the plane of dimensionless coordinates. Thus, for a given shape Γ we deals with a one-parameter problem, the size of contour γ playing the part of this parameter in eq. (7.2.16).

In the following we consider the axially symmetric shapes of equilibrium of the circular membrane. They are described by the equation

$$\frac{d^2 w}{d\rho^2} + \frac{1}{\rho}\frac{dw}{d\rho} - \frac{1}{w^2} = 0, \tag{7.2.17}$$

where $w = 1 - v$. The boundary conditions are as follows: $w(\kappa R) = 1$ and $w(0)$ is limited (R is the membrane radius). Equation (7.2.17) presents an unstudied version of the Emden-Fowler equation (the detailed analysis of such equation will be carried out in Section 7.6).

The one-dimensional analogue of problem (7.2.16)

$$\ddot{v} + \frac{1}{(1-v)^2} = 0, \quad v(0) = v(\kappa L) = 0 \tag{7.2.18}$$

describes the equilibrium of ferromagnetic string-strip, that is, a stretched flexible strip whose width b is much smaller than length l. However b is much greater than the initial distance between the string and magnet h and displacement u. The string is included in the magnetic circuit according to Fig. 7.2. In eq. (7.2.18) a dot denotes differentiation with respect to τ, $v = v(\tau)$, $\tau = \kappa x$, where x is the coordinate along the string, $\kappa^2 = \mu_0 b I^2/2Th^3$ and T is the string tension.

Assuming a nonplanar surface of magnet near the membrane or string, instead of eqs. (7.2.15) and (7.2.18) we obtain the equations

$$\triangle v + \frac{1}{(f-v)^2} = 0, \quad \ddot{v} + \frac{1}{(f-v)^2} = 0, \tag{7.2.19}$$

where $f = h/h_0$ is a prescribed function of the coordinates and h_0 is a constant.

The equations of bending of thin ferromagnetic plate and beam are written by analogy

$$\triangle\triangle v - \frac{1}{(1-v)^2} = 0, \quad v^{IV} - \frac{1}{(1-v)^2} = 0. \tag{7.2.20}$$

Generally speaking, the load (7.2.13) can be applied to different elastic bodies which gives rise to many nonlinear boundary-value problems analogous to the above-mentioned.

In the cases when the body is subjected to external forces in addition to the electromagnetic forces the problems of the "interaction" of these two factors arises. Substituting $v_* = 1 - f + v$ in eq. (7.2.19) we obtain the equation

$$\triangle v_* + \frac{1}{(1-v_*)^2} = -\triangle f. \tag{7.2.21}$$

It follows from the latter equation and eq. (7.2.16) that applying external load is somehow equivalent to distortion of the magnet.

Let us now consider the cases when the ferromagnet can not be considered as ideal one. For two-dimensional and, especially, for one-dimensional bodies one can often use approximate methods of determining the field by methods of the theory of magnetic circuits etc. The magnetomotive forces are assumed to be given and this is equivalent to prescribing the scalar potential. That is why we consider the problem of magnetostatics to be reduced to determining φ and write down the approximate equation for distribution of the potential in thin bodies.

Let us first solve the following problem. Let us find the solution of Laplace equation for an elongate rectangle $0 \leq x \leq l, 0 \leq y \leq h_0, h_0 \ll l$ with the boundary conditions: $\varphi = I$ at $x = 0$; $\partial\varphi/\partial x = 0$ at $x = l$; $\partial\varphi/\partial y = 0$ at $y = h_0$ and $\varphi = a\partial\varphi/\partial y$ at $y = 0$. This problem corresponds to the problem of determining the field in a rather broad (in the direction perpendicular to the plane of the drawing in Fig. 7.3) layer of ferromagnet separated by a thin gap from the surface of ideal ferromagnetic body. We assume that $\varphi = 0$ on this surface. The problem can be viewed as a plane one in the regions far from the edges. We will clarify as to how to reduce it to a one-dimensional problem by taking advantage of the fact that the rectangle is elongated. According to the above-said, parameter $a = h\mu/\mu_0 \sim l^2/h_0$.

Using Fourier's method we find the solution for the rectangle

$$\varphi(x,y) = 2I \sum_{k}^{\infty} \frac{\sin \nu_k}{\nu_k} \frac{\cos((1 - y/h_0)\nu_k)}{1 + \dfrac{\cos^2 \nu_k}{a\nu_k^2/h_0}} \frac{\cosh\left((1 - x/l)\nu_k l/h_0\right)}{\cosh(\nu_k l/h_0)}, \quad (7.2.22)$$

where $\nu_k, k = 1,\ldots$ are the roots of the equations

$$\cos \nu = (a/h_0)\nu \sin \nu. \qquad (7.2.23)$$

If $a/h_0 \gg 1$ we have $\nu_1 \cong (k-1)\pi, k = 2,\ldots$. Therefore all terms of series (7.2.22), except for the first, are small. This yields the approximate equality

$$\varphi(x,0) = I\frac{\cosh(\nu(1 - x/l))}{\cosh \nu}, \quad \nu = \frac{l}{h_0}\sqrt{\frac{h_0}{a}}. \qquad (7.2.24)$$

Equation (7.2.24) can be obtained without exact solution. Let us consider an area between the straight lines parallel to axis y and passing through the points with abscissas x and $x + dx$. Let us assume that the induction is independent of y and write down the relationship expressing the balance of magnetic fluxes through the "lower" and "lateral" surfaces of the element

$$h_0 B(x) - h_0 B(x + dx) - B_0 dx = 0. \qquad (7.2.25)$$

Having expressed the induction in terms of the scalar potential we obtain the equation

$$\varphi'' - \frac{\mu_0}{\mu}\frac{1}{hh_0}\varphi = 0 \quad \left(\varphi' = \frac{d\varphi}{dx}\right). \qquad (7.2.26)$$

The solution of differential equation (7.2.26) under the boundary conditions $\varphi(0) = I, \varphi'(l) = 0$ is given by eq. (7.2.24). This justifies using the equations of such sort.

In the problems of magnetoelasticity for one-dimensional bodies it is necessary to replace the value h in eq. (7.2.26) by $h - u(x)$ where u is the deflection. As a result, we obtain the system of two nonlinear equations for $u(x)$ and $\varphi(x)$. For example, for the beam in magnet circuit (Fig. 7.2) which is attracted to an ideal ferromagnetic body these equations are given by

$$u^{IV} - \frac{1}{2}\frac{\mu_0 b}{EJ}\frac{\varphi^2}{(h - u)^2} = 0,$$

$$\varphi^{IV} - \frac{1}{2}\frac{\mu_0}{\mu h_0}\frac{\varphi}{(h - u)^2} = 0, \qquad (7.2.27)$$

where b is the beam width and EJ denotes the bending rigidity.

The equations describing distribution of the potential in two-dimensional case are derived by analogy. For example, let us consider a thin plate provided that field lines shown in Fig. 7.3 pass the gap between a body with potential φ_0 and a plate.

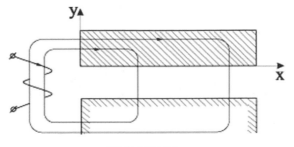

FIGURE 7.3.

We cut out an elementary cylinder with the base area S and write down the relationship of the balance of magnetic fluxes

$$h_0 \int_\Gamma B_n dl + \int_S H_0 \mu_0 dS = 0. \tag{7.2.28}$$

Here Γ denotes the contour of the cylinder base, B_n is the projection of the induction on the outward normal to the contour. Proceeding to limit and using eq. (7.2.5), we obtain the following equation

$$\triangle \varphi - \frac{\mu_0}{\mu h_0} \frac{\varphi - \varphi_0}{h} \frac{1}{u} = 0, \tag{7.2.29}$$

where $\triangle$ denotes Laplace operator in the plane.

7.3   Boundary-value problems for equilibrium of conductors with currents

Let us now consider the equilibrium of closely adjacent thin rods with currents and write down the expression for the load acting on the conductors by keeping only the lowest terms in h/l (which is adequate to the accuracy of the bending equations).

We first study the following auxiliary problem. We consider a linear conductor (that is, the conductor with the negligible dimensions of the cross-section) and a point P, the distance $|\mathbf{r}_0|$ between P and the conductor being small. We look for the field at this point caused by the current in the conductor under the assumption that $\mu = \mu_0$ in whole space. We start from the well-known formula

$$\mathbf{H}(P) = \frac{I}{4\pi} \int \frac{\boldsymbol{\tau} \times \mathbf{r}}{r^3} ds, \tag{7.3.1}$$

where $\boldsymbol{\tau}$ is the unit vector tangential to the contour and directed along the current, I is the current and $\mathbf{r}$ is the vector from a contour point to P.

Integration in eq. (7.3.1) is performed along a closed path of the current. We introduce s such that $\mathbf{r}(0) = \mathbf{r}_0, \boldsymbol{\tau}(0) = \boldsymbol{\tau}_0$ and $-l_1 \leq s \leq l - l_1, \mathbf{r}(-l_1) = \mathbf{r}(l - l_1)$. We transform eq. (7.3.1) to the form

$$\mathbf{H}(P) = \frac{I}{4\pi} \int_{l_1}^{l-l_1} \frac{\boldsymbol{\tau}_0 \times \mathbf{r}_0 + \ldots}{(h^2 + s^2 + \ldots)^{3/2}} ds = \frac{I}{2\pi h} \mathbf{b_0} + \Delta\mathbf{H}, \qquad (7.3.2)$$

where

$$\mathbf{b_0} = \frac{1}{h}(\boldsymbol{\tau}_0 \times \mathbf{r}_0).$$

The term $I\mathbf{b_0}/2\pi h$ in eq. (7.3.2) is obtained if the integrand contains only the written-out components. The addend $\Delta\mathbf{H}$ is such that $h|\Delta\mathbf{H}| \to 0$ if $h \to 0$ as it is clear from the following argumentation. Clarifying the behaviour of $h|\Delta\mathbf{H}|$ at $h \to 0$ it is sufficient to perform integration in eq. (7.3.2) over such interval in the vicinity of point $s = 0$ where $\mathbf{r}$ and $\boldsymbol{\tau}$ can be presented by power series in terms of s. Then $\Delta\mathbf{H}$ is presented by a sum of the integrals of the type of the principal term in eq. (7.3.2) however with the higher degrees of s in the numerator. All these integrals multiplied by h tend to zero as $h \to 0$. Strictly speaking, the value $|\Delta\mathbf{H}|$ incorporates the logarithm $\ln(l/h)$ which increases as $h \to 0$. This is the reason why the accuracy of the following equations is lower than that of the equations in Section 7.2.

Only the lowest term in $\mathbf{H}(P)$, which is $I\mathbf{b_0}/2\pi h$, is retained in what follows. That means that for computation of the lowest term in the expression of the field intensity near the linear conductor this conductor can be replaced by an infinite straight conductor of the vanishing cross-section.

Assuming small sizes of the cross-section of a thin conductor in comparison with the length and the characteristic radius of the curvature (nevertheless the sizes can distinctly vary only on the distances of order l) we find now the field near the conductor with the above accuracy. Let point P lie in cross-section σ_1 of the conductor, then we have

$$\mathbf{H}(P) = \frac{1}{2\pi} \int_{\sigma_1} \frac{\mathbf{j_1}(Q) \times \mathbf{r}(P, Q)}{r^2(P, Q)} d\sigma_1. \qquad (7.3.3)$$

Here Q is a point in σ_1, $\mathbf{r}(P, Q)$ is the vector from Q to P and $\mathbf{j_1}(Q)$ is the vector of the current density. The difference in the direction of $\mathbf{j_1}$ at various points in σ_1 is not considered and all these directions are assumed to be tangential to the conductor axis. It is also assumed that $r(P, Q)$ is of the order of the characteristic dimension of the cross-section. Then the conductor can be replaced by an infinite right cylinder with a generatrix parallel to the axis tangent and this yields eq. (7.3.3).

Let us now suppose that P is a point in cross-section σ_2 of another conductor. We denote current density as $\mathbf{j_2}(P)$. As the conductors are located near each other the difference in direction of their axis (and thus vectors $\mathbf{j}_1$ and $\mathbf{j}_2$) should be small. Therefore we can assume that σ_1 and σ_2 lie in the same plane perpendicular to $\mathbf{j}_1$ and $\mathbf{j}_2$. The density of the body forces generated by the field of first conductor in the second one is given by

$$
\mathbf{f}(P) = \frac{\mu_0}{2\pi} \int_{\sigma_1} \frac{\mathbf{j}_2(P) \times [\mathbf{j}_1(Q) \times \mathbf{r}(P,Q)]}{r^2(P,Q)} d\sigma_1 =
$$

$$
= \frac{\gamma\mu_0}{2\pi} \int_{\sigma_1} \frac{j_2(P)j_1(Q)\mathbf{r}(P,Q)}{r^2(P,Q)} d\sigma_1. \tag{7.3.4}
$$

Here $\gamma = -1$ if the vectors $\mathbf{j}_1$ and $\mathbf{j}_2$ are parallel, and $\gamma = 1$ if they are antiparallel.

Only load $\mathbf{q}$ obtained by integrating $\mathbf{f}$ over the cross-section σ_2

$$
\mathbf{q} = \frac{\gamma\mu_0}{2\pi} \int_{\sigma_2} \int_{\sigma_1} \frac{j_2(P)j_1(Q)\mathbf{r}(P,Q)}{r^2(P,Q)} d\sigma_1 d\sigma_2 \tag{7.3.5}
$$

is of importance for the technical theory of bending. The interaction of currents in the same conductor also causes deformations however its influence on bending can be neglected. The body force $\mathbf{f}$ generates, along with load $\mathbf{q}$, distributed bending moments and torques, however their sum in a distance of the order of l has the order of fh^3l while the order of the bending moment due to load $\mathbf{q}$ is fh^2l^2. For this reason, the distributed moments can be neglected. Finally, while determining $\mathbf{r}(P,Q)$ one should take into account only the displacements of the cross-sections rather than their rotations caused by bending and twisting. Indeed, in the linear theory of thin rods the value ϑl (ϑ is the characteristic rotational displacement) should be of the order of h or smaller. Therefore the displacements of points on the cross-sectional contour due to the rotation are of the order of h^2/l, that is, they are small as compared with the deflection.

In the problems of bending of rods with currents the load $\mathbf{q}$ should be introduced into the equilibrium equations of the rod. The current density is calculated for the undeformed state however $\mathbf{r}(P,Q)$ should be taken by accounting for the elastic displacements. Since the dependence of $\mathbf{r}$ on the displacement is prescribed then $\mathbf{q}$ is a prescribed function of $\mathbf{r}$. As a result we obtain a nonlinear boundary-value problem for the mechanical unknowns entering in the equations of statics for thin rods.

Expression (7.3.5) is essentially simplified for strings or threads. In these cases the cross-sectional sizes of the conductor are small compared with h, r and vector $\mathbf{r}$ can be considered as being independent of P and Q. Then we obtain

$$
\mathbf{q} = \frac{\gamma\mu_0 I_1 I_2}{2\pi} \frac{\mathbf{r}}{r^2}, \tag{7.3.6}
$$

where I_1 and I_2 are the currents in the interacting conductors.

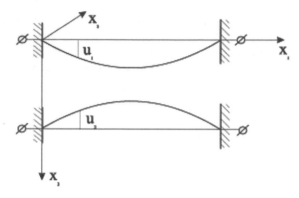

FIGURE 7.4.

Relationship (7.3.5) can also be used for determining the load acting on a current-carrying rod in the neighborhood of the surface of ferromagnet. In this case under σ_2 one should understand a figure lying in the plane of σ_1. This figure should be symmetric with respect to σ_1 about the straight line obtained by intersection of the plane σ_1, σ_2 and the tangential plane to the ferromagnet surface. The tangential plane should be passed through the point on the surface close to σ_1. There is an arbitrariness in choice of the tangency point as the orientation of the plane admits an error of order h/l. The current density $j_2(P)$ is also prescribed as being symmetric about $j_1(Q)$.

As an example of using relationship (7.3.6) we derive the equilibrium equations for two strings which are parallel in the undeformed state. Let us introduce axes x_1, x_2, x_3 , cf. Fig. 7.4, the unit vectors of the corresponding axes $\mathbf{i}_1, \mathbf{i}_2, \mathbf{i}_3$ and the displacements $\mathbf{u}_1 = v_1\mathbf{i}_2 + w_1\mathbf{i}_3$, $\mathbf{u}_2 = v_2\mathbf{i}_2 + w_2\mathbf{i}_3$ of the first and second strings, respectively (the ends of first string lie on axis x_1). The initial distance between the strings and the string tension are respectively denoted by h and T. Then we have

$$\frac{d^2\mathbf{u}_1}{dx_1^2} + \frac{\gamma I_1 I_2}{2\pi T} \frac{\mathbf{u}_1 - \mathbf{u}_2 - h\mathbf{i}_2}{|\mathbf{u}_1 - \mathbf{u}_2 - h\mathbf{i}_2|^2} = 0,$$

$$\frac{d^2\mathbf{u}_2}{dx_1^2} - \frac{\gamma I_1 I_2}{2\pi T} \frac{\mathbf{u}_1 - \mathbf{u}_2 - h\mathbf{i}_2}{|\mathbf{u}_1 - \mathbf{u}_2 - h\mathbf{i}_2|^2} = 0. \qquad (7.3.7)$$

Adding equations in (7.3.7) we obtain $\mathbf{u}_1'' + \mathbf{u}_2'' = 0$ and $\mathbf{u}_1'' = d^2\mathbf{u}_1/ds^2$. From the boundary conditions we have $\mathbf{u}_1 \equiv -\mathbf{u}_2$. Let us introduce the following denotations

$$\kappa^2 = \frac{\mu_0 I_1 I_2}{\pi T h^2}, \quad \tau = \kappa x_1, \quad 2v_1/h = v, \quad 1 - v = w, \quad 2w_1/h = z. \quad (7.3.8)$$

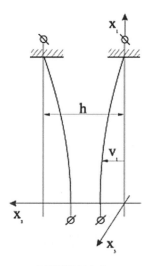

FIGURE 7.5.

Removing vector $\mathbf{u}_2$ from eq. (7.3.7), projecting the result onto the axes and introducing the nondimensional variables we arrive at the system

$$\ddot{w} + \gamma \frac{w}{w^2 + z^2} = 0, \quad \ddot{z} + \gamma \frac{z}{w^2 + z^2} = 0, \quad \left(\dot{w} = \frac{dw}{d\tau} \right) \qquad (7.3.9)$$

with boundary conditions $w(0) = w(\kappa l) = 1$; $z(0) = z(\kappa l) = 0$. Let us recall that $\gamma = -1$ for attracted strings and $\gamma = 1$ for repelled ones.

Equations (7.3.9) have the form of equations of motions for particle in a central force field since these force are inversely proportional to the distance. The motion takes place in the plane with the Cartesian coordinates w and z, whereas the center of attraction or repulsion is at the origin of the coordinate system. The orbit is projection of the string on this plane.

Let us construct the equilibrium equations for vertically suspended massive threads, Fig. 7.5. We will limit our consideration to analysis of the plane forms in plane x_1, x_3. As above $v_1 \equiv v_2$. Setting

$$\frac{2v_1}{h} = v, \quad 1 - v = w, \quad \rho = \kappa x_1, \quad \kappa = \frac{\mu_0 I_1 I_2}{\pi dgh^2}, \qquad (7.3.10)$$

where dg denotes the weight of the unit length of the thread we obtain the equation

$$\frac{d}{d\rho} \left(\rho \frac{dw}{d\rho} \right) + \gamma \frac{1}{\rho} = 0 \qquad (7.3.11)$$

with the boundary condition $w(\kappa l) = 1$ and the condition that $w(0)$ is limited. Similar to the above equation (7.2.17) for equilibrium of the ferromagnetic membrane equation (7.3.11) presents a variant of the Emden-Fowler equation.

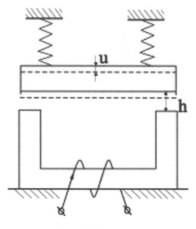

FIGURE 7.6.

7.4   Equilibrium of elastically suspended rigid bodies

Let us determine the equilibrium of the rigid ferromagnetic body which is elastically suspended and attracted by the electromagnet, Fig. 7.6. We assume that the body can move only translatory and both gaps between the body and the magnet poles are equal. Let $h - u$ denote the gap size where u and h denote respectively the displacement and the gap for undeformed springs. Force Q acting on the body is obtained by integrating expression (7.2.13) over surfaces of the poles. Both the electromagnet poles and the body surface are assumed to be flat. As a result we obtain the well-known Maxwell formula for the attraction force of the electromagnet

$$Q = \frac{\mu_0 I^2 S}{4 (h - u)^2},\qquad(7.4.1)$$

where $I = ni$ denotes the total current (or the magnetomotive force of the winding), n is the number of coils, i is the current and S is the pole area. The equilibrium equation is as follows

$$cu = \frac{\mu_0 I^2 S}{4 (h - u)^2},\qquad(7.4.2)$$

where c denotes the total stiffness of the springs.

Having introduced the nondimensional quantities

$$v = u/h, \quad \alpha^2 = \frac{\mu_0 I^2 S}{4ch^3}$$

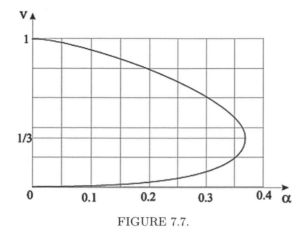

FIGURE 7.7.

we arrive at the cubic equation for v with a single parameter α

$$v = \frac{\alpha}{(1-v)^2}. \tag{7.4.3}$$

The solutions have a physical meaning only for $v < 1$. With the help of curve $v(\alpha)$, Fig. 7.7, referred to as the equilibrium curve, we find that for a particular value of parameter α the system can have either two equilibrium positions or one equilibrium at the limiting point or no equilibrium at all. It is worth noting that there is a linear series of equilibrium positions (upper branch) which tends, as $\alpha \to 0$, to the position of contact of the body and the magnet rather than to the undeformed state.

Following the results of Section 7.2 of the stability analysis we can consider the current in winding as an independent parameter and take advantage of Poincare's results of changing stability on the equilibrium curves.

System described by equation (7.4.3) has the force function

$$V = \Pi - W = \frac{1}{2}v^2 - \frac{\alpha^2}{1-v}, \tag{7.4.4}$$

where Π and W are nondimensional potential energy and the power of magnetic field respectively. The equilibrium is stable if $d^2V/dv^2 > 0$ for $v = v_r$ where v_r is the root of equation (7.4.3). Let us take solution $v = 0$ at $\alpha = 0$. For it $d^2V/dv^2 = 1$, thus this solution is stable (clearly as it corresponds to the undeformed state without any field). Hence the entire linear series of equilibria is stable from $\alpha = 0, v = 0$ to the limiting point. At the limiting point the upper branch is no longer stable.

Thus, in the considered system the displacement is only stable if it is less than or equal to one third of the initial gap h (Fig. 7.7). The permitted values of parameter α are also limited ($\alpha \le 2\sqrt{3}/9$).

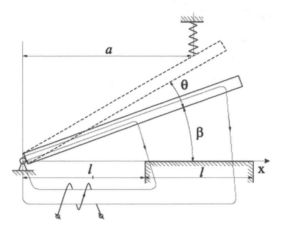

FIGURE 7.8.

When the parameter exceeds the limiting value the equilibrium is no more possible. This means that the elastically suspended body collides the magnet.

Equation (7.4.3) describes also equilibrium of the body rotating about the axis as shown in Fig. 7.8. Let ϑ and b denote the angle of rotation and the width of the body. The moment of the ponderomotive forces about the axis is equal to

$$M = \int_{l_1}^{l+l_1} xbqdx = \frac{1}{2}\mu_0 I^2 b \ln(1 + l/l_1)\frac{1}{(\beta - \vartheta)^2}. \qquad (7.4.5)$$

Putting

$$v = \vartheta/\beta, \quad \alpha^2 = \frac{1}{2}\mu_0 b \frac{I^2}{ca\beta^3} \ln(1 + l/l_1) \qquad (7.4.6)$$

we arrive at eq. (7.4.3).

Let us now consider the system whose centerline passes through the body midpoint, cf. Fig. 7.9. The moment of the ponderomotive forces is given by

$$M = \frac{1}{2}\mu_0 I^2 b \int_{-l/2}^{l/2} \frac{xdx}{(h - \vartheta x)^2} =$$

$$= \frac{1}{2}\mu_0 I^2 b \frac{1}{\vartheta} \left(\frac{hl}{h^2 - \frac{1}{4}l^2\vartheta^2} + \frac{1}{\vartheta^2} \ln \frac{h - 1/2\vartheta l}{h + 1/2\vartheta l} \right). \qquad (7.4.7)$$

Introducing the nondimensional quantities

$$\vartheta = \frac{1}{2}\vartheta l/h, \alpha^2 = \frac{1}{8}\mu_0 I^2 l^2 b/ch^3 \qquad (7.4.8)$$

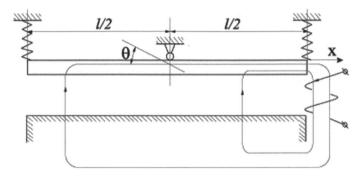

FIGURE 7.9.

yields the equilibrium equations

$$v = \alpha^2 p(v), \quad p(v) = \frac{1}{v}\left(\frac{1}{1-v^2} + \frac{1}{2v}\ln\frac{1-v}{1+v}\right). \tag{7.4.9}$$

For any α this equation has a solution $v = 0$ corresponding to the undeformed system. The remaining solutions are obtained from the equation

$$v^2\left(\frac{1}{1-v^2} + \frac{1}{2v}ln\frac{1-v}{1+v}\right)^{-1} - \alpha^2. \tag{7.4.10}$$

The corresponding equilibrium curve is displayed in Fig 7.10. In order to find a bifurcation point one needs to calculate the limit of the left hand side of eq. (7.4.10) at $v \to 0$ which yields $\alpha = \sqrt{3/2}$. The tangent to branch (7.4.10) is vertical at $v = 0$, cf. Fig. 7.10. Hence the bifurcation point is simultaneously a limiting point. Since the branch $v \equiv 0$ between point $\alpha = 0$ and the bifurcation point is stable we find that the remaining branches are unstable.

Similar to the previous example this system can be in equilibrium only for restrictive values of the parameters. However now it is caused by the loss of stability rather than vanishing the solutions (solution $v = 0$ exists for all α).

7.5 Equilibrium of string subjected to magnet force and prescribed external load

Let us consider a thin ferromagnetic string attracted by the electromagnet and loaded by uniformly distributed load q. We assume that the permeability of the string, magnet and magnetic circuit is infinitely large. The induction lines are assumed to be closed, see Fig. 7.2, and embraced by the same total current I. Let us introduce the following denotations: b and l are

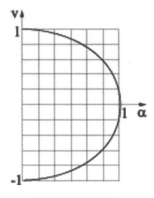

FIGURE 7.10.

respectively the width and length of the string, h is the distance between the undeformed string and the magnet and $u(x)$ denotes the displacement of the string cross-section. It is also assumed that $h \ll b \ll l$ and u is of the order of the string deflection Δ. We retain the lowest terms in u/b in the expression for ponderomotive force and arrive at the following boundary-value problem

$$\ddot{v} + \frac{1}{(1-v)^2} + \gamma = 0, \quad \dot{v} = \frac{dv}{d\tau}, \quad v(0) = v(\alpha) = 0,$$

$$v = \frac{u}{h}, \quad \tau = \frac{\alpha}{l}x, \quad \alpha^2 = \frac{\mu_0 b l^2 I^2}{2T\Delta^3}, \quad \gamma = \frac{2qh^2}{\mu_0 b I^2}. \qquad (7.5.1)$$

Here v and τ are the nondimensional displacement and coordinate respectively, T is the tension force and μ_0 is the medium permeability.

Let us reduce the problem to quadratures, determine the number of solutions and their dependence on the parameters and analyse their stability[2].

The same boundary-value problem is obtained if one looks for the equilibrium in the field of a curved cylinder in absence of load under the condition $h(x) = h_0 \pm x(l-x)h_1/l^2$ where h_1 is comparable with h_0 (in this case the magnet has the shape of a parabolic cylinder).

The solution of boundary-value problem (7.5.1) is essentially dependent on the sign of γ.

Let $\gamma > 0$, that is, the load is directed toward the magnet. The equation in (7.5.1) has no singular points and its first integral is given by

$$w^2 = 2(v_m - v)\left[(1-v)^{-1}(1-v_m)^{-1} + \gamma\right], \quad w = \dot{v}. \qquad (7.5.2)$$

Here v_m denotes the value of v at the point of intersection of the phase trajectory and axis v. The solution of problem (7.5.1) corresponds to the

---

[2]The solution is obtained together with I.Z. Shtirelman [53].

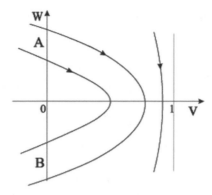

FIGURE 7.11.

piece AB of the phase trajectory which the phase point runs in the "time interval" $\tau - \alpha$, cf. Fig. 7.11. Since line AB is symmetric about $0v$, the string shape is symmetric about the axis passing through the string mid-point. The value v_m is equal to the maximum dimensionless displacement achieved in the middle of string.

Integrating eq. (7.5.2) with account for condition $v(0) = 0$ we obtain

$$\tau = \int_0^v \left\{ 2(v_m - z) \left[\frac{1}{(1 - v_m)(1 - z)} + \gamma \right] \right\}^{-1/2} dz,$$

$$0 \leq \tau \leq \alpha/2. \tag{7.5.3}$$

The shape of string is determined by relationship (7.5.3) and condition $v(\alpha - \tau) = v(\tau)$ up to a constant v_m, the latter being found from the condition $v_m = v(\alpha/2)$. This determines the dependence $v_m = v_m(\alpha, \gamma)$. Having replaced the integration variable in integral obtained from eq. (7.5.3) for $\tau = \alpha/2, v = v_m$, we reduce the integral to the form

$$\frac{\alpha}{\sqrt{2}} = \sqrt{v_m} \int_0^1 \left\{ (1 - z) \left[\frac{1}{(1 - v_m)(1 - v_m z)} + \gamma \right] \right\}^{-1/2} dz. \tag{7.5.4}$$

Functions $\tau(v)$ and $\alpha(v_m, \gamma)$ are expressed in terms of combinations of the elliptical integrals of the first and second kind and the elementary functions. However the inversion of these functions is difficult and for this reason we begin with the expressions with quadratures.

It is clear that $\alpha(0, \gamma) = \alpha(1, \gamma) = 0$ for all $\gamma > 0$. We have

$$\frac{\partial \alpha}{\partial v_m} = \frac{\sqrt{2}}{2} \int_0^1 \frac{dz}{\sqrt{1-z}} [g_1(z, v_m) - g_2(z, v_m) g_3(z, v_m)], \tag{7.5.5}$$

$$g_1 = \frac{1}{\sqrt{v_m}} \left[\frac{1}{(1-v_m)(1-v_m z)} + \gamma \right]^{-1/2}, \quad g_2 = \frac{1+z-2v_m z}{\sqrt{(1-v_m)(1-v_m z)}},$$

$$g_3 = \sqrt{v_m} [1 + \Gamma(1-v_m)(1-v_m z)]^{-3/2}.$$

Function g_1, g_2 and g_3 are non-negative, and for $0 \le z \le 1$ function g_1 is a decreasing function of v_m whereas g_2 and g_3 are monotonically increasing functions of v_m. Thus, the integrand in eq. (7.5.5) decreases monotonically as v_m increases. Therefore the derivative $\partial \alpha / \partial v_m$ has not more than one zero in the interval $0 < v_m < 1$. As $\partial \alpha / \partial v_m \to \infty$ as $v_m \to 0$ and $\partial \alpha / \partial v_m \to -\infty$ as $v_m \to 1$ there exists a value v_m for which $\partial \alpha / \partial v_m = 0$. Thus for any fixed value of γ function $\alpha(v_m)$ has only one maximum, cf. Figs. 7.12 and 7.13. Correspondingly for $\alpha < \alpha_l(\gamma)$ the string has two forms of equilibrium for any set of parameters, for $\alpha = \alpha_l$ it has only one form and for $\alpha > \alpha_l$ no equilibrium is possible.

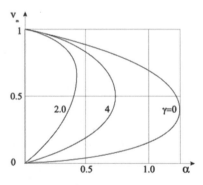

FIGURE 7.12.

Let $\gamma < 0$, i.e. the load is directed in opposition to the magnet. The eq. (7.5.1) has a singular point $v = 1 - (-\gamma)^{1/2}$, $w = 0$ in the half-plane $v, w, v < 1$. The first integral of this equation is

$$w^2 = 2\gamma (v_{m1} - v)(v_{m2} - v)(1-v)^{-1}, \tag{7.5.6}$$

where

$$(v_{m1} - v)(v_{m2} - v) = -1/\gamma \tag{7.5.7}$$

determines closed phase trajectories crossing axis v at $v = v_{m1}$ and v_{m2}. Hence the singular point is a center in the interval $0 < v < 1$ for $\gamma < -1$ and in the interval $-\infty < v < 0$ for $\gamma > -1$, see Figs. 7.14 and 7.15.

The solution of the boundary-value problem (7.5.1) describes two sets of parts of the phase trajectories which begin and end on axis w: segment AB and segment BA, cf. Figs. 7.14 and 7.15. Segment AB determines the single-extremum forms of equilibrium with positive displacements (as in the case when the load is directed to the magnet) while segment AB determines those with the negative displacements. These form are symmetric and the displacement with the maximum absolute value is achieved in the string middle and is equal to $v_{m1} > 0$ and $v_{m2} < 0$ for the "positive" and "negative" forms respectively. In addition to this, there can be the equilibrium forms corresponding to the passages like ABA and BAB, $ABAB$ etc. Among these segments one should choose those which the phase point runs in "time" $\tau = \alpha$.

The difference in geometry of position of the center in the cases $\gamma < -1$ and $\gamma > -1$ has a crucial influence on the character of solutions, that is, it is necessary to consider these cases separately.

Let $\gamma < -1$. Let us first consider the "positive" single-extremum forms of equilibrium. Relationships (7.5.3) and (7.5.4) are valid for them however the limits of possible change in v_m are now $v_p \leq v_m < 1$ where $v_p = 1+1/\gamma$. The latter condition expresses two requirements: firstly, the integrand in eqs. (7.5.3) and (7.5.4) must be real-valued for all $0 \leq z \leq 1$ and, secondly, the considered phase trajectory must embrace the trajectory passing through the origin of the phase plane since only such trajectories cross axis w and "satisfy" the boundary conditions.

Let us show that function $\alpha(v_m)$ decreases monotonically to zero if v_m increases from $v_m = v_p$ to $v_m = 1$. Let us put $v_m = v_p + \varepsilon$ and introduce a new variable $t = (1 + \varepsilon/v_p)^{-1} z$ in eq. (7.5.4). Then we obtain

$$\alpha(v_m) = \sqrt{2} \int\limits_0^{v_p} [(v_p - t) f_1(t, \varepsilon) + f_2(t, \varepsilon) f_3(t, \varepsilon)]^{-1/2} dt,$$

$$f_1 = t (1 - v_p - \varepsilon)^{-1} [1 - (1 + \varepsilon/v_p) t]^{-1},$$

$$f_2 = (1 - v_p)^{-1} (1 - v_p - \varepsilon)^{-1} \quad , f_3 = \varepsilon (1 + \varepsilon/v_p)^{-1}. \qquad (7.5.8)$$

Functions f_1, f_2 and f_3 are non-negative and increase monotonically with growth of ε. Hence the integrand in eq. (7.5.8) is a monotonically increasing function of ε and $\alpha(\varepsilon)$ decreases monotonically. It follows from eq. (7.5.5) that $\partial \alpha / \partial v_m \to -\infty$ as $v_m \to 1$ or $v_m \to v_p$, see Fig. 7.13. Thus the positive single-extremum forms exist in the region $0 \leq \alpha \leq \alpha_p(\gamma)$ and this form is unique for a given α.

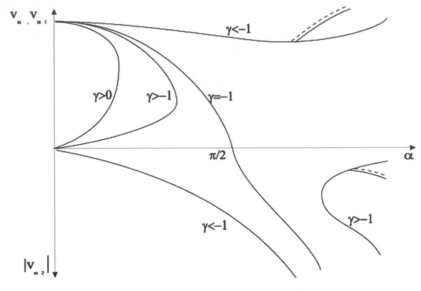

FIGURE 7.13.

Let us consider the negative single-extremum forms. The dependence $\alpha(|v_{m2}|)$ is given by

$$\frac{\alpha}{\sqrt{2}} = \int_0^1 \left[\frac{|v_{m2}|(1 + |v_{m2}|)(1 + v_{m2}z)}{(1 - z)[-1 - \gamma(1 + |v_{m2}|)(1 + |v_{m2}|z)]} \right]^{1/2} dz. \qquad (7.5.9)$$

The derivative of the integrand in eq. (7.5.9) with respect to $|v_{m2}|$ is positive for any $|v_{m2}|$. Hence, $\alpha(v_{m2})$ exists for any $v_{m2} < 0$ and increases monotonically with increase in $|v_{m2}|$. Therefore for a given value of α a negative single-extremum form exists and is unique for any $0 \leq \alpha < \infty$, besides, $\alpha(0) = 0$ and $\partial\alpha/\partial v_{m2} \to -\infty, v_{m2} \to 0$, see Fig. 7.13.

At $\alpha = \alpha_p(\gamma)$ one branch of the "positive" single-extremum form is split into three branches of multi-extremum forms. Two of these forms are asymmetric, correspond to the paths ABA and BAB of the phase point, cf. Fig. 7.14, and are symmetric about the plane passing through the string midpoint perpendicular to axis x. The third form is symmetric and corresponds to the transition $BABA$.

All these forms can be found in the following way. Let us consider a "positive" single-extremum form for some value $\alpha = \alpha_1$, then we obtain $v_{m1}(\alpha_1)$. Let us find an "adjoint" value v_{m2} from eq. (7.5.7). We determine α_2 from the relationship $v_{m2} = v_{m2}(\alpha_2)$ for "negative" single-extremum forms. Then value $\alpha = \alpha_1 + \alpha_2$ corresponds to double-extremum form ABA. The string possessing such a form is split into two parts: in the first part the form coincides with the single-extremum form at $\alpha = \alpha_1$, while in

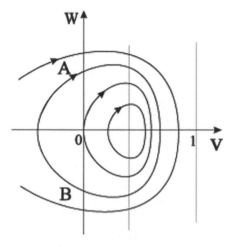

FIGURE 7.14.

the second one $\alpha = \alpha_2$. The forms $ABABA$ appearing at $\alpha = 2\alpha_p$ etc. are determined by analogy.

Let $\gamma > -1$. We consider the "positive" single-extremum forms. They are defined by relationships (7.5.3) and (7.5.4), and $0 \le v_m < 1$ similar to the case $\gamma > 0$. Let us show that similar to the case $\gamma > 0$ the dependence $\alpha = \alpha(v_m)$ has only one maximum. To this end, we show that the bifurcations of the extremum line $\alpha(v_m)$ are impossible under change of γ. As above, $\partial\alpha/\partial v_m \to \infty$ as $v_m \to 0$ and $\partial\alpha/\partial v_m \to -\infty$ as $v_m \to 1$, namely $\alpha(v_m)$ has at least one maximum. Let us assume that at some value $\gamma = \gamma_*$ one maximum is split into two maxima and one minimum. Let us assume that for $\gamma = \gamma_*$ the maximum of $\alpha(v_m)$ is achieved at point $v_m = v_{m*}$. Then for $\gamma = \gamma_*, v_m = v_{m*}$ we should have

$$\frac{\partial\alpha}{\partial v_m} = 0, \qquad \frac{\partial^2\alpha}{\partial v_m \partial\gamma} = 0. \qquad (7.5.10)$$

Let us now assume that at $\gamma = \gamma_*$ a pair "maximum - minimum" appears. The form of line $\alpha(v_m)$ at $\gamma = -1$ assumes that for certain $\gamma = \gamma_{**}$, $\gamma_* > \gamma_{**} > -1$ these extrema must either merge and disappear or merge with the above maximum. In the latter case the equalities in eq. (7.5.10) hold at $\gamma = \gamma_{**}$. In the first case the values $v_m(\gamma)$ corresponding to the maximum of $\alpha(v_m)$ satisfy the relationship $\partial v_m(\gamma)/\partial\gamma = \pm\infty$ at $\gamma = \gamma_*$ and $\gamma = \gamma_{**}$ respectively. Hence there exists such a value of γ in the interval (γ_*, γ_{**}) where $\partial v_m/\partial\gamma = 0$ which is possible only in the case when equalities (7.5.10) are met. The same equalities must hold for any other way of appearance of the new extrema.

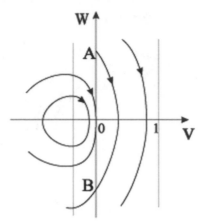

FIGURE 7.15.

Differentiating eq. (7.5.5) we obtain

$$\frac{\partial \alpha}{\partial v_m} = \frac{\sqrt{2}}{2} \int_0^1 \frac{dz}{\sqrt{v_m(1-z)}} \left(F^{-1/2} - GF^{-3/2} \right),$$

$$F = [(1-v_m)(1-v_m z)]^{-1} + \gamma,$$

$$G = v_m(1+z-2v_m z)[(1-v_m)(1-v_m z)]^{-2}, \qquad (7.5.11)$$

$$\frac{\partial^2 \alpha}{\partial v_m \partial \gamma} = \frac{\sqrt{2}}{4} \int_0^1 \frac{dz}{\sqrt{v_m(1-z)}} (-F^{-3/2} + 3GF^{-5/2}). \qquad (7.5.12)$$

Let us show that the equalities is eq. (7.5.10) are incompatible. To this end, we show that one can find such a number n that

$$(F^{-1/2} - GF^{-3/2}) + n(-F^{-3/2} + 3GF^{-5/2}) > 0 \qquad (7.5.13)$$

for all z. Since $F, G > 0$ it is sufficient that n satisfies the system of inequalities $F - n > 0$, $-F + 3n > 0$. Let us show that there is a number n satisfying to a more strong system of inequalities

$$\min_z F - n > 0, \quad -\max_z F + 3n > 0, \qquad (7.5.14)$$

$$\min_z F = (1-v_m)^{-1} + \gamma, \quad \max_z F = (1-v_m)^{-2} + \gamma. \qquad (7.5.15)$$

It is evident that such n exists provided that $\min F > 1/3 \max F$ or, what is the same, if

$$(2 - 3v_m)(1 - v_m)^{-2} > -2\gamma. \qquad (7.5.16)$$

Let us prove validity of the latter inequality. Relationship (7.5.9) can be brought to the form

$$\frac{\alpha}{\sqrt{2}} = \sqrt{v_m(1 - v_m)}\psi(v_m),$$

$$\psi(v_m) = \int\limits_0^1 \left(\frac{1 - v_m z}{1 - z}\right)^{1/2} [1 + \gamma(1 - v_m)(1 - v_m z)]^{-1/2} dz. \qquad (7.5.17)$$

Function ψ is positive and monotonically decreasing function of v_m. Function $[v_m(1 - v_m)]^{1/2}$ in the half-interval $1/2 < v_m \leq 1$ monotonically decreases, too. Thus $\alpha(v_m)$ can have an extremum only in $0 < v_m < 1/2$. For such v_m the following inequality

$$(2 - 3v_m)(1 - v_m)^{-2} > 2 \qquad (7.5.18)$$

holds true. On the other hand, it should be $-2\gamma < 2$ for $\gamma > -1$. Thus inequality (7.5.16) is valid, a required number n exists, equalities (7.5.10) are incompatible, the generation of new extrema for $0 \geq \gamma > -1$ is impossible and the line $\alpha(v_m)$ has only one maximum, cf. Fig. 7.13.

For $\gamma > -1$ there are also "negative" single-extremum forms of equilibrium. In this case, dependence $\alpha(|v_{m2}|)$ is defined as before by relationship (7.5.9), in which $|v_{m2}| \geq v_p = |1 + 1/\gamma|$. Let us find the derivative $\partial\alpha/\partial|v_{m2}|$. It has a form analogous to that given by eq. (7.5.11). At $|v_{m2}| = v_p$ the function analogous to g_2 in eq. (7.5.5) becomes equal to $-z^{-3/2}/v_p$ and the corresponding integral diverges, namely $\partial\alpha/\partial|v_{m2}| \to -\infty$ as $|v_{m2}| \to v_p$. On the other hand, as $|v_{m2}| \to \infty$ function analogous to $g_1(z, v_m)$ decreases as $|v_{m2}|^{-1/2}$ and the analogue of the product g_2g_3 decreases as $|v_{m2}|^{-5/2}$. Hence, for a certain value of $|v_{m2}|$ the derivative $\partial\alpha/\partial|v_{m2}|$ is positive. Thus, the branch of the "negative" single-extremum forms possesses a limiting point, cf. Fig. 7.13.

Also multi-external forms exist in the considered case. The simple ones originate from that branch of the "negative" single-extremum forms which is split into three branches at point (α_p, v_p), $\alpha_p = \alpha(v_p)$. One of them corresponds to the three-extremum symmetric forms $ABAB$ whereas two others correspond to three-extremum asymmetric forms ABA and BAB, see Fig. 7.13. These branches can be found by the above method of "addition".

There remains to consider the case $\gamma = -1$. In this case the load generated by the electromagnet under the undeformed string is equal in magnitude and oppositely directed to the direction of the given external load. Therefore for any α there is the solution $v \equiv 0$ corresponding to the undeformed state.

Let us find another solutions. Putting $\gamma = -1$ in eq. (7.5.4) and cancelling $\sqrt{v_m}$ we obtain

$$\frac{\alpha}{\sqrt{2}} = \sqrt{1 - v_m} \int\limits_0^1 \sqrt{\frac{1 - v_m z}{1 - z}} \frac{dz}{\sqrt{1 + (1 - v_m z)}}. \tag{7.5.19}$$

For $v_m > 0$ relationship (7.5.19) describes the branch of curve $v_m(\alpha)$ which, at $\gamma \to -1+0$, becomes the branch of the single-extremum "positive" forms $\gamma > -1$ and located between the limiting point and point $v_m = 1$. Another ("lower") branch of these forms containing point $\alpha = 0$, $v_m = 0$ becomes an interval of axis α between point $\alpha = 0$ and the point of intersection of line (7.5.19) with axis α. Assuming $v_m = 0$ in eq. (7.5.19) we obtain $\alpha = \pi/\sqrt{2}$. The limiting point of the branch of "positive" single-extremum arrives at point $v_m = 0$, $\alpha = \pi/\sqrt{2}$ if $\gamma \to -1+0$ such that the tangent to line $v_m(\alpha)$ at this point is vertical, see Fig. 7.15. This can be shown with the help of eq. (7.5.19).

The limiting point of the branch of "negative" single-extremum forms arrives at the same point $v_m = 0$, $\alpha = \pi/\sqrt{2}$ as $\gamma \to -1+0$. As $\gamma \to -1+0$ the part of this branch between the limiting point and the bifurcation point passes into the interval $(\pi/\sqrt{2}, \pi\sqrt{2})$ of axis α whereas its infinite part passes into the infinite branch of negative single-extremum forms described by expression (7.5.19) provided that v_m is replaced by $-|v_{m2}|$.

Additionally, there are more multi-external symmetric and asymmetric forms. They branch off from the undeformed state at the values α which are multiple to $\pi/\sqrt{2}$ and form an infinite number of branches.

Let us now proceed to the stability analysis of equilibrium configurations of the string.

Equation (7.5.1) can be derived from the following variational principle

$$\delta V = 0, \quad V = \int\limits_0^\alpha \left(\frac{1}{2}\dot{v}^2 - \frac{1}{1 - v} - \gamma v\right) d\tau, \tag{7.5.20}$$

where the allowed functions $v(\tau)$ are such that $v(0) = 0$, $v(\alpha) = 0$, $\dot{v} \in L_2(0, \alpha)$ and $v(\tau) < 1$. As above, we judge the stability by the sign of the second variation of functional V for the solution of problem (7.5.1) and use Poincare's results about changing stability under change of parameters.

Let us first analyse stability of the undeformed state at $\gamma = -1$. The second variation V for the solution $v \equiv 0$ is given by

$$\delta^2 V = \int\limits_0^\alpha \left(\frac{1}{2}\dot{\zeta}^2 - \zeta^2\right) d\tau, \tag{7.5.21}$$

where $\zeta(\tau)$ satisfies the conditions $\zeta(0) = \zeta(\alpha) = 0, \dot{\zeta} \in L_2(0,\alpha)$. The inequality [39]

$$\int_0^\alpha \dot{\zeta}^2 d\tau \geq \frac{\pi^2}{\alpha^2} \int_0^\alpha \zeta^2 d\tau \qquad (7.5.22)$$

holds for these values of $\zeta(\tau)$. It is also known that if a greater coefficient is taken in front of the integral in the right hand side of eq. (7.5.22), then one can find such $\zeta(\tau)$ from the indicated class that the opposite inequality will be fulfilled. Therefore $\delta^2 V > 0$ for $\alpha < \pi/\sqrt{2}$, that is, the undeformed state up to the first bifurcation point is stable. At the bifurcation point $\alpha = \pi/\sqrt{2}$ the undeformed state becomes unstable and the branch of "negative" single-extremum forms becomes stable. Consider the corresponding eigenvalue problem one can see that at the following bifurcation point $\alpha = \pi\sqrt{2}$ the undeformed state is again unstable as well at the departing branches similar to the problem of axial bending.

Now we consider the "positive" single-extremum forms for $\gamma > -1$. We shall take some form $v(\tau, \alpha, \gamma)$ on the branch with smaller values of v_m. At $\alpha = $ const and $\gamma \to -1$ this form continuously goes over into a stable undeformed state. As this transformation does not affect the stability the original form is stable. By means of the curve $v_m = v_m(\alpha)$ at $\gamma = $ const we conclude that at the limiting point the stability is lost and the branch with greater values of v_m is unstable.

The same reasoning leads to the result that the branch of "negative" single-extremum forms outgoing from the limit point to infinity is stable for $0 > \gamma > -1$. Also the "negative" single-extremum forms are stable for $\gamma < -1$. The remaining forms are unstable.

It is worth noting that for $0 > \gamma > -1$ there are two non-crossing stable branches, the equilibrium forms of the negative branch having a point of inflection due to a change of increase and decrease in $w = \dot{v}$ with growth of τ. For comparison we mention that in some problems of Euler elastica the forms with inflections are always unstable.

In the problem of equilibrium of the string without external load but in the field of a curved magnet the displacement is determined according to eq. (7.2.19) as a sum of the above function $v(\tau)$ and a quadratic polynomial of τ. The results is only "positive" forms without inflection points, similar to the case where the load retains its sign for any displacement.

7.6 The Emden-Fowler equation. Reduction to autonomous system

As indicated in Section 7.2, the problem of equilibrium of a round ferromagnetic membrane and massive treads with currents are reduced to determining the bounded solution of the Emden-Fowler equation, the exponent of power of the unknown function in the nonlinear term being equal to minus two and minus one. For these problems we find the dependence of the solutions on the parameter and the number of solutions.

To begin with, we first generalise the problem. The reason is that the analysis of the Emden-Fowler equation and eq. (7.2.17) or (7.3.11) is equally difficult. Besides, the theory of Emden-Fowler equation presents a certain interest regardless of applications to the problems of magnetoelasticity, thus it makes sense to study the case of a negative degree n of the unknown function[3].

Let us consider the general form of the Emden-Fowler equation

$$\frac{d}{d\rho}\left(\rho^k \frac{dw}{d\rho}\right) \pm \rho^m w^n = 0 \qquad (7.6.1)$$

with three parameters $k, m, n \, (n < 0)$. We seek only the solution $w > 0$. If n is a rational quotient with an odd denominator (i.e. when a real-valued solution $w < 0$ exists) the case $w < 0$ is reduced to the previous one by replacing $w_1 = -w$. For the same reason we can take $\rho > 0$.

Equation (7.6.1) is integrated in quadratures only for certain combinations of k and m. However it can be reduced to a system without explicit argument and this simplifies the study. We shall introduce[4] new variables: the argument τ and the unknown function η by the relationships $\rho = a \exp(-\beta\tau), w = b\eta \exp(-\delta\tau)$. Then we arrive at the equation

$$\eta'' - (2\delta - \beta + k\beta)\eta' + \delta(\delta - \beta + k\beta)\eta \pm$$
$$\pm a^{2+m-k}b^{n-1}\beta^2\eta^n \exp\{[\delta(1-n) - \beta(2+m-k)]\tau\} = 0,$$
$$(\eta' = d\eta/d\tau). \qquad (7.6.2)$$

Let us assume

$$a^{2+m-k}b^{n-1}\beta^2 = 1, \quad \delta = 1, \quad (2+m-k)\beta = 1-n \qquad (7.6.3)$$

(the case $2 + m - k = 0$ will be considered later). Then τ does not appear explicitly in eq. (7.6.2) and instead of eq. (7.6.2) we can write the following system of equations

$$\eta' = \vartheta, \quad \vartheta' = (c+1)\vartheta - c\eta + \gamma\eta^n,$$
$$\gamma = \pm 1, \quad c = 1 - \beta + k\beta. \qquad (7.6.4)$$

---

[3] The Emden-Fowler equation was first obtained for the problem of the temperature distribution in a gas sphere [30]. Such problems lead only to the case $n > 0$ and only this case is studied for the general equation. For $n > 0$ the theory of the Emden-Fowler equation is set out e.g. in books by R. Bellmann [12] and J. Sansone [103].

[4] R. Ackerberg studied a particular case of eq. (7.6.1) by using another substitution [1]. The same case was studied by the above method in [47, 48].

Here $\gamma = -1$ provided that in eq. (7.6.1) a plus sign is taken in front of the nonlinear term and $\gamma = 1$ for a minus sign.

In what follows we use the relationships obtained from eq. (7.6.4)

$$(c-1)\eta(\tau) = (c\eta_0 - \vartheta_0)e^\tau - (\eta_0 - \vartheta_0)e^{c\tau} + \gamma \int_0^\tau \left[e^{c(\tau-\xi)} - e^{\tau-\xi} \right] \eta^n(\xi)d\xi,$$

$$(c-1)\vartheta(\tau) = (c\eta_0 - \vartheta_0)e^\tau - c(\eta_0 - \vartheta_0)e^{c\tau} + \gamma \int_0^\tau \left[ce^{c(\tau-\xi)} - e^{\tau-\xi} \right] \eta^n(\xi)d\xi$$

$$\eta_0 = \eta(0), \quad \vartheta_0 = \vartheta(0). \tag{7.6.5}$$

We will consider some combinations of signs of c and γ, determine the form of the integral curve of system (7.6.4) and estimate the rate of increase in $\eta(\tau), \vartheta(\tau)$ as $\tau \to \infty$.

1. Let $c > 0$ and $\gamma = 1$. Then system (7.6.4) has a singular point $\eta = \eta_c, \vartheta = 0, \eta_c^{n-1} = c$. Let us put $\zeta = \eta - \eta_c$ and linearise eq. (7.6.3) in the neighbourhood of the singular point. We obtain

$$\zeta' = \vartheta, \quad \vartheta' = (c+1)\vartheta - c(1-n)\zeta. \tag{7.6.6}$$

The character of the singular point is determined by the coefficients of system (7.6.6). Let us consider a subcase when

$$(c-1)^2 + 4nc < 0 \tag{7.6.7}$$

and the singular point is an unstable focal point. Let a straight line L have the equation $(c+1)\vartheta = c\eta + \gamma\eta^n$ and show the derivative

$$d\vartheta/d\eta = \vartheta^{-1}\left[(c+1) - c\eta + \gamma\eta^n\right] \tag{7.6.8}$$

of the integral curve $\vartheta = \vartheta(\eta)$ in the half-plane $0\eta\vartheta$ by arrows in Fig. 7.16. The orientation of the arrows corresponds to increase in τ. We also find the second derivative

$$\frac{d^2\vartheta}{d\eta^2} = -\frac{1}{\vartheta^3}\left[\left(c - \gamma n\eta^{n-1}\right)\vartheta^2 - (c+1)(c\eta - \gamma\eta^n)\vartheta + (c\eta - \gamma\eta^n)^2\right]. \tag{7.6.9}$$

The trinomial in ϑ in the square brackets in eq. (7.6.8) is positive if $\eta < \eta_g$, $\eta_g^{n-1} = -1/4(c-1)^2/n$. Hence axis 0ϑ can not be an asymptote for the integral curve (however it is possible for other values of c and γ).

It follows from eq. (7.6.7) that $\eta_g^{n-1} < \frac{1}{4}4nc/n = c$ and $\eta_g > \eta_c$. Let us write down the expression for the roots of the trinomial for $\gamma = 1$

$$\vartheta_{1,2} = \frac{\vartheta_L(\eta)}{2(c - n\eta^{n1})}\left[(c+1)^2 \pm (c+1)\sqrt{(c+1)^2 - 4(c - n\eta^{n-1})}\right]. \tag{7.6.10}$$

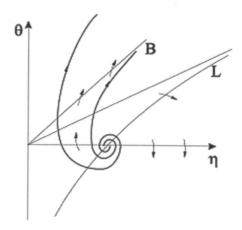

FIGURE 7.16.

Here $\vartheta_L(\eta)$ denotes the value of ϑ in L. Since the roots $\vartheta_{1,2}$ exist only for $\eta \geq \eta_g > \eta_c$ we have $\vartheta_L > 0$. In addition to this, $\vartheta_{1,2} > 0$. For $c - n\eta^{n-1} \leq 1/4(c+1)^2$ we have

$$(c+1)^2 - (c+1)\sqrt{(c+1)^2 - 4(c - n\eta^{n-1})} - 2(c - n\eta^{n-1}) > 0, \quad (7.6.11)$$

which is easy to show. Therefore the smaller root is $\vartheta_2 > \vartheta_l$. Thus the trinomial is positive in L, below L and for $\vartheta < 0$.

The straight lines B and C having respectively the equations $\vartheta = \eta$ and $\vartheta = c\eta$ are crossed by the integral curve from the zone below the straight line into the zone above it. Indeed, $d\vartheta/d\eta = 1 + \eta^n/\vartheta > 1$ in B and $d\vartheta/d\eta = c + \eta^n/\vartheta > c$ in B.

Taking into account the signs of $d\vartheta/d\eta$ and $d^2\vartheta/d\eta^2$ we conclude that the integral curve crossing L for $\vartheta > 0$ will then cross axis 0η. For $n \neq -1$ all curves in the region $\vartheta < 0$ arrive at zone $\vartheta > 0$ in the interval $0, \eta_c$. For $n > -1$ we have curves having either the origin or the end in axis 0ϑ.

Let $c > 1$. We will show that in this case there is an integral curve which asymptotically approaches B from below. On the lines crossing B or 0η for $\eta > \eta_c$, η and ϑ increase monotonically with growth in τ. We take η_0, ϑ_0 on the "monotonic" part of this curve and make use of relationship (7.6.5). Because of the monotony we have $\eta(\tau) > \eta_0$ in $\tau > 0$, that is

$$(c - 1)\vartheta(\tau) < (c\eta_0 - \vartheta_0)e^\tau - c(\eta_0 - \vartheta_0)e^{c\tau} +$$

$$+ \int_0^\tau [ce^{c(\tau-\xi)} - e^{\tau-\xi}]\eta_0^n d\xi = (c\eta_0 - \vartheta_0 - \eta_0^n)e^\tau + (c\vartheta_0 - c\eta_0 + \eta_0^n)e^{c\tau}.$$

$$(7.6.12)$$

If $c\vartheta_0 - c\eta_0 + \eta_0^n < 0$, then $\vartheta(\tau)$ decreases for some values of τ. But it is impossible as we consider a "monotonic" part of the separating curve. Since any point in this part can be taken as an initial one we have $0 < \eta - \vartheta < \eta^n/c$ and $\eta - \vartheta \to 0$ for $\eta \to \infty$ everywhere in this part.

In what follows we show that there exists only one separating curve. Let us assume the contrary and consider the branches of two separating curves which are monotonically tending in infinity. Let one curve have the equation $\vartheta = \vartheta_1(\eta)$ and the second curve be given by $\vartheta = \vartheta_2(\eta)$, where $\vartheta_1(\eta_0) > \vartheta_2(\eta_0), \eta_0 > \eta_c$. Let us denote $\Delta\vartheta = \vartheta_1 - \vartheta_2$ then we have

$$\frac{d(\Delta\vartheta)}{d\eta} = (c\eta - \eta^n)\frac{\Delta\vartheta}{\vartheta_2(\vartheta_2 + \Delta\vartheta)}. \qquad (7.6.13)$$

Since the curves $\vartheta_1(\eta)$ and $\vartheta_2(\eta)$ have the same asymptote then $\Delta\vartheta$ must tend to zero as $\eta \to \infty$. At the same time it should be $\Delta\vartheta > 0$. However the positive solution of eq. (7.6.13) for $\eta > \eta_c$ increases with growth of η. Hence two curves $\vartheta_1(\eta)$ and $\vartheta_2(\eta)$ can not exist.

Thus, in the case of $n \neq -1$ and increasing τ all integral curves, except for the singular point $\eta = \eta_c$ and the separating curve, make an infinite number of turns about the focus, then cross B and go to infinity. The separating curve leaves the focus and asymptotically approaches B from below, Fig. 7.16. In the case of $n > -1$ there are more curves coming from the focus to $O\vartheta$ and tending to infinity.

The most interesting question in theory of the Emden-Fowler equation is that about the behaviour of solutions for $\rho \to 0$ and $\rho \to \infty$ which is important, in particular, for investigation of the boundary-value problems. Dealing with system (7.6.4) it is necessary to clear up the features of solutions for $\tau \to \pm\infty$. In the case of $\tau \to -\infty$ they are evident. In the case $\tau \to \infty$ the integral curves go to infinity, the nonlinear term in eq. (7.6.4) decreases and the behaviour of solutions is mostly defined by the linear part. Let us take any curve going to infinity and different from the singular curve $\eta \equiv \eta_c$ and the separating one. Multiplying both parts of eq. (7.6.5) by $e^{c\tau}$ and proceeding to limit we obtain that there are finite limits

$$\lim_{\tau \to \infty} \eta(\tau)e^{-c\tau} = (c-1)^{-1}\left[\vartheta_0 - \eta_0 + \int_0^\infty e^{-c\xi}\eta^n(\xi)d\xi\right] > 0,$$

$$\lim_{\tau \to \infty} \vartheta(\tau)e^{-c\tau} = c\lim_{\tau \to \infty} \eta(\tau)e^{-c\tau}. \qquad (7.6.14)$$

Let us consider the separating line and make use of the identity

$$\eta(\tau) \equiv \eta_0 e^\tau + \int_0^\tau e^{\tau-\xi}[\vartheta(\xi) - \eta(\xi))]d\xi. \qquad (7.6.15)$$

Since $\eta(\tau) - \vartheta(\tau) \to 0$ as $\tau \to \infty$ then the integral

$$\int_0^\infty e^{-\xi}[\vartheta(\xi) - \eta(\xi)]d\xi \qquad (7.6.16)$$

converges and, thus, there are finite limits

$$\lim_{\tau\to\infty} \eta(\tau)e^\tau = \eta_0 - \int_0^\infty e^{-\xi}[\eta(\xi) - \vartheta(\xi)]d\xi,$$

$$\lim_{\tau\to\infty} \vartheta(\tau)e^\tau = \lim_{\tau\to\infty} \eta(\tau)e^{-\tau}. \tag{7.6.17}$$

To study the boundary-value problems it is necessary to know as to the indicated limits change under change in the initial point on the integral curve. Let us consider the limits of functions $f(\tau)\eta_1(\tau)$ and $f(\tau)\eta_2(\tau)$, where $\eta_1(\tau)$ and $\eta_2(\tau)$ correspond to the same integral curve but have different values at $\tau = 0$. Let $\eta_1(0) = \eta_{10}, \vartheta_1(0) = \vartheta_{10}, \eta_2(0) = \eta_{20}$ and $\vartheta_2(0) = \vartheta_{20}$. Then $\eta_2 = \eta_1(\tau + \tau_{12})$ where τ_{12} is such that $\eta_1(\tau_{12}) = \eta_{20}, \vartheta_1(\tau_{12}) = \vartheta_{20}$. Here $\tau_{12} > 0$ provided that the transition from point $(\eta_{10}, \vartheta_{10})$ to point $(\eta_{20}, \vartheta_{20})$ takes place in the direction of increasing τ and $\tau_{12} < 0$ in the case of a decrease. We have

$$\lim_{\tau\to\infty} f(\tau)\eta_1(\tau) = \lim_{\tau\to\infty} \frac{f(\tau + \tau_{12})}{f(\tau)} \lim_{\tau\to\infty} f(\tau)\eta_2(\tau) \tag{7.6.18}$$

under the condition that these limits exist. Therefore for all curves but the separating one

$$\lim_{\tau\to\infty} \eta_1(\tau)e^{-c\tau} = e^{-c\tau_{12}} \lim_{\tau\to\infty} \eta_2(\tau)e^{-c\tau}. \tag{7.6.19}$$

On the separating curve

$$\lim_{\tau\to\infty} \eta_1(\tau)e^{-\tau} = e^{-\tau_{12}} \lim_{\tau\to\infty} \eta_2(\tau)e^{-\tau}. \tag{7.6.20}$$

The case $0 < c < 1$ is similar to the previous one, i.e. the separating curve asymptotically approaches the straight line C. On this curve the limit

$$\lim_{\tau\to\infty} \eta(\tau)e^{-c\tau} = \frac{1}{c} \lim_{\tau\to\infty} \vartheta(\tau)e^{-c\tau} \tag{7.6.21}$$

is finite whereas on all remaining integral curve the following limit

$$\lim_{\tau\to\infty} \eta(\tau)e^{-\tau} = \lim_{\tau\to\infty} \vartheta(\tau)e^{-\tau} \tag{7.6.22}$$

is finite. Finally, for $c = 1$ on separating curve the limit

$$\lim_{\tau\to\infty} \eta(\tau)e^{-\tau} = \lim_{\tau\to\infty} \vartheta(\tau)e^{-\tau} \tag{7.6.23}$$

is finite whereas on all remaining integral curve the following limit

$$\lim_{\tau\to\infty} \eta(\tau)e^{-\tau}\tau^{-1} = \lim_{\tau\to\infty} \vartheta(\tau)e^{-\tau}\tau^{-1} \tag{7.6.24}$$

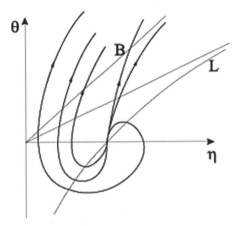

FIGURE 7.17.

is finite. This case considered in [47, 48, 1] covers the equilibrium problem of the ferromagnetic membrane and the attraction of threads with currents, as well as the problem of equilibrium of the electrostatically charged drops, cf. [1, 116].

In the second subcase, when $(c-1)^2 + 4nc > 0$, the singular point of system (7.6.4) is an unstable node, Fig. 7.17. Let us estimate the slope of the integral curves on the half-line $\vartheta = z_1(\eta - \eta_c), \eta > \eta_c$ where

$$z_1 = \frac{1}{2}(c+1) + \sqrt{\frac{1}{4}(c-1)^2 + nc}. \qquad (7.6.25)$$

The curve η^n for $\eta > 0, n < 0$ is convex downward and, thus, lies on one side from any of its tangents. For the tangent passing through point $\eta = \eta_c$ we obtain

$$\eta^n > \eta_c^n + n\eta_c^{(n-1)}(\eta - \eta_c). \qquad (7.6.26)$$

Applying inequality (7.6.26) to the expression for $d\vartheta/d\eta$ on the considered half-line we obtain

$$\frac{d\vartheta}{d\eta} = c + 1 - \frac{c\eta}{z_1(\eta - \eta_c)} + \frac{\eta^n}{z_1(\eta - \eta_c)} >$$

$$> c + 1 - \frac{c\eta}{z_1(\eta - \eta_c)} + \frac{\eta_c^n + n\eta_c^{(n-1)}(\eta - \eta_c)}{z_1(\eta - \eta_c)} = z_1.$$

Hence the integral curve lying left from the straight line $\vartheta = z_1(\eta - \eta_c)$ for $\vartheta > 0$ cannot move right and, thus, can not cross axis 0η. For this reason the integral curve can have not more than two crossing axes (apart from the nodal point). There exist curves which do not cross axis 0η at all. In particular, the separating curve is such a curve. Indeed, $\min(c, 1) < z_1$, that is, for $c > 1$ the straight line crosses B, and for $c < 1$ it crosses C. Thus,

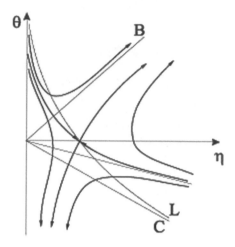

FIGURE 7.18.

if the separating curve could traverse 0η, then it should remain above the straight line $\vartheta = z_1(\eta - \eta_c)$ and could not have straight line B or C as an asymptote. For $n \leq -1$ the integral curves are displayed in Fig. 7.17. The rate of increasing the solutions for $\tau \to \infty$ is coincident with that for the case of $(c-1)^2 + 4nc < 0$.

Finally, the case $(c-1)^2 + 4nc = 0$ differs from the case $(c-1)^2 + 4nc > 0$ only in that all integral curves are tangential to the same straight line $\vartheta = z_1(\eta - \eta_c)$.

2. Let $c < 0$ and $\gamma = -1$. The singular point of system (7.6.4) is $\eta = \eta_c, \vartheta = 0, \eta_c^{n-1} = -c$. Linearising eq. (7.6.4) by analogy with derivation of eq. (7.6.6) we obtain that for $c < 0$ the singular point is a saddle.

Let $-1 < c < 0$. Observing directions of the integral curve (shown by arrows in Fig. 7.18) we conclude that for $n \leq -1$ there are curves asymptotically approaching axis 0η. An unbounded value of ϑ is achieved on these curves within a finite "time". Indeed, let us assume that $\eta \to 0, \vartheta \to \infty$ as $\tau \to \infty$. Then

$$\eta(\tau) = \eta_0 + \int_0^\tau \vartheta(\xi)d\xi, \quad \lim_{\tau \to \infty} \eta(\tau) = 0, \quad \int_0^\infty \vartheta(\xi)d\xi = -\eta_0, \qquad (7.6.27)$$

which is impossible since the latter integral diverges. Analogously if $\eta \to 0$ and $\vartheta \to \infty$ as τ decreases, then τ has a lower bound.

As above, we obtain from eq. (7.6.5) that on the parts of the curve where $\eta, \vartheta \to \infty$ at $\tau \to \infty$, the limits $\lim \eta(\tau)e^{-\tau}$ and $\lim \vartheta(\tau)e^{-\tau}$ are bounded, and on the parts where $\eta \to \infty, \vartheta \to -\infty$ at $\tau \to -\infty$, the limits $\lim \eta(\tau)e^{-c\tau}$ and $\lim \vartheta(\tau)e^{-c\tau}$ are bounded. Provided that $\eta(\tau) \to \infty$ for

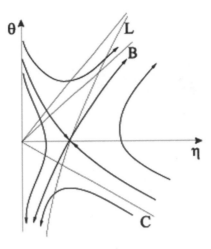

FIGURE 7.19.

$\tau \to \pm\infty$, we have

$$\lim_{\tau \to \infty} \int_0^\tau e^{c(\tau-\xi)}\eta^n(\xi)d\xi = 0, \qquad \lim_{\tau \to -\infty} \int_0^\tau e^{\tau-\xi}\eta^n(\xi)d\xi = 0. \qquad (7.6.28)$$

For this reason, on the above mentioned parts $\vartheta - \eta \to 0$ and $\vartheta - c\eta \to 0$ respectively, cf. Fig 7.18.

If $c < -1$, then L and the integral curves are transformed according to Fig. 7.19. For $c = -1$ curve L degenerates in a straight line $\eta = \eta_c$, whereas system (7.6.4) and the original Emden-Fowler equation are integrated in quadratures. The solution behaviour remains the same.

3. Provided that $c > 0, \gamma = -1$ system (7.6.4) has no singular points, L does not cross axis 0η and reaches minimum for $\eta = \eta_m, \eta_m^{n-1} = -c/n$. It is clear that there exists integral curves which either do not cross L or cross it twice, Fig. 7.20, axis 0η being a "two-sided" asymptote for these curves for $n \leq -1$. Let us show that there are curves crossing L only once. To this end, we find $d\vartheta/d\eta$ at $\vartheta = 1/2(c+1)\eta$ and obtain that $d\vartheta/d\eta > 1/2(c+1)$ at $\eta > \eta_l, \eta_l^{n-1} = 1/4(c-1)^2$, that is, in the region $\eta > \eta_l, \vartheta > 1/2(c+1)\eta$ the integral curve can not leave it and must go to infinity.

Let $c > 1$. From eq. (7.6.5) we obtain

$$\vartheta < (c-1)^{-1}[(\vartheta_0 - \eta_0)e^{c\tau} + (c\eta_0 - \vartheta)e^\tau]. \qquad (7.6.29)$$

This means that a curve below B crosses L. Therefore the integral curves can move to infinity ($\eta, \vartheta \to \infty$) only lying above B. Among them there is only one curve which asymptotically approaches B (separating curve). Let us prove it. Among the integral curves there exists a "first" curve which does not cross B. We take η_0, ϑ_0 on that part of the curve where $c\eta_0 > \vartheta_0$

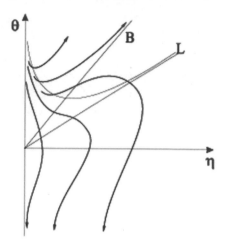

FIGURE 7.20.

and replace $\eta(\xi)$ by η_0 in the integrand in the second relationship in eq. (7.6.5) . Taking into account that $\eta(\xi)$ increases with growth of τ we obtain

$$\vartheta(\tau) > f(\tau, \eta_0, \vartheta_0) = \frac{(c\eta_0 - \vartheta_0 + \eta_0^n)\,e^\tau + (c\vartheta_0 - c\eta_0 - \eta_0^n)e^{c\tau}}{c - 1}. \qquad (7.6.30)$$

Let us assume that there is point at which $\vartheta_0 - \eta_0 > \eta_0^n/c$. Then there exists such $\varepsilon > 0$ that $\vartheta_0 - \varepsilon - \eta_0 > \eta_0^n/c$. Let us take a curve passing through point $(\eta_0, \vartheta_0 - \varepsilon)$. We understand $\eta_0, \vartheta_0 - \varepsilon$ as the initial point and consider a part of the latter curve up to the crossing with 0η. On this part $\eta(\tau)$ increases and thus $\vartheta(\tau, \eta_0\vartheta_0 - \varepsilon) > f(\tau, \eta_0, \vartheta_0 - \varepsilon)$, while $\vartheta(0, \eta_0, \vartheta_0 - \varepsilon) = f(0, \eta_0, \vartheta_0 - \varepsilon) = \vartheta_0 - \varepsilon$. The latter inequality is however impossible. Indeed, $f(\tau, \eta_0, \vartheta_0 - \varepsilon)$ is an increasing function of τ whereas for certain values of τ function $\vartheta(\tau, \eta_0, \vartheta_0 - \varepsilon)$ must decrease because the curve with the initial point $\eta_0, \vartheta_0 - \varepsilon$ "begins" below the "first" of the curves which do not cross B and L. Thus, on the "first" of the curves which do not cross B the inequality $0 < \vartheta - \eta < \eta^n/c$ is valid and this means that this curve has B as an asymptote.

Having written an equation, analogous to eq. (7.6.13), we can convince ourselves that there is only one separating curve.

Similar to the case $c > 0, \gamma = 1$ the limit $\lim \eta(\tau)e^{-\tau} > 0$ for $\tau \to \infty$ is bounded on the separating curve. For $c < 1$ the separating curve asymptotically approaches C etc. which is also analogous to the case $c > 0, \gamma = 1$.

The case $c > 0, \gamma = -1$ considered above is also appropriate for the problem of repulsion of threads with currents.

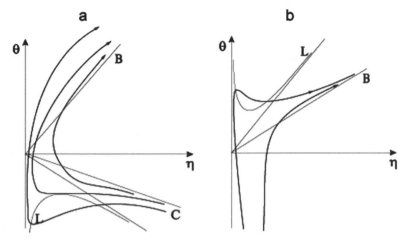

FIGURE 7.21.

4 When $c < 0, \gamma = 1$ we arrive at Fig. 7.21a (for $c > -1$) or Fig. 7.21b (for $c < -1$). The limits

$$\lim_{\tau \to \infty} \eta(\tau)e^{-\tau} = \lim_{\tau \to \infty} \vartheta(\tau)e^{-\tau}, \quad \lim_{\tau \to -\infty} \eta(\tau)e^{-c\tau} = \frac{1}{c}\lim_{\tau \to -\infty} \vartheta(\tau)e^{-c\tau}$$

$$(7.6.31)$$

are always bounded. This result is also valid for the case $c = -1$ when system (7.6.4.) and original Emden-Fowler equation are integrated in quadratures.

5. If $c = 0$ we have Fig. 7.22a $(\gamma = 1)$ and Fig. 7.22b $(\gamma = -1)$. The features of the solutions on the curves asymptotically approaching 0ϑ and B are coincident with those in other analogous cases. Let us consider the curves asymptotically approaching axis 0η. For $\gamma = -1$ such curve is unique. Instead of relationships (7.6.5) we have

$$\vartheta = \vartheta_0 e^{\tau} - \int_0^{\tau} e^{\tau - \xi}\eta^n(\xi)d\xi,$$

$$\eta = \eta_0 - \vartheta_0 + \vartheta_0 e^{\tau} - \int_0^{\tau}(e^{\tau-\xi} - 1)\eta^n(\xi)d\xi. \qquad (7.6.32)$$

It follows from eq. (7.6.32) and condition $\vartheta \to 0$ that for $\tau \to \infty$ $\eta(\tau)$ increases as an integral of $\eta^n(\tau)$, namely $\eta \cong \tau^{\nu}, \nu = (1 - n)^{-1}$. For the case shown in Fig. 7.22b, the same behaviour is typical for $\eta(\tau)$ at $\tau \to -\infty$.

6. For $2 + m - k = 0$ one should put $\delta = 0$ in eq. (7.6.2). Taking additionally that $k\beta - \beta = 1$ we arrive at system (7.6.4) in the form of the previous case $(c = 0)$ however with another relation between η and w.

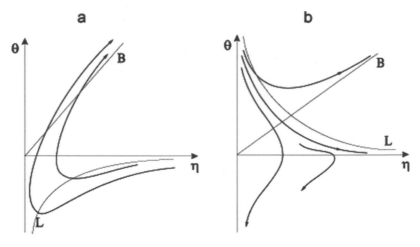

FIGURE 7.22.

7.7 Equilibrium of ferromagnetic membrane and threads with currents

If we know the behaviour of $\eta(\tau)$ for $-\infty < \tau < \infty$ we can establish the features of $w(\rho)$ for $0 < \rho < \infty$, specifically, we can find the property of function $w(\rho)$ as $\rho \to 0$ and $\rho \to \infty$. It allows one to determine the number and some features of solutions of the boundary-value problems subjected to conditions $w(\alpha) = 1$ and $w(0)$ or $w(\infty)$ is limited. The above problems of magnetoelasticity and equilibrium of the electrostatically charged drops are reduced to the first problem [1, 116].

Let $\beta > 0$. Then $\rho \to 0$ corresponds to $\tau \to \infty$, and $\rho \to \infty$ corresponds to $\tau \to -\infty$. Let us put in eqs. (7.6.2) and (7.6.3)

$$a = \alpha, \quad b = (\beta^{-2}\alpha^{k-m-2})^{1/(n-1)}. \tag{7.7.1}$$

Then $\rho = \alpha$ corresponds to $\tau = 0$ and the boundary condition $w(\alpha) = 1$ yields $\eta(0) = 1/b$. Therefore the relation between ρ, w and τ, η takes the form $\rho = ae^{\beta\tau}, w = \eta_0^{-1}\eta e^{-\tau}$. We will consider the problem with the condition: $w(0)$ is limited. It requires existence of a bounded limit $\lim \eta(\tau)e^{-\tau}$ as $\tau \to \infty$.

The most interesting result is obtained in the case of $c \geq 1, \gamma = 1$. In this case the product $\eta(\tau)e^{-\tau}$ is bounded if functions $\eta(\tau), \vartheta(\tau)$ provide one with a parametric representation for the separating curve. The latter condition and the condition $\eta_0 = 1/b$ determine the sought-for solutions of the boundary-value problem. For a prescribed value of α the number of solutions is coincident with the number of crossing the straight line $\eta = \eta_0 = b^{-1}(\alpha)$ by the separating curve.

Let $(c-1)^2 + 4nc < 0$. Let $\eta_{*1}, \eta_{*2}, \ldots$ denote the values of η at the points of intersection of the separating curve and axis 0η. Let their numbering correspond to moving along the separating curve as τ decreases. For $\eta_0 < \eta_{*1}$ the boundary-value problem has no solutions, for $\eta_0 > \eta_{*2}$ there is only one solution, and for $\eta_{*1} < \eta_0 < \eta_{*3}$ there are two solutions etc. There exist the intervals of η_0 (or α) where there are m solutions, m being any positive integer. For $\eta_0 = \eta_c$ the problem has an infinite number of solutions (more exactly a countable set).

Since

$$\frac{dw}{d\rho} = b\beta^{-1}\alpha^{-1}e^{(\beta-1)\tau}(\eta - \vartheta) \tag{7.7.2}$$

$w(\rho)$ decreases monotonically from a certain w_m to a unity as ρ increases from zero to α, and $w_m = \eta_0^{-1}\lim\limits_{\tau\to\infty}\eta(\tau)e^{-\tau}$.

Let us consider the dependence $w_m(\alpha)$. For $\alpha \to 0$ we have $b^{-1} = \eta_0 \to \infty$ and

$$\lim_{\eta_0\to\infty} w_m = \lim_{\eta_0\to\infty}[\eta_0^{-1}\lim_{\tau\to\infty}\eta(\tau,\eta_0)e^{-\tau}] =$$

$$= 1 - \lim_{\eta_0\to\infty}\frac{1}{\eta_0}\int_0^{\overset{\text{\tiny$w$}}{}}e^{-\xi}[\eta(\xi) - \vartheta(\xi)]d\xi =$$

$$- 1 - \lim_{\tau_{12}\to\infty}\frac{1}{\eta_0}\int_0^\infty e^{-\xi}[\eta(\xi+\tau_{12},\eta_{U*}) - \vartheta(\xi+\tau_{12},\eta_{U*})]d\xi =$$

$$= 1 - \lim_{\tau_{12}\to\infty}\frac{e^{\tau_{12}}}{\eta_0}\int_{\tau_{12}}^\infty e^{-\xi}[\eta(\xi,\eta_{0*}) - \vartheta(\xi,\eta_{0*})]d\xi = 1, \tag{7.7.3}$$

see eq. (7.6.17). Here $\eta(\xi,\eta_0)$ has a variable η_0 and is expressed in terms of some function $\eta(\xi,\eta_{0*})$ with a fixed initial value η_{0*} by the relationship

$$\eta(\xi,\eta_0) = \eta(\xi+\tau_{12},\eta_{0*}). \tag{7.7.4}$$

Let us find the derivative

$$\frac{dw_m}{d\alpha} = \frac{d\eta_0}{d\alpha}\frac{d}{d\eta_0}\left\{-\frac{1}{\eta_0}\int_0^\infty e^{-\xi}[\eta(\xi,\eta_0) - \vartheta(\xi,\eta_0)]d\xi\right\} =$$

$$= \frac{d\eta_0}{d\alpha}\left\{\frac{1}{\eta_0}(1 - w_m) - \frac{1}{\eta_0}\int_0^\infty\frac{1}{\vartheta_0}\left[\vartheta(\xi,\eta_0) - \frac{d}{d\xi}\vartheta(\xi,\eta_0)\right]e^{-\xi}d\xi\right\} =$$

$$= -\frac{1}{\alpha\beta}\frac{\eta_0 - \vartheta_0}{\vartheta_0}w_m. \tag{7.7.5}$$

Here we used the following relationships

$$\frac{d\eta(\xi,\eta_0)}{d\eta_0} = \frac{1}{\vartheta_0}\frac{d\eta(\xi,\eta_0)}{d\xi} = \frac{1}{\vartheta_0}\vartheta(\xi,\eta_0) \tag{7.7.6}$$

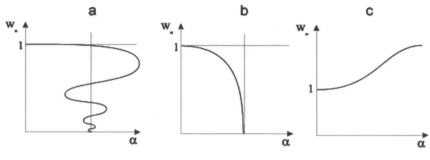

FIGURE 7.23.

and the term in eq. (7.7.5) with $d\vartheta/d\xi$ was integrated by parts.

Since $\eta > \vartheta$ on the separating curve, the signs of ϑ_0 and derivative $dw_m/d\alpha$ are opposite, see eq. (7.7.5). Let us consider a set of such solutions that point (η_0, ϑ_0) lies on the part of separating curve going from point $(\eta_{*1}, 0)$ to infinity. For these solutions $\vartheta_0 > 0$ and $dw_m/d\alpha < 0$, i.e. w_m decreases with increase of α in the interval $0 < \alpha < \alpha_{*1})$, where α_{*1} is such that $b^{-1}(\alpha_{*1}) = \eta_{*1}$. For $\alpha_{*2} < \alpha < \alpha_{*1}$ there are such solutions that point (η_0, ϑ_0) lies on the scroll of the separating curve linking η_{*1} and η_{*2}. It is typical for these solutions that $\vartheta_0 < 0$ and w_m increases with increasing α. Considering the solutions on the next scrolls of the separating curve we obtain a curve $w_m(\alpha)$ of a rather rare form, cf. Fig. 7.23a. This curve winds about the straight line $\alpha = \alpha_c = (\beta^2 \eta_c^{1-n})^{1/(k-m-2)}$, that is, there is an infinite number of solutions for $\alpha = \alpha_c$.

The equilibrium equation for the ferromagnetic membrane (7.2.17) is reduced to the form (7.6.1) after multiplication by ρ. As a result we obtain $k = 1, m = 1, n = -2, \beta = 3/2, b = (9/4\alpha^2)^{1/3}, c = 1$, that corresponds to the case $c > 0, \gamma = 1, (c-1)^2 + 4nc < 0$. Hence, the dependence of the nondimensional deflection in the centre of membrane v_m from parameter α can be obtained from Fig. 7.23a by the replacement $v_m = 1 - w_m$. Taking into account that $\eta_c = 1, \alpha_c = 2/3$, we arrive at the equilibrium curve $v_m(\alpha)$ depicted in Fig. 7.24.

It was established above that $0 < \eta - \vartheta < \eta^n/c$ on the monotonic branch of the separating curve. In this case we obtain $0 < \eta - \vartheta < 1/\eta^2$. This yields $0 < (\eta - \vartheta)e^{2\tau} < (\eta e^{-\tau})^{-2}$ and, thus, the quantity $(\eta - \vartheta)e^{2\tau}$ is limited at $\tau \to \infty$. From eq. (7.7.2) we find that $e^{(\beta-1)\tau}(\eta - \vartheta) = e^{\tau/2}(\eta - \vartheta) \to 0$ at $\tau \to \infty$ and $dw/d\rho \to 0$ if $\rho \to 0$, i.e. for any equilibrium form the membrane has in the centre a tangential plane which is parallel to the contour plane. An exception is the form of equilibrium corresponding to $\eta \equiv \eta_c$ and its equation is $v(\rho) = 1 - (3\rho/2)^{2/3}$. Such membrane looks like a crater with a cusp in the centre. It is natural that this solution (and close functions on the equilibrium curve) should be considered only

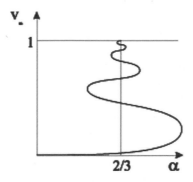

FIGURE 7.24.

as formal ones, because such a great value of $dv/d\rho$ is in conflict with the assumptions made for derivation of the original equations.

Figure 7.24 yields only a qualitative form of the equilibrium curve. In order to obtain the numerical values it is necessary to develop other methods, for example, a numerical integration. Some numerical results were obtained by R. Ackerberg [1], in particular, he found out that the first limiting point is achieved at $\alpha \simeq 0.9, v_m \simeq 0.35$, that is, only relatively small deflections are possible.

A certain segment of the equilibrium curve, which is close to the undeformed state, contains stable forms. However one can not assert that the entire first branch is stable up to the limiting point since this branch can give rise to nonsymmetric forms for values of α smaller that the limiting one. A final assessment about the stability is made in the next Section with the help of more general results.

In the problem of attraction of the threads with currents (cf. eq. (7.3.11)) we have $k = 1, m = 0, n = -1, \beta = 2, b = 2\sqrt{\alpha}, c = 1$. This corresponds to the case of the membrane. Replacing $\rho = 1/4s^2$ we can go over from eq. (7.3.11) to the equation in which m and k are equal to the same values in the problem of membrane. The equilibrium curve has the same form (Fig. 7.24) however now $\alpha_c = 1/4$. In the case of attraction the nonplanar forms are not possible which is proved in the following way. We write out the equilibrium equations, analogous to the equations of attracted strings

$$\frac{d}{d\rho}\left(\rho\frac{dw}{d\rho}\right) - \frac{w}{w^2 + z^2} = 0, \quad \frac{d}{d\rho}\left(\rho\frac{dz}{d\rho}\right) - \frac{z}{w^2 + z^2} = 0,$$
$$w(\alpha) = 1, \quad z(\alpha) = 0. \tag{7.7.7}$$

Let us put $p(\rho) = (w^2 + z^2)^{-1}$. It is evident that for any non-negative $p(\rho)$ the boundary-value problem

$$\frac{d}{d\rho}\left(\rho\frac{dz}{d\rho}\right) - p(\rho)z = 0, \quad z(\alpha) = 0, |z(0)| < \infty \tag{7.7.8}$$

has solution only at $z \equiv 0$.

The case $c \geq 1, \gamma = 1, (c-1)^2 + 4nc > 0$ can be considered by analogy. As a result we arrive at the curve $w_m(\alpha)$ shown in Fig. 7.23b. Interestingly enough, the curve $v_m = 1 - w_m$ approaches point $v_m = 1$ having no singularities for smaller values of α. It is however unclear whether there exists such a physical system corresponding to this curve. A more general question is also of interest: does a system of ideally ferromagnetic bodies exist such that all values of the maximum displacement up to a contact of the closely adjacent surfaces are possible without loss of stability.

In the case $c \geq 1, \gamma = -1$ the curve $w_m(\alpha)$ is displayed in Fig. 7.23c. In this case the solution of the boundary-value problem exists and is unique for any α. The latter is relevant, for instance, for the problem of the planar form of equilibrium of the repelled threads provided that the investigation is limited to determination of the forms without threads' contact. The branch of these forms does not have bifurcation solutions and, thus, does not cross the branches of the nonplanar forms. The same statement can also be proved for repelled strings. Additionally, there exist formal solutions describing planar forms with a cusp similar to string's forms with a cusp. Provided that "usual" forms describe motions with the initial value $\tau = 0$, $\eta = \eta_0$ on the separating curve, the forms with cusps correspond first to motion along the curve where $\vartheta \to -\infty, \eta \to 0$ (Fig. 7.20), then they jump on the separating curve and move along it from the value $\eta = 0, \vartheta = \infty$ to the value $\eta, \vartheta \to \infty$. In this case the values $\eta = 0, w = 0, v = 1$ are achieved, i.e. the threads touch each other.

The curves in Fig. 7.23 belong to the case $c \geq 1$. For $c < 1$ regardless of sign γ, equation (7.6.1) admits two-parameter family of solutions $\eta(\tau)$ such that $\lim \eta(\tau)e^{-\tau}$ is finite for $\tau \to \infty$. Therefore the boundary-value problem with the conditions $w(\alpha) = 1, |w(0)| < \infty$ has a single-parameter family of solutions. By analogy, the boundary-value problem with the condition "w is limited for $\rho \to \infty$" has either no solutions or a family of solutions. For $\beta < 0$ the above conclusion about the problem with the boundedness condition at zero is valid for the problem with boundedness condition at infinity and vice versa.

7.8   On the stable equilibrium forms of conductors with currents

The boundary-value problem of the forms of equilibrium of conductors with currents was formulated in Section 7.3. The equilibrium equations of the initially parallel conductors was reduced to the following boundary-value problem

$$\ddot{u} - \gamma \frac{\lambda^2 u}{u^2 + w^2} = 0, \quad \ddot{w} - \gamma \frac{\lambda^2 w}{u^2 + w^2} = 0 \quad \left(\dot{u} = \frac{du}{ds} \right) \tag{7.8.1}$$

with the denotations of Section 7.3

$$s = \frac{x}{l}, \quad \lambda^2 = \kappa^2 l^2, \quad u = w, \quad w = z, \quad \left(\dot{u} = \frac{du}{ds} \right).$$

Equations (7.8.1) coincide with the equations of a particle in the field of a central force whose value is inverse proportional to the distance. The motion takes place in the plane where u, w are the Cartesian coordinates and the center of attraction or repulsion is placed in origin of the coordinate system. The orbit is projection of the string on this plane. We look for the trajectory which begins at $s = 0$ at the point $u = 1, w = 0$ and returns to this point at $s = l$.

When the strings are attracted then $\gamma = 1$ which corresponds to motion of the particle under the repelling force. In this case there are no closed or self-intersecting trajectories. Therefore the body can return to the initial position only if it moves along a straight line passing through the centre and its velocity was initially directed toward the centre. Only such motions can correspond to the solution of the considered boundary-value problem. Because the initial point $u = 1, w = 0$ lies on axis Ou passing through the centre we obtain $w = 0$. Instead of system (7.8.1) we have a single equation

$$\ddot{u} - \frac{\lambda^2}{u} = 0, \quad u(0) = u(1) = 1. \tag{7.8.2}$$

Next we introduce the variable $v = 1 - u$ which is the string deflection from the undeformed state. Then

$$\ddot{v} + \frac{\lambda^2}{1 - v} = 0 \quad v(0) = v(1) = 0. \tag{7.8.3}$$

This equation admits the first integral $\frac{1}{2}\dot{v} + \lambda^2 \ln(1 - v) = h$, that is, the integral curves in the plane $v, \dot{v}$ are symmetric about axis v. Therefore the string form should be symmetric about axis passing through its midpoint and the maximum displacement v_m is reached at the midpoint. For this reason the integration constant in the first integral is equal to $\lambda^2 \ln(1 - v_m^2)$. A further integration with account for the condition $v(0) = 0$ yields

$$s = \frac{1}{\sqrt{2\lambda^2}} \int\limits_0^v \frac{dv}{\sqrt{\ln(1 - v) - \ln(1 - v_m)}} = \sqrt{\frac{2}{\lambda^2}(1 - v)} \int\limits_{\varphi(v)}^{\varphi(0)} e^{z^2} dz, \tag{7.8.4}$$

where

$$\varphi_1(v) = \sqrt{\ln(1 - v) - \ln(1 - v_m)}, \quad \varphi_1(0) = \sqrt{-\ln(1 - v_m)}.$$

For $1/2 \le s \le 1$ the dependence $v(s)$ is defined by the equality $v(s) = v(1 - s)$. The form of the string is now found up to a constant v_m. In order

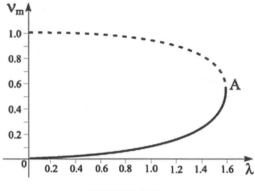

FIGURE 7.25.

to determine it one needs to utilise the second boundary condition or the equivalent relationship $v(1/2) = v_m$. Supposing $s = 1/2, v = v_m$, in eq. (7.8.4) we obtain

$$1 = F(v_m) = \frac{2\sqrt{2}}{\lambda}(1 - v_m) \int_0^{\sqrt{-\ln(1-v_m)}} e^{z^2} dz. \qquad (7.8.5)$$

This determines the dependence of the deflection v_m on the only parameter λ. By virtue of curve $v_m = v_m(\lambda)$, Fig. 7.25, referred to as the equilibrium curve, we find that for a certain value of this parameter the string can have either two equilibrium forms or only one (the corresponding point A on the curve is termed a limiting point) or have no equilibrium at all.

The maximum value of λ^2 for which there exists a solution of eq. (7.8.5) is equal to 2.34189 and the corresponding value of $v_m = 0.5745$ (point A in Fig. 7.25). As λ increases from zero up to λ_m the values $v_{m\,\min}$ and $v_{m\,\max}$ begin to merge. At $\lambda = \lambda_m$ we have $v_{m\,\min} = v_{m\,\max}$. Let us analyse the stability of the obtained branches of equilibrium. We write down the homogeneous problem in variations

$$\xi'' + \lambda^2 a(s, \lambda)\xi = 0, \quad \xi(0) = \xi(1),$$

where

$$a(s, \lambda) = \frac{1}{(1 - v_0)^2}. \qquad (7.8.6)$$

This linear equation has either only one solution or a family of solutions, the latter being dependent on arbitrary constants when λ is equal to one of the eigenvalues μ of the problem

$$\varphi'' + \mu^2 a(s, \lambda)\varphi = 0, \quad \varphi(0) = \varphi(1) = 0. \qquad (7.8.7)$$

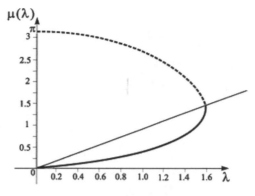

FIGURE 7.26.

We reduce this problem to the Fredholm integral equation

$$\varphi = \mu^2 \int_0^1 K(s, \eta) a(\eta, \lambda) \varphi(\eta) d\eta, \qquad (7.8.8)$$

with the kernel

$$K(s, \eta) = \begin{cases} \eta(1 - s), & \eta < s, \\ (1 - \eta)s, & s < \eta. \end{cases}$$

Next we prescribe λ and determine $v_0(s, \lambda)$ from eq. (7.8.4), and the solution of equation in variations and dependence $\mu(\lambda)$ (Fig. 7.26) is obtained from eq. (7.8.8). The value $\lambda = \mu$ is the limiting point. The lower branch is stable while the upper one is unstable.

In order to analyse the stability of the found forms of attraction if they leave the plane we vary the second equation in (7.8.1) by setting $\eta = u - u_0, w = 0$

$$\eta'' - \frac{\lambda^2}{u_0^2} \eta = 0, \quad \eta(0) = \eta(1) = 0. \qquad (7.8.9)$$

Clearly, this equation has only trivial solution for any λ, i.e. the planar forms do not lose stability.

The case $\gamma = -1$ corresponds to repulsion of the strings or attraction to the centre. Let us find the planar forms of equilibrium. They correspond to motion along a straight line passing through the centre.

For the initial velocity directed from the centre, the body first moves from the centre, then to the centre and finally returns to the initial position (the body can not go to infinity by virtue of the inequality $h - \ln u \geq 0$, where h is a constant energy). This motion (from the origin to the return) can be described by the solution of the original problem for which $w = 0$. Integrating the obtained equation for u

$$\ddot{v} - \frac{\lambda^2}{1 - v} = 0, \quad v(0) = v(1) = 0$$

we find

$$s = \sqrt{\frac{2}{\lambda^2}(1 - v_m)} \int\limits_{\varphi_2(v)}^{\varphi_2(0)} e^{-z^2} dz, \quad 0 \le s \le \frac{1}{2}, \tag{7.8.10}$$

where

$$\varphi_2(v) = \sqrt{\ln(1 - v_m) - \ln(1 - v)}, \quad \varphi_2(0) = \sqrt{\ln(1 - v_m)}.$$

The dependence $v_m(\lambda)$ is shown in Fig. 7.25. The obtained series of forms exists for all λ and $\lambda(v_m)$ is one-to-one dependence. The examples of forms $v_0(\lambda, s)$ are shown in Fig. 7.26.

These series of forms serves as a continuation of the lower branch (7.8.5) with respect to parameter $\gamma\lambda$ however it is not sufficient for judging its stability. For treatment of stability of the repulsion forms we write down a homogeneous problem in variations which is obtained by varying eq. (7.8.1) for $u_0 = 1 - v_0, w_0 = 0$

$$\xi'' - \frac{\lambda^2}{u_0^2}\xi = 0, \quad \xi(0) = \xi(1) = 0,$$

$$\eta'' + \frac{\lambda^2}{u_0^2}\eta = 0, \quad \eta(0) = \eta(1) = 0. \tag{7.8.11}$$

For any λ the first equation has only trivial solution, that is, the planar form of repulsion cannot lose stability without escaping from the plane. The second equation has either a trivial solution or a family of solutions. The family of solutions appears if λ is equal to one of the eigenvalues μ of the problem

$$\varphi'' + \frac{\mu^2}{u_0^2}\varphi = 0, \quad \varphi(0) = \varphi(1) = 0. \tag{7.8.12}$$

Let us show that it is impossible. We have two equations

$$\varphi'' + \frac{\mu^2}{u_0^2}\varphi = 0, \quad u_0'' + \frac{\lambda^2}{u_0} = 0. \tag{7.8.13}$$

We multiply the first one by u_0 and the second by φ, then we add them and integrate from 0 to 1. Integration by parts yields

$$\varphi'(0) - \varphi'(1) + (\lambda^2 - \mu^2) \int\limits_0^1 \frac{\varphi}{u_0} ds = 0. \tag{7.8.14}$$

Subtracting eq. (7.8.14) from eq. (7.8.13) we have

$$\lambda^2 = \mu^2 \left\{ 1 - \int\limits_0^1 \frac{\varphi}{u_0^2} ds \left[\int\limits_0^1 \frac{\varphi}{u_0} ds \right]^{-1} \right\}. \tag{7.8.15}$$

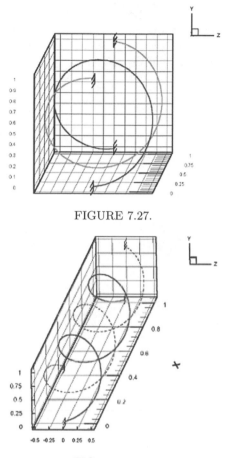

FIGURE 7.27.

FIGURE 7.28.

Let $\mu = \mu_1, \varphi = \varphi_1$ denote the first eigenfrequency and eigenfunction respectively. Function $\varphi_1(s)$ can be taken as being positive whereas $v_0(s)$ is negative, i.e. $u_0 = 1 - v_0$ is positive. Hence both integrals in the latter equation are positive and $\mu_1^2 > \lambda^2$ for any λ. As the inequality $\mu = \lambda$ is not possible, the considered series of the planar forms do not have bifurcational solutions. Since this series is a continuation of the series of stable forms in the problems of attracting strings we conclude that the series of the form of repulsion is stable, too.

The nonplanar forms can not be considered here. We note at this place that they exist. The forms corresponding to a circular orbit are elementary determined. In this case $u = \cos \lambda s$ and $w = \sin \lambda s$. From the boundary conditions $u(0) = u(1) = 1, w(0) = w(1) = 0$ we obtain that $\lambda = 2\pi n$, that is, the string form is a helix with n winds. Such forms exist in pairs: in the form of right and left twisted spirals which corresponds to two directions of rotation of a particle about the centre. The distance between the corresponding points of the conductors ρ is constant within the whole length and equal to one. The examples of these helixes for $n = 1$ and $n = 2$ are displayed in Fig. 7.27 and 7.28 respectively.

Appendix A

Averaging quasilinear systems with several fast variables

A.1 Linear equation with coefficients given by asymptotical series

Let us consider a system

$$\dot{y} = A(x)y + f(x,t) \quad (\dot{y} = dy/dt), \tag{A.1.1}$$

where y denotes the sought vector of dimension m, A is a $m \times m$ matrix, x is a vector of dimension n which is a function of fast time t, slow time $\tau = \varepsilon t$ and parameter ε. Vector x is assumed to be given by the asymptotic series

$$x \sim \xi(\tau) + \varepsilon u_1(\tau, t) + \varepsilon^2 u_2(\tau, t) + \dots \tag{A.1.2}$$

with the prescribed coefficients $\xi(\tau)$ and $u_i(\tau, t)$.

By definition of the asymptotic series for $t \in T_\varepsilon$, $T_\varepsilon = [t_0, t_0 + L/\varepsilon]$, $0 < \varepsilon \leq \varepsilon_0$, the following inequality $|x - x^{(k-1)}| \leq \varepsilon^k C_k$ holds where $C_k = \text{const}$ is independent of ε and

$$x^{(k-1)} = \xi(\tau) + \varepsilon u_1(\tau, t) + \dots + \varepsilon^{(k-1)} u_k(\tau, t). \tag{A.1.3}$$

In what follows we make the following assumptions: A and f are analytical functions of x provided that x belongs to a certain region G and all values of x $\left(x^{(k)} \in G\right)$, ξ and u_i are infinitely differentiable with respect to τ in $\tau_0 \leq \tau \leq \tau_0 + L$. Function x is assumed to be continuously differentiable with respect to t, and the inequality $|\dot{x}| \leq \varepsilon M = \text{const}$ is valid for $x \in G$, $t \in T_\varepsilon$, $0 < \varepsilon \leq \varepsilon_0$; $|f| \leq F = \text{const}$ for $x \in G$, $t_0 \leq t < \infty$; $|u_i| \leq U_i =$

const for $\tau_0 \leq \tau \leq \tau_0 + L$, $t_0 \leq t < \infty$. Analogous inequalities are valid for all derivatives of f and u_i with respect to x or τ. The roots $\lambda_\nu(x)$, $\nu = 1, \cdots, m$, of the equation

$$|A(x) - \lambda E| = 0 \tag{A.1.4}$$

satisfy the condition $\mathrm{Re}\,\lambda_\nu(x) < -\beta$, $\beta = \mathrm{const} > 0$ for all $x \in G$. Of special interest is the case when f and u_i are periodic or quasi-periodic functions of t having the same period (or frequency basis) which is independent of x or τ. Under a quasi-periodic function we understand a linear combination of $\cos \omega_i t$ and $\sin \omega_i t$, $i = 1, \ldots, N$, when at least two numbers ω_k, ω_j are incommensurable.

Let $U(t, t_1, \varepsilon)$ be the fundamental matrix of the solutions of the homogeneous equation $\dot{v} = A(x)v$ such that $U(t_1, t_1, \varepsilon) = E$, where E is the identity matrix and $t_1 \in T_\varepsilon$. In what follows we use the following

Lemma 1. *Under the above assumptions there exists such ε_* that for $0 < \varepsilon \leq \varepsilon_*$ the resolving matrix $U(t, \theta, \varepsilon)$ for $t_0 \leq \theta \leq t$, $t \in T_\varepsilon$, fulfils the condition*

$$\|U(t, \theta, \varepsilon)\| \leq N e^{-\gamma(t-\theta)}, \tag{A.1.5}$$

where the constants γ, $N > 0$ are independent of ε.

This lemma presents Coppel's theorem [128] however the inequality $|\dot{x}| \leq \varepsilon M$ holds in the interval $t \in T_\varepsilon$, that is, inequality (A.1.5) is valid in the above interval rather than in the infinite interval, as in Coppel's theorem. In the following we assume that $\varepsilon \leq \varepsilon_*$.

Let us describe the process of construction of a formal series defining a partial solution φ of systems (A.1.1)

$$\varphi \sim \varphi_0(\tau, t) + \varepsilon \varphi_1(\tau, t) + \ldots . \tag{A.1.6}$$

Functions $A(x)$ and $f(x, t)$ are represented by asymptotic series (for matrices, the difference $A - A^{(k)}$ is estimated by means of the norm)

$$A(x) \sim A(\xi) + \varepsilon A_1(\tau, t) + \ldots, \quad f(x, t) \sim f(\xi, t) + \varepsilon f_1(\tau, t) + \ldots, \tag{A.1.7}$$

which can be obtained by substitution of eq. (A.1.2) into the expansion of functions A and f in the neighbourhood of point $x = \xi$, while the series coefficients are infinitely differentiable with respect to τ. Substituting eqs. (A.1.2) and (A.1.7) into (A.1.1) and equating the coefficients at the powers of ε we obtain the system

$$\frac{\partial \varphi_i}{\partial t} = A(\xi)\varphi_i + g_i, \quad g_i = f_i + A_i \varphi_0 + \cdots + A_1 \varphi_{i-1} - \frac{\partial \varphi_{i-1}}{\partial \tau}, \quad g_0 = f_0. \tag{A.1.8}$$

By virtue of the infinite differentiability of A_i, f_i and ξ with respect to τ, functions g_i are determined for all k as functions of t, τ, φ_0 etc. Equations (A.1.8) are successively integrable for ξ, $\tau = \mathrm{const}$. When $\varphi_0, \varphi_1, \ldots, \varphi_{i-1}$ have been found, function g_i is a known function of t and τ.

Let φ_0 be some partial solution of the first equation in (A.1.8). Since function g_0 is bounded at $t_0 \leq t < \infty$, and $\operatorname{Re} \lambda_\nu < 0$ function φ_0 is also bounded in $t_0 \leq t < \infty$, $\xi \in G$. Let us denote $\psi_0 = \partial \varphi_0 / \partial \tau$. From the first equation in (A.1.8) we obtain

$$\frac{\partial \psi_0}{\partial t} = A(\xi)\psi_0 + \frac{\partial g_0}{\partial \tau} + \left(\frac{\partial}{\partial \tau} A(\xi) \right) \varphi_0. \qquad (A.1.9)$$

Under the made assumptions "the inhomogeneous part" of eq. (A.1.9) is bounded in $t \geq t_0$ therefore function ψ_0 is also bounded. Hence function g_1, and in turn φ_1 are bounded, too. Having chosen φ_1, we find functions g_2, φ_2 etc., all of them being bounded for $x \in G$, $\tau_0 \leq \tau \leq \tau + L$, $t_0 \leq t < \infty$.

When f and u_i are periodic or quasi-periodic functions of t having the same period (or frequency basis) which is independent of x and τ, any periodic or quasi-periodic solution of the first equation in (A.1.8) can be understood as φ_0. Then function g_1 will be periodic or quasi-periodic, in a similar manner one can take φ_1, φ_2 etc.

Theorem 1. *A formal series (A.1.6) asymptotically converges to a partial solution of equation (A.1.1).*

Let $\varphi_k(t_0)$ denote the value $\varphi_k(t)$ for $t = t_0$, and $\varphi(t_0, \varepsilon)$ denote one of the functions ε such that

$$\varphi(t_0, c) \sim \varphi_0(t_0) + \varepsilon\varphi_1(t_0) + \dots . \qquad (A.1.10)$$

Let us consider a partial solution $\varphi(t, \varepsilon)$ of system (A.1.1) coinciding with $\varphi(t_0, \varepsilon)$ at $t = t_0$ and denote $d_k = \varphi(t, \varepsilon) - \varphi^{(k)}(t, \varepsilon)$. The function $\varphi^{(k)} = \varphi_0 + \varepsilon\varphi_1 + \dots + \varepsilon^k \varphi_k$ satisfies the equation

$$\dot{\varphi}^{(k)} = A^{(k)} \varphi^{(k)} + f^{(k)} + \varepsilon^{k+1} \left[\frac{\partial \varphi_k}{\partial \tau} - \right.$$
$$\left. - A_1 \varphi_k - \dots - A_k \varphi_1 - \varepsilon(A_2 \varphi_k + \dots + A_k \varphi_2) \dots \right]. \qquad (A.1.11)$$

For d_k we have the following equation

$$\dot{d}_k = A(x)d_k + (f - f^{(k)}) + (A - A^{(k)})\varphi^{(k)} + \varepsilon^{k+1} \left[\frac{\partial \varphi_k}{\partial \tau} + A_1 \varphi_k + \dots \right.$$
$$\left. \dots + A_k \varphi_1 + \varepsilon(A_2 \varphi_k + \dots + A_k \varphi_2) + \dots + \varepsilon^{k-1} A_k \varphi_k \right] \qquad (A.1.12)$$

and the initial conditions $d_k(t_0) = \varphi(t_0, \varepsilon) - \varphi^{(k)}(t_0, \varepsilon)$. The solution of eq. (A.1.12) can be written in the form

$$d_k = U \left(\varphi(t_0, \varepsilon) - \varphi^{(k)}(t_0, \varepsilon) \right) + \int\limits_{t_0}^{t} U \rho_k d\theta, \qquad (A.1.13)$$

where ρ_k designates the sum of all terms in the right hand side of eq. (A.1.12) beginning with $f - f^{(k)}$. It follows from eq. (A.1.13) that

$$|d_k| \le \|U\| |\varphi(t_0, \varepsilon) - \varphi^{(k)}(t_0, \varepsilon)| + \int\limits_{t_0}^{t} \|U\| |\rho_k| d\theta. \qquad (A.1.14)$$

Taking into account eq. (A.1.5) we obtain

$$|d_k| \le N_0 |\varphi(t_0, \varepsilon) - \varphi^{(k)}(t_0, \varepsilon)| e^{\gamma(t-t_0)} + N \int\limits_{t_0}^{t} e^{-\gamma(t-\theta)} |\rho_k| d\theta. \quad (A.1.15)$$

The definition (A1.1.3) of the asymptotic series yields

$$|\varphi(t_0, \varepsilon) - \varphi^{(k)}(t_0, \varepsilon)| \le \varepsilon^{k+1} C_{k+1}, \qquad |f - f^{(k)}| \le \varepsilon^{k+1} C'_{k+1},$$

where the constants C_{k+1}, C'_{k+1} are independent of ε.

Since functions $\varphi_k(t, \tau)$ are bounded (see above) for $t_0 \le t < \infty$, $\tau_0 \le \tau \le \tau_0 + L$ we have $|\varphi^{(k)}| \le \Phi^{(k)} = \sum\limits_i \varepsilon_0^i \Phi_i$ where Φ_i are independent of ε. The estimates for all values in square brackets in eq. (A.1.12) are obtained by analogy. Then we have

$$|(A - A^{(k)})\varphi^{(k)}| \le \|A - A^{(k)}\| |\varphi^{(k)}| \le \varepsilon^{k+1} \Phi'_{k+1}.$$

Collecting the estimates of values entering in the expression for ρ_k, we obtain $|\rho_k| \le \varepsilon^{k+1} R_{k+1}$ and

$$|d_k| \le \varepsilon^{k+1} \left| N_0 C_{k+1} e^{-\gamma(t-t_0)} + \frac{N R_{k+1}}{\gamma} (1 - e^{\gamma(t-t_0)}) \right| \le \varepsilon^{k+1} D_{k+1},$$
$$(A.1.16)$$

where D_{k+1} is independent of ε. This proves the asymptotic convergence of series (A.1.6) to the partial solution of equation (A.1.1).

In order to construct the general integral of the homogeneous system we look for the fundamental matrix $U(t, \tau, \varepsilon)$ in the following form

$$U(t, \tau, \varepsilon) = U_0(t, \tau) + \varepsilon U_1(t, \tau) + \dots, \qquad (A.1.17)$$

where U_i satisfies the matrix equations

$$\frac{\partial U_0}{\partial t} = A U_0, \quad \frac{\partial U_1}{\partial t} = A U_1 + A_1 U_0 - \frac{\partial U_0}{\partial \tau} \qquad (A.1.18)$$

etc. and the initial conditions $U_0(t_0, \tau) = E$, $U_i(t_0, \tau) = 0$, $i = 1, 2, \dots$. One determines U_0 from the first equation in (A.1.18). Substituting matrix U_0 into the second equation in eq. (A.1.18) we determine U_1 and so on.

Theorem 2. *Sequence (A.1.17) asymptotically converges to the fundamental matrix of the homogeneous system.*

It is evident that derivatives $\partial U_i/\partial \tau$, $i = 0, 1, \ldots$, are bounded by norm by the functions $C_i t^k e^{-\gamma t}$, $\gamma > 0$, where k is integer and $C_i = $ const. In addition to this, all A_1, A_2, ... are bounded. The convergence is proved similar to the proof of theorem 1.

The introduced matrices contain secular terms which are products of exponential functions and powers of t. For this reason, in the case when the roots λ_ν are real-valued, negative and simple (or a multiple root with simple divisors) whereas u_i are periodic or quasi-periodic functions (with period or frequency basis independent of τ) another algorithm is preferable. We will seek the partial solutions y_ν of the homogeneous system in the form $y_\nu = \zeta_\nu z_\nu$ where ζ_ν is a scalar functions and z_ν is a vector function, both being governed by the following differential equations

$$\dot{\zeta}_\nu = \mu_\nu \zeta_\nu \quad \dot{z}_\nu + \mu_\nu z_\nu = A z_\nu. \tag{A.1.19}$$

We look for the asymptotic series for z_ν and μ_ν in the form

$$z_\nu \sim z_{\nu 0}(\tau) + \varepsilon z_{\nu 1}(\tau, t) + \ldots, \quad \mu_\nu \sim \mu_{\nu 0}(\tau) + \varepsilon \mu_{\nu 1}(\tau) + \ldots. \tag{A.1.20}$$

Inserting eqs. (A.1.20) and (A.1.7) into eq. (A.1.19) and equating the terms with the equal powers of ε we obtain the system

$$A(\xi)z_{\nu 0} - \mu_{\nu 0} z_{\nu 0} = 0,$$

$$\frac{\partial z_{\nu 1}}{\partial t} = (A(\xi) - \mu_{\nu 0} E) z_{\nu 1} + h_{\nu 1}, \quad h_{\nu 1} = A_1 z_{\nu 0} - \frac{\partial z_{\nu 0}}{\partial \tau} - \mu_{\nu 1} z_{\nu 0} \tag{A.1.21}$$

and so on. Let us assume that $\mu_{\nu 0}(\xi)$ is equal to one of the roots of the equation

$$|A(\xi) - \mu E| = 0. \tag{A.1.22}$$

Then $z_{\nu 0}$ is one of eigenvectors of matrix $A(\xi)$. It is determined up to a scalar factor which is an arbitrary infinitely differentiable function of τ, a specific form of this factor being out of interest. The equations for $z_{\nu 1}$, $z_{\nu 2}$ etc. are integrated at ξ, $\tau = $ const.

The roots of the characteristic equation

$$|A(\xi) - (\mu_{\nu 0} + \mu)E| = 0 \tag{A.1.23}$$

are the differences $\mu_\alpha - \mu_{\nu 0}$, where μ_α is a root of eq. (A.1.22). All these differences are real-valued, only one among them being identically equal to zero. Let us choose $\mu_{\nu i}$ by means of the following condition

$$\lim_{T \to \infty} \frac{1}{T} \int_{t_0}^{t_0+T} (h_{\nu k}, z_{\nu 0}^*) dt = 0, \tag{A.1.24}$$

where $z_{\nu 0}^*$ denotes the eigenvector of matrix $A^* - \mu_{\nu 0} E$ corresponding to zero root of the characteristic equation. Equalities (A.1.24) are simple linear equations allowing one to determine successively $\mu_{\nu 1}$, $\mu_{\nu 2}$ ... These equations are resolvable since the coefficient at $\mu_{\nu k}$ in $k - th$ equation is equal to $(z_{\nu 0}, z_{\nu 0}^*)$ and does not vanish.

If u_j are periodic, then function $h_{\nu i}$ will be also periodic provided that $z_{\nu 1}$, ..., $z_{\nu i-1}$ are periodic functions. Under the condition (A.1.24) which can be replaced by the equality that mean value averaged over the period is equal to zero, function $z_{\nu i}$ can be understood as a periodic one. But $h_{\nu i}$ is periodic, thus all $z_{\nu i}$ can be taken as being periodic.

Functions $h_{\nu i}$ are polynomials of u_j whereas $h_{\nu i}$ is a linear form in $z_{\nu 1}$, ..., $z_{\nu i-1}$. For this reason, for quasi-periodic functions u_j, $z_{\nu 1}$, ..., $z_{\nu i-1}$ the function $h_{\nu i}$ will be quasi-periodic and under condition (A.1.24) all $z_{\nu i}$ can be chosen as quasi-periodic.

Taking periodic or quasi-periodic functions $z_{\nu i}$ and having found some number of values $\mu_{\nu i}$, we can calculate the corresponding approximations to ζ_ν

$$\zeta_\nu^{(k)} = \exp\left(\int_{t_0}^t \mu_\nu^{(k)} \xi(\varepsilon\sigma) d\sigma\right). \tag{A.1.25}$$

The sequence $y_\nu^{(k)} = \zeta_\nu^{(k)} z_\nu^{(k)}$ converges asymptotically to a partial solution of the homogeneous system. For different ν the solutions are linearly independent and the general integral can be constructed with the help of these solutions.

A.2   The case when function $\xi(\tau, \varepsilon)$ is given by an asymptotic sequence

In the previous analysis $\xi(\tau)$ and $u_i(\tau, t)$ were considered as prescribed functions of τ and t. Let us now consider the case when u_i are given functions of ξ and t, analytical with respect to ξ while function ξ is assumed to be given by an asymptotic sequence $\xi^{(i)}(\tau, \varepsilon)$, $i = 0, 1, \ldots$, and the derivatives of functions $\xi^{(i)}$ are expressed in terms of these functions by the relationships

$$\frac{d\xi^{(i)}}{dt} = \varepsilon \Xi_1(\xi^{(i)}) + \cdots + \varepsilon^{i+1} \Xi_{i+1}(\xi^{(i)}), \tag{A.2.1}$$

where functions Ξ_1, Ξ_2, ... are analytical. As for $\xi(\tau, \varepsilon)$ we assume that for $0 < \varepsilon \leq \varepsilon_0$, $\tau_0 \leq \tau \leq \tau_0 + L$ there exists the derivative $\partial\xi/\partial\tau$ which is the limit of the asymptotically converging sequence $\partial\xi^{(k)}/\partial\tau$.

In this case the approximations $y^{(k)}$ can be expressed in terms of $\xi^{(k)}(\tau)$ and $\Xi_i(\xi^{(i)})$. We seek $\varphi^{(k)}$ in the following form

$$\varphi^{(k)} = \varphi_0(\xi^{(k)}, t) + \varepsilon\varphi_1(\xi^{(k)}, t) + \cdots + \varepsilon^k\varphi_k(\xi^{(k)}, t), \qquad (A.2.2)$$

and instead of $d\varphi^{(k)}/dt$ we use the sum

$$\frac{\partial\varphi_0}{\partial t} + \varepsilon\left[\frac{\partial\varphi_0}{\partial\xi^{(k)}}\Xi_1 + \frac{\partial\varphi_1}{\partial t}\right] + \varepsilon^2\left[\frac{\partial\varphi_0}{\partial\xi^{(k)}}\Xi_2 + \frac{\partial\varphi_1}{\partial\xi^{(k)}}\Xi_1 + \frac{\partial\varphi_2}{\partial t}\right] + \ldots \quad (A.2.3)$$

The approximation to x up to the term of ε^{k+1} is given by

$$x^{(k)} = \xi + \varepsilon u_1(\xi, t) + \cdots + \varepsilon^k u_k(\xi, t). \qquad (A.2.4)$$

Since $\xi^{(k)}$ approximates ξ with the accuracy $O(\varepsilon^{k+1})$ then

$$x^{(k)} = \xi^{(k)} + \varepsilon u_1(\xi^{(k)}, t) + \cdots + \varepsilon^k u_k(\xi^{(k)}, t) \qquad (A.2.5)$$

approximates $x(t, \varepsilon)$ with the same accuracy. Similar to eq. (A.1.7), we can replace ξ by $\xi^{(k)}$, to have

$$A^{(k)} = A(\xi^{(k)}) + \varepsilon A_1(\xi^{(k)}, t) + \cdots + \varepsilon^k A_k(\xi^{(k)}, t),$$
$$f^{(k)} = f(\xi^{(k)}) + \varepsilon f_1(\xi^{(k)}, t) + \cdots + \varepsilon^k f_k(\xi^{(k)}, t). \qquad (A.2.6)$$

Let us insert these approximations into eq. (A.1.1) and equate the coefficients in front of the powers of ε. We arrive at the system

$$\frac{\partial\varphi_0}{\partial t} = A(\xi^{(k)})\varphi_0 + g_0, \quad g_0 = f(\xi^{(k)}, t),$$
$$\frac{\partial\varphi_1}{\partial t} = A(\xi^{(k)})\varphi_1 + g_1, \quad g_1 = f(\xi^{(k)}, t) + A_1\varphi_0 - \frac{\partial\varphi_0}{\partial\xi^{(k)}}\Xi_1, \quad \ldots \quad (A.2.7)$$

To prove the asymptotic convergence of sequence $\varphi^{(k)}$ to the solution of equation (A.1.1) we construct an equation for $\varphi^{(k)}$. We multiply equations (A.2.7) by the corresponding degree of ε and sum the results, then we obtain

$$\dot{\varphi}^{(k)} = A^{(k)}\varphi^{(k)} + f^{(k)} + \varepsilon^{k+1}[A_1\varphi_k + \cdots + A_k\varphi_1 + \varepsilon(A_2\varphi_k + \cdots +$$
$$+ A_k\varphi_2) + \cdots + \varepsilon^{k-1}A_k\varphi_k + \frac{\partial\varphi_k}{\partial\xi^{(k)}}(\Xi_1 + \cdots + \varepsilon\Xi_{k+1})], \qquad (A.2.8)$$

that differs from eq. (A.1.8) in the terms of order ε^{k+1} and higher, as well as that the argument of y, $A^{(k)}$, $f^{(k)}$ etc. is $\xi^{(k)}$ rather than ξ. It is however insignificant and therefore the convergence is proved by analogy with Theorem 1.

The fundamental matrix U is obtained from eq. (A.1.18) where $\partial U_0 / \partial \tau$ is replaced by $(\partial U_0 / \partial \xi^{(k)}) \Xi_1$ etc. The asymptotic convergence is proved as above.

Finally, the $k - th$ approximation to $y = \varphi + UC$ is as follows

$$y^{(k)} = \varphi_0 + \varepsilon \varphi_1 + \cdots + (U_0 + \varepsilon U_1 + \ldots)C = y_0 + \varepsilon y_1 + \ldots, \qquad \text{(A.2.9)}$$

where C denotes the vector of the arbitrary constants determined from the initial conditions, whereas φ_0, φ_1, U_0, U_1, ... are prescribed functions of $\xi^{(k)}$, Ξ_1 etc.

Having determined $y^{(k)}$ we can also take $x^{(k)}$ in the following form

$$x^{(k)} = \xi^{(k)} + \varepsilon u_1(\xi^{(k-1)}, t) + \cdots + \varepsilon^k u_k(\xi^0, t). \qquad \text{(A.2.10)}$$

Analogously, we can write formulae for $A^{(k)}$ and $f^{(k)}$. The argument $\xi^{(k)}$ in eq. (A.2.7) should be replaced by the lowest approximations, namely in the equation for $\varphi^{(i)}$ it is necessary to replace $\xi^{(k)}$ by $\xi^{(k-i)}$. The remaining equations will be not changed and the obtained solution approximates the exact solution with the accuracy $O(\varepsilon^{k+1})$.

A.3   Separating slow variables in quasi-linear systems

Let us consider the system

$$\begin{aligned} \dot{x} &= \varepsilon X(x, y, t, \varepsilon) = \varepsilon X_1(x, y, t) + \varepsilon^2 X_2(x, y, t) + \ldots, \\ \dot{y} &= A(x)y + f(x, t). \end{aligned} \qquad \text{(A.3.1)}$$

The quasi-linear system (A.3.1) presents a particular case of systems with many fast variables studied by V.M. Volosov [128] who indicated a replacement of variables allowing separation of slow and fast variables. However the fact that system (A.3.1) is quasi-linear suggests a method of separation different from Volosov's method. The present method relies on combination of the averaging method with the method of asymptotic integration of linear equations explained in Sections A1 and A2.

Let us assume that $X(x, y, t, \varepsilon)$ is an analytical function of x, y, ε for $x \in G$, $y \in G_1$, $0 \le \varepsilon \le \varepsilon_0$, and $A(x)$ and $f(x, t)$ are analytical functions of x. The other assumptions will be listed in what follows.

We look the $k-th$ approximation to $x(t, \varepsilon)$ ($k = 0, 1, \ldots$) in the following form

$$x^{(k)}(t, \varepsilon) = \xi^{(k)} + \varepsilon u_1(\xi^{(k)}, t) + \cdots + \varepsilon^k u_k(\xi^{(k)}, t), \qquad \text{(A.3.2)}$$

where $\xi^{(k)}(\tau, \varepsilon)$ satisfies the equation

$$\frac{d\xi^{(k)}}{dt} = \varepsilon \Xi_1(\xi^{(k)}) + \cdots + \varepsilon^{k+1} \Xi_{k+1}(\xi^{(k)}), \qquad \text{(A.3.3)}$$

and express $k - th$ approximation to $y(t, \varepsilon)$ in terms of $\xi^{(k)}$ and Ξ_i with the help of relationships (A.2.2).

We replace $X_i(x^{(k)}, y^{(k)}, t)$, $i = 1, \ldots, k + 1$, by the sums

$$X_i(x^{(k)}, y^{(k)}, t) = X_i(\xi^{(k)} + \varepsilon u_1 + \ldots, y_0 + \varepsilon y_1 + \ldots, t) \tag{A.3.4}$$

$$= X_i(\xi^{(k)}, y_0, t) + \varepsilon \left[\frac{\partial X_i}{\partial x} u_1 + \frac{\partial X_i}{\partial y} y_1 \right] + \cdots + \varepsilon^{k+1} [\ldots],$$

keeping in each sum the terms with powers of ε not higher than $k - i + 1$ (the derivatives in eq. (A.3.4) are calculated at $x = \xi^{(k)}$, $y = y_0$).

Let us take a derivative in eq. (A.3.2), replace the derivative of $\xi^{(k)}$ by means of eq. (A.3.3) and keep the terms with powers of ε not higher than $k + 1$. We replace $\dot{x}$ in the first equation in (A.3.1) by the obtained expressions for the derivative, omit the terms beginning with $O(\varepsilon^{k+2})$ in the right hand side of this equation and replace X_i according to eq. (A.3.4). Then we obtain the following formal equality

$$\varepsilon \Xi_1 + \cdots + \varepsilon^k \Xi_k + \varepsilon \frac{\partial u_1}{\partial t} + \varepsilon \frac{\partial u_1}{\partial \xi^{(k)}} (\varepsilon \Xi_1 + \cdots + \varepsilon^{k-1} \Xi_{k1}) + \varepsilon \frac{\partial u_k}{\partial t} =$$

$$= \varepsilon X_1(\xi^{(k)}, y_0, t) + \varepsilon^2 \left[\frac{\partial X_1}{\partial x} u_1 + \frac{\partial X_1}{\partial y} y_1 \right] + \cdots + \varepsilon^{k+1} [\ldots]. \tag{A.3.5}$$

In equation (A.3.5) we equate the terms at the powers of ε and obtain

$$\frac{\partial u_1}{\partial t} = X_1(\xi^{(k)}, y_0(\xi^{(k)}, t), t) - \Xi_1, \tag{A.3.6}$$

$$\frac{\partial u_2}{\partial t} = X_2(\xi^{(k)}, y_0(\xi^{(k)}, t), t) + \frac{\partial X_1}{\partial x} u_1 + \frac{\partial X_1}{\partial y} y_1 - \frac{\partial u_1}{\partial \xi^{(k)}} \Xi_1 - \Xi_2$$

etc. Altogether we obtain $k + 1$ equations.

It is typical for the asymptotic methods that u_i are required to be bounded functions of t. To this end, it is necessary and sufficient that the following mean values

$$\langle W_i \rangle = \lim_{T \to \infty} \frac{1}{T} \int_{t_0}^{t_0 + T} W_i(\xi, t) dt, \qquad i = 1, \ldots, k + 1 \tag{A.3.7}$$

exist, where W_i is the sum of all terms in the right hand sides of eq. (A.3.6), except for Ξ_i, and the integrals

$$\int_{t_0}^{t} [W_i(\xi, \theta) - \langle W_i \rangle] d\theta, \qquad i = 1, \ldots, k + 1 \tag{A.3.8}$$

should be bounded functions of time.

We point out two important cases in which these conditions are satisfied.

1. Functions X_i and f are periodic functions of t with period T which is independent of x. The mean values (A.3.7) are assumed to be independent of C.

2. Functions X_i and f are quasi-periodic functions of t with a frequency basis independent of x, and X_i are the polynomials in the components of y. In this case y_0 and W_1 are sums of the terms with attenuating exponential functions and a quasi-periodic function. Therefore for $\Xi_1 = \langle W_1 \rangle$ function u_1 has the form of W_1. But u_1 appears linearly in eq. (A.1.8) (in terms of A_1 and f_1) therefore φ_1 and y_1 are sums of attenuating functions and a quasi-periodic function. Function W_2 is the same character. In general, since u_i appear in eq. (A.1.8) in the form of u_i^m, m being positive and integer, all y_i are the sums of exponentially attenuating and quasi-periodic functions. Therefore under the conditions

$$\Xi_i(\xi) = \langle W_i(\xi, t) \rangle, \qquad i = 1, 2, \dots, \tag{A.3.9}$$

u_i will have the form of W_i, and, thus, they are bounded.

The second case occurs, for example, in mechanics and the theory of electromechanical systems. Then X_i are quadratic forms in the components of y.

Provided that the integral (A.3.8) is bounded, one determines Ξ_i from eq. (A.3.9) whereas u_i is found by quadratures in terms of W_i up to an arbitrary function of ξ.

Finally we arrive at the following algorithm.

We take the initial conditions for $\xi(\tau)$ in the form of $\xi(\tau_0) = x(t_0)$ and choose $u_i(\xi(\tau), t)$ such that $u_i(\xi(\tau_0), t_0) = 0$. Then we find the solution of the first equation (A.2.7) under the condition $\varphi_0(t_0) = y(t_0)$. This condition can be satisfied since the value of $\xi(\tau_0)$ is known. Then we introduce the obtained function $\varphi_0(t) = y_0$ in eq. (A.3.6), average the result according to eq. (A.3.7) and find $\Xi_1(\xi)$. Integrating the first equation in eq. (A.3.6) we find u_1 by choosing the integration function of ξ such that $u_1(\xi(\tau_0), t_0) = 0$. Next, from the second equation in (A.2.7) we find $\varphi_1 = y_1$ under the condition $\varphi_1(t_0) = 0$. Introducing the result into the second equation in eq. (A.3.6) and constructing W_2 we find function Ξ_2 with the help of eq. (A.3.7). Then integrating the second equation in eq. (A.3.6) we find u_2 by virtue of $u_2(\xi(\tau_0), t_0) = 0$ and so on.

The choice of the conditions $\xi(\tau_0) = x(t_0)$, $u_i(\xi(\tau_0), t_0) = 0$, $\varphi_0(t_0) = y(t_0)$, $\varphi_i(t_0) = 0$ is not necessary. The arbitrary functions in u_i can be chosen from another consideration. Then the starting values $\xi(\tau_0)$, $\varphi_i(t_0)$ are found in terms of the initial conditions $x(t_0)$, $y(t_0)$ and the initial values $u_i(\xi(\tau_0), t_0)$.

In what follows we use the theorem [130] on higher approximations for the systems in standard form

$$\dot{x} = \varepsilon X(x, t, \varepsilon), \tag{A.3.10}$$

which states the following.

Let X have a sufficient number of the derivatives and $|X| \leq M = \text{const}$ for $x \in G$, $t \in T_\varepsilon$, $0 < \varepsilon \leq \varepsilon_0$. Let us assume that we have the initial conditions for eq. (A.3.10) and the improved $k - th$ approximation $x^{(k)} = \xi + \varepsilon u_1(\xi, t) + \cdots + \varepsilon^k u_k(\xi, t)$, where ξ is the solution of equation in the form (A.3.3). Let u_i, $i = 1, \ldots, k$, be bounded function of time satisfying the above initial conditions. Let ξ remains in G for $t \in T_\varepsilon$, $0 < \varepsilon \leq \varepsilon_0$. Then there exists such $\varepsilon_1 \leq \varepsilon_0$ that for $0 < \varepsilon \leq \varepsilon_1$ and the same initial conditions, equation (A.3.10) admits a solution remaining in G for $t \in T_{\varepsilon_1}$ and satisfying the inequality $|x - x^{(k)}| \leq \varepsilon^{k+1} \cdot \text{const}$.

We will also use the following

Lemma 2. *Let us consider two systems*

$$\dot{x}_1 = \varepsilon X_1(x_1, t, \varepsilon), \qquad \dot{x}_2 = \varepsilon X_2(x_2, t, \varepsilon), \qquad (A.3.11)$$

where $|X_1(x, t, \varepsilon) - X_2(x, t, \varepsilon)| \leq \varepsilon^{k+1} K$, $K = \text{const in } x \in G$, $t \in T_\varepsilon$. *Then under the assumption that the condition of the previous theorem is valid , we have* $|x_1 - x_2| \leq \varepsilon^{k+1} \cdot \text{const}$, $\left| x_1 - x_2^{(k)} \right| \leq \varepsilon^{k+1} \cdot \text{const for } t \in T_\varepsilon$ *and the same initial conditions.*

The proof follows from the inequalities

$$|x_1 - x_2| \leq \varepsilon \int_{t_0}^{t} |X_1(x_1, t, \varepsilon) - X_2(x_2, t, \varepsilon)| dt \leq \varepsilon \int_{t_0}^{t} |X_1(x_1, t, \varepsilon) -$$

$$- X_1(x_2, t, \varepsilon)| + |X_1(x_2, t, \varepsilon) - X_2(x_2, t, \varepsilon)| dt \leq$$

$$\leq \varepsilon C \int_{t_0}^{t} |x_1 - x_2| dt + \varepsilon^{k+1} K \leq \varepsilon^{k+1} K e^{CL},$$

$$|x_1 - x_2^{(k)}| \leq |x_1 - x_2| + |x_2 - x_2^{(k)}| \leq \varepsilon^{k+1} \cdot \text{const}. \qquad (A.3.12)$$

Here we used the lemma condition, Lipschits condition for X_1, inequality (A.3.12) and the theorem on systems in standard form.

Theorem 3. *Let for* $t \in T_\varepsilon$, $0 < \varepsilon \leq \varepsilon_0$ *the curves* $\xi^{(0)}(\tau)$, $y_0(t, \tau)$ *satisfy* $\xi^{(0)}(\tau_0) = x(t_0)$, $y_0(t_0, \tau_0) = y_0(t_0)$ *and remain, along with their vicinities, in* G *and* G_1, *respectively. Then there exists such* $\varepsilon_1 \leq \varepsilon_0$ *that for* $0 < \varepsilon \leq \varepsilon_1$ *the solution* $x(t, \varepsilon)$, $y(t, \varepsilon)$ *of system (A.3.1) becomes* $x(t_0)$, $y(t_0)$ *at* $t = t_0$ *and remains in* G, G_1 *for* $t \in T_\varepsilon$, *i.e.* $x \in G$, $y \in G_1$. *The approximations* $x^{(k)}$, $y^{(k)}$ *obtained under the conditions* $x^{(k)}(t_0, \varepsilon) = x(t_0, \varepsilon)$, $y^{(k)}(t_0, \varepsilon) = y(t_0, \varepsilon)$ *remain in* G, G_1 *for* $t \in T_\varepsilon$ *and satisfy the inequalities* $|x - x^{(k)}| \leq \varepsilon^{k+1} \cdot \text{const}$, $|y - y^{(k)}| \leq \varepsilon^{k+1} \cdot \text{const}$, *where the constants depend only on* k.

Let functions u_i, y_i, Ξ_i, $i = 1, \ldots, k$, be constructed. We resolve eq. (A.3.2) for $\xi^{(k)}$, to obtain

$$\xi^{(k)} = x^{(k)} + \varepsilon v_1(x^{(k)}, t) + \cdots + \varepsilon^k v_k(x^{(k)}, t) + \varepsilon^{k+1}\delta\xi_k,$$

$$v_1 = -u_1(x^{(k)}, t), \quad v_2 = -u_2(x^{(k)}, t) + \left(\frac{\partial u_1}{\partial \xi^{(k)}}\right) u_1(\xi^{(k)}, t), \ldots,$$

$$\text{(A.3.13)}$$

which is possible only for sufficiently small ε. Let us keep in eq. (A.3.13) the terms up to ε^k included and substitute them in the expression for $y^{(k)}(\xi^{(k)}, t, \varepsilon)$ and keep the terms of the same order. Then we obtain

$$y_*^{(k)} = y_0(x^{(k)}, t) + \varepsilon y_{*1}(x^{(k)}, t) + \cdots + \varepsilon^k y_{*k}(x^{(k)}, t). \qquad \text{(A.3.14)}$$

Now we introduce eq. (A.3.14) into the first equation in (A.3.1) and arrive at the system in standard form

$$\dot{x}_{*k} = \varepsilon X(x_{*k}, y_*^{(k)}(x_{*k}, t, \varepsilon), t, \varepsilon). \qquad \text{(A.3.15)}$$

A function satisfying this formally constructed system is denoted as x_{*k}. We consider $i - th$ ($i \leq k$) approximation to the solution of this system obtained by the averaging method

$$x_{*k}^{(i)} = \xi_*^{(i)} + \varepsilon u_{*1}(\xi_*^{(i)}, t) + \cdots + \varepsilon^i u_{*i}(\xi_*^{(i)}, t),$$

$$\dot{\xi}_*^{(i)} = \varepsilon \Xi_{*1}(\xi_*^{(i)}) + \cdots + \varepsilon^{i+1}\Xi_{*i+1}(\xi_*^{(i)}), \qquad \text{(A.3.16)}$$

whose solution being independent of k.

Let us prove that functions $u_{*j}(\xi, t)$, $\Xi_{*j}(\xi)$ coincide with $u_j(\xi, t)$, $\Xi_j(\xi)$ in eqs. (A.3.2) and (A.3.3). Let us assume that this is correct for u_{*1}, Ξ_{*1}, ..., u_{*i-1}, Ξ_{*i1}. Substituting $x_{*k}^{(i)}$ from eq. (A.3.16) in the expression $y_*^{(k)}$, instead of $x^{(k)}$ we obtain a function of $\xi_*^{(i)}$, t, ε, whose coefficients of the expansion in terms of ε up to ε^{i-1} included are coincident with the coefficients $y^{(k)}(\xi^{(k)}, t, \varepsilon)$ provided that $\xi^{(k)}$ is replaced $\xi_*^{(k)}$ in the latter expression. Therefore function W_{*i} in the equation

$$\frac{\partial u_{*i}}{\partial t} = W_{*i} - \Xi_{*i} \qquad \text{(A.3.17)}$$

determined by functions u_{*1}, ..., u_{*i-1} is coincident with W_i after replacing the argument $\xi_*^{(i)}$ by $\xi^{(k)}$. Hence, Ξ_{*i} and Ξ_i, coincide and under an appropriate choice of the arbitrary functions ξ after integration u_{*i} and u_i will be coincident, too.

Let us consider the first approximation. We have

$$\dot{x}_{*k} = \varepsilon X_1(x_{*k}, y_0(x_{*k}, t), t) + \ldots; \qquad W_{*1} = X_1(\xi_*^{(i)}, y_0(\xi_*^{(i)}, t), t),$$

that is, u_{*1} coincides with u_1 after replacing the arguments. By induction we can conclude that all u_{*j}, Ξ_{*j} coincide with the corresponding u_j, Ξ_j up to $j = k + 1$.

Let us now assume that there exists such ε_1 and such initial conditions that for $0 < \varepsilon \leq \varepsilon_1$ the solution of system (A.3.1) and all $i-th$, $i = 0, 1, \ldots$, approximations satisfying the initial conditions remain in G, G_1 for $t \in T_\varepsilon$.

We shall consider the solution and approximation to it under these initial conditions.

Function $y_0(x, t)$ satisfies the equation

$$\dot{y}_0 = A(x)y_0 + f(x, t) + \varepsilon \frac{\partial y_0}{\partial x}(X_1 + \varepsilon X_2 + \ldots). \tag{A.3.18}$$

The coefficient at ε in eq. (A.3.18) is bounded under the considered values of the variables. By analogy with Section A.1 we can show that $|y_0(x, t) - y| \leq \varepsilon \cdot \text{const}$.

Let us now consider the equation

$$\dot{x}_{*0} = \varepsilon X_1(x_{*0}, y_0(x_{*n}, t), t). \tag{A.3.19}$$

From the estimate $|y - y_0| \leq \varepsilon \cdot \text{const}$ and the boundedness of function $X_2, \ldots$ we conclude by virtue of Lemma 2 that $|x - x_{*0}| \leq \varepsilon \cdot \text{const}$ and $|x - \xi^{(0)}| \leq \varepsilon \cdot \text{const}$.

Let us assume that

$$|x - x^{(k-1)}(\xi^{(k-1)}, t, \varepsilon)| \leq \varepsilon^k \cdot \text{const}, \quad |y - y^{(k-1)}(\xi^{(k-1)}, t, \varepsilon)| \leq \varepsilon^k \cdot \text{const}$$

and consider the function $\xi^{(k-1)}(x^{(k1)}, t, \varepsilon)$ obtained from eq. (A.3.2) by replacing k by $k - 1$. According to the theorem on implicit functions this function exists for sufficiently small ε (say for $\varepsilon \leq \varepsilon_1$) and is an analytical function of parameter ε and bounded functions of t. Hence

$$\xi^{(k-1)} = x^{(k-1)} + \varepsilon v_1 + \cdots + \varepsilon^{k-1} v_{k-1} + \varepsilon^k \delta \xi_{k-1},$$

where $|\delta \xi_{k-1}|$ is a bounded function in the domain under consideration. Therefore we have $|y^{(k-1)}(\xi^{(k-1)}, t, \varepsilon) - y_*^{(k-1)}(x^{(k1)}, t, \varepsilon)| \leq \varepsilon^k \cdot \text{const}$ and $|y - y_*^{(k-1)}(x^{(k-1)}, t, \varepsilon)| \leq \varepsilon^k \cdot \text{const}$. Hence

$$|X(x, y_*^{(k-1)}(x, t, \varepsilon), t, \varepsilon) - X(x, y, t, \varepsilon)| \leq \varepsilon^k \cdot \text{const}$$

and system (A.3.1) may be rewritten in the following form

$$\dot{x} = \varepsilon X(x, y_*^{(k1)}(x, t, \varepsilon), t, \varepsilon) + \varepsilon^{k+1} \delta X_k, \tag{A.3.20}$$

where δX_k is a bounded function. By virtue of the theorem on the systems in standard form there exists a function $\xi^{(k)}(\tau, \varepsilon)$ such that

$$x^{(k)} = \xi^{(k)} + \varepsilon u_1(\xi^{(k)}, t) + \cdots + \varepsilon^k u_k(\xi^{(k)}, t), \tag{A.3.21}$$

approximates x with the accuracy $O(\varepsilon^{k+1})$, while $u_1, \ldots, u_k$ are independent of the terms of the order ε^{k+1} in eq. (A.3.20). Thus under this choice $\xi^{(k)}$ $x = x^{(k)} + \varepsilon^{k+1}\delta x_k$ and $y = y^{(k)}(\xi^{(k)}, t, \varepsilon) + \varepsilon^{k+1}\delta y_k$, where $|\delta x_k|$ and $|\delta y_k|$ are bounded.

All $\xi^{(k)}(\tau, \varepsilon)$ satisfying this condition are expressed in terms of x by the relationship

$$\xi^{(k)} = x + \varepsilon v_1(x, t) + \cdots + \varepsilon^k v_k(x, t) + \varepsilon^{k+1}\delta\xi_k, \qquad (A.3.22)$$

where the latter term in the right hand side of eq. (A.3.22) is also bounded. Let us construct function $y_*^{(k)}(x, t, \varepsilon)$ by using eq. (A.3.22). For this function we have $|y - y_*^{(k)}(x, t, \varepsilon)| \leq \varepsilon^{k+1} \cdot \mathrm{const}$. Therefore the first equation in (A.3.1) can be rewritten in the following form

$$\dot{x} = \varepsilon X(x, y_*^{(k)}(x, t, \varepsilon), t, \varepsilon) + \varepsilon^{k+2}\delta X_{k+1}, \qquad (A.3.23)$$

where δX_{k+1} is bounded. Omitting the last term in eq. (A.3.23) we arrive at the system whose solution, under these initial conditions, approximates the exact solution with the accuracy $O(\varepsilon^{k+1})$. After replacing x by x_{*k} this system coincides with that in eq. (A.3.15), and $k - th$ approximation to its solution coincides with $x^{(k)}$. Thus, provided that $(k - 1) - th$ approximation approximates the exact solution with accuracy $O(\varepsilon^k)$, then $k - th$ approximation approximates it with accuracy $O(\varepsilon^{k+1})$. However for $x^{(0)} = \xi^{(0)}$, $y^{(0)} = y_0(\xi^{(0)}, t)$ the approximation has been derived above, thus the approximation of order $k + 1$ is proved for all k.

It remains to show that for sufficiently small values of ε, $\varepsilon > 0$ functions $x \in G$ and $y \in G_1$, provided that $\xi^{(0)}$, y_0 remain in G, G_1 along with some vicinities. We have

$$(\xi^{(k)} - \xi^{(0)})^{\cdot} = \varepsilon(\Xi_1(\xi^{(k)}) - \Xi_1(\xi^{(0)})) + \cdots + \varepsilon^{k+1}\Xi_{k+1}(\xi^{(k)}) \quad (A.3.24)$$

or after the integration

$$|\xi^{(k)} - \xi^{(0)}| \leq \varepsilon \int_{t_0}^{t} |\Xi_1(\xi^{(k)}) - \Xi_1(\xi^{(0)})|dt + \varepsilon C_1 \leq \qquad (A.3.25)$$

$$\leq \varepsilon C_2 \int_{t_0}^{t} |\xi^{(k)} - \xi^{(0)}|dt + \varepsilon C_1 \leq \varepsilon C_1 e^{C_2 L}, \qquad C_1, C_2 = \mathrm{const}.$$

Here we used the inequality of eq. (A.3.12). Hence for a sufficiently small ε function $\xi^{(k)} \in G$ for $t \in T_\varepsilon$. Since u_i are bounded functions of time which are differentiable with respect to $\xi^{(k)}$ then $y^{(k)} \in G_1$ for sufficiently small ε and $x^{(k)} \in G, y^{(k)} \in G_1$. Finally, the proved inequalities for $x - x^{(k)}$, $x - x_*$ suggest that $x, x_* \in G$, $y \in G_1$ for $t \in T_\varepsilon$ and sufficiently small values of ε.

A.4   Remarks

In practice it is often sufficient to construct only the equations for slow variables. Then for periodic or quasi-periodic X_i and f it is sufficient to find only periodic or quasi-periodic partial solution for y, while u_i will be periodic or quasi-periodic.

When $X_i(x, y, t)$ and $f(x, t)$ are periodic or quasi-periodic functions and λ_ν are simple and real-valued then one can use the algorithm of Section A.1 and obtain the approximations having only periodic or quasi-periodic functions and exponential functions rather than the secular terms. To this end we should use the following representation

$$y^{(k)} = y_0 + \varepsilon \delta y_1 + \cdots + \varepsilon^k \delta y_k, \qquad \delta y_i = \left(y^{(i)} - y^{(i-1)} \right) / \varepsilon^i, \qquad \text{(A.4.1)}$$

where the fractions are of the order of one, and to replace everywhere y_i by $(y^{(i)} - y^{(i-1)})/\varepsilon^i$.

The method substantiated above for the quasilinear systems of general form was originally proposed by K.Sh. Khodzhaev for the system describing slow motions in electromechanical systems and considered in [124]. In this case, in the first approximation the equations for slow variables describe the conservative system, and the higher approximations are required for a qualitative assessment of the motion. The corresponding computations were carried out by M.M. Vetyukov and K.Sh. Khodzhaev with the help of Volosov's method [124] and with the help of the above method. The approximations obtained by both methods were proved to be coincident however the computations by the suggested method turned out to be simpler.

Having studied the case in which f and X_i are periodic functions of t, R.F. Nagaev proposed to construct, instead of functions $\varphi^{(k)}(\xi, t, \varepsilon)$, the function $\varphi^{(k)}(x, t, \varepsilon)$ which are periodic with respect to t by replacing $\dot{x}$ by the expansion εX. Then $\varphi_0(x, t)$ is obtained from $\varphi_0(\xi, t)$ if ξ is replaced by x. For $\varphi_1(x, t)$ one obtains the equation

$$\frac{\partial \varphi_1}{\partial t} = A(x)\varphi_1 + f(x, t) - \frac{\partial \varphi_0}{\partial x} X_1(x, \varphi_0, t) \qquad \text{(A.4.2)}$$

and so on. Having found $\varphi^{(k)}$ and replaced y in first equation (A.3.1) by $\varphi^{(k)}$ one arrives at the system in standard form. It can be shown that equations (A.3.3) and (A.3.6) are obtained after the averaging. Analogously the general solution $y^{(k)}(x, t, \varepsilon)$ can be constructed and there is no need to consider only the periodic functions f and X_i.

Appendix B

Systems integrable in the first approximation of the averaging method

The higher approximations of the averaged equations obtained by the Krylov-Bogolyubov method are known to be not uniquely constructed since the corresponding replacement is performed up to some arbitrary functions of slow variables. It is natural to try to choose these functions such that the averaged equations are simplified. As shown in the present Section, such a possibility appears, in particular, in the case in which the general integral of the equation of the first approximation is known. In this case the choice of these arbitrary functions allows one to ensure that all highest terms of the averaged equations turn to zero, that is, for any approximation the averaged equations coincide with the equations of the first approximation.

Let us consider the system in standard form

$$\dot{x} = \varepsilon X(x, t, \varepsilon) = X_0(x, t) + \varepsilon X_1(x, t) + \dots, \qquad (\text{B.1})$$

where x and X are $n \times 1$ column-vectors. Function $X(x, t, \varepsilon)$ is assumed to be periodic function of the explicit time t with a given period independent of x and ε, have all partial derivatives and these derivatives are uniformly bounded for $|\varepsilon| < \varepsilon_0$ and x from a certain region. The approximate solution x_m of $m - th$ approximation of the averaging method is defined by the relationship

$$x_m = \xi + \varepsilon u_1(\xi, t) + \varepsilon^2 u_2(\xi, t) + \dots + \varepsilon^{m-1} u_{m-1}(\xi, t). \qquad (\text{B.2})$$

Here functions $u_1(\xi, t), \dots, u_{m-1}(\xi, t)$ are periodic with respect to t. The slow variables ξ are dependent on the number of approximation m, however for the sake of simplicity we will write ξ instead of ξ_m.

Generally speaking, the system for ξ is given by

$$\dot{\xi} = \varepsilon\Xi_0(\xi) + \varepsilon^2\Xi_1(\xi) + \ldots + \varepsilon^m\Xi_{m-1}(\xi). \tag{B.3}$$

Functions $u_1, \ldots, u_{m-1}$ and, respectively, functions $\Xi_1, \ldots, \Xi_{m-1}$ are determined not uniquely since any arbitrary function of ξ can be added to u_i, however it does not affect the approximation of the accurate solutions x by the approximate solutions x_m, cf. [83].

Let us assume that the general integral of the averaged system of the first approximation

$$\dot{\xi} = \varepsilon\Xi_0(\xi) \tag{B.4}$$

be obtained. Let us show that in this case one can determine the arbitrary corrections to u_i such that all functions $\Xi_1, \ldots, \Xi_{m-1}$ are identically equal to zero. Then the averaged equations of any approximation coincide with the equation of the first approximation and can be integrated.

Let us consider the following relationships for determining functions u_1, u_2, Ξ_0, Ξ_1

$$\Xi_0(\xi) + \frac{\partial u_1(\xi, t)}{\partial t} = X_0(\xi, t),$$

$$\Xi_1(\xi) + \frac{\partial u_1(\xi, t)}{\partial t}\Xi_0(\xi) + \frac{\partial u_2(\xi, t)}{\partial t} = \frac{\partial X_0(\xi, t)}{\partial \xi}u_1(\xi, t) + X_1(\xi, t). \tag{B.5}$$

Function $\Xi_0(\xi)$ is determined uniquely

$$\Xi_0(\xi) = \langle X_0(\xi, t)\rangle, \tag{B.6}$$

where the broken brackets denote time-averaging over the period. The expression for u_1 can be set in the form

$$u_1 = \int (X_0(\xi, t) - \Xi_0)dt + u_{c1}(\xi) = u_{\nu 1}(\xi, t) + u_{c1}(\xi). \tag{B.7}$$

Here and in what follows index ν denote the quadratures whose mean value over the period is equal to zero and u_{c1} is yet an arbitrary function of ξ.

Function Ξ_1 is determined from the condition of periodicity of u_2

$$\Xi_1(\xi) = \left\langle -\frac{\partial u_1(\xi, t)}{\partial \xi}\Xi_0 + \frac{\partial X_0(\xi, t)}{\partial \xi}u_1 + X_1(\xi, t)\right\rangle. \tag{B.8}$$

Introducing eq. (B.7) into eq. (B.8) we obtain

$$\Xi_1 = -\frac{\partial u_{c1}}{\partial \xi}\Xi_0 = \frac{\partial \Xi_0}{\partial \xi}u_{c1} + G_1(\xi), \tag{B.9}$$

where

$$G_1(\xi) = \left\langle \frac{\partial X_0}{\partial \xi}u_{\nu 1} + X_1\right\rangle. \tag{B.10}$$

Let us select function u_{c1} such that $\Xi_1(\xi) \equiv 0$. This requirement leads to the equation for u_{c1}

$$\frac{\partial u_{c1}}{\partial \xi}\Xi_0 = \frac{\partial \Xi_0}{\partial \xi}u_{c1} + G_1(\xi). \tag{B.11}$$

Strictly speaking, for the vector-function u_{c1} we obtained a system of differential equations in partial derivatives of the first order with the identical principal part [26].

The characteristic system for eq. (B.11) is given by the following system, cf. [26]

$$\frac{\partial \xi^{(k)}}{\partial \xi^{(n)}} = \frac{\Xi_0^{(k)}}{\Xi_0(n)}, \quad k = 1, \ldots, n-1,$$

$$\frac{\partial u_{c1}}{\partial \zeta^{(n)}} = \frac{\partial \Xi_0}{\partial \xi}\frac{u_{c1}}{\Xi_0^{(n)}} + \frac{G_1}{\Xi_0^{(n)}}. \tag{B.12}$$

Index k in the first group of equations in eq. (B.12) denotes $k - th$ component of the column-vectors $(\xi^{(1)}, \ldots, \xi^{(n)})^{\mathrm{T}}$ and $\Xi_0 = (\Xi_0^{(1)}, \ldots, \Xi_0^{(n)})^{\mathrm{T}}$. The unknowns in eq. (B.12) are $n - 1$ functions $\zeta^{(1)}(\zeta^{(n)}), \ldots, \zeta^{(n-1)}(\zeta^{(n)})$ and n components of vector $u_{c1}(\xi^{(n)})$. Any other component of vector ξ can be taken as the argument in system (B.12).

Provided that the general integral of equations (B.4) is prescribed in the form $F(\xi, \tau) = C$, where $\tau = \varepsilon t$, we can find the general integral of the first group of equations (B.12). To this end, the slow time τ should be eliminated from n relationships $F(\xi, \tau) = C$ and we obtain $n - 1$ relationships of the form $F_*(\xi) = C_*$, where C_* is a $(n-1) \times 1$ column-vector. Another constant contained in the general integral of system (B.4) is removed under elimination of τ since system (B.4) is autonomous and one constant is contained additively with τ. Here we solve the problem without the general integral of equations (B.4) as it is sufficient to know their $n-1$ autonomous integrals.

The second group of equations in eq. (B.12) is also integrated since it is obtained from the system of equations in variations for equation (B.4). In order to obtain the equations of this group it is necessary to add the inhomogeneity $G_1(\xi)$ to the variational equations and take the argument $\xi^{(n)}$ instead of time.

Let us consider the inhomogeneous system with the argument τ and the unknown variable u_{c1}

$$\frac{du_{c1}}{d\tau} = \frac{\partial \Xi_0(\xi)}{\partial \xi}u_{c1} + G_1(\xi). \tag{B.13}$$

Its general solution is given by

$$u_{c1} = \frac{\partial \xi(\tau + C^{(n)}, C_*)}{\partial C} A + \frac{\partial \xi(\tau + C^{(n)}, C_*)}{\partial C} \times$$

$$\times \int_{\tau_0}^{\tau + C^{(n)}} \left(\frac{\partial \xi(\vartheta + C^{(n)}, C_*)}{\partial C} \right)^{-1} G_1(\xi(\vartheta + C^{(n)}, C_*)) d(\vartheta + C^{(n)}). \quad (B.14)$$

Here A denotes $n \times 1$ column-vector of the arbitrary constants, and functions $\xi(\tau + C^{(n)}, C_*)$ are obtained by "inversion" of the general integral $F(\xi, \tau) = C$, where $C^{(n)}$ denotes the constant appearing in the solution of system (B.4) additively with τ.

In order to obtain the general solution of the second group of equations (B.12), it is necessary to substitute $\tau + C^{(n)} = f(\xi^{(n)}, C_*)$ into eq. (B.14). This substitution is determined in terms of the dependences

$$\xi^{(1)}(\xi^{(n)}, C_*), \ldots, \xi^{(n-1)}(\xi^{(n)}, C_*), \tau + C^{(n)} = f(\xi^{(n)}, C_*)$$

derived from n equations $F(\xi, \tau) = C$.

Now we can write down the general integral of system (B.12). Instead of eq. (B.14) we have

$$\left(\frac{\partial \xi(\tau + C^{(n)}, C_*)}{\partial C} \right)^{-1} u_{c1} - \int_{\tau_0}^{\tau + C^{(n)}} \left(\frac{\partial \xi(\vartheta + C^{(n)}, C_*)}{\partial C} \right)^{-1} \times$$

$$\times G_1(\xi(\vartheta + C^{(n)}, C_*)) d(\vartheta + C^{(n)})|_{\tau + C^{(n)} = f(\xi^{(n)}, C_*)} = A. \quad (B.15)$$

Then one should replace C_* on $F_*(\xi)$ in eq. (B.15). Then we arrive at the relationship of the form $H_1(u_{c1}, \xi) = A$. Along with the relationship $F_*(\xi) = C$ it comprises the general integral of system (B.12).

The general solution of system (B.11) is now obtained in the following way. We take n arbitrary differentiable functions of $2n-1$ arguments (these functions can be viewed as the components of vector Ψ) and write down n functional equations

$$\Psi(F_*(\xi), H_1(u_{c1}, \xi)) = 0. \quad (B.16)$$

The vector-function Ψ should satisfy the condition that equations (B.16) are resolvable for u_{c1}. Then the dependence $u_{c1}(\xi)$ ensuring the equality $\Xi_1 = 0$ is found from eq. (B.16).

It is evident that when we use this method the "degree of arbitrariness" in the choice of functions u_{c1} is less than in the general case, however it is not clear how it is possible to decrease this "degree".

For the estimation of functions $u_{c2}, \ldots, u_{cm-1}$ we obtain the following equations

$$\frac{\partial u_{ci}}{\partial \xi} \Xi_0(\xi) = \frac{\partial \Xi_0}{\partial \xi} u_{ci} + G_i(\xi), \quad (B.17)$$

which differ from equation (B.11) in the form of the "inhomogeneous" part of G_i, function G being a prescribed function of ξ at $i - th$ stage of the investigation. For this reason, u_{ci} will be determined in the same manner as u_{c1}, with the only difference that G_1 is replaced by G_i.

Instead of relationship (B.2) we can look for an approximate solution in the following form

$$x_m = \xi + \varepsilon u_1(\xi, \tau, t) + \ldots + \varepsilon^{m-1} u_{m-1}(\xi, \tau, t). \tag{B.18}$$

In general case, if we do not try to eliminate functions $\Xi_1, \ldots, \Xi_{m-1}$ these replacements depending explicitly upon the slow time τ, are clearly not rational, because the averaged equations for the higher approximations are not autonomous and more difficult than those under standard replacement. When the suggested method is used, the non-autonomous (with respect to τ) replacement leads even to simpler relationships than the autonomous one.

Provided that the replacement of variables with the explicit slow time τ is used, the averaged system takes the form

$$\dot{\xi} = \varepsilon \Xi_0(\xi) + \varepsilon^2 \Xi_1(\xi, \tau) + \ldots + \varepsilon^{m-1} \Xi_{m-1}(\xi, \tau),$$
$$\dot{\tau} = \varepsilon. \tag{B.19}$$

As above, function $\Xi_0(\xi)$ is defined by relationship (B.6). Let us take u_1 in the following form $u_1 = u_{\nu 1}(\xi, t) + u_{c1}(\xi, \tau)$ where function $u_{\nu 1}$ is given by eq. (B.7). For $u_2(\xi, \tau, t)$ we obtain the equation

$$\Xi_1(\xi, \tau) + \frac{\partial u_1(\xi, \tau, t)}{\partial \xi} \Xi_0(\xi) + \frac{\partial u_1}{\partial \tau} + \frac{\partial u_2(\xi, \tau, t)}{\partial t} =$$
$$= \frac{\partial X_0(\xi, t)}{\partial \xi} u_1(\xi, \tau, t) - X_1(\xi, t). \tag{B.20}$$

The condition of periodicity of u_2 with respect to t yields

$$\Xi_1 = \left\langle -\frac{\partial u_1}{\partial \xi} \Xi_0 - \frac{\partial u_1}{\partial t} + \frac{\partial X_0}{\partial \xi} u_1 + X_1 \right\rangle. \tag{B.21}$$

Let us require that $\Xi_1 \equiv 0$, then we arrive at the equation for u_{c1}, analogous to eq. (B.11)

$$\frac{\partial u_{c1}}{\partial \xi} \Xi_0 + \frac{\partial u_{c1}}{\partial \tau} = \frac{\partial \Xi_0}{\partial \xi} u_{c1} + G_1(\xi), \tag{B.22}$$

where

$$G_1 = \left\langle \frac{\partial X_0(\xi, t)}{\partial \xi} u_{\nu 1} + X_1(\xi, t) \right\rangle. \tag{B.23}$$

Under this assumption, the corresponding characteristic system takes the following form

$$\frac{d\xi}{d\tau} = \Xi_0(\xi),$$

$$\frac{du_{c1}}{d\tau} = \frac{\partial \Xi_0(\xi)}{\partial \xi} u_{c1} + G_1(\xi). \tag{B.24}$$

Basically speaking, the general integral $F(\xi, \tau) = C$ of system (B.4) can deliver the dependences $\xi(\tau, C)$. For this reason, the solution of the second equation in eq. (B.24) is as follows

$$u_{c1} = \frac{\partial \xi(\tau, C)}{\partial C} A + \frac{\partial \xi(\tau, C)}{\partial C} \int_{\tau_0}^{\tau} \left(\frac{\partial \xi(\vartheta, C)}{\partial C} \right)^{-1} G(\xi(\vartheta, C)) d\vartheta. \tag{B.25}$$

From this equation we determine the general integral analogous to that in eq. (B.15). Replacing in this integral C by $F(\xi, \tau)$ we arrive at the relationship of the form $H(u_{c1}, \xi, \tau) = A$. By analogy with the previous analysis we introduce n arbitrary differentiable functions which form vector Ψ and construct the equations

$$\Psi\left(F(\xi, \tau), H(u_{c1}, \xi, t)\right) = 0. \tag{B.26}$$

Provided that functions Ψ are chosen in such a way that equations (B.26) are resolvable for u_{c1}, from eq. (B.26) one can obtain the dependences $u_{c1}(\xi, \tau)$ guaranteeing the equality $\Xi_1 \equiv 0$.

For the following functions $u_{c2}, \ldots, u_{cm-1}$ we obtain the equations of the form of eq. (B.22), where however functions G_i are dependent on ξ and τ. Functions $u_{\nu 2}, \ldots, u_{\nu m-1}$ will also depend upon τ.

The suggested method can be applied to the systems with a single fast phase

$$\dot{x} = \varepsilon X(x, \varphi, \varepsilon),$$

$$\dot{\varphi} = \omega(x) + \varepsilon \Phi(x, \varphi, \varepsilon), \quad \omega > 0. \tag{B.27}$$

Indeed, by introduced a new argument φ we arrive at the system in the standard form.

Finally, this method can be also utilised for averaging the systems with many fast variables different from the phases [128], that is,

$$\dot{x} = \varepsilon X(x, y, t, \varepsilon),$$

$$\dot{y} = Y(x, y, t, \varepsilon). \tag{B.28}$$

Specifically, the following quasi-linear system presents some interest

$$\dot{x} = \varepsilon X(x, y, t, \varepsilon),$$

$$\dot{y} = A(x)y + F(x, t, \varepsilon). \tag{B.29}$$

In this case we can use the expansions described in [96]. An alternative approach is to eliminate the fast variables by determining the coefficients of the following expansion

$$y = y_0(t, x) + \varepsilon y_1(t, x) + \ldots \tag{B.30}$$

When the expression in eq. (B.30) is differentiated, the derivatives $\dot{x}$ are substituted by means of the first equation in (B.29). After substitution of eq. (B.30) into the first equation (B.29) we obtain a system in the standard form.

Appendix C

Higher approximations of the averaging method for systems with discontinuous variables

Let us consider a system whose state is characterised by n-dimensional vector x which is a discontinuous function of time. Let the system be described by the equations in standard form

$$\dot{x} = \varepsilon X(x, t, \varepsilon) \tag{C.1}$$

within the time intervals between the discontinuities. When the phase point in the extended phase space x, t reaches the surface $F(x, t, \varepsilon)$ the unknown variables x becomes discontinuous. Let t_j denote the time instant of the discontinuity. The relation between the value of variable x before and after the discontinuity is described by the formula

$$x(t_j + 0) - x(t_j - 0) = \varepsilon \Delta \left(x(t_j - 0), t_j, \varepsilon \right). \tag{C.2}$$

We assume that function Δ given on the surface $F(x, t\varepsilon)$ is such that the phase point crosses the surface of discontinuity, i.e. it transits from the subspace where $F(x, t, \varepsilon) > 0$ (or $F(x, t, \varepsilon) < 0$) into the subspace, where $F(x, t, \varepsilon) < 0$ (or $F(x, t, \varepsilon) > 0$). The sliding regimes with repetitive crossings, which occur in the small time interval of the order of ε, are not possible. It is also assumed that for $|\varepsilon| < \varepsilon_0$ functions X, F and Δ are continuous with respect to t and determined in certain regions of the corresponding spaces, have continuous derivatives with respect to x, ε of the necessary order and uniformly bounded in these regions along with the above-mentioned derivatives.

The objective of the forthcoming analysis is to substantiate and apply the analogue of the averaging method developed for the systems in standard form to the systems under consideration at any approximation.

The first approximation of the averaging method for the systems with discontinuous variables was proposed and substantiated in the works by A.M. Samoilenko (see for example [101]). In these works the substantiation of the method is presented under weaker requirements (than those in the present Section) to smoothness of the functions, in particular, the functions in the right hand sides of the equation of motions between the discontinuities. Under these conditions the substantiation of the averaging method in the first [3] and higher [130] approximations turns out to be essentially more difficult even for systems with continuous variables than in the case when there is a required number of bounded derivatives (in the latter case the substantiation of the averaging method at the first approximation is suggested, for example, in [5]). Hence under the assumptions of the present work it would be essentially simpler to substantiate the method for the systems with discontinuous variables at the first approximation than in [3] and [130]

According to the suggested method, $m - th$ approximation is sought in the following form

$$x_m = \xi_m + \varepsilon u_1^{(m)}(\xi_m, t, \varepsilon) + \ldots + \varepsilon^{m-1} u_{m1}^{(m)}(\xi_m, t, \varepsilon), \qquad (C.3)$$

where $u_r^{(m)}$ is a $2\pi/\omega-$periodic function of time having discontinuities at $t = t_{im}, i = 1, \ldots, h$. Function ξ_m is continuous and one can construct its equation in the form of a "standard" averaging method

$$\dot{\xi}_m = \varepsilon \Xi_0(\xi_m) + \ldots + \varepsilon^m \Xi_{m-1}(\xi_m). \qquad (C.4)$$

The time instants of discontinuities are defined in the form of following expansion

$$t_{im} = t_{i0}(\xi_m) + \varepsilon \Delta t_{i1}(\xi_m) + \ldots + \varepsilon^{m-1} \Delta t_{im-1}(\xi_m). \qquad (C.5)$$

For determining t_{im} we use the equation $F(x, t, \varepsilon) = 0$. For the sake of simplicity we assume in what follows that for all x, ε from the region under consideration this equation in the half-interval $0 \leq t < 2\pi/\omega$ has the same number h of the simple roots $t_i(x, \varepsilon)$.

Because functions $u_r^{(m)}$ are $2\pi/\omega-$periodic with respect to time it is sufficient to consider only the roots from the indicated half-interval.

The approximate value of the instants of discontinuity can be sought in parallel with functions u_r. In order to keep the description of the method down it is more convenient to express the expansion coefficients (C.5) and the right hand sides of equations (C.4) in terms of $t_{im}(\xi_m, \varepsilon)$ rather than determine these functions.

Decompositions (C.5) will be found in the end of calculations.

Let us denote the coefficients in eq. (C.3) and the second terms of equations in (C.4) expressed in terms of the yet unknown functions $t_{im}(\xi_m, \varepsilon)$ respectively through $u^{(r)}(\xi_m, t, \varepsilon)$ and $\Xi^{(r)}(\xi_m, \varepsilon)$, functions $u^{(r)}$ and $\Xi^{(r)}$

being dependent on ε only in terms of $t_{1m}, \ldots, t_{hm}$ (there is no need to note such a dependence on m by an index). Hence if we wish to find, for example, a $(m+1)-th$ approximation, one can utilise functions $u^{(r)}$, found under the calculation of $m-th$ approximation by replacing the index of ξ_m and t_{im} by $m+1$. The same is valid for functions $\Xi^{(r)}$ and Ξ_r, but $u_r^{(m)}$, whose discontinuity times are assumed to be computed according to eq. (C.5). Instead of (C.3), (C.4) we obtain

$$x_m = \xi_m + \varepsilon u^{(1)}(\xi_m, t, \varepsilon) + \ldots + \varepsilon^{m-1} u^{(m-1)}(\xi_m, t, \varepsilon) \qquad (C.6)$$

$$\dot{\xi}_m = \varepsilon \Xi^{(0)}(\xi_m, \varepsilon) + \ldots + \varepsilon^{m-1} \Xi^{(m-1)}(\xi_m, \varepsilon). \qquad (C.7)$$

Functions Ξ_r in eq. (A2.4) are obtained by a "re-expansion" of the right hand sides of eq. (C.7) in terms of ε after the expansion (C.5) has been found and inserted into eq. (C.7), and u_r has been obtained from $u^{(r)}$ by substitution of eq. (C.5).

For computation of functions $u^{(r)}$ within the time intervals between the discontinuities one should substitute (C.6) in eq. (C.1), differentiate $u^{(r)}$ with respect to t taking into account that ξ_m are functions of time, replace ξ_m by the right hand side of eq. (C.7), expand X in a power series in terms of ε by using Taylor's formula with the remainder of order of ε^m and equate the coefficients at the identical powers of ε in the left and right hand sides.

While the expressions $(\partial u^{(r)} / \partial \xi_m) \dot{\xi}_m$ are calculated one should take into account dependence $u^{(r)}$ on ξ_m in terms of $t_{1m}, \ldots, t_{hr}$. However for balance of terms with the identical powers of ε one should take into account only the powers of ε in front of functions $\Xi^{(r)}, u^{(r)}$ and their derivatives, however these functions are not expanded in power series in terms of ε. As a result one obtains equations for $u^{(r)}$ coinciding formally with those of the "standard" method of averaging.

Having substituted eq. (C.6) into eq. (C.2) and using Taylor's formula for function Δ we obtain the relationships which determine the discontinuities of functions $u^{(r)}$. These relationships have the form

$$\Delta_{i1} = u^{(1)}(\xi_m, t_{im} + 0, \varepsilon) - u^{(1)}(\xi_m, t_{im} - 0, \varepsilon) = \Delta(\xi_m, t_{im}, 0),$$

$$\Delta_{i2} = u^{(2)}(\xi_m, t_{im} + 0, \varepsilon) - u^{(2)}(\xi_m, t_{im} - 0, \varepsilon) =$$

$$= \left(\frac{\partial \Delta}{\partial x}\right)_i u^{(1)}(\xi_m, t_{im} - 0, \varepsilon) + \left(\frac{\partial \Delta}{\partial \varepsilon}\right)_i \qquad (C.8)$$

and so on. In eq. (C.8) derivations of function Δ are calculated for $x = \xi_m, \varepsilon = 0, t = t_{im}$. The dependence of $t_{1m}, \ldots, t_{im}$ on ε are not taken into account in the expansion in terms of ε.

It is essential that the discontinuities of functions u_1 can be found from eq. (C.8) without knowledge of this function. Let us proceed to equation for $u^{(1)}$ in the time interval between the discontinuities

$$\Xi^{(0)}(\xi_m, \varepsilon) + \frac{\partial u^{(1)}}{\partial t} = X(\xi_m, t, 0). \qquad (C.9)$$

From the requirement of $2\pi/\omega$—periodicity of the discontinuous function $u^{(1)}$ it is possible to determine $\Xi^{(0)}(\xi_m, \varepsilon)$

$$\Xi^{(0)} = \langle X(\xi_m, t, 0)\rangle + \frac{\omega}{2\pi}\sum_{i=1}^{h}\Delta_{i1},$$

$$\Delta X(\xi_m, t, 0)\rangle = \frac{\omega}{2\pi}\int_{0}^{2\pi/\omega} X(\xi_m, t, 0)dt. \qquad (C.10)$$

The mean value $\langle X(\xi_m, t, 0)\rangle$ is also dependent on ε since it is calculated for function having discontinuities at $t_{im}(\xi_m, \varepsilon), \ldots, t_{hm}(\xi_m, \varepsilon)$. Then from eq. (C.9) and the first relationship (C.8) we find

$$u^{(1)} = \int_{0}^{t}(X(\xi_m, t, 0) - \Xi^{(0)})dt + \sum_{i=1}^{h}\Delta_{i1}\sigma(t - t_{im}). \qquad (C.11)$$

Here σ is Heaviside function, $\sigma(t - t_{im}) = 0, t < t_{im}, \sigma(t - t_{im}) = 1, t > t_{im}$ and values σ at $t = t_{im}$ are not considered. Typical for the averaging method is that an arbitrary function of ξ_m can be added to the right hand side of eq. (C.11) (for brevity, such functions are omitted here and in what follows). Having obtained function $u^{(1)}$ one can determine the discontinuities of functions $u^{(2)}$ from the second relationship (C.8) and write down the equation for $u^{(2)}$ between the discontinuities. From the condition of $2\pi/\omega$—periodicity of $u^{(2)}$ we obtain the following expression for $\Xi^{(1)}$

$$\Xi^{(1)} = \left\langle \left(\frac{\partial X}{\partial x}\right)_0 + \left(\frac{\partial X}{\partial \varepsilon}\right)_0 - \frac{\partial u^{(1)}}{\partial \xi_m}\Xi^{(0)}\right\rangle + \frac{\omega}{2\pi}\sum_{i=1}^{h}\Delta_{i2}. \qquad (C.12)$$

Then it is possible to find $u^{(2)}$

$$u^{(2)} = \int_{0}^{t}\left[\left(\frac{\partial X}{\partial x}\right)_0 u^{(1)} + \left(\frac{\partial X}{\partial \varepsilon}\right)_0 - \frac{\partial u^{(1)}}{\partial \xi_m}\Xi^{(0)} - \Xi^{(1)}\right]dt +$$

$$+ \sum_{i=1}^{h}\Delta_{i2}\sigma(t - t_{im}). \qquad (C.13)$$

While computing $\partial u^{(1)}/\partial \xi_m$ and in the similar cases one should take into account that u_1 is dependent on ξ_m in terms of $t_{im}(\xi_m, \varepsilon)$. This derivative is a function defined everywhere in the region under consideration, except for the time instants $t = t_{im}$ and is $2\pi/\omega$—periodic with respect to t.

Continuing the process we can calculate the discontinuities of functions $u^{(3)}$ in the time intervals between the discontinuities, determine $\Xi^{(2)}$ from the conditions of $2\pi/\omega-$periodicity of $u^{(3)}$, then determine $u^{(3)}$ and so on. By analogy with the "standard" averaging method in order to estimate function $\Xi^{(m-1)}$ it is necessary to add the term $\varepsilon^m u^{(m)}(\xi_m, t, \varepsilon)$ in eq. (C.3), find the discontinuities Δ_{im} of functions $u^{(m)}$ and write down the equation for $u^{(m)}$ between the discontinuities. There is no need to calculate function $u^{(m)}$ itself since it is required only for construction of the improved $m-th$ approximation.

Next one should substitute eq. (C.3) with the already obtained functions $u^{(r)}(\xi_m, t)$ in equation $F(x, t, \varepsilon) = 0$ and use eq. (C.5). Then we have

$$F(\xi_m + \varepsilon u^{(1)}(\xi_m, t_{im} - 0, \varepsilon) + ... + \varepsilon^{(m-1)} u_{m-1}(\xi_m, t_{im} - 0, \varepsilon), t_{im}, \varepsilon) = 0. \tag{C.14}$$

The values $u^{(r)}(\xi_m, t_{im} - 0, \varepsilon) = u^{(r)}(\xi_m, t_{i0} + ... + \varepsilon^{m1}\Delta t_{i,m-1} - 0, \varepsilon)$ are dependent on $t_{10}, ..., t_{h0}, ..., \Delta_{th}, m - 1$ and their derivative with respect to ξ_m. They can be expanded in a series in terms of ε by keeping the necessary number of conditions. The same should be done for functions $\Xi^{(r)}(\xi_m, \varepsilon)$. Then function F of the indicated arguments can be expanded, too. Equating the coefficients at the powers of ε to zero we arrive at the equations for t_{i0}, Δ_{i1} and so on.

Considering the non-small terms we obtain

$$F(\xi_m, t_{i0}, 0) = 0, \tag{C.15}$$

yielding h roots $t_{i0}(\xi_m), i = 1, ..., h$.

Let us consider the first order terms. They contain the following expression

$$u^{(1)}(\xi_m, t_{im} - 0, \varepsilon) = \frac{\omega}{2\pi} \int\limits_0^{t_{i0}+...+\varepsilon^{m-1}\Delta t_{i,m-1}} (X(\xi_m, t, 0) - \Xi^{(0)})dt + \sum_{j=1}^{i-1}\Delta_{j1} \tag{C.16}$$

which should be expanded in series in terms of ε. This expansion is written in the following form

$$u^{(1)}(\xi_m, t_{im} - 0) = u_{1i,0} + \varepsilon u_{1i,1} + ... + \varepsilon^{m-1} u_{1i,m1} + ..., \tag{C.17}$$

where dots denote the remainder in Taylor's formula. The first terms of the expansion is as follows

$$u_{1i,0}(\xi_m, t_{i0}) = \int\limits_t^{t_{i0}} (X - \Xi_0^{(0)})dt + \sum_{j=1}^h \Delta(\xi_m, t_{j0}, 0), \tag{C.18}$$

where

$$\Xi_0^{(0)} = \frac{\omega}{2\pi} \int\limits_0^{2\pi/\omega} X(\xi_m, t, 0)dt + \frac{\omega}{2\pi} \sum_{j=1}^h \Delta(\xi_m, t_{i0}, 0). \tag{C.19}$$

For estimation Δt_{i1} we have the linear equations

$$\left(\frac{\partial F}{\partial x}\right)_{i0} u_{i1,0} + \left(\frac{\partial F}{\partial t}\right)_{i0} \Delta t_{i1} + \left(\frac{\partial F}{\partial \varepsilon}\right)_{i0} = 0, \quad i = 1, \ldots, h. \quad (C.20)$$

The derivatives with index "$i0$" are calculated at $x = \xi_m, t = t_{i0}, \varepsilon = 0$.

In order to show the structure of quantities obtained at the expansion in terms of powers of ε we will consider an additional equation for Δt_{i2}. We obtain

$$\frac{1}{2}\left[u_{i1,0}^T \left(\frac{\partial^2 F}{\partial x^2}\right)_{i0} u_{i1,0} + \left(\frac{\partial^2 F}{\partial t^2}\right)_{i0} \Delta t_{i1}^2 + \left(\frac{\partial^2 F}{\partial \varepsilon^2}\right)_{i0}\right] +$$

$$+ \left(\frac{\partial^2 F}{\partial x \partial t}\right)_{i0} u_{i1,0} \Delta t_{1i} + \left(\frac{\partial^2 F}{\partial t \partial \varepsilon}\right)_{i0} \Delta t_{i1} + \left(\frac{\partial^2 F}{\partial x \partial \varepsilon}\right)_{i0} u_{1i,0} +$$

$$+ \left(\frac{\partial F}{\partial x}\right)_{i0} u_{1i,1} + \left(\frac{\partial F}{\partial t}\right)_{i0} \Delta t_{i2} = 0. \quad (C.21)$$

The expression for $u_{1i,1}$ is of interest

$$u_{1i,1} = -\Xi_1^{(0)} t_{i0} + \left(X(\xi_m, t_{(i0)}, 0) - \Xi_0^{(0)} \right) \Delta t_{i1} + \sum_{j=1}^{i-1} \left(\frac{\partial \Delta}{\partial t}\right)_{i0} \Delta t_{i1},$$

$$\Xi_1^{(0)} = \frac{\omega}{2\pi} \sum_{i=1}^{h} \left(\frac{\partial \Delta}{\partial t}\right)_{i0} \Delta t_{i1}. \quad (C.22)$$

In the following approximation it is necessary to expand the repeated integrals of the discontinuous functions in series in powers of ε however it is necessary to take into account that ε appears also in the integration limits. As one can see from eqs. (C.20) and (C.21) it is essential that the absolute term and the coefficient at Δt_{ir} in the linear equation for Δt_{ir} are known provided that $\Delta t_{i1}, \ldots, \Delta t_{ir-1}$ are calculated.

Corrections to the time instants of discontinuity can be sought alternatively. Let us consider the equation $F(x, t, \varepsilon) = 0$. It has h solutions $t^{(i)}(x, \varepsilon)$. Substituting x from eq. (C.7) in these solutions we obtain that $t_{im}(\xi_m, \varepsilon) = t^{(i)}(\xi_m + \varepsilon u^{(1)} + \ldots + \varepsilon^{m-1} u^{(m-1)}, \varepsilon)$. Using eq. (C.5) we find Δt_{ir} which will be expressed in terms of functions of $t^{(i)}$ and their derivatives.

The described computations yield the expansions of functions $\Xi^{(r)}(\xi_m, \varepsilon)$. Inserting these expansions into eq. (C.7) and keeping the required powers of ε we obtain eq. (C.4). In the first approximation it is necessary to find only values $t_{i0}, \Delta(\xi_m, t_{i0}, 0)$ and function $\Xi = \Xi_0^{(0)}$. Up to the designations we obtain the results reported e.g. in [101].

Under computation of the discontinuities of solution x_m

$$x_m(t_i + 0) - x_m(t_i - 0) = \varepsilon\Delta(\xi_m, t_{im}, 0) + \varepsilon^2 \Delta_{i2} + \ldots + \varepsilon^{m-1}\Delta_{im1} \quad (C.23)$$

we obtain the terms with the "redundant" powers of ε because $t_{im}, \ldots, t_{hm}$ are also functions of ε. These "redundant" powers of ε should be eliminated by a re-expansion with the help of eq. (C.5).

As a result all values which are independent of the explicitly entering time t will be presented by their expansions in terms of ε. Only functions u_r will be not decomposed.

Let the initial condition $x(t_0)$ be given for the original problem. From eq. (C.3) we can find the corresponding initial condition for ξ_m. Let the solution $\xi = f(t)$ of equation (C.4) be found for these initial conditions. The approximate solution of the original problem is obtained provided that $\xi_m = f(t)$ is substitute into eq. (C.3). Then the argument of the Heaviside function in the expressions for u_r is the difference $t - t_{im}(f(t), \varepsilon)$. Hence we can calculate the time instances of the discontinuities $t_m^{(ik)}$ as solutions of the equations $t_m^{(ik)} = T_{im}\left(f(t_m^{(ik)}), \varepsilon\right)$. Under $T_{im}(\xi_m, \varepsilon)$ we understand an infinite-valued function which is obtained if one takes h dependences $t_{im}(\xi_m, \varepsilon)$ in the half-interval $0 \le t < 2\pi/\omega$ and continues them $2\pi/\omega$-periodically along axis T_{im}. Index k takes values $k = k_1, k_1 + 1, \ldots,$ where k_1 is dependent on t_0. It is also possible to find values ξ_m at discontinuities from the equation $\xi_m^{(ik)} = f\left(T_{im}(\xi_m^{(ik)}, \varepsilon)\right)$ and then find $t_m^{(ik)}$.

It is also possible to use a variant of the method when the corrections to the time instants of discontinuity Δ_{ik} are sought together with functions $u_r^{(k)}, k = 1, \ldots, m,$ appearing in the required approximations. Let us assume that t_{i0} can be found without knowledge of functions $u_{(k)}$. If we know t_{i0} we can found $u_1^{(1)}$ in the form of eq. (C.10) however with the discontinuities at $t = t_{i0}$. Having such functions we can find the corrections Δt_{i1} and construct function $u_1^{(2)}$ with discontinuities at $t = t_{i0} + \varepsilon \Delta_{i1}$. i.e. this variant of the method is concerned with the functions having discontinuities first at t_{i0} and then at $t_{i0} + \varepsilon \Delta_{i1}$ and so on.

Appendix D

On qualitative investigation of motion by the asymptotic methods of nonlinear mechanics

The general theorem on the asymptotic methods for the systems in standard form [83] or systems with many fast and slow variables [128] allows one to judge whether the exact and approximate solutions are close in a finite interval of time of the order of $1/\varepsilon$ where ε is a small parameter. For study of the features of solutions of these systems in an infinite time interval, one uses, firstly, the theorems on existence of the exact solutions of the original equation in a certain form (for example, quasi-periodic solutions) obtained by the method of integral manifolds [84], secondly, the theorems on approximation of the exact solutions by approximate ones in the infinite time interval. Banfi's theorem [31, 38] and its generalization to the systems with many fast variables [31] were proved to be useful among the theorems of the second group.

It should be mentioned that these results are applicable only under rather hard restrictions imposed on the solutions of the averaged system. These restrictions are uniform asymptotic stability in the case of Banfi's theorem, existence of a limit cycle for the equation of first approximation in the case of the theorem about quasi-periodic solutions etc. Specifically, one can not handle the case in which the averaged system is "indifferent" in the first approximation, for example, it is conservative whereas the attenuation and the limit cycle are exposed only in higher approximations. Then the approximation does not provide one with a correct approximate solution of the original equation in the infinite time interval. It can be seen already in the case of uniform exponential stability detected in higher approximations [96].

Similar situation is observed at study of slow motions of a rigid body in an alternating magnetic field. In the most interesting case of "pure mechanical" potential generalized forces, for example the moments of gravity force, the first approximation in the averaged equations of motion describes a conservative system whose motions which can be qualitatively changed by the forces of higher order of smallness. Therefore the second approximations are necessary for deriving the averaged equations describing various motions such as attenuating or divergent ones.

In that follows we suggest a simple method of purely qualitative investigation of motions with the help of the asymptotic methods of nonlinear mechanics under the circumstances when the exact solutions of the initial system and the approximate solutions obtained by the averaging method are close only in a finite time interval. The qualitative comparison of the accurate and approximate solutions will be carried out for the systems in standard form and quasi-linear systems with many fast variables.

Let $x(t)$ denote the sought $n \times 1$ column-vector governed by the system in standard form

$$\dot{x} = \varepsilon X(x, t, \varepsilon), \quad \varepsilon \geq 0 \tag{D.1}$$

and $x^{(m)}(\xi^{(m)}, t, \varepsilon)$ denote $m - th$ approximation to $x(t)$ obtained by the averaging method

$$x^{(m)} = \xi^{(m)} + \varepsilon u_1(\xi^{(m)}, t) + \ldots + \varepsilon^{m-1} u_{m-1}(\xi^{(m)}, t). \tag{D.2}$$

The equation for $\xi^{(m)}$ is given by

$$d\xi^{(m)}/dt = \varepsilon \Xi_1(\xi^{(m)}) + \ldots + \varepsilon^m \Xi_m(\xi^{(m)}) = \varepsilon \Xi^{(m)}(\xi^{(m)}, \varepsilon). \tag{D.3}$$

It is assumed that for $t \geq t_*, \varepsilon \leq \varepsilon_0$ and for x from a certain region D function X is continuous with respect to t and uniformly bounded along with the derivative of order m with respect to x and ε, and functions $\Xi_1, \ldots, \Xi_m$ and $u_1, \ldots, u_{m-1}$ are uniformly bounded along with the first order derivatives with respect to $\xi^{(m)}$ respectively for $\xi^{(m)} \in D$ and $\xi^{(m)} \in D, t \geq t_*$.

Let us consider the function $\xi_m(t, \varepsilon)$ defined by the relationship

$$x = \xi_m + \varepsilon u_1(\xi_m, t) + \ldots + \varepsilon^{m-1} u_{m-1}(\xi_m, t). \tag{D.4}$$

The results will be the same if we introduce $\xi_m(t, \varepsilon)$ by a similar relationship with the component $O(\varepsilon^m)$.

The following estimate of closeness of functions ξ_m and $\xi^{(m)}$ is known. Let $x(t_0) \in D_\alpha$ where region D_α is such that its α-vicinity, $\alpha = O(\varepsilon)$, coincides with D. We will find $\xi_m(t_0)$ with the help of eq. (D.4), then $\xi_m(t_0) \in D_\alpha$ for sufficiently small values of ε. Let the solution of eq. (D.3) for the initial condition $\xi^{(m)}(t_0) = \xi_m(t_0)$ remains in D_α within the time interval $t_0 \leq t \leq t_0 + T/\varepsilon$. Then for functions $\xi^{(m)}, \xi_m$ obeying the above initial conditions the relationship

$$|\xi_m(t) - \xi^{(m)}(t)| \leq c_m \varepsilon^m, \quad t_0 \leq t \leq t_0 + T/\varepsilon$$

is valid for sufficiently small $\varepsilon \leq \varepsilon_*$, constants c_m and T being independent of ε.

In what follows we point out some cases when some features of function ξ_m in an infinite time interval can be established by means of the analogous features of functions $\xi^{(m)}$.

Let us consider a bounded domain D_1 and its δ–vicinity $D_\delta, \delta = d\varepsilon^{m-1}$, $d = \text{const} > 0$, where $D_\delta \subset D_\alpha$, and a single-valued scalar function $V(\xi, \varepsilon)$ defined in $D_\delta, \varepsilon \leq \varepsilon_*$. It is assumed that $|\partial V/\partial \xi| \neq 0$ in $\xi \leq D_\delta, \varepsilon \leq \varepsilon_*$ and there exist

$$V_M = \sup_{\xi \in D_\delta, \varepsilon \leq \varepsilon_*} V(\xi, \varepsilon), \quad F = \sup_{\xi \in D_\delta, \varepsilon \leq \varepsilon_*} \left| \frac{\partial V}{\partial \xi} \right|. \tag{D.5}$$

Theorem 1. *Let us assume that under the above assumptions there exists such a function $V(\xi, \varepsilon)$ that for any solution $\xi^{(m)}(t)$ of eq. (D.3) remaining in D_δ within some time interval $t^{(0)} \leq t \leq t^{(0)} + T/\varepsilon$ the following inequality*

$$V(\xi^{(m)}(t^{(0)} + T/\varepsilon)) \geq V(\xi^{(m)}(t^{(0)})) + \varepsilon^{m-1}W_0 \tag{D.6}$$

holds for the same $W_0 > 0$ for all $t^{(0)} \geq t_*, \varepsilon \leq \varepsilon_*$ and all $\xi^{(m)}(t)$.

Then there exists no solution $\xi_m(t)$ remaining in D_1 for all $t \geq t_*$.

Remarks. 1. Condition (D.6) is automatically satisfied if there exists a function $V(\xi, \varepsilon)$ in D_δ whose derivative satisfies the relationship $\dot{V} \geq \varepsilon^m w_0, w_0 = \text{const} > 0$, by virtue of eq. (D.3). Then it can be accepted that $W_0 = w_0 T$.

2. It immediately follows from condition (D.6) that there exists no solution $\xi^{(m)}(t)$ remaining in D_δ for all $t \geq t_*$. Indeed, let $\xi^{(m)}(t_0) \in D_\delta$. Let us consider the time interval $t_0 \leq t \leq t_0 + kT/\varepsilon$, where k is an integer. After this time interval function V gains an increment $V(\xi^{(m)}(t_0+kT/\varepsilon)) - V(\xi^{(m)}(t_0)) \geq k\varepsilon^{m-1}W_0$ for any solution remaining in D_δ. As a result, for a sufficiently great k the value of function $V(\xi, \varepsilon)$ for the solution $\xi^{(m)}(t)$ exceeds V_M, which is impossible.

Proof. Let us compare the sequence of "approximate" solutions $\xi_j^{(m)}(t)$, $j = 0, 1, \ldots$, defined by the conditions

$$\xi_0^{(m)}(t_0) = \xi_m(t_0),$$
$$\xi_1^{(m)}(t_0 + T/\varepsilon) = \xi_m(t_0 + T/\varepsilon), \ldots,$$
$$\xi_j^{(m)}(t_0 + jT/\varepsilon) = \xi_m(t_0 + jT/\varepsilon)$$

with the "exact" solution $\xi_m(t), \xi_m(t_0) \in D_1$. Let us prove that it is not possible that $\xi_j^{(m)}(t) \in D_\delta$ for all arbitrary large j in the interval $t_0 + jT/\varepsilon \leq t \leq t_0 + (j+1)T/\varepsilon$. Let $\xi_j^{(m)}(t) \in D_\delta, t_0 + jT/\varepsilon \leq t \leq t_0 + (j+1)T/\varepsilon$ for all j. Then $\xi_m(t_0 + jT/\varepsilon) \in D_\delta$ for all j. For the solution $\xi_0^{(m)}(t)$ at $t_0 \leq t \leq t_0 + T/\varepsilon$ function $V(\xi, \varepsilon)$ obtains the increment

$$V\left(\xi_0^{(m)}(t_0 + T/\varepsilon)\right) - V\left(\xi_0^{(m)}(t_0)\right) \geq \varepsilon^{m-1}W_0 \qquad (D.7)$$

by virtue of the theorem conditions.

Let us now consider the value $V(\xi_m(t_0 + T/\varepsilon))$. According to eq. (D.5) we have

$$\left| V\left(\xi_m(t_0 + T/\varepsilon)\right) - V\left(\xi_0^{(m)}(t_0 + T/\varepsilon)\right) \right| \leq$$
$$\leq F\left| \xi_m(t_0 + T/\varepsilon) - \xi_0^{(m)}(t_0 + T/\varepsilon) \right| \leq Fc_m\varepsilon^m. \qquad (D.8)$$

Hence

$$V(\xi_m(t_0 + T/\varepsilon)) \geq V(\xi_0(t_0 + T/\varepsilon)) - \varepsilon^m Fc_m. \qquad (D.9)$$

Taking into account that $V(\xi_0^{(m)}(t_0)) = V(\xi_m(t_0))$ and denoting $W = W_0(1 - \varepsilon_* Fc_m/W_0)$ we obtain the following estimate from eqs. (D.7) and (D.9)

$$V(\xi_m(t_0 + T/\varepsilon)) - V(\xi_m(t_0)) \geq \varepsilon^{m-1}W. \qquad (D.10)$$

By analogy with the increment $V(\xi_1^{(m)}(t))$ in the time interval $t_0 + T/\varepsilon \leq t \leq t_0 + 2T/\varepsilon$ one can estimate $V(\xi_m(t_0 + 2T/\varepsilon))$. Analogously to eq. (D.10) we obtain

$$V(\xi_m(t_0 + 2T/\varepsilon)) - V(\xi_m(t_0 + T/\varepsilon)) \geq \varepsilon^{m-1}W.$$

From this equation and eq. (D.10) we have

$$V(\xi_m(t_0 + 2T/\varepsilon)) - V(\xi_m(t_0)) \geq 2\varepsilon^{m-1}W. \qquad (D.11)$$

Using $j + 1$ functions $\xi_0^{(m)}, \ldots, \xi_j^{(m)}$ yields

$$V(\xi_m(t_0 + (j+1)T/\varepsilon)) - V(\xi_m(t_0)) \geq (j+1)\varepsilon^{m-1}W. \qquad (D.12)$$

As a result value $V(\xi_m(t_0 + (j+1)T/\varepsilon))$ exceeds V_M for a sufficiently great j which is impossible.

Thus, there exists such k and $t_1, t - 0 + kT/\varepsilon \leq t_1 \leq t_0 + (k+1)T/\varepsilon$ that $\xi_k^{(m)}(t_1) \notin D_1$. In a similar manner one can show that $\xi_m(t)$ does not leave not only D_1 but also any $\theta\delta$-vicinity of D_1, where $\theta < 1$ and θ is independent of ε.

It is helpful to estimate time Δt which is needed for solution $\xi_m(t)$ to leave region D_1

$$\Delta t \leq \frac{V_M}{\varepsilon^m W} = \frac{T_1}{\varepsilon^m}.$$

Theorem 2. *Let function V satisfy the conditions of Theorems 1 and besides the derivative $\dot{V}$ computed by virtue of equation (D.3) is nonnegative everywhere in D_δ. Let region D_1 be bounded by the surfaces $V = C_1, V = C_2, C_2 > C_1$ and all surfaces of the family $V = C$ are closed. Then any solution $\xi_m(t), \xi_m(t_0) \in D_1$ after some time will cross the surface $V = C_2$ and leave region D_1 forever.*

Remark. Let surface $V = C$ be denoted as $S(C)$. It follows from the above features of the derivative $\partial V/\partial\xi$ that family $S(C)$ has no singularities in D_δ. Since the surfaces are closed they cover each another. Let us assume that for all $C' < C''$ the surface $S(C')$ covers $S(C'')$ and this enables using the expressions "a region outside $S(C_1)$", "a region inside $S(C_2)$" etc. The case when $S(C'')$ covers $S(C')$ is considered by analogy.

Proof. First we show that $\xi_m(t)$ will necessarily fall in $S(C_2)$. An "approximate" solution $\xi_0^{(m)}(t)$ can leave D_δ only by crossing $S(C_2)$. If $\xi_0^{(m)}(t)$ leaves D_δ in the time interval $t_0 \le t \le t_0 + T/\varepsilon$ then $\xi_m(t)$ leaves D_1 in the same time interval through $S(C_2)$ (if $\xi_0^{(m)}(t)$ lies near the boundary $S(C_1)$ then $\xi_m(t)$ can leave D_1 through this boundary however it is not important).

Let $\xi_0^{(m)}(t) \in D_\delta$ for $t_0 \le t \le t_0 + T/\varepsilon$. The point $\xi_0^{(m)}(t_0 + T/\varepsilon)$ lies in $S(C_1)$ on the distance ρ from it. For ρ we have an estimate (analogous to (D.8)) $\rho \ge \varepsilon^{m-1}W_0/F$. By virtue of (D.5) point $\xi_m(t_0 + T/\varepsilon)$ is also in $S(C_1)$ on the distance $\rho_1 = O(\rho)$ from it. Considering function $\xi_1^{(m)}(t)$ we obtain that point $\xi_m(t_0 + 2T/\varepsilon)$ is in $S(C_1)$ on the distance $\rho_2 \ge \rho_1 + \varepsilon^{m-1}W/F$ from the boundary. In the second interval the point $\xi_m(t)$ can not be outside $S(C_1)$. Then we obtain $\rho_3 \ge \rho_1 + 2\varepsilon^{m-1}W/F$ and so on. Thus point $\xi_m(t)$ which must leave D_1 according to Theorem 1, leaves D_1 through $S(C_2)$.

Let now $\xi_m(t_1)$ lies in $S(C_2)$. Let us assume that some number of solutions $\xi_j^{(m)}(t)$, $\xi_0^{(m)}(t_1)$ etc. remain in D_δ. Then already in the second time interval $t_1 + T/\varepsilon \le t \le t_1 + 2T/\varepsilon$ the solution $\xi_m(t)$ can not leave $S(C_2)$ just like the solution $\xi_m(t)$ could not leave $S(C_1)$ in the corresponding intervals. There remains the case in which solution $\xi_k^{(m)}(t)$ escapes D_δ. Let C_M denote the largest value of C for which $S(C)$ lies completely in D_δ, then $C_M - C_2 = O(\varepsilon^{m-1})$. In the time interval $t_1 + kT/\varepsilon \le t \le t_1 + (k+1)T/\varepsilon$ point $\xi_m(t)$ can lie outside $S(C_M)$ only on the distance $O(\varepsilon^m)$ from the boundary. It is now clear that $\xi_m(t)$ can not reach D_1 in the next interval irrespective whether $\xi_{k+1}^{(m)}(t)$ enters D_δ or escapes it.

From Theorem 2 follows

Theorem 3. *Let surface $S(C)$ collapse into a point $C \to C_*$. Assume that for any infinitesimally small $\eta > 0$ there exists such $\varepsilon(\eta)$ that for $\varepsilon \le \varepsilon(\eta) \le \varepsilon_*$ and $C_2 - C_* = \eta$ Theorem 2 holds true. Then for sufficiently small ε beginning with a certain $t = t(\eta)$ the solution $\xi_m(t)$ stays forever in the infinitesimally small $\eta-$vicinity of the point $V = C_*$.*

According to eq. (D.4), solution $x(t)$ of the initial system (D.1) for sufficiently great values of t remains in $(\eta + O(\varepsilon))$–vicinity of the point $V = C_*$. Let point $V = C_*$ correspond to the equilibrium position of a mechanical system. Then the vibrations described by the initial system (D.1) will qualitatively present the superposition of (D.4) of a slow evolutional motion tending "almost" to the equilibrium position and small fast vibrations. For sufficiently great t the motion reduces to the small fast vibrations about the mean position, which probably is slowly wandering in a small vicinity of the equilibrium position. Such vibrations slightly differ from the quasi-static ones.

In the case when surface $S(C'')$ covers $S(C')$ at $C'' > C'$ it follows from Theorem 2 that solution $\xi_m(t)$ leaves surface $S(C_2)$ forever. Provided that the condition of Theorem 2 are met for sufficiently large or even arbitrary large values of C_2 then $x(t)$ describes the superposition of small vibrations and the "outgoing" motion.

After the evident modifications in the formulations of Theorems 1-3 it is possible to rely on existence of decreasing functions V rather than increasing ones. For example, in Theorem 1 we can take the condition

$$V\left(\xi^{(m)}(t_0 + T/\varepsilon)\right) - V\left(\xi^{(m)}(t^{(0)})\right) \leq -\varepsilon^{m-1}W_0$$

assuming that $\inf V$ exists in D_δ.

Function V can be found in the case in which eq. (D.3) admits first integral $V(\xi, \varepsilon) = \text{const}$ at the first approximation or in the several lowest approximations and in the following approximation this integral vanishes and sign of $\dot{V}$ can be determined. Clearly, in this case $\dot{V} = O(\varepsilon^m)$, m denoting the number of approximation the integral vanishes for the first time. The simplest case is that when $V = \text{const}$ is the energy integral and the relationship $\dot{V} = 0$ is violated because of the dissipation in higher approximations.

The above theorems can be generalised on the systems different from the systems in standard form provided that the exact and approximate solutions are proved to be close in the interval T/ε and the determination of the approximate solution is reduced to integration of the autonomous system. This is the case of quasi-linear system with many fast variables different from the phases

$$\dot{x} = \varepsilon X(x, y, t, \varepsilon),$$
$$\dot{y} = A(x)y + f(x, t). \tag{D.13}$$

Here x, y denote $n \times 1$ and $l \times 1$ vectors-columns and matrix $A(x)$ is such that its eigenvalues $\lambda_1(x), \ldots, \lambda_l(x)$ satisfy the condition $\text{Re}\,\lambda_i \leq -\mu < 0$, $\mu = \text{const}$.

For system (D.13) we have [82, 96]

$$|\xi_m - \xi^{(m)}| \le c_m \varepsilon^m, \quad |x - x_m| \le c^{(m)}\varepsilon^m,$$
$$|x - x_m^{(m)}| \le c_m^{(m)}\varepsilon^m, \quad |y - y^{(m-1)}(\xi^{(m)}, t, \varepsilon)| \le b_m \varepsilon^m. \quad \text{(D.14)}$$

Here

$$y^{(m-1)}(\xi_m, t, \varepsilon) = \phi^{(m-1)}(x(\xi_m, t, \varepsilon), t, \varepsilon) = \sum_{i=0}^{m1} \varepsilon^i \phi_i(x, t, \varepsilon) \quad \text{(D.15)}$$

and functions $\phi_0, \phi_1, \ldots, \phi_{m-1}$ are determined from the linear equations

$$\dot{\phi}_0 = A\phi_0 + f,$$
$$\dot{\phi}_1 = A\phi_1 - \frac{\partial \phi_0}{\partial x} X(x, \phi_0, t, 0) \quad \text{(D.16)}$$

etc. which are integrated if $x = $ const and under the initial conditions $\phi_0(x(t_0), t_0) = y(t_0), \phi_i(x(t_0), t_0) - 0$. Next, x_m in eq. (D.14) denotes the solution of the system in standard form

$$\dot{x}_m = \varepsilon X(x_m, \phi^{(m-1)}(x_m, t, \varepsilon), t, \varepsilon), \quad \text{(D.17)}$$

where $x_m^{(m)}$ is $m - th$ approximation to x_m in the form of eq. (D.2) and $\xi_m, \xi^{(m)}$ are introduced for system (D.17) by analogy with (D.1). For $\xi^{(m)}$ one obtains the autonomous system in the form of (D.3), then $\xi^{(m)}$ and ξ_m can be qualitatively compared with the help of above theorems. The corresponding properties of the sought functions $x(t)$ and $y(t)$ can be proved, too.

References

[1] Ackerberg R. On a nonlinear differential equation of electrohydrodynamics. Proc. Royal Society A, 1969, vol. 312, No. 1508.

[2] Alekseev I.I., Kropotkin A.P. Geomagnetism and aeronomia, No. 10,
1970, p. 953.

[3] Alekseeva M.M. Machine generators of higher frequency (in Russian).
Leningrad, Energiya, 1967.

[4] Alper I.Ya., Terzyan A.A. Inductor generators (in Russian). Moscow,
Energiya, 1970.

[5] Arnold V.I. Mathematical methods of classical mechanics (in Russian). Moscow, Nauka, 1974.

[6] Artemeva M.S., Skubov D.Yu. Dynamics of conducting bodies of pendulum type in the high-frequency magnetic field (in Russian). Transaction of Russian Academy of Sciences, Mechanics of Solids, 2001,
No. 4, pp. 29-39.

[7] Banfi C. Sull'approssimazione di processi non stazionari in mecanica
non lineare. Boll. Unione mat. Ital., 1967, vol.22, No. 4.

[8] Barbashin E.A., Tabueva V.A. Dynamic systems with cylindrical
phase space (in Russian). Moscow, Nauka, 1969.

[9] Bazhenichev A.V., Khodzhaev K.Sh., Chirkov A.G. On motion of
the non-relativistic charged particle in electromagnetic field having

a constant and a high-frequency components (in Russian). In: Analysis and synthesis of nonlinear mechanical oscillating systems, St. Petersburg, vol. 2, 1998, pp. 424-433.

[10] Bautin N.N., Leontovich E.A. Methods and approaches of qualitative investigation of dynamic systems in plane (in Russian). Moscow, Nauka, 1990, 487p.

[11] Beletskiy V.V., Hentov A.A. Rotatory motion of magnetized satellite (in Russian). Moscow, Nauka, 1985.

[12] Bellman R. Theory of stability of solutions of differential equations (in Russian). Moscow, Inostrannaya Literatura, 1959.

[13] Belyustina L.N., Belih V.N. Qualitative investigation of dynamic system on cylinder (in Russian). Ordinary Differential Equations, 1973, vol. IX, No. 3.

[14] Blekhman I.I., Ladizhenskiy L.A., Polyakov V.I., Popova I.A., Khodzhaev K.Sh. Vibrations of the resonant multi-drive vibration devices (in Russian). Transaction of USSR Academy of Sciences MTT (Mechanics of Solids), 1967, No. 5.

[15] Blekhman I.I. Synchronization of dynamical systems (in Russian). Moscow, Nauka, 1971.

[16] Blekhman I.I., Polyakov V.I., Khodzhaev K.Sh. Synthesis of the modes of forced vibrations and tuning of multi-drive resonance vibration machines (in Russian). Transaction of USSR Academy of Sciences MTT (Mechanics of Solids), 1967, No. 6.

[17] Bogolyubov N.N., Zubarev D.N. Ukranian Matematic Journal No. 7, 1955, p. 5.

[18] Bogolyubov N.N., Mitropolskiy Yu.A. Asymptotic methods in theory of nonlinear oscillations (in Russian). Moscow, Nauka, 1974.

[19] Bolotnik N.N. Optimisation of the shock absorption systems (in Russian). Moscow, Nauka, 1983.

[20] Braginskiy S.I. To the theory of motion of charged particles in magnetic field (in Russian). Ukranian Mathematical Journal, vol. 8, No. 2, 1956, p. 119.

[21] Braunbeck W. Freies Schweben diamagnetischer Korper im Magnetfeld. Z. Phis., 1939, 112, H. 9, pp. 764-769.

[22] Burov A.A., Subkhankulov G.I. About motion of a rigid body in magnetic field (in Russian). PMM (Applied Mathematics and Mechanics), 1989, vol. 50, No. 6, pp. 960-966.

[23] Burshtein E.L., Solov'ev L.S. Hamilton function of the averaged motion (in Russian). Transactions of the USSR Academy of Science, 1961, vol. 139, No. 4, pp. 855-858.

[24] But D.A. Non-contact electric machines (in Russian). Moscow, Vysshaya Shkola, 1990.

[25] Bykov Yu.M., Gutkin B.M., Grigorash A.I. et al. Analogic-digital modelling of systems with valve transformers (in Russian). Electricity, No. 11, pp. 40-44.

[26] Courant R. Differential equations with partial derivatives (in Russian). Moscow, Nauka, 1965.

[27] Daletskiy Yu.D., Krein M.G. Stability of solutions of differential equations in Banach space (in Russian). Moscow, Nauka, 1970.

[28] Dombur L.E. Axial inductor generators (in Russian). Riga, Zinatne, 1984.

[29] Eliseev V.V. Mechanics of elastic bodies (in Russian). Publishers of St. Petersburg State Polytechnic University, 1999.

[30] Emden R. Gaskugeln. Leipzig-Berlin, 1907.

[31] Filatov A.N. Asymptotic methods in the theory of differential and integro-differential equations (in Russian). Tashkent, Fan, 1974.

[32] Filer Z.E. On dynamics of electromagnetic vibrator (in Russian). Izvestia Vuzov, Electromechanics, 1965, No. 10.

[33] Gegerin R.P. Inductor generators (in Russian). Moscow, Gosenergoizdat, 1961.

[34] Geker I.R. Interaction of strong magnetic fields with plasma (in Russian). Moscow, Atomizdat, 1978, 312 p.

[35] Gelig A.H., Leonov G.A., Yakubovich V.A. Stability of nonlinear systems with non-unique position of equilibrium (in Russian). Moscow, Nauka, 1978.

[36] Glebov I.A., Kasharskiy E.G., Rutberg F.G. Synchronous generators in electrophysical equipment (in Russian). Leningrad, Nauka, 1977.

[37] Gruzdev A.Yu., Sablin A.D. Dynamics of synchronous machine of brief action with inductive storage of energy (in Russian). Electricity, 1993, No. 5, pp. 27-33.

[38] Halanay A. Stability problems for synchronous machines. VII Internationale Konferenz über nichtlineare Schwingungen. Berlin, 1977, No. 5, S. 407-421.

[39] Hardy G. Inequalities (in Russian). Moscow, Gosizd. Inostrannaya Literatura, 1948.

[40] Ioffe B.A., Kalnin R.K. Orientation of details by means of electromagnetic field (in Russian). Riga, Zinatne, 1972, 300 p.

[41] Khodzhaev K.Sh. Vibrations in the system with many electromagnetic exciters (in Russian). Transaction of USSR Academy of Sciences, Mechanics of Solids, 1966, No. 2, pp. 23-29.

[42] Khodzhaev K.Sh. Synchronization of the mechanical vibrators related to the linear oscillating system (in Russian). Transaction of the USSR Academy of Sciences, Mechanics of Solids, 1967, No. 4, pp. 14-24.

[43] Khodzhaev K.Sh. Vibrations excited by electromagnets in linear mechanical systems (in Russian). Transaction of the USSR Academy of Sciences, Mechanics of Solids, No. 5, pp. 11-26.

[44] Khodzhaev K.Sh. Integral criterion of stability for systems with quasi-cyclic coordinates and energy relations under vibration of current conductors (in Russian) PMM (Applied Mathematics and Mechanics), vol. 33, 1969, No. 1, pp. 85-100.

[45] Khodzhaev K.Sh. On influencing of nonlinearity in ferromagnetics on the vibrations excited by electromagnets (in Russian). Transaction of the USSR Academy of Sciences, MTT (Mechanics of Solids), 1973, No. 6, pp. 36-46.

[46] Khodzhaev K.Sh. On stability of stationary motions of systems with quasi-cyclic coordinates and mechanical equilibrium in magnetic field (in Russian). PMM (Applied Mathematics and Mechanics), vol. 37, 1973, No. 3, pp. 400-406.

[47] Khodzhaev K.Sh. Nonlinear problems of deforming elastic bodies by magnetic field (in Russian). PMM (Applied Mathematics and Mechanics), 1970, vol. 34, No. 4, pp. 653-671.

[48] Khodzhaev K.Sh. On equation of electrohydrodynamics and magnetoelasticity (in Russian). PMM (Applied Mathematics and Mechanics), vol. 35, 1971, No. 6, pp. 1038-1046.

[49] Khodzhaev K.Sh., Chirkov A.G., Shatalov S.D. Theory of motion of the relativistic charged particle in strongly heterogeneous crossed electric and magnetic fields (in Russian). Journal of Technical Physics, vol. 52, 1982, No. 8, pp. 1493-1499.

[50] Khodzhaev K.Sh., Chirkov A.G., Shatalov S.D. On motion of the charged particle in magnetic and electric fields at the strong heterogeneity of magnetic field (in Russian). Journal of Applied Mechanics and Technical Physics, 1981, No. 4, pp. 3-6.

[51] Khodzhaev K.Sh., Shatalov S.D. On qualitative research of motions by asymptotic methods of nonlinear mechanics (in Russian). PMM (Applied Mathematics and Mechanics), vol. 44, 1980, No. 5, pp. 802-810.

[52] Khodzhaev K.Sh., Shatalov S.D. On slow motions of a conducting body in magnetic field (in Russian). Transactions of the USSR Academy of Sciences, Mechanics of Solids, 1981, No. 2, pp. 175-182.

[53] Khodzhaev K.Sh., Shtilerman I.Z. On a boundary-value problem of magnetoelasticity (in Russian). PMM (Applied Mathematics and Mechanics), vol. 36, 1972, No. 5, pp. 952-956.

[54] Kiselev P.V., Lev O.M. Synchronisation of two generators of brief action (in Russian). In: Materials of the Scientific-Practice Conference, Pskov, NTO, 1982, pp. 94-95.

[55] Kiselev P.V., Lev O.M., Khodzhaev K.Sh. Asymptotic separation of transitional processes of synchronous machine (in Russian). Transaction of the USSR Academy of Sciences, Energy and Transport, 1983, No. 3, pp. 76-83.

[56] Kiselev P.V., Khodzhaev K.Sh. Equations of nonstationary processes of synchronous generator feeding loading via rectifier (in Russian). Electricity, 1983, No. 4, pp. 33-38.

[57] Kiselev P.V., Khodzhaev K.Sh. Averaging equations of transitional processes in synchronous machines (in Russian). Transaction of the USSR Academy of Sciences, Power Engineering and Transport, 1987, No. 4, pp. 69-80.

[58] Kiselev P.V., Khodzhaev K.Sh. Dynamics of the turbogenerator of brief action operating into plasmotron of alternating current (in Russian). Electricity, 1988, No. 3, pp. 47-53.

[59] Kiselev P.V., Sablin A.D., Khodzhaev K.Sh. Control of excitation of turbogenerators of brief action (in Russian). Transaction of Russian Academy of Sciences, Power Engineering, 1992, No. 6, pp. 50-57.

[60] Kitsenko A.B., Ponkratov I.M., Stepanov K.N. Nonlinear stage of excitation of monochromatic vibrations of plasma in magnetic field by a flux of charged particles (in Russian). Journal of Experimental and Theoretical Physics, 1974, vol. 66, No. 1, pp. 166-175.

[61] Kitsenko A.B., Ponkratov I.M., Stepanov K.N. Nonlinear cyclotron resonance in plasma (in Russian). Journal of Experimental and Theoretical Physics, 1974, vol. 67, No. 5 (11), pp. 1728-1737.

[62] Kovilin Yu.Ya., Basov S.A. Interaction of electromagnetic vibrator with the source of sinusoidal voltage (in Russian). Transaction of the Siberian Division of the USSR Academy of Sciences, Physical-Technical Problems of Mineral Processing, 1965, No. 4.

[63] Kozlov V.V. To the problem of rotation of a rigid body in magnetic field (in Russian). Transaction of the USSR Academy of Sciences. MTT (Mechanics of Solids), 1985, No. 6, pp. 28-33.

[64] Kozorez V.V. Dynamic systems of magnetically interacting free bodies (in Russian). Kiev, Naukova Dumka, 1981.

[65] Kolovskiy M.Z. On condition of existence of periodic solutions of system of differential equations with discontinuous right parts containing a small parameter (in Russian). PMM (Applied Mathematics and Mechanics), 1960, vol. 24, No. 4.

[66] Kononenko V.O. Oscillating systems with limited supply (in Russian). Moscow, Nauka, 1964.

[67] Korablev S.S. To the theory of electromechanical vibration extinguisher (in Russian). Applied Mechanics, 1968, vol. 4, No. 1.

[68] Kotkin G.L., Serbo V.G. Collection of problems in classical mechanics (in Russian). Moscow, Nauka, 1979, 431p.

[69] Kuvikin V.I. Magnetic friction in non-contact suspensions (in Russian). Nizhniy Novgorod, Publishers Itelservis, 1997.

[70] Lavrov B.P. Spatial problem of synchronisation of mechanical vibrators (in Russian). Transaction of the USSR Academy of Sciences, OTN, Mechanics and Mechanical Engineering, 1961, No. 5.

[71] Landau L.D., Lifshits E.M. Electrodynamics of continuous media (in Russian). Moscow, Nauka, 1955.

[72] Landau L.D., Lifshits E.M. Theory of field (in Russian). Moscow, Nauka, 1967.

[73] Landau L.D., Lifshits E.M. Mechanics (in Russian). Moscow, Nauka, 1973.

[74] Litvak A.G., Miller M.A., Shelehov N.V. Refinement of the averaged equation of motion of the charged particle in the field of standing electromagnetic wave (in Russian). Izvestiya Vuzov, Radiophysics, 1962, vol. 5, No. 6, pp. 1160-1174.

[75] Loginov S.I., Valkov V.I. Investigation of electrodynamic model of the regime of operation of the feeding aggregate (in Russian). In: Powerful generators of low temperature plasma and methods of research of their parameters, Leningrad, VNIIElektromash, 1970, pp. 166-171.

[76] Lurie A.I. Analytical mechanics. Springer-Verlag, Berlin-Heidelberg, 2002, 864 p.

[77] Lvovich A.Yu. Electromechanical systems (in Russian). Publishers of Leningrad State University, 1989.

[78] Malkin I.G. Some problems of theory of nonlinear vibrations (in Russian). Moscow, GITTL, 1956.

[79] Martinenko Yu.G. Motion of conducting rigid body near an immobile point in magnetic field (in Russian). Transaction of the USSR Academy of Sciences, MTT (Mechanics of Solids), 1977, No. 4, p. 36.

[80] Martinenko Yu.G. Motion of the solid in electric and magnetic fields (in Russian). Moscow, Nauka, 1988.

[81] Miller M.A. Motion of charged particles in high-frequency electromagnetic fields (in Russian). Izvestiya Vuzov, Radiophysics, 1959, vol. 1, No. 3, pp. 110-123.

[82] Mirkina A.S., Khodzhaev K.Sh. Approximation of non-stationary processes in infinite time domain under exponential stability of slow motions (in Russian). PMM (Applied Mathematics and Mechanics), 1979, vol. 43, No. 2.

[83] Mitropolskiy Yu.A. Method of averaging in nonlinear mechanics (in Russian). Kiev, Naukova Dumka, 1971.

[84] Mitropolskiy Yu.A., Likova O.B. Integral manifolds in nonlinear mechanics (in Russian). Moscow, Nauka, 1973.

[85] Merkin D.R. Introduction to the theory of stability of motion (in Russian). Moscow, Nauka, 1987.

[86] Metlin V.B. Magnetic and magnetohydrodynamic supports (in Russian). Moscow, Nauka, 1967, 519 p.

[87] Morton K.W. Phys. Fluids, No. 7, 1964, p.1800.

[88] Nagaev R.F. Periodic solutions of the piecewise continuous systems with small parameter (in Russian). PMM (Applied Mathematics and Mechanics), 1972, vol. 36, No. 6.

[89] Nagaev R.F. Dynamics of synchronising systems. Springer-Verlag, Berlin-Heidelberg, 2002. 326 p.

[90] Neimark Yu.I., Fufaev N.A. Dynamics of nonholonomic systems (in Russian). Moscow, Nauka, 1964, 519 p.

[91] Neimark Yu.I., Shilnikov L.P. On application of method of small parameter to the systems of differential equations with discontinuous right parts (in Russian). Transaction of the USSR Academy of Sciences, OTN, Mechanics and Mechanical Engineering, 1959, No. 6.

[92] Nortrop T. Adiabatic theory of motion of the charged particles (in Russian). Moscow, Atomizdat, 1967.

[93] Othman H.A. Lesieutre B.C., Sauer P.W. Averaging Theory in Reduced-Order Synchronous Machine Modelling. In: Proc. of 19th Annual North American Symposium, NAPS 87, Edmonton, N.Y., 1987.

[94] Park R.H. Two reaction theory of synchronous machines. AIEE Trans, 1933, Pt. 2, pp. 352-355.

[95] Razmadze Sh.M. Transforming circuits and systems (in Russian). Moscow, Visshaya Shkola, 1967.

[96] Reimers N.A., Khodzhaev K.Sh. Averaging quasi-linear systems with many fast variables (in Russian). Differential Equations, 1978, vol. 14, No. 8.

[97] Reimers N.A., Khodzhaev K.Sh., Shatalov S.D. Analysis and synthesis of the mechanical oscillating systems (in Russian). Publishers of Institute of Problems in Mechanical Engineering, St. Petersburg, 1998, vol. 2, pp. 88-97.

[98] Rodin V.I., Zagorskiy A.E., Shakaryan Yu.G. Controlled electric generators at the variable frequency (in Russian). Moscow, Energy, 1978, 152 p.

[99] Rumyantsev A.A. Superadiabatic acceleration of the charged particles and flashes in the Sun and stars (in Russian). Transaction of the USSR Academy of Sciences, Series Physics, vol. 41, No. 9.

[100] Sablin A.D., Khodzhaev K.Sh. Control of the transient process in synchronous generator of brief action (in Russian). Electricity, 1982, No. 12, pp. 33-37.

[101] Samoylenko A.M. Method of averaging in the systems with pushes (in Russian). In: Mathematical Physics, Republican Interuniversity Collection. Kiev, 1971, No. 9, pp. 107-117.

[102] Samsonov V.A. On rotation of body in magnetic field (in Russian). Transaction of the USSR Academy of Sciences, MTT (Mechanics of Solids), 1984, No. 4, pp. 32-34.

[103] Sansone J. Ordinary differential equations (in Russian). Moscow, Inostrannaya Literatura, vol. 2, 1954.

[104] Sermons G.Ya. Dynamics of bodies in magnetic field (in Russian). Riga, Zinatne, 1974, 247 p.

[105] Skubov D.Yu. Electrodynamic forces at the eccentric motion of conducting rotor of implicit-pole machine (in Russian). Transaction of the USSR Academy of Sciences, Power Engineering and Transport, 1989, No. 6, pp. 82-89.

[106] Skubov D.Yu., Khodzhaev K.Sh. Asymptotic transformation of equations of synchronous many-contour machine with a superconducting excitation winding (in Russian). Transaction of the USSR Academy of Sciences, Power Engineering, 1994, No. 5, pp. 80-88.

[107] Skubov D.Yu., Khodzhaev K.Sh. Equations of electromechanical processes in a single-phase inductor generator operating into active-inductive loading (in Russian). Electricity, 1994, No. 9, pp. 33-40.

[108] Skubov D.Yu., Khodzhaev K.Sh. Equations for transient processes in three-phase inductor generators (in Russian). Electricity, 1995, No. 11, pp. 29-36.

[109] Skubov D.Yu., Khodzhaev K.Sh. Systems with electro-magnetic vibration extinguishers (in Russian). Transaction of the Russian Academy of Sciences, MTT (Mechanics of Solids), 1996, No. 2, pp. 64-74.

[110] Skubov D.Yu., Khodzhaev K.Sh. Application of asymptotic methods of analysis of dynamical processes in synchronous electric machines. In: Proceedings of Second European Nonlinear Oscillations Conference, Prague, 1996.

[111] Skubov D.Yu., Khodzhaev K.Sh. Asymptotic and qualitative methods in the theory of synchronous electric machines (in Russian). Publishers of St. Petersburg State Polytechnic University, 1999, 154 p.

[112] Skubov D.Yu., Shumakovich I.V. Stability of rotor of asynchronous machine in magnetic field of current windings (in Russian). Transaction of the Russian Academy of Sciences, MTT (Mechanics of Solids), 1999, No. 4, pp. 36-50.

[113] Sperling L. Beitrag zur allgemeinen Theorie der Selbstsynchronisation umlaufender Unwuchtmassen im Nichtresonanzfall. Wissenschaftliche Zeitschrift der Technischen Hochschule Otto von Guericke, Magdeburg, 1967, 11.

[114] Stoker J. Nonlinear vibrations in mechanical and electric systems. Moscow, SILT, 1952.

[115] Tamm I.E. Basics of the theory of electricity (in Russian). Moscow, Nauka, 1989.

[116] Taylor C. The coalescence of closely spaled drops when they are at different electric potentials. Proc. Royal Society A, 1968, vol. 306, No. 1487.

[117] Trikomi F. Differential equations (in Russian). Moscow, SILT, 1962.

[118] Urman Yu.M. Drift of the moments caused by a non-spherical rotor in suspension with axisymmeric field (in Russian). Transaction of the USSR Academy of Sciences, MTT (Mechanics of Solids), 1973, No. 1, pp. 24-31.

[119] Vazhnov A.I. Transient processes in the alternate current machines (in Russian). Leningrad, Energy, 1980.

[120] Vainberg M.M., Trenogin V.A. Theory of solving nonlinear equations (in Russian). Moscow, Nauka, 1969.

[121] Vasilev V.N., Toptigin I.N., Chirkov A.G. Interaction of energetic particles with the front of shock wave in turbulent medium (in Russian). Geomagnetism and Aeronomy, No. 13, 1978, p. 415-422.

[122] Venikov V.A . Transient processes in the electric systems (in Russian). Moscow, Vysshaya Shkola, 1978.

[123] Vetyukov M.M., Khodzhaev K.Sh. Excitation of shock vibrations by electromagnets (in Russian). Transaction of the USSR Academy of Sciences, MTT (Mechanics of Solids), 1976, No. 4, pp. 71-78.

[124] Vetyukov M.M., Khodzhaev K.Sh. Equations of slow motions of the systems with quasi-cyclic coordinates and electromechanical systems (in Russian). In: Dynamics of Systems, Publishers of Gorky University, Gorky, 1976, No. 9, pp. 92-106.

[125] Vlasov E.N., Lev O.M., Khodzhaev K.Sh. Asymptotic separation of transient processes in synchronous generator (in Russian). In: Mechanics and Control Processes, Publishers of Saratov University, Saratov, 1981, p. 111-125.

[126] Vlasov E.N., Sablin A.D., Khodzhaev K.Sh. Equations of slow transient processes of synchronous machine (in Russian). Electricity, 1980, No. 9, pp. 41-44.

[127] Vlasov E.N., Khodzhaev K.Sh. Nonstationary processes in synchronous generator feeding an inductive storage of energy (in Russian). Electricity, 1982, No. 11, pp. 19-24.

[128] Volosov V.M., Morgunov B.I. Method of averaging in theory of nonlinear oscillating systems (in Russian). Moscow, Publishers of Moscow State University, 1971.

[129] Yanko-Trinitskiy A.A. New method of analysis of work of synchronous motor under sharply changing loads (in Russian). Leningrad, Gosenergoizdat, 1958.

[130] Zabreyko P.P., Ledovskaya I.B. To substantiation of the method by N.M. Bogolyubov and N.M. Krylov for ordinary differential equations (in Russian), Differential Equations, 1969, vol. 5, No. 2.

Index

Foundations of Engineering Mechanics

Series Editors: Vladimir I. Babitsky, Loughborough University, UK
                    Jens Wittenburg, Karlsruhe University, Germany

Further volumes of this series can be found on our homepage: springer.com

(Continued from page ii)

Printing: Krips bv, Meppel, The Netherlands
Binding: Stürtz, Würzburg, Germany